Springer Proceedings in Physics

Volume 234

Indexed by Scopus

The series Springer Proceedings in Physics, founded in 1984, is devoted to timely reports of state-of-the-art developments in physics and related sciences. Typically based on material presented at conferences, workshops and similar scientific meetings, volumes published in this series will constitute a comprehensive up-to-date source of reference on a field or subfield of relevance in contemporary physics. Proposals must include the following:

- name, place and date of the scientific meeting
- a link to the committees (local organization, international advisors etc.)
- scientific description of the meeting
- list of invited/plenary speakers
- an estimate of the planned proceedings book parameters (number of pages/ articles, requested number of bulk copies, submission deadline).

More information about this series at http://www.springer.com/series/361

Anjan Giri · Rukmani Mohanta
Editors

16th Conference on Flavor Physics and CP Violation

FPCP 2018

Springer

Editors
Anjan Giri
Department of Physics
IIT Hyderabad
Hyderabad, Telangana, India

Rukmani Mohanta
School of Physics
University of Hyderabad
Hyderabad, Telangana, India

ISSN 0930-8989 ISSN 1867-4941 (electronic)
Springer Proceedings in Physics
ISBN 978-3-030-29621-6 ISBN 978-3-030-29622-3 (eBook)
https://doi.org/10.1007/978-3-030-29622-3

This Springer imprint is published by the registered company Springer Nature Switzerland AG
The registered company address is: Gewerbestrasse 11, 6330 Cham, Switzerland

To our parents.

Contributions

Akshay Chatla, University of Hyderabad
Alakabha Datta, University of Mississippi
Anirban Karan, Institute of Mathematical Sciences, Chennai
Aritra Biswas, Indian Association for the Cultivation of Science (IACS)
Atasi Ray, University of Hyderabad
Atika Fatima, Aligarh Muslim University
Avelino Vicente, IFIC—CSIC/University of Valencia
Bharti Kindra, Physical Research Laboratory, Ahmedabad
Bhupal Dev, Washington University In St. Louis, USA
Biswaranjan Behera, IIT Hyderabad
Bob Kowalewski, University of Victoria (CA)
Brij Kishor Jashal, Tata Institute of Fundamental Research, Mumbai
Debjyoti Bardhan, Ben Gurion University of the Negev
Dhananjaya Thakur, Indian Institute of Technology Indore
Diganta Das, University of Delhi
Dinesh Kumar, University of Rajasthan, Jaipur
Emilie Bertholet, Centre National De La Recherche Scientifique (FR)
Eryk Czerwiski, Jagiellonian University, Marian Smoluchowski Institute of Physics
Gauhar Abbas, Indian Institute of Technology (BHU), Varanasi, India
Girish Kumar, Tata Institute of Fundamental Research, Mumbai
Indumathi D., The Institute of Mathematical Sciences, Chennai
Jurgen Engelfried, Universidad Autonoma De San Luis Potosi (MX)
Jusak Tandean, National Taiwan University
Justin Albert, University of Victoria (CA)
Jyoti Saini, Indian Institute of Technology Jodhpur
Kajari Mazumdar, TIFR, Mumbai
Kavita Lalwani, Malaviya National Institute of Technology, Jaipur
Konstantin Gizdov, The University of Edinburgh (GB)
Lei Xia, University of Science and Technology of China
Manish Kumar, MNIT, Jaipur
Mark Smith, Imperial College (GB)

Marta Moscati, Karlsruhe Institute of Technology
Minakshi Nayak, Tel Aviv University
Muhammad Muhammad Alibordi, Indian Institute of Technology Madras, Chennai
Nairit Sur, Tata Institute of Fundamental Research, Mumbai
Nakul Soni, The Maharaja Sayajirao University of Baroda
Nejc Konik, J. Stefan Institute, University of Ljubljana
Nimmala Narendra, Indian Institute of Technology, Hyderabad
Nora Patricia Estrada Tristan, Universidad Autonoma De San Luis Potosi
Prasanth Krishnan K. P., TIFR, Mumbai
Pratishruti Saha, Harish-Chandra Research Institute, Allahabad
Preeti Dhankher, Indian Institute of Technology
Raghunath Sahoo, Indian Institute of Technology Indore (IN)
Rajarshi Bhattacharya, Saha Institute of Nuclear Physics, Kolkata
Rajeev Kumar, Punjab Agricultural University
Rajeev N., NIT Silchar
Resmi P. K., IIT Madras
Ria Sain, Institute of Mathematical Sciences, Chennai
Rudra Majhi, University of Hyderabad
Satoshi Mihara, KEK, JAPAN
Saurabh Sandilya, University of Cincinnati
Sean Bensen, CERN
Seema Choudhury, IIT Hyderabad
Shivaramakrishna Singirala, University of Hyderabad
Shohei Nishida, KEK
Smruti Patel, Veer Narmad South Gujarat University, Surat
Soram Robertson Singh, University of Hyderabad
Soumya C., Institute of Physics, Bhubaneswar
Sruthilaya M., University of Hyderabad
Steven Gottlieb, Indiana University
Subhasmita Mishra, IIT Hyderabad
Suchismita Sahoo, University of Hyderabad
Suman Kumbhakar, IIT Bombay
Teppei Kitahara Karlsruhe, Institute of Technology
Vindhyawasini Prasad, University of Science and Technology of China, Hefei, China
Vipin Bhatnagar, Panjab University, Chandigarh
Vipin Gaur Virginia, Polytechnic Institute and State University
Vishal Bhardwaj, IISER Mohali
Zubair Ahmad Dar, Aligarh Muslim University

Editorial Board

Sponsoring Institutions

SERB, Government of India
UPE-II, University of Hyderabad
IIT Hyderabad

Organization

16th Conference on Flavor Physics and CP Violation (FPCP 2018) was jointly organized by the University of Hyderabad and IIT Hyderabad, Hyderabad, India

International Advisory Committee

Fabrizio Bianchi, Torino
Guennadi Borissov, Lancaster
Roy Briere, Carnegie Mellon
Tom Browder, Hawaii
Zdenek Dolezal, Charles University
Ulrik Egede, London
Aida El-Khadra, UIUC
Marcello Giorgi, Pisa
Anjan Giri, Hyderabad
Benjamin Grinstein, UCSD
Xiao-Gang He, NTU & SJTU
David Hitlin, Caltech
Toru Iijima, Nagoya—**Chair**
C. S. Kim, Yonsei
Bob Kowalewski, Victoria
Zoltan Ligeti, Berkeley
Renaud Le Gac, CPPM Marseille
F. Di Lodovivo, London
Leandro de Paula, UFRJ Rio
Abner Soffer, Tel Aviv
Sheldon Stone, Syracuse
T. Yamanaka, Osaka

Local Organizing Committee

P. Behera, Chennai
A. Dighe, Mumbai
A. Giri, Hyderabad—**Chair**
J. Libby, Chennai
N. Mahajan, Ahmedabad
B. Mawlong, Hyderabad
R. Mohanta, Hyderabad—**Co-chair**
G. Mohanty, Mumbai
N. Sahu, Hyderabad
R. Sinha, Chennai
S. Swain, Bhubaneswar
S. Umasankar, Mumbai

Preface

In continuation of the practice of publishing the proceedings for previous editions, we decided to publish the proceedings for the 16th Conference on Flavor Physics and CP Violation (FPCP 2018). Publication of the proceedings constitutes an important aspect of any conference and the conference under consideration (FPCP 2018) is no way different. It is very important that we spread the scientific ideas, new results, and discussions during the meeting through the proceedings of the conference.

Flavor Physics and CP Violation (FPCP) conference is held every year, each time at a different location of the world, and is considered to be one of the premiere conferences organized in the area of high energy physics. Around 150 physicists and researchers participated in FPCP 2018 to discuss the latest advancements in the fields of interests. In the previous editions, we used to have only plenary invited lectures and poster contributions. Therefore, to give opportunities to younger scientists to present their work in FPCP, we decided to introduce few parallel sessions of short talks (based on the submitted abstracts). These include B physics & CP violation, Charm Physics, Exoctic composites, Kaon physics, Neutrino Physics and leptonic CP violation, Leptogenesis and Baryogenesis, Flavour and Dark sector connection, important results from various experiments like BABAR, BELLE, CLEO, LHCb, CMS, ATLAS, ALICE, BES-III, NOVA, T2K, DAYA BAY, KOTO, and future experiments.

The 16th edition of the conference, FPCP 2018, was held at the University of Hyderabad, Hyderabad, India during 14–18 July, 2018. The conference was jointly organized by the University of Hyderabad and IIT Hyderabad. This volume includes manuscripts from both invited, contributed talks, and poster contributions from Flavor Physics and CP violation, BSM physics, lattice gauge theory, high p-T physics, heavy ion physics, lepton flavor violation, etc. Some new results in the flavor sector were also presented, including review talks on the new developments during the past 1 year covering almost all areas connected to FPCP conference.

We are going through a very important and exciting phase of scientific development, where we have completely verified the standard model of particle physics. There are still some discrepancies and other observations, which we cannot explain within the framework of the standard model and we believe that there is physics beyond it but the nature of the same is not known so far. The same is truly reflected in this volume. The book is intended for both young as well as advanced researchers of the field who are actively following the exciting time that we are going through when we are expecting something new to show up either at the energy frontier or may be at the intensity and cosmic frontiers.

Hyderabad, India

Anjan Giri
Rukmani Mohanta

Acknowledgements

Organizing a conference like FPCP 2018 (16th Conference on Flavor Physics and CP violation) and bringing out the proceedings of the conference is undoubtedly a difficult job, and it would not have been possible to do so without the support and active participation of our colleagues and many students in both the organizing Institutes. The cooperation we received from the International Advisory Committee (IAC) members during the organization of the conference was truly exemplary. We thank all the members of the IAC from the bottom of our heart for their help in making FPCP 2018 a grand success. We take this opportunity to specially thank Toru Iijima (Chair), Zoltan Ligeti, Zdenek Dolezal, Ben Grinstein, Abi Soffer, Aida El-Khadra, Ulrik Egede, Sheldon Stone, Fabrizio Bianchi, David Hitlin, Roy Briere, H. Di Lodovico, C. S. Kim, Xiao-Gang He, and Tom Browder for their valuable suggestions regarding various organizational aspects.

We thank all the members of the Local Organizing Committee for their constant support and encouragement for making the organization of the conference smooth and enjoying. We would also like to thank the members of our Editorial board for their constant support in reviewing the abstracts, finalizing the programs, and finally reviewing the proceedings drafts. Our special thanks to Gagan Mohanty and James Libby for all the support extended to us at critical times. We must also take this opportunity to thank our colleagues in School of Physics, University of Hyderabad, and Department of Physics, IIT Hyderabad. It was indeed a pleasure to work with you all!

We were lucky to have a large pool of research scholar volunteers who worked relentlessly during the conference. It would have been a herculean task to organize this event without the active support of our students, in particular, Mitesh Behera, Rudra Majhi, Atasi Ray, Aishwarya Bhatt, Soram Robertson, Shivaramkrishna Singirala, Suchismita Sahoo, Akshay Chatla, Manas Mohapatra, Subhasmita Mishra, and Seema Choudhury. Thank you very much for providing the helping hand. We would also thank the Vice-Chancellor, University of Hyderabad and the

Director, IIT Hyderabad for their constant support and encouragement and the staff members both at School of Physics, University of Hyderabad, and Department of Physics, IIT Hyderabad for providing the assistance whenever needed. Needless to say, the conference would not have been possible without the financial support from the Science and Engineering Research Board (SERB), DST, Government of India, UPE-II, University of Hyderabad, and IIT Hyderabad.

We would like to thank Ms. Sabine Lehr, Associate Editor, Springer-Verlag, GmbH and Mr. Nareshkumar Mani (Springer Nature, India) for constantly working with us from the beginning to bring out this volume.

Hyderabad, India

Anjan Giri
Rukmani Mohanta

Contents

Part II Charm, Kaon and Heavy Leptons

Part III Flavour Mixing, CP and CPT

Contributors

Gauhar Abbas Department of Physics, Indian Institute of Technology (BHU), Varanasi, India

H. Aihara Department of Physics, University of Tokyo, Tokyo, Japan

J. Albert University of Victoria, Victoria, BC, Canada

Muhammad Alibordi Indian Institute of Technology Madras, Chennai, Tamil Nadu, India

Ashutosh Kumar Alok Indian Institute of Technology Jodhpur, Jodhpur, India

T. Aziz Tata Institute of Fundamental Research, Mumbai, India

S. Bacher The Henryk Niewodniczanski Institute of Nuclear Physics, Krakow, Poland

S. Bahinipati Indian Institute of Technology Bhubaneswar, Satya Nagar, India

Bindu A. Bambah School of Physics, University of Hyderabad, Hyderabad, India

E. Barberio School of Physics, University of Melbourne, Melbourne, VIC, Australia

Debjyoti Bardhan Ben Gurion University of the Negev, Be'er Sheva, Israel

T. Baroncelli School of Physics, University of Melbourne, Melbourne, VIC, Australia

A. K. Basith Indian Institute of Technology Madras, Chennai, India

G. Batignani Dipartimento di Fisica, Università di Pisa, Pisa, Italy;
INFN Sezione di Pisa, Pisa, Italy

A. Bauer Institute of High Energy Physics, Austrian Academy of Sciences, Vienna, Austria

Damir Bečirević Laboratoire de Physique Théorique (Bât. 210), Université Paris-Sud and CNRS, Orsay-Cedex, France

Biswaranjan Behera IIT Hyderabad/Fermilab, Hyderabad, India

P. K. Behera Indian Institute of Technology Madras, Chennai, India

Sean Benson Nikhef National Institute for Subatomic Physics, Amsterdam, The Netherlands

V. Bertacchi Dipartimento di Fisica, Università di Pisa, Pisa, Italy;
INFN Sezione di Pisa, Pisa, Italy

Emilie Bertholet LPNHE, Sorbonne Universite, Paris Diderot Sorbonne Paris Cite, CNRS/IN2P3, Paris, France

S. Bettarini Dipartimento di Fisica, Università di Pisa, Pisa, Italy;
INFN Sezione di Pisa, Pisa, Italy

Vishal Bhardwaj Indian Institute of Science Education and Research Mohali, Punjab, India

Sanjeeda Bharti Das Department of Physics, Ramanujan Junior College, Nagaon, India

V. Bhatnagar Department of Physics, Panjab University, Chandigarh, India

Rajarshi Bhattacharya Saha Institute of Nuclear Physics, HBNI, Kolkata, India

Tanvi Bhavsar Government Science College, Veer Narmad South Gujarat University, Songadh, Gujarat, India

P. S. Bhupal Dev Department of Physics and McDonnell Center for the Space Sciences, Washington University, St. Louis, MO, USA

B. Bhuyan Indian Institute of Technology Guwahati, Assam, India

T. Bilka Faculty of Mathematics and Physics, Charles University, Prague, Czech Republic

Aritra Biswas Indian Association for the Cultivation of Science, Kolkata, India

Monika Blanke Institut für Kernphysik, Karlsruhe Institute of Technology, Eggenstein-Leopoldshafen, Germany

F. Bosi INFN Sezione di Pisa, Pisa, Italy

L. Bosisio Dipartimento di Fisica, Università di Trieste, Trieste, Italy;
INFN Sezione di Trieste, Trieste, Italy

A. Bozek The Henryk Niewodniczanski Institute of Nuclear Physics, Krakow, Poland

F. Buchsteiner Institute of High Energy Physics, Austrian Academy of Sciences, Vienna, Austria

G. Caria School of Physics, University of Melbourne, Melbourne, VIC, Australia

G. Casarosa Dipartimento di Fisica, Università di Pisa, Pisa, Italy;
INFN Sezione di Pisa, Pisa, Italy

M. Ceccanti INFN Sezione di Pisa, Pisa, Italy

D. Červenkov Faculty of Mathematics and Physics, Charles University, Prague, Czech Republic

Akshay Chatla School of Physics, University of Hyderabad, Hyderabad, India

S. Choudhury IIT Hyderabad, Hyderabad, India

T. Czank Department of Physics, Tohoku University, Sendai, Japan

Eryk Czerwiński Faculty of Physics, Astronomy and Applied Computer Science, Institute of Physics, Jagiellonian University, Kraków, Poland

Soumya Dansana Department of Physical Sciences, Indian Institute of Science Education and Research, Kolkata, India

Zubair Ahmad Dar Department of Physics, Aligarh Muslim University, Aligarh, India

Diganta Das Department of Physics and Astrophysics, University of Delhi, Delhi, India

N. Dash Indian Institute of Technology Bhubaneswar, Satya Nagar, India

Alakabha Datta Department of Physics and Astronomy, University of Mississippi, Oxford, MS, USA

Sudipan De Discipline of Physics, School of Basic Sciences, Indian Institute of Technology Indore, Indore, India

M. De Nuccio Dipartimento di Fisica, Università di Pisa, Pisa, Italy;
INFN Sezione di Pisa, Pisa, Italy

P. Dhankher Indian Institute of Technology Bombay, Mumbai, Maharashtra, India

Z. Doležal Faculty of Mathematics and Physics, Charles University, Prague, Czech Republic

Ilja Doršner University of Split, Faculty of Electrical Engineering, Mechanical Engineering and Naval Architecture in Split (FESB), Split, Croatia

Rupak Dutta National Institute of Technology Silchar, Silchar, India

Jurgen Engelfried Instituto de Física, Universidad Autónoma de San Luis Potosí, San Luis Potosi, Mexico

Nora Patricia Estrada-Tristán Universidad de Guanajuato, Guanajuato, Mexico; Universidad Autónoma de San Luis Potosi, San Luis Potosi, Mexico

Svjetlana Fajfer Jožef Stefan Institute, Ljubljana, Slovenia;
Faculty of Mathematics and Physics, University of Ljubljana, Ljubljana, Slovenia

Darius A. Faroughy Jožef Stefan Institute, Ljubljana, Slovenia

A. Fatima Department of Physics, Aligarh Muslim University, Aligarh, India

F. Forti Dipartimento di Fisica, Università di Pisa, Pisa, Italy;
INFN Sezione di Pisa, Pisa, Italy

M. Friedl Institute of High Energy Physics, Austrian Academy of Sciences, Vienna, Austria

Shireen Gangal Center for Neutrino Physics, Department of Physics, Virginia Tech, Blacksburg, USA

V. Gaur Virginia Polytechnic Institute and State University, Blacksburg, VA, USA

A. Giri IIT Hyderabad, Hyderabad, India

Anjan Giri IIT, Hyderabad, India

Anjan K. Giri Department of Physics, IIT Hyderabad, Kandi, India

Konstantin Gizdov University of Edinburgh, Edinburgh, UK

B. Gobbo INFN Sezione di Trieste, Trieste, Italy

Steven Gottlieb Indiana University, Bloomington, IN, USA

J. A. M. Grimaldo Department of Physics, University of Tokyo, Tokyo, Japan

K. Hara High Energy Accelerator Research Organization (KEK), Tsukuba, Japan

Chandan Hati Laboratoire de Physique de Clermont, CNRS/IN2P3 - UMR 6533, Aubière, France

T. Higuchi Kavli Institute for the Physics and Mathematics of the Universe (WPI), University of Tokyo, Kashiwa, Japan

D. Indumathi The Institute of Mathematical Sciences, Chennai, India

C. Irmler Institute of High Energy Physics, Austrian Academy of Sciences, Vienna, Austria

A. Ishikawa Department of Physics, Tohoku University, Sendai, Japan

Brij Kishor Jashal Instituto de Física Corpuscular (IFIC—University of Valencia and CSIC), València, Spain;
Tata Institute of Fundamental Research, Mumbai, India

H. B. Jeon Department of Physics, Kyungpook National University, Daegu, Korea

C. Joo Kavli Institute for the Physics and Mathematics of the Universe (WPI), University of Tokyo, Kashiwa, Japan

M. Kaleta The Henryk Niewodniczanski Institute of Nuclear Physics, Krakow, Poland

N. Kambara High Energy Accelerator Research Organization (KEK), Tsukuba, Japan

J. Kandra Faculty of Mathematics and Physics, Charles University, Prague, Czech Republic

K. H. Kang Department of Physics, Kyungpook National University, Daegu, Korea

Anirban Karan The Institute of Mathematical Sciences, HBNI, Chennai, India

Daljeet Kaur Department of Physics, Delhi University, Delhi, India

Bharti Kindra Physical Research Laboratory, Ahmedabad, India;
Indian Institute of Technology, Gandhinagar, India

T. Kitahara Institute for Advanced Research, Nagoya University, Nagoya, Japan;
Kobayashi-Maskawa Institute for the Origin of Particles and the Universe, Nagoya University, Nagoya, Japan;
Physics Department, Technion–Israel Institute of Technology, Haifa, Israel

P. Kodyš Faculty of Mathematics and Physics, Charles University, Prague, Czech Republic

T. Kohriki High Energy Accelerator Research Organization (KEK), Tsukuba, Japan

S. Koike High Energy Accelerator Research Organization (KEK), Tsukuba, Japan

I. Komarov INFN Sezione di Trieste, Trieste, Italy

Nejc Košnik Jožef Stefan Institute, Ljubljana, Slovenia;
Faculty of Mathematics and Physics, University of Ljubljana, Ljubljana, Slovenia

Bob Kowalewski University of Victoria, Victoria, BC, Canada

Prasanth Krishnan Tata Institute of Fundamental Research, Mumbai, India

Dinesh Kumar Department of Physics, University of Rajasthan, Jaipur, India

Girish Kumar Tata Institute of Fundamental Research, Mumbai, India

Jacky Kumar Department of High Energy Physics, Tata Institute of Fundamental Research, Mumbai, India

M. Kumar Malaviya National Institute of Technology, Jaipur, Jaipur, India

R. Kumar Punjab Agricultural University, Ludhiana, India

Rajeev Kumar Department of Mathematics, Statistics and Physics, Punjab Agri University, Ludhiana, India

Sanjeev Kumar Department of Physics, Delhi University, Delhi, India

Abinash Kumar Nayak The Institute of Mathematical Sciences, HBNI, Chennai, India

Suman Kumbhakar Indian Institute of Technology Bombay, Mumbai, India

W. Kun Department of Physics, University of Tokyo, Tokyo, Japan

P. Kvasnička Faculty of Mathematics and Physics, Charles University, Prague, Czech Republic

K. Lalwani Malaviya National Institute of Technology Jaipur, Jaipur, India

L. Lanceri Dipartimento di Fisica, Università di Trieste, Trieste, Italy;
INFN Sezione di Trieste, Trieste, Italy

J. Y. Lee Department of Physics and Astronomy, Seoul National University, Seoul, Korea

S. C. Lee Department of Physics, Kyungpook National University, Daegu, Korea

J. Libby Indian Institute of Technology Madras, Chennai, India

C. La Licata Dipartimento di Fisica, Università di Trieste, Trieste, Italy;
INFN Sezione di Trieste, Trieste, Italy

David London Physique des Particules, Université de Montréal, Montréal, QC, Canada

T. Lueck Dipartimento di Fisica, Università di Pisa, Pisa, Italy;
INFN Sezione di Pisa, Pisa, Italy

Namit Mahajan Physical Research Laboratory, Ahmedabad, India

Rudra Majhi School of Physics, University of Hyderabad, Hyderabad, India

P. Mammini INFN Sezione di Pisa, Pisa, Italy

A. Martini Dipartimento di Fisica, Università di Pisa, Pisa, Italy;
INFN Sezione di Pisa, Pisa, Italy

Barilang Mawlong School of Physics, University of Hyderabad, Hyderabad, India

S. N. Mayekar Tata Institute of Fundamental Research, Mumbai, India

Kajari Mazumdar Tata Institute of Fundamental Research, Mumbai, India

Satoshi Mihara KEK, J-PARC Center, and SOKENDAI, Tsukuba, Ibaraki, Japan

Subhasmita Mishra IIT, Hyderabad, India

R. Mohanta University of Hyderabad, Hyderabad, India

Rukmani Mohanta School of Physics, University of Hyderabad, Hyderabad, India

G. B. Mohanty Tata Institute of Fundamental Research, Mumbai, India

T. Morii Kavli Institute for the Physics and Mathematics of the Universe (WPI), University of Tokyo, Kashiwa, Japan

Marta Moscati Institut für Theoretische Teilchenphysik, Karlsruhe Institute of Technology, Karlsruhe, Germany

Md. Naimuddin Department of Physics, Delhi University, Delhi, India

K. R. Nakamura High Energy Accelerator Research Organization (KEK), Tsukuba, Japan

Nimmala Narendra Indian Institute of Technology Hyderabad, Sangareddy, Telangana, India

Z. Natkaniec The Henryk Niewodniczanski Institute of Nuclear Physics, Krakow, Poland

Minakshi Nayak Tel Aviv University, Tel Aviv-Yafo, Israel

Ulrich Nierste Institut für Theoretische Teilchenphysik, Karlsruhe Institute of Technology, Karlsruhe, Germany

Shohei Nishida High Energy Accelator Research Organization (KEK) and Sokendai University, Tsukuba, Japan

Y. Onuki Department of Physics, University of Tokyo, Tokyo, Japan

Jean Orloff Laboratoire de Physique de Clermont, CNRS/IN2P3 - UMR 6533, Aubière, France

W. Ostrowicz The Henryk Niewodniczanski Institute of Nuclear Physics, Krakow, Poland

Arantza Oyanguren Campos Instituto de Física Corpuscular (IFIC—University of Valencia and CSIC), València, Spain

A. Paladino Kavli Institute for the Physics and Mathematics of the Universe (WPI), University of Tokyo, Kashiwa, Japan

J. N. Pandya Applied Physics Department, Faculty of Technology and Engineering, The Maharaja Sayajirao University of Baroda, Vadodara, Gujarat, India

E. Paoloni Dipartimento di Fisica, Università di Pisa, Pisa, Italy;
INFN Sezione di Pisa, Pisa, Italy

H. Park Department of Physics, Kyungpook National University, Daegu, Korea

Smruti Patel Government Science College, Veer Narmad South Gujarat University, Songadh, Gujarat, India

Sudhanwa Patra Indian Institute of Technology Bhilai, GEC Campus, Raipur, Chhattisgarh, India

Vindhyawasini Prasad State Key Laboratory of Particle Detection and Electronics, Department of Modern Physics, University of Science and Technology of China, Hefei, China

K. Prasanth Tata Institute of Fundamental Research, Mumbai, India

A. Profeti INFN Sezione di Pisa, Pisa, Italy

N. Rajeev National Institute of Technology Silchar, Silchar, India

K. K. Rao Tata Institute of Fundamental Research, Mumbai, India

Soumya Rao National Centre for Nuclear Research, Warsaw, Poland

I. Rashevskaya INFN Sezione di Trieste, Trieste, Italy;
TIFPA - INFN, Trento, Italy

Atasi Ray School of Physics, University of Hyderabad, Hyderabad, India

P. K. Resmi Indian Institute of Technology Madras, Chennai, India

G. Rizzo Dipartimento di Fisica, Università di Pisa, Pisa, Italy;
INFN Sezione di Pisa, Pisa, Italy

Soram Robertson Singh School of Physics, University of Hyderabad, Hyderabad, India

M. Rozanska The Henryk Niewodniczanski Institute of Nuclear Physics, Krakow, Poland

Pratishruti Saha Harish-Chandra Research Institute, Jhunsi, Allahabad, India

D. Sahoo Tata Institute of Fundamental Research, Mumbai, India

Raghunath Sahoo Discipline of Physics, School of Basic Sciences, Indian Institute of Technology Indore, Indore, India

Suchismita Sahoo Theoretical Physics Division, Physical Research Laboratory, Ahmedabad, India

Narendra Sahu Indian Institute of Technology Hyderabad, Sangareddy, Telangana, India

Ria Sain The Institute of Mathematical Sciences, Chennai, India;
Homi Bhabha National Institute Training School Complex, Mumbai, India

Jyoti Saini Indian Institute of Technology Jodhpur, Jodhpur, India

M. Sajjad Athar Department of Physics, Aligarh Muslim University, Aligarh, India

S. Sandilya University of Cincinnati, Cincinnati, Ohio, USA

S. Uma Sankar Indian Institute of Technology Bombay, Mumbai, India

J. Sasaki Department of Physics, University of Tokyo, Tokyo, Japan

N. Sato High Energy Accelerator Research Organization (KEK), Tsukuba, Japan

S. Schultschik Institute of High Energy Physics, Austrian Academy of Sciences, Vienna, Austria

C. Schwanda Institute of High Energy Physics, Austrian Academy of Sciences, Vienna, Austria

Ruchi Sharma Indian Institute of Technology Jodhpur, Jodhpur, India

Sujay Shil Institute of Physics, Bhubaneswar, Odisha, India;
Homi Bhabha National Institute Training School Complex, Anushakti Nagar, Mumbai, India

S. K. Singh Department of Physics, Aligarh Muslim University, Aligarh, India

Shivaramakrishna Singirala School of Physics, University of Hyderabad, Hyderabad, India

N. R. Soni Applied Physics Department, Faculty of Technology and Engineering, The Maharaja Sayajirao University of Baroda, Vadodara, Gujarat, India

C. Soumya Institute of Physics, Bhubaneswar, India

M. Sruthilaya University of Hyderabad, Hyderabad, India

J. Stypula The Henryk Niewodniczanski Institute of Nuclear Physics, Krakow, Poland

Olcyr Sumensari Dipartimento di Fisica e Astronomia "G. Galilei", Università di Padova, Padova, Italy;
Istituto Nazionale Fisica Nucleare, Sezione di Padova, Padova, Italy

Nairit Sur Tata Institute of Fundamental Research, Mumbai, India

J. Suzuki High Energy Accelerator Research Organization (KEK), Tsukuba, Japan

S. Tanaka High Energy Accelerator Research Organization (KEK), Tsukuba, Japan

Jusak Tandean Department of Physics, National Taiwan University, Taipei 106, Taiwan;
Physics Division, National Center for Theoretical Sciences, Hsinchu 300, Taiwan

H. Tanigawa Department of Physics, University of Tokyo, Tokyo, Japan

G. N. Taylor School of Physics, University of Melbourne, Melbourne, VIC, Australia

Ana M. Teixeira Laboratoire de Physique de Clermont, CNRS/IN2P3 - UMR 6533, Aubière, France

Dhananjaya Thakur Discipline of Physics, School of Basic Sciences, Indian Institute of Technology Indore, Indore, India

R. Thalmeier Institute of High Energy Physics, Austrian Academy of Sciences, Vienna, Austria

K. Trabelsi High Energy Accelerator Research Organization (KEK), Tsukuba, Japan

T. Tsuboyama High Energy Accelerator Research Organization (KEK), Tsukuba, Japan

P. Urquijo School of Physics, University of Melbourne, Melbourne, VIC, Australia

Avelino Vicente Instituto de Física Corpuscular (CSIC-Universitat de València), Valencia, Spain

P. C. Vinodkumar Government Science College, Veer Narmad South Gujarat University, Songadh, Gujarat, India

L. Vitale Dipartimento di Fisica, Università di Trieste, Trieste, Italy;
INFN Sezione di Trieste, Trieste, Italy

M. Watanabe Department of Physics, Niigata University, Niigata, Japan;
Nippon Dental University, Niigata, Japan

Ryoutaro Watanabe Physique des Particules, Université de Montréal, Montréal, QC, Canada

S. Watanuki Department of Physics, Tohoku University, Sendai, Japan

I. J. Watson Department of Physics, University of Tokyo, Tokyo, Japan

J. Webb School of Physics, University of Melbourne, Melbourne, VIC, Australia

J. Wiechczynski The Henryk Niewodniczanski Institute of Nuclear Physics, Krakow, Poland

S. Williams School of Physics, University of Melbourne, Melbourne, VIC, Australia

Lei Xia University of Science and Technology of China, Hefei, People's Republic of China;
State Key Laboratory of Particle Detection and Electronics, Hefei, People's Republic of China

H. Yin Institute of High Energy Physics, Austrian Academy of Sciences, Vienna, Austria

L. Zani Dipartimento di Fisica, Università di Pisa, Pisa, Italy;
INFN Sezione di Pisa, Pisa, Italy

Part I
Heavy Flavour Physics

Chapter 1
Leptonic B Decays at the e^+e^- B Factories

Bob Kowalewski

Abstract Leptonic B decays provide theoretically clean probes that are sensitive to physics beyond the standard model. These decays can be accessed at high luminosity e^+e^- colliders operating near the $\Upsilon(4S)$ resonance, where kinematic constraints and the nearly 4π solid angle coverage of the Belle and BaBar detectors allow for clean measurements of final states that involve non-interacting particles. Charged B decays to lepton pairs will be reviewed, with emphasis on two recent Belle measurements. In addition, a search for the flavour-changing neutral-current decay $B^- \to \Lambda\overline{p}\nu\overline{\nu}$ from BaBar will be presented.

1.1 Introduction

The e^+e^- B meson factories at SLAC and KEK produced collisions at center-of-mass (CM) energies near $\sqrt{s} = 10.58\,\text{GeV}$, where the $\Upsilon(4S)$ resonance is produced. The Belle [1] and BaBar [2] detectors collected large samples of B mesons from $\Upsilon(4S)$ decays to B^+B^- or $B^0\overline{B}^0$ pairs, where each B meson has a known energy in the CM frame. The detectors reconstruct charged particles and photons in >90% of the solid angle and have dedicated particle identification systems to distinguish amongst $e^{\pm}$, $\mu^{\pm}$, $\pi^{\pm}$, $K^{\pm}$ and $p^{\pm}$. The performance of the detectors is well understood, and the detailed detector simulations are accurate in most respects. These properties enable the study of final states involving neutrinos, for which selection criteria include the use of vetoes on additional charged tracks or neutral calorimeter energy.

The pure leptonic decay of a B^+ meson (charge conjugation is implied throughout this paper) is dominated by an annihilation diagram through which a W boson produces a charged-lepton–neutrino pair. The standard model (SM) decay rate to lowest order is

---

Bob Kowalewski—On behalf of the Belle and BaBar collaborations.

B. Kowalewski (✉)
University of Victoria, Victoria, BC, Canada
e-mail: kowalews@uvic.ca

A. Giri and R. Mohanta (eds.), *16th Conference on Flavor Physics and CP Violation*, Springer Proceedings in Physics 234,
https://doi.org/10.1007/978-3-030-29622-3_1

$$\Gamma\left(B^+ \to \ell^+\nu_\ell\right) = \frac{G_F^2 m_B^3 f_B^2}{8\pi} |V_{ub}|^2 x^2(1-x)^2, \tag{1.1}$$

where $x = m_\ell/m_B$ and f_B is the B meson decay constant. The appearance of x^2 in the decay rate reflects the helicity suppression due to the $V - A$ nature of the charged weak interaction. The decay rate can be calculated using a non-perturbative method; a recent lattice QCD calculation [3] gives 186 ± 4 MeV, while a calculation based on QCD sum rules [4] gives 207^{+17}_{-9} MeV. Using the lattice QCD value along with a recent determination [5] $|V_{ub}| = (3.74 \pm 0.14) \times 10^{-3}$, one obtains branching fraction predictions

$$\mathrm{BF}(B^+ \to \tau^+\nu_\tau) = (8.45 \pm 0.70) \times 10^{-5} \text{ (SM)}, \tag{1.2}$$

$$\mathrm{BF}(B^+ \to \mu^+\nu_\mu) = (3.80 \pm 0.31) \times 10^{-7} \text{ (SM)}, \tag{1.3}$$

$$\mathrm{BF}(B^+ \to e^+\nu_e) = (8.89 \pm 0.73) \times 10^{-12} \text{ (SM)}. \tag{1.4}$$

The suppression of these decays in the SM makes them interesting probes for beyond-the-standard-model (BSM) physics. Examples of BSM physics that would affect these decay rates are charged Higgs bosons, leptoquarks, and R-parity-violating supersymmetry (SUSY). With much larger datasets, as are expected at Belle II, the muonic and tauonic decays can, in the absence of BSM physics, be used to determine $|V_{ub}|$ with small theoretical uncertainties.

Decays with a radiated photon in the final state, $B^+ \to \ell^+\nu_\ell\gamma$, have predicted branching fractions of $\sim 10^{-6}$ for both electronic and muonic decays. The SM decay rate is given by

$$\Gamma\left(B \to \ell\nu_\ell\gamma\right) = \alpha \frac{G_F^2 m_B^5}{48\pi^3} |V_{ub}|^2 y^3(1-y)\left(f_A^2(y) + f_V^2(y)\right) \tag{1.5}$$

where α is the fine structure constant, $y = 2E_\gamma/m_B$ in the B rest frame and $f_{A,V}$ are the axial-vector and vector form-factors for the $B \to \gamma$ transition. These form factors have large theoretical uncertainties, and measurements of these decays will help test form-factor calculations.

The experimental status of leptonic decays is tabulated in the Review of Particle Properties [6]. Of particular interest for this talk are the tauonic and muonic decay channels, for which the world-average results are

$$\mathrm{BF}(B^+ \to \tau^+\nu_\tau) = (10.9 \pm 2.4) \times 10^{-5}, \tag{1.6}$$

$$\mathrm{BF}(B^+ \to \mu^+\nu_\mu) < 1.0 \times 10^{-6} \text{ at 90\% CL}, \tag{1.7}$$

$$\mathrm{BF}(B^+ \to e^+\nu_e) < 9.8 \times 10^{-7} \text{ at 90\% CL}. \tag{1.8}$$

For the radiative leptonic decays, the most sensitive current determination [7] gives $\mathrm{BF}\left(B^+ \to \ell^+\nu\gamma\right) < 3.5 \times 10^{-6}$ at 90% CL.

1.2 Analyses at $\Upsilon(4S)$

The e^+e^- B factories run at center-of-mass (CM) energy $\sqrt{s} = 10.58\,\text{GeV}$, where they produce the $\Upsilon(4S)$ resonance which decays almost exclusively $B\overline{B}$ pairs. Each B in these events has momentum $|\mathbf{p}_B| = \sqrt{s/4 - m_B^2} \simeq 325\,\text{MeV}$ in the CM frame. As a result, the decay products of the B mesons are not spatially segregated in the detector. For 2-body B decay modes involving a neutrino (e.g. $B^+ \to \mu^+\nu$), one can select single charged tracks with CM momentum near $m_B/2$, smeared out by the B motion, to achieve a high selection efficiency $\epsilon \sim \mathcal{O}(50\%)$. However, analyses of decays involving more than one neutrino require the assignment of particles to a B meson parent through its full reconstruction in a recognized decay mode. This process is known as "B tagging" and results in a much smaller efficiency. Having reconstructed one B meson, the remaining charged tracks and neutral calorimeter energy deposits are presumed to come from the second B meson, and are compared with the particles expected from the signal decay under study (e.g. $B^+ \to \tau^+\nu \to \pi^+\overline{\nu}\nu$). The two main modalities for tag reconstruction are *hadronic* tags, where the B meson decays into visible particles and where the tag reconstruction efficiency, summed over all decay modes, is $\sim$0.5%, and *semileptonic* tags, where the B meson decays into a D or D^* meson, an electron or muon, and a corresponding unseen neutrino, with a tag reconstruction efficiency of $\sim$1%. The use of hadronic tags provides a determination of the B meson momentum vector, but this does not lead to a useful kinematic constraint if the signal has more than one neutrino. In fact, the lower multiplicity of interacting particles when using semileptonic tags can be an advantage when looking for decays like $B^+ \to \tau^+\nu$ due to the vetoes placed on additional activity in the event selection criteria.

1.2.1 Preliminary Measurement of $B^+ \to \mu^+\nu$ from Belle

The Belle collaboration recently presented an analysis [5] of this decay mode using an untagged method on their full dataset ($7.7 \times 10^8\, B\overline{B}$ events). Candidate events are selected by requiring an identified muon with CM momentum in the range 2.60–2.84 GeV. The invariant mass of the sum of the remaining particles, calculated using the mass hypothesis per track determined from particle identification information, must satisfy $M_{bc} > 5.1\,\text{GeV}$, where $M_{bc} = \sqrt{E_{beam}^2 - |\sum_i \mathbf{p}_i|^2}$. Further requirements are placed on the direction of the muon candidate relative to the thrust axis, and no more than one additional lepton and one K^0 cluster in the outer detector are allowed. Additional discrimination between signal and background is achieved using a neural network based on 14 input parameters that are uncorrelated with the muon momentum. The network output variable, o_{nn}, and the muon momentum, p_μ^*, are used in a binned 2-dimensional fit for the normalizations of each of the following contributions: $B^+ \to \mu^+\nu$, $B \to \pi\ell^+\nu$, $B \to \rho\ell^+\nu$, other $B \to X_u\ell^+\nu$,

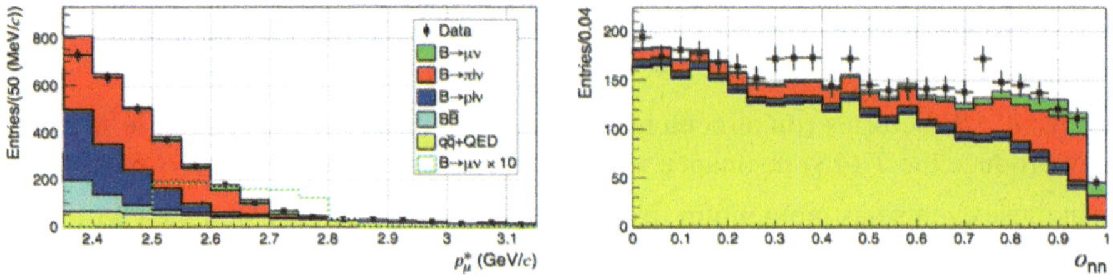

Fig. 1.1 Projections of the data and fitted distributions on the p^*_μ axis for $o_{nn} > 0.84$ and on the o_{nn} axis for $2.6 < p^*_\mu < 2.85$ GeV for the Belle analysis of the decay $B^+ \to \mu^+\nu$. (taken from [5])

other $B\overline{B}$ and $e^+e^- \to c\overline{c}$, $q\overline{q}$ $(q = u, d, s)$, $\tau^+\tau^-$ and $e^+e^-\mu^+\mu^-$ (two-photon production of muon pairs). Additional small backgrounds are estimated using simulation. Projections of the fit output onto the p^*_μ and o_{nn} axes are show in Fig. 1.1.

The fit results in 195 ± 67 signal events and determines the ratio $R = N(B^+ \to \mu^+\nu)/N(B \to \pi\ell^+\nu) = (1.66 \pm 0.57) \times 10^{-2}$. The double ratio R_{data}/R_{MC} is used to extract the ratio of branching fractions

$$\frac{\mathrm{BF}(B^+ \to \mu^+\nu)}{\mathrm{BF}(B \to \pi\ell^+\nu)} = (4.45 \pm 1.53_{stat}) \times 10^{-3}. \tag{1.9}$$

The most significant sources of systematic uncertainty are from the shapes of the $B \to \rho\ell^+\nu$ and $B^+ \to \mu^+\nu$ distributions and of the continuum $c\overline{c}$ and $q\overline{q}$ $(q = u, d, s)$ distributions. After accounting for all uncertainties, the fitted yield corresponds to a 2.4σ indication of a signal with branching fraction

$$\mathrm{BF}(B^+ \to \mu^+\nu) = (6.5 \pm 2.2_{stat} \pm 1.6_{sys}) \times 10^{-7}\ . \tag{1.10}$$

The corresponding 90% CL limits are determined using a frequentist prescription [8]: $\mathrm{BF}(B^+ \to \mu^+\nu) \in [2.9, 10.7] \times 10^{-7}$, consistent with the SM prediction in (1.3).

1.2.2 Belle Measurement of $B^+ \to \tau^+\nu$ Using Semileptonic Tags

The leptonic decay $B^+ \to \tau^+\nu$ results in two or three neutrinos in the final state, depending on the tau decay mode. This makes the use of semileptonic B tags well suited to the analysis of this decay at e^+e^- B factories. The Belle collaboration uses their full data set ($7.7 \times 10^8 B\overline{B}$ events) and selects events where one B meson is reconstructed in a decay $B^- \to D^{(*)0}\ell^-\overline{\nu}$ and the remaining particles in the event are consistent with coming from a τ^+ decay to $e^+\nu\overline{\nu}$, $\mu^+\nu\overline{\nu}$, $\pi^+\overline{\nu}$ or $\rho^+\overline{\nu}$. The tag B is selected using a NeuroBayes multivariate selection (MVS) that considers 10 D^0 decay modes and both $D^{*0} \to D^0\pi^0$ and $D^{*0} \to D^0\gamma$ transitions. Requirements on the MVS output and on $\cos\theta_{B,D^{(*)}\ell}$[1] are optimized separately for each τ^+ decay channel. An important discriminant between signal and background events is E_{ECL}, the sum of all calorimeter energy deposits (above a noise threshold) that are not associated with either B candidate. This quantity is sensitive to both beam-related backgrounds and the detailed modeling of the detector. The simulation of E_{ECL} is validated using a sample of events where the tag B is reconstructed as described above and the other B is reconstructed in the decay $B^+ \to \overline{D}^0\pi^+$, and this comparison is used to estimate systematic uncertainties due to the modeling of this quantity.

Events satisfying the $B^- \to D^{(*)0}\ell^-\overline{\nu}$, $B^+ \to \tau^+\nu$ decay followed by a $\tau^+ \to$ 1-prong decay enter an unbinned maximum-likelihood fit that takes as input the p^*_{sig} and E_{ECL} values from each event. The fit determines the branching fraction for $B^+ \to \tau^+\nu$ along with the background normalization in each tau decay channel. The shapes of the fitted components are fixed. Positive signals are seen in each channel; the combined yield is 222 ± 50 $B^+ \to \tau^+\nu$ decays. After accounting for systematic uncertainties, which are dominated by the description of the background composition and component shapes and by the determination of the selection efficiency, the resulting measurement is

$$\mathrm{BF}(B \to \tau^+\nu) = (1.25 \pm 0.28 \pm 0.27) \times 10^{-4}, \tag{1.11}$$

consistent with the SM expectation, (1.2). The significance of the observed signal yield is 3.8σ. A clear, 5σ observation of this decay mode will have to await data from Belle II.

1.2.3 Preliminary Search for $B^- \to \Lambda\overline{p}\nu\overline{\nu}$ from BaBar

This decay is the analogue to $B \to K\nu\overline{\nu}$ with baryons in the final state. The leading SM diagrams are shown in Fig. 1.2. There are no previously reported experimental studies of this decay. Its predicted [9] branching fraction is

$$\mathrm{BF}(B^- \to \Lambda\overline{p}\nu\overline{\nu}) = (7.9 \pm 1.9) \times 10^{-7}\ (\mathrm{SM}). \tag{1.12}$$

Given the small SM rate, this decay mode is a sensitive probe for BSM physics processes that produce flavor-changing neutral-currents (FCNC).

Due to the presence of two neutrinos in the signal decay, one B in the event must be tagged. The analysis reconstructs the tag B by considering thousands of hadronic decay chains following the procedure described in [10]. The remaining particles are required to be consistent with the signal topology, namely a $p\pi^-$ combination that forms a good Λ candidate and an additional identified anti-proton. This combination

[1] $\cos\theta_{B,D^{(*)}\ell} = (2E_{beam}E_{D^{(*)}\ell} - m_B^2 - m^2_{D^{(*)}\ell})/(2p^*_B p^*_{D^{(*)}\ell})$ where all quantities are in the CM frame and p^*_B is the nominal B meson momentum calculated from the beam energy E_{beam} and the nominal mass m_B.

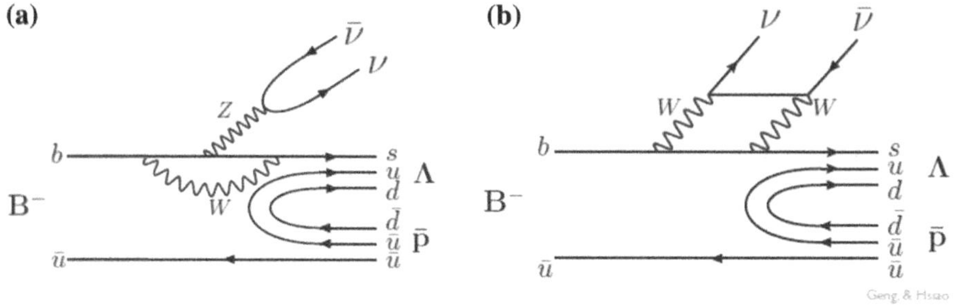

Fig. 1.2 The leading diagrams in the standard model for the decay $B^- \to \Lambda\overline{p}\nu\overline{\nu}$ (taken from [9])

of a tag B and identified baryons strongly suppresses backgrounds. Further reductions in background come from a multivariate discriminant that reduces the contribution from $e^+e^- \to q\overline{q}$ continuum interactions and a requirement that the total calorimeter energy unassociated with any tag- or signal-B decay particles satisfies $E_{extra} <$ 0.4 GeV. The final selection, frozen prior to unblinding the data, has an estimated signal efficiency of $(3.42 \pm 0.08 \pm 0.80) \times 10^{-4}$ with an expected background of $2.3 \pm 0.7 \pm 0.6$ events. The observed number of events in the signal region was three, consistent with the background expectation. The resulting branching fraction is

$$\mathrm{BF}(B^- \to \Lambda\overline{p}\nu\overline{\nu}) = (0.4 \pm 1.1 \pm 0.6) \times 10^{-5} \tag{1.13}$$

which corresponds to a 90% CL frequentist upper limit [11] of 3.0×10^{-5}. The larger datasets expected from Belle II will allow significant gains in sensitivity for this channel.

1.3 Summary

Leptonic B decays are sensitive to BSM physics, and e^+e^- B factories have unique sensitivities for decays that involve one or more neutrinos or other invisible particles. The current analyses based on the full Belle and BaBar datasets allow the $B^+ \to \tau^+\nu$ decay to be measured with a significance of approximately 4σ. A new Belle analysis has seen the first indications of the decay $B^+ \to \mu^+\nu$ and set an upper limit on the branching fraction of less than three times the SM prediction. These results are consistent with other measurements [6] of these decay modes.

A rare FCNC decay involving two neutrinos, $B^- \to \Lambda\overline{p}\nu\overline{\nu}$, has been searched for by BaBar. An upper limit was determined on the branching fraction for the first time, albeit at a level well above that expected in the SM.

These analyses have not revealed any evidence for departures from the SM. With the large dataset expected from Belle II, leptonic B decays will be sensitive to a variety of BSM physics, including charged Higgs bosons, leptoquarks and R-parity-violating SUSY, and may provide a complementary way to measure $|V_{ub}|$.

References

1. A. Abashian et al. (Belle Collab.), Nucl. Instrum. Methods A **479**, 117 (2002)
2. B. Aubert et al. (BaBar Collab.), Nucl. Instrum. Methods A **729**, 615 (2013); *ibid* **479**, 1 (2002)
3. R.J. Dowdall et al. (HPQCD Collab.), Phys. Rev. Lett. **110**, 222003 (2013)
4. P. Gelhausen et al., Phys. Rev. D **88**, 014015 (2013); Errata *ibid* **89**, 099901 (2014); **91**, 099901 (2015)
5. A. Sibidanov et al. (Belle Collab.), Phys. Rev. Lett. **121**, 031801 (2018)
6. M. Tanabashi et al. (Particle Data Group), Phys. Rev. D **98**, 030001 (2018)
7. A. Heller et al. (Belle Collab.), Phys. Rev. D **91**, 112009 (2015)
8. G.J. Feldman, R.D. Cousins, Phys. Rev. D **57**, 3873 (1998)
9. C.Q. Geng, Y.K. Hsiao, Phys. Rev. D **85**, 094019 (2012)
10. J.P. Lees et al. (BaBar Collab.), Phys. Rev. D **88**, 072012 (2013)
11. R. Barlow, Comput. Phys. Commun. **149**, 97 (2002)

Chapter 2
Status of $B_{d,s} \to \mu^+\mu^-$ at the LHC

Kajari Mazumdar

Abstract Salient features of the measurements of $B_{d,s} \to \mu^+\mu^-$ by LHC experiments are reported including results for the branching fractions and effective lifetime for $B_s \to \mu^+\mu^-$ decay.

2.1 Introduction

The field of high energy physics is knocking at the heaven's door. The direct searches at the CERN-LHC have not yet yielded any positive indication for existence of beyond Standard Model (SM) physics at TeV energy scale. The search for New Physics (NP) has become a long drawn-out effort, which is sustained by the exciting deviations in several B physics measurements compared to the SM expectations. Surely with only about few % of data delivered at the LHC till now, the hunt for NP will continue and, probably, more innovative strategy is required. In recent times, search for NP in heavy flavour sector, has become very interesting, due to several deviations observed in data compared to expectations from the SM. However the interpretations remains inconclusive awaiting more measurements. In general NP can become manifest in several ways in heavy flavor sector: (i) via enhancement or suppression of decay rates, (ii) via introduction of a new source of CP violation, or, (iii) via modification of the angular distribution of final state particles. $B_{d,s} \to \mu^+\mu^-$ processes can be subtle players in this endeavour. These studies complement the searches for NP via the processes $b \to s\gamma$, $\mu \to e\gamma$ etc.

The decays of $B_{d,s} \to \mu^+\mu^-$ have very clean experimental signatures and have been searched extensively during last several decades. Theoretically, the rates are very low in the SM, since flavour changing neutral current is forbidden at the tree level. Furthermore there is a helicity suppression factor of m_μ^2/M_B^2. The transitions

Kajari Mazumdar—On behalf of ATLAS, CMS and LHCb Collaborations.

K. Mazumdar (✉)
Tata Institute of Fundamental Research, Mumbai 400005, India
e-mail: mazumdar@tifr.res.in

A. Giri and R. Mohanta (eds.), *16th Conference on Flavor Physics and CP Violation*, Springer Proceedings in Physics 234,
https://doi.org/10.1007/978-3-030-29622-3_2

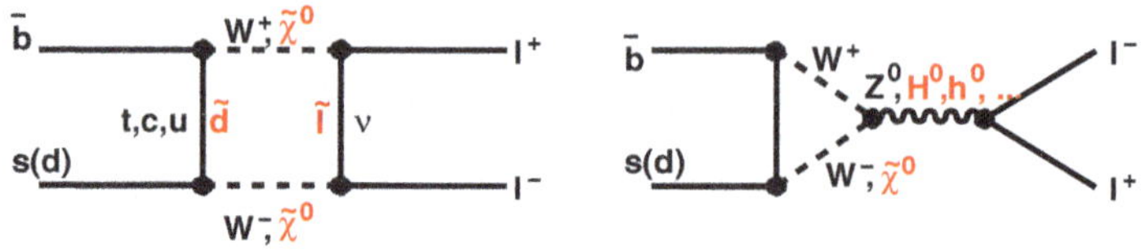

Fig. 2.1 Representative Feynman diagrams for $B_s \to \mu^+\mu^-$ transition. Possible contributions from beyond SM physics is indicated in terms of the particles in red

proceed via loops as shown by the representative Feynman diagrams in Fig. 2.1. These processes are excellent probes for NP, since new, heavy particles envisaged in scenarios of NP, can appear in the loops. The additional contributions may lead to distinct modifications in the individual rates.

In SM the branching fractions for both B_d and B_s have been calculated accurately, taking into account the best value of top quark mass as well as higher order contributions from electroweak (next-to-leading order) and strong interactions (next-to-next-to-leading order). The predictions are [1]:

$$\mathcal{B}(B_d \to \mu^+\mu^-) = (1.06 \pm 0.09) \times 10^{-10} \ \& \ \mathcal{B}(B_s \to \mu^+\mu^-) = (3.57 \pm 0.17) \times 10^{-9}$$

The lower value of the branching fraction for B_d compared to that of B_s is due to the fact that $V_{td} < V_{ts}$. The theoretical uncertainties are dominated by those in CKM matrix elements as well as in the relevant decay constants f_{B_d} or f_{B_s}. In the ratio of the branching fractions, some of the systematic uncertainties get cancelled and hence it is also an interesting quantity to be measured in the experiment with the estimated value of $R = 0.0295^{+0.0028}_{-0.0025}$ in the SM.

2.1.1 New Physics Effects

As evident from Fig. 2.1 the decays typically involve at least two different energy scales: the electroweak scale, characterized by the weak vector boson mass, which determines the transition at the quark level, and the scale of strong interactions Λ_{QCD}, related to the hadron formation. Hence the decay amplitudes contain large logarithms of the type $\log(m_W/\Lambda_{QCD})$ and perturbation theory cannot be applied. Here m_W is the mass of W boson. However by integrating out the heavy SM fields (W and Z bosons, as well as the top quark) at the electroweak scale, a low-energy effective theory can be constructed in terms of only the light SM fields. Thus the weak effective Lagrangian has local operators of dimension six or higher, involving light SM fermions, photon and gluon fields, and is suppressed by inverse powers of m_W.

In effective theory formalism via operator product expansion method, the nature of possible new interaction can be deciphered in a model-independent way in terms of effective Hamiltonian $\mathcal{H}_{\text{eff}}$ given by

$$\mathcal{H}_{\text{eff}} = -\frac{4G_F}{\sqrt{2}} V_{tb} V_{ts}^* \Sigma_i \left[\mathcal{C}_i(\mu)\mathcal{O}_i(\mu) + \mathcal{C}_i'(\mu)\mathcal{O}_i'(\mu) \right].$$

where V_{tb}, V_{ts}^* are the CKM matrix elements, and the non-perturbative, long distance aspects of QCD, are indicated by the operators $\mathcal{O}_i(\mu)$ and $\mathcal{O}_i'(\mu)$, though the latter is suppressed in the SM. The Wilson coefficients $C_i(\mu)$ correspond to perturbative, short distance physics, sensitive at an energy higher than the electroweak scale and carry the important information about the interaction. Here $i = 1, 2$ indicate tree level processes while $i = 3$–6 and 8 refer to gluon penguin, $i = 7$ to photon penguin, $i = 9, 10$ to electroweak penguin, $i = S$ to Higgs (scalar) penguin, and, $i = P$ for pseudoscalar penguin. Interestingly, the purely leptonic exclusive decays can be only due to the axial-current operator in the SM, C_{10}, and the short-distance contribution is not affected by the mixing with four-quark operators. In principle, NP can modify Wilson coefficients, as well as induce new operators.

For $B_s \to \mu^+\mu^-$ in the SM, there is only the axial-vector contribution, C_{10}. Thus the partial width is expressed as

$$\Gamma(B_s \to \mu^+\mu^-) \sim \frac{G_F^2\alpha^2}{64\pi^3} m_{B_s}^2 f_{B_s}^2 |V_{tb}V_{ts}|^2 |2m_\mu C_{10}|^2$$

Here the accurate value of f_{B_s} comes as input from lattice calculations and V_{tb} is to be determined experimentally. In the presence of an extended Higgs sector, the terms C_s and C_p contribute additionally and the partial width is given by

$$\Gamma(B_s \to \mu^+\mu^-) \sim |V_{tb}V_{tq}| \left[\left(1 - \frac{4m_\ell^2}{m_B^2}\right) |C_s - C_s'|^2 + |(C_P - C_P') + \frac{2m_\ell}{m_B}(C_{10} - C_{10}')|^2 \right]$$

Thus in the models for minimal flavour violation (MFV) the rates can be significantly higher. Hence accurate measurement of $B_s \to \mu^+\mu^-$ provides strong constraints on NP via the Wilson coefficients.

2.2 B Physics at the LHC

The decay $B_s \to \mu^+\mu^-$ has been searched for almost 30 years before observation at the LHC. Until recently, the B-factory machines fell short of the required energy to produce B_s mesons. Having a very large production cross section of about 300 and 700μb at centre-of-mass energy $\sqrt{s} = 7$ and 13 TeV respectively, the LHC provides an unique opportunity to study the rare processes. There can be three modes of production of an inclusive $b\bar{b}$ pair, corresponding to the processes of flavor creation,

Table 2.1 Comparison of salient features of LHC experiments

Parameter	ATLAS/CMS	LHCb
Instantaneous luminosity (cm^{-2} s^{-1})	1×10^{34}	2×10^{32}
Avg. interactions/crossing	50	0.5
$b\bar{b}$ events/10^7 s	$5 \times 10^{13} \times$ accept	$1 \times 10^{12} \times$ accept
Track measurement (mrad)	$\theta > 220$	$10 < \theta < 300$
p_T threshold for trigger (GeV)	4 (3)	1.5
$m_{\mu^+\mu^-}$ mass resolution (MeV)	32–75 (η dependent)	25
Proper time resolution (fs)	77	36

flavor excitation and gluon splitting. Only about 10% of the produced b-quarks hadronize to B_s meson. The b-hadrons have typically large boost in the forward-backward direction, leading to an average total momentum of a B meson to be about 100 GeV at $\sqrt{s} = 13$ TeV. Further, in a large fraction of events, b and $\bar{b}$ are produced back-to-back. In terms of capabilities, there is significant complementarity between LHCb, the dedicated b physics related experiment using forward production in one arm and CMS as well as ATLAS, which are multi-purpose experiments designed originally for high-p_T physics. The b-quark lifetime being about 1.6 ps, it is extremely important to measure the secondary vertex with precision in the experiment. The LHC experiments are equipped with excellent tracking detectors which play a key role in providing a high resolution measurement of the secondary vertices of b decay and the derived observables. A comparison between the two types of detectors has been made in Table 2.1 considering the scenario of Run-2 operation of the LHC during 2015–2018. It is to be noted that for LHCb the b event selection efficiency using a low p_T trigger threshold is much higher than for CMS/ATLAS.

2.3 Measurement of Branching Fractions

Several interesting measurements have been performed [2] by the experiments using Run-1 proton-proton collision data at $\sqrt{s} = 7$ and 8 TeV. For LHCb the data volume correspond to an integrated luminosity $\mathcal{L}$ of 1 and 2 fb^{-1} respectively while for CMS it is $\mathcal{L} = 25$ fb^{-1} at $\sqrt{s} = 8$ TeV. The results have yielded 6.2 (3.0)σ significance observation for $B_s \to \mu^+\mu^-$ ($B_d \to \mu^+\mu^-$) decay. The measured branching fractions are within 1.2 (2.2)σ of the SM predictions and are given by

$$\mathcal{B}(B_s \to \mu^+\mu^-) = \left(2.8^{+0.7}_{-0.6}\right) \times 10^{-9} \;\&\; \mathcal{B}(B_d \to \mu^+\mu^-) = \left(3.9^{+1.6}_{-1.4}\right) \times 10^{-9}$$

The signal strength, defined as the ratio of observation wrt the SM expectation, is $0.76^{+0.20}_{-0.18}$ ($3.9^{+1.6}_{-1.4}$) for $B_s \to \mu^+\mu^-$ ($B_d \to \mu^+\mu^-$). Further the ratio of branching

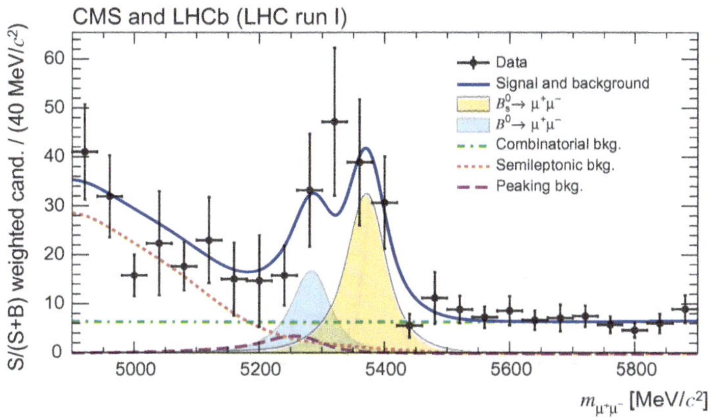

Fig. 2.2 Invariant mass distribution for $B_s \to \mu^+\mu^-$ and $B_d \to \mu^+\mu^-$ candidates (yellow and blue shaded components respectively); data points are shown in black [2]

fractions B_d/B_s is measured to be $R = 0.14^{+0.06}_{-0.08}$ which is within 2.3σ of SM. The dimuon invariant mass distribution for combined LHCb and CMS data is displayed in Fig. 2.2.

A preliminary measurement by ATLAS Collaboration using part of Run-2 data corresponding to $\sqrt{s} = 13$ TeV and $\mathcal{L} = 23.6$ fb^{-1} yields a branching ratio value of $\mathcal{B}(B_s \to \mu^+\mu^-) = (3.2^{+1.1}_{-1.0}) \times 10^{-9}$ and an upper limit for $\mathcal{B}(B_d \to \mu^+\mu^-) < 4.3 \times 10^{-10}$ at 95% confidence level. After combination with Run-1, the results from ATLAS Collaboration [3] are $\mathcal{B}(B_s \to \mu^+\mu^-) = (2.8^{+0.8}_{-0.7}) \times 10^{-9}$ and $\mathcal{B}(B_d \to \mu^+\mu^-) < 2.1 \times 10^{-10}$, which is within 2.4σ of the SM expectation.

The LHCb analysis of Run-2 data corresponding to $\sqrt{s} = 13$ TeV and $\mathcal{L} = 1.4$ fb^{-1} has been combined with that of Run-1 data amounting to 3 fb^{-1} leading to the first observation of $B_s \to \mu^+\mu^-$ process by a single experiment with a significance of 7.8σ. However there is no evidence of $B_d \to \mu^+\mu^-$ and the corresponding upper limit on the branching fraction is $\mathcal{B}(B_d \to \mu^+\mu^-) < 3.4 \times 10^{-10}$ based on CLs method; this is smaller than the Run-1 measurement.

The analysis of Run-2 data has been performed with an improved strategy resulting in better rejection of background compared to Run-1 analysis. The resolution of dimuon invariant mass $m_{\mu^+\mu^-}$ is improved due to the use of new signal isolation criteria and enhanced rejection of di-hadron background due to better particle identification criteria. Hence a narrower window of $m_{\mu^+\mu^-}$ has been considered along with a new multivariate BDT analysis based on kinematics, geometrical and isolation variables for the dimuon system. Figure 2.3 presents the mass distribution of selected $B_s \to \mu^+\mu^-$ candidates in LHCb. The signal peak position is calibrated with the decay modes $B_S \to K^+K^-$ and $K^\pm\pi^\mp$. The normalization channel is $B^+ \to J/\psi K^+$. The background has been estimated using a data driven method combined with Monte Carlo samples as well as theoretical inputs. The fraction of hadronization f_d/f_s is estimated using the ratio of $B^+ \to J/\psi K^+$ to $B_s \to J/\psi\phi$;

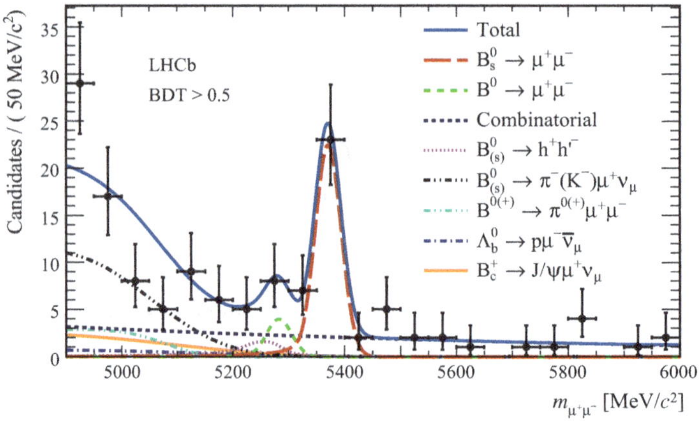

Fig. 2.3 Invariant mass distribution for $B_s \to \mu^+\mu^-$ candidates along with the fit result and different components [5]

it depends on $\sqrt{s}$ and has increased by 6.8% from Run-1 to Run-2. In the finally selected dimuon sample 62 (7) events of type $B_s(B_d) \to \mu\mu$ are expected over the whole BDT range. Branching fractions have been determined from an unbinned maximum likelihood fit in high BDT region, while signal and exclusive background fractions are constrained to their expected values.

2.4 Measurement of the Effective Lifetime

$B_s - \bar{B}_s$ oscillation leads to CP-even and CP-odd mass eigenstates with the difference in decay widths between the heavy and the light states given by $\Delta\Gamma = \Gamma(B_{s,H} \to \mu^+\mu^-) - \Gamma(B_{s,L} \to \mu^+\mu^-) = 0.082 \pm 0.007\,\text{ps}$. In the SM only the heavy state decays to dimuon final state. This is in contrast to the situation in presence of large CP violation in B_s system in certain scenarios of NP. The effective lifetime for the dimuon decay, defined as

$$\tau_{\mu^+\mu^-} = \frac{\int_0^\infty t \cdot \Gamma(B_s(t) \to \mu^+\mu^-)dt}{\int_0^\infty \Gamma(B_s(t) \to \mu^+\mu^-)dt}$$

is a complementary probe for NP. Here,

$$\Gamma(B_s(t) \to \mu^+\mu^-) \equiv \frac{1}{2}[\Gamma(B_s(t) \to \mu^+\mu^-) + \Gamma(\bar{B}_s(t) \to \mu^+\mu^-)]$$

Presence of NP affects the decay widths $\Gamma(B_{s,H})$ and $\Gamma(B_{s,L})$ differently and hence the asymmetry between them is defined as [4]

$$A_{\Delta\Gamma}^{\mu^+\mu^-} = \frac{\Gamma(B_{s,H} \to \mu^+\mu^-) - \Gamma(B_{s,L} \to \mu^+\mu^-)}{\Gamma(B_{s,H} \to \mu^+\mu^-) + \Gamma(B_{s,L} \to \mu^+\mu^-)}$$

Interestingly, $A_{\Delta\Gamma}^{\mu^+\mu^-}$ can assume any value in the range [−1,+1] for NP and is equal to +1 in the SM, since only $B_{s,H}$ decays to $\mu^+\mu^-$. It has been shown that the effect of NP on this asymmetry is orthogonal to the effect on the branching fraction. The effective lifetime of B_s decay involves this asymmetry:

$$\tau_{\mu^+\mu^-} = \frac{\tau_{B_s}}{1 - y_s^2}\left[\frac{1 + 2A_{\Delta}^{\mu^+\mu^-} y_s + y_s^2}{1 + 2A_{\Delta}^{\mu^+\mu^-} y_s}\right]$$

Here $y_s \equiv \tau_{Bs}\Delta\Gamma/2 = 0.062 \pm 0.006$ and $\tau_{B_s} = 1/\Gamma_s$, the average decay width. Thus an accurate measurement of $\tau_{\mu^+\mu^-}$ will potentially indicate the presence of NP.

The effective lifetime is estimated by LHCb Collaboration from signal weighted decay lifetime fit as [5]

$$\tau_{\mu^+\mu^-} = 2.04 \pm 0.44 \pm 0.05 \text{ ps}$$

The fit, as shown in Fig. 2.4, is performed in 2 steps, first to dimuon invariant mass distribution in a tighter mass window avoiding the peak region of $B_d \to \mu^+\mu^-$. Using $\mathcal{S}$-plot technique of calculating event weight, the background is subtracted from data before fitting the distribution to extract $\tau_{\mu^+\mu^-}$. The acceptance function for the decay time has been modeled on simulated events of $B_s \to \mu^+\mu^-$ and is validated using $B_d \to K\pi$ events.

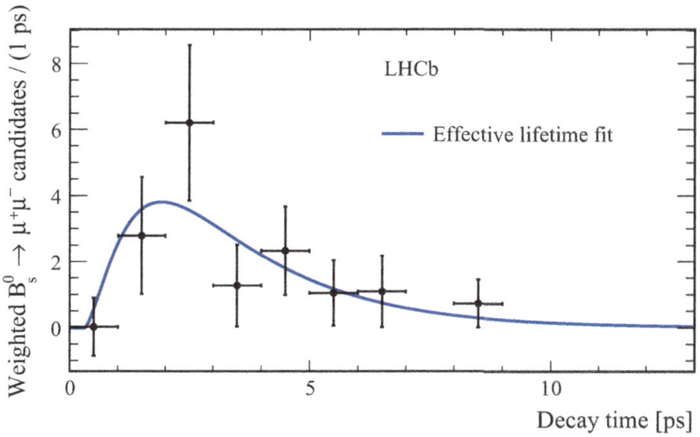

Fig. 2.4 Background subtracted $B_s \to \mu^+\mu^-$ decay time distribution with the fit result imposed [5]

This value can be compared with the SM prediction of 1.510 ± 0.005 ps. This is also consistent with $A_{\Delta\Gamma} = 1(-1)$ at $1.0(1.4)\sigma$ level. It is expected that with a data sample corresponding to $\mathcal{L}$ of about $50\,\text{fb}^{-1}$ to be collected by LHCb experiment during Run-3 (2021–2023) the effective lifetime can be measured with an accuracy of about 5%.

2.5 Conclusion

Results of the measurement of $B_{d,s} \to \mu^+\mu^-$ from LHC experiments have been presented. It is noted, that at present the theoretical error due to uncertainties in V_{CKM} and f_{Bs} are well below the statistical error for the measurement of branching fraction. Analyses based on larger samples of Run-2 data and the eventual combinations from all experiments is expected to yield results of much higher precision leading to interesting interpretations. Analysis of data to measure the equally important process $B_s \to \mu^+\mu^-\gamma$ is going on at present and the result of the branching fraction will be very interesting to compare with the expectation from the SM which predicts a value comparable to the rate for $B_s \to \mu^+\mu^-$. However direct measurement is highly challenging.

References

1. C. Bobeth, M. Gorbahn, T. Hermann, M. Misiak, E. Stamou, M. Steinhauser, $B_{s,d} \to \ell^+\ell^-$ in the standard model with reduced theoretical uncertainty. Phys. Rev. Lett. **112**, 101801 (2014). https://doi.org/10.1103/PhysRevLett.112.101801
2. CMS and LHCb Collaborations, Observation of the rare $B_s \to \mu^+\mu^-$ decay from the combined analysis of CMS and LHCb data. Nature **522** (2015) 68. https://doi.org/10.1038/nature14474
3. ATLAS Collaboration, Study of the rare decays of B_s and B_d into muon pairs from data collected during 2015 and 2016 with the ATLAS detector, in *ATLAS-CONF-2018-046*. http://cds.cern.ch/record/2639673
4. K.D. Bruyn, R. Fleischer, R. Knegjens, P. Koppenburg, M. Marcel, A. Pellegrino, N. Tuning, Probing new physics via the $B_s \to \mu^+\mu^-$ effective lifetime. Phys. Rev. Lett. **109**, 041801 (2012). https://doi.org/10.1103/PhysRevLett.109.041801
5. LHCb Collaboration, Measurement of the $B_s \to \mu^+\mu^-$ branching fraction and effective lifetime and search for $B_d \to \mu^+\mu^-$ decays. Phys. Rev. Lett. **118**, 191801 (2017). https://doi.org/10.1103/PhysRevLett.118.191801

Chapter 3
Radiative and Electroweak Penguin Decays at e^+e^- B-Factories

S. Sandilya

Abstract B-meson decays involving radiative and electroweak penguin processes are sensitive probes to new physics beyond the standard model. The Belle experiment recently reported measurements of the inclusive radiative decay $B \to X_s\gamma$, and exclusive radiative decays $B \to K^*\gamma$ and $B \to K_S^0\eta\gamma$. A lepton-flavor dependent measurements of angular observables for the decays $B \to K^*\ell\ell$ by Belle hinted at possible deviation from lepton-flavor-universality. Any departure from lepton flavor universality is essentially accompanied by lepton flavor violation. Recently, lepton-flavor-violating decays $B^0 \to K^{*0}\mu^{\pm}e^{\mp}$ are searched for at Belle, and stringent limits on their branching fractions are set. The BaBar experiment has searched for the decay $B^+ \to K^+\tau^+\tau^-$, which comprises third generation of the lepton family. The decays $B \to h\nu\bar{\nu}$ are searched for at Belle and obtained upper limits for these decays are close to the standard model predictions.

3.1 Introduction

The B-factories, Belle and BaBar experiments, were located at the interaction region of e^+e^- asymmetric colliders of KEKB and PEP-II, respectively. These B-factories had about a decade long very successful operational period, and recorded a combined data sample over 1.5 ab^{-1}, which corresponds to more than 1.2×10^9 B-meson pairs. In these experiments electron and positron beams collide at the $\Upsilon(4S)$ resonance, which leads to a clean sample of quantum correlated pairs of B-mesons and makes analyses with missing final states straightforward. Also, a low background environment at the B-factories enables an efficient reconstruction of the neutral particles.

In the standard model (SM), flavor changing neutral current (FCNC) processes are forbidden at tree level and proceed via penguin loop or box-diagrams at lowest order. In these loops, non-SM heavy particles can also enter. Thus, FCNC processes involving $b \to s$ quark-level transition are among the most sensitive probes for the

S. Sandilya (✉)
University of Cincinnati, Cincinnati, OH 45221, USA
e-mail: saurabhsandilya@gmail.com

A. Giri and R. Mohanta (eds.), *16th Conference on Flavor Physics and CP Violation*, Springer Proceedings in Physics 234,
https://doi.org/10.1007/978-3-030-29622-3_3

new physics (NP) beyond the SM. Recently, B-factories provided important measurements involving radiative and electroweak penguin B decays. The measurements from Belle and BaBar experiments reviewed here, are based on the total recorded data sample of 711 and 424 fb^{-1}, respectively.

3.2 Measurements of $B \to X_s\gamma$ and $B \to K^*\gamma$

The FCNC transition $b \to s\gamma$ proceeds dominantly through electromagnetic penguin diagrams and, is sensitive to NP. The branching fraction (BF) of the decay $B \to X_s\gamma$ [1], $\mathcal{B}(B \to X_s\gamma)$ [2, 3] is consistent with the SM prediction and constrains NP. The uncertainty in the SM prediction of $\mathcal{B}(B \to X_s\gamma)$ is about 7% [4], which is close to the current experimental uncertainties. The upcoming Belle II experiment is expected to further improve the uncertainties in measurement to about 3% [5]. On the other hand, the dominant uncertainty in the SM prediction is due to non-pertubative effects, and it is related to the isospin asymmetry (Δ_{0-}) in the decay $B \to X_s\gamma$ [6]. If Δ_{0-} is found to be zero, then it will lead to the reduction in uncertainty in the SM prediction of $\mathcal{B}(B \to X_s\gamma)$. Another interesting observable sensitive to NP is the difference of direct CP asymmetries between the B^+ and B^0 mesons: $\Delta A_{CP} = A_{CP}(B^+ \to X_s^+\gamma) - A_{CP}(B^0 \to X_s^0\gamma)$ [7]. Any significant deviation of ΔA_{CP} from zero will indicate the presence of NP [7–9].

Recently, Belle reported measurements of Δ_{0-} and ΔA_{CP} for the decay $B \to X_s\gamma$, where X_s is reconstructed from 38 exclusive final states [10]. Among these reconstructed modes, 11 are flavor non-specific modes, which are only used for Δ_{0-} measurements. The result for $\Delta_{0-} = (-0.48 \pm 1.49 \pm 0.97 \pm 1.15)\%$ is found to be consistent with zero, where the first uncertainty is statistical, the second is systematic and the third is due to uncertainty in the BF ratio of the $\Upsilon(4S) \to B^+B^-$ and $\Upsilon(4S) \to B^0\bar{B}^0$ decays. This measured value of Δ_{0-} will be important in improving the theoretical uncertainty in $\mathcal{B}(B \to X_s\gamma)$. The obtained ΔA_{CP} value is $(+3.69 \pm 2.65 \pm 0.76)\%$, which is also consistent with zero as well as with the SM prediction, and hence can be used to constrain NP.

In another analysis of the exclusive decay $B \to K^*\gamma$, Belle reported measurements of Δ_{0+} and A_{CP} [11]. These BF ratios Δ_{0+} and A_{CP} provide a strong constraint on NP, as form-factor related uncertainties in the theoretical prediction cancel out [12]. In this analysis, the first evidence of isospin violation is reported with a significance of 3.1 standard deviations (σ), with a value of $\Delta_{0+} = (+6.2 \pm 1.5 \pm 0.6 \pm 1.2)\%$, where the third uncertainty is again due to the BF ratio of the $\Upsilon(4S) \to B^+B^-$ and $\Upsilon(4S) \to B^0\bar{B}^0$ decays. The results for $A_{CP}(B \to K^*\gamma) = (-0.4 \pm 1.4 \pm 0.3)\%$ and, $\Delta A_{CP} = (+2.4 \pm 2.8 \pm 0.5)\%$ are also reported. All these measurements are consistent with the SM.

3.3 Measurement of Time-Dependent *CP* Asymmetries in $B^0 \to K_S^0\eta\gamma$

According to the SM, the photon polarization in the $b \to s\gamma$ transition is predominantly left-handed. Right-handed currents can, however, enter in the loop through various NP models and enhancing the right-handed photon polarization. A promising avenue to observe such NP scenarios is the measurement of time-dependent *CP* violation in a decay of the form $B \to P_1 P_2\gamma$, where P_1 and P_2 are scalar or pseudoscalar mesons and the $P_1 P_2$ system is a *CP* eigenstate [13]. A small mixing-induced *CP* violation parameter ($\mathcal{S}$) is generated via interference between the $\bar{B}^0 \to P_1 P_2\gamma_{L(R)}$ and $B^0 \to P_1 P_2\gamma_{L(R)}$ decays. Thus, the value of $\mathcal{S}$ could be enhanced by the NP related right-handed currents.

Belle and BaBar, have measured *CP* violation parameters in the decays $B^0 \to K_S^0\pi^0\gamma$ (including $K_S^0\pi^0$) [14, 15], $B^0 \to K_S^0\eta\gamma$ [16], $B^0 \to K_S^0\rho^0\gamma$ [17, 18], and $B^0 \to K_S^0\phi\gamma$ [19]. These results are consistent with SM predictions [20–22]. First measurement of time-dependent *CP* asymmetries in the decay $B^0 \to K_S^0\eta\gamma$ is reported by Belle [23]. The obtained parameters are

$$\mathcal{S} = -1.32 \pm 0.77(\text{stat.}) \pm 0.36(\text{syst.}),$$
$$\mathcal{A} = -0.48 \pm 0.41(\text{stat.}) \pm 0.07(\text{syst.}),$$

that lie outside the physical boundary, defined by $\mathcal{S}^2 + \mathcal{A}^2 = 1$, shown in Fig. 3.1. These measurements are consistent with the null-symmetry hypothesis within 2σ as well as with SM predictions.

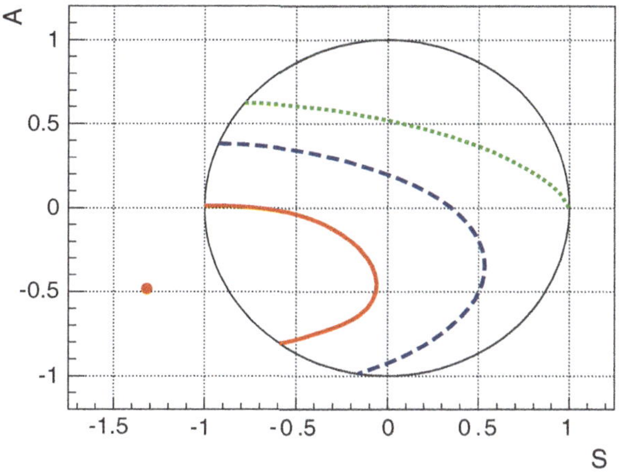

Fig. 3.1 The solid red, dashed blue and dotted green curves show the 1σ, 2σ and 3σ confidence contours, respectively. The red dot shows the Belle result, which is is consistent with a null asymmetry within 2σ [23]. The physical boundary $\mathcal{S}^2 + \mathcal{A}^2 = 1$ is drawn with a thin solid black curve

3.4 Angular Analysis of $B \to K^*\ell^+\ell^-$

The decay $B \to K^*\ell^+\ell^-$ involves a quark-level $b \to s\ell^+\ell^-$ FCNC transition and proceeds via electroweak penguin or a box diagrams. NP particles may enter in these loops, and thus can alter the BF and angular distributions of the final-state particles [24]. Interestingly, in the recent years, several measurements have shown possible deviations from the SM for a number of decays involving $b \to s\ell^+\ell^-$ transition [25–27]. Global fits are performed including experimental and theoretical correlations, and these fits hint at possible lepton flavor universality (LFU) violation [28]. The angular observables P_i' for the $B \to K^*\ell^+\ell^-$ is introduced in [29], which are considered mostly to be free from form-factor related uncertainities [30]. And, also if the differences of P_i' between the muon and the electron modes, $Q_i = P_i'^{\mu} - P_i'^{e}$ ($i = 4, 5$), deviates from zero, it would be a signature of NP [31].

Belle has reported a measurement of angular observables, P_i' for both lepton flavors separately and Q_i, in the decay $B \to K^*\ell^+\ell^-$ [32]. In this measurement, the $B^+ \to K^{*+}\ell\ell$ decays are reconstructed along with $B^0 \to K^{*0}\ell\ell$ decays, where K^{*+} is reconstructed from $K^+\pi^0$ or $K_S^0\pi^+$ and K^{*0} from $K^+\pi^-$. The analysis is performed in the four independent bins of q^2 (invariant mass square of the two leptons). Comprehensively, the results are compatible with SM predictions [31]. The largest deviation of 2.6σ from the SM is observed for P_5' of the muon modes in the $q^2 \in (4.0, 8.0)$ GeV^2/c^2 bin, in the same bin electron modes deviate by 1.3σ, and both combined the deviation is about 2.5σ.

The $Q_{4,5}$ observable is shown in Fig. 3.2, where no significant deviation from zero is observed. Global fits including this Belle result hint at LFU violation [33, 34].

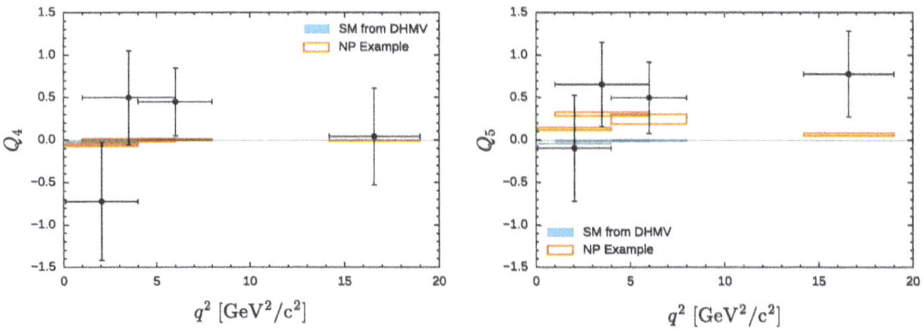

Fig. 3.2 Q_4 (left) and Q_5 (right) observables compared with SM and NP scenario [31], shown by the cyan filled and brown open boxes [32], respectively

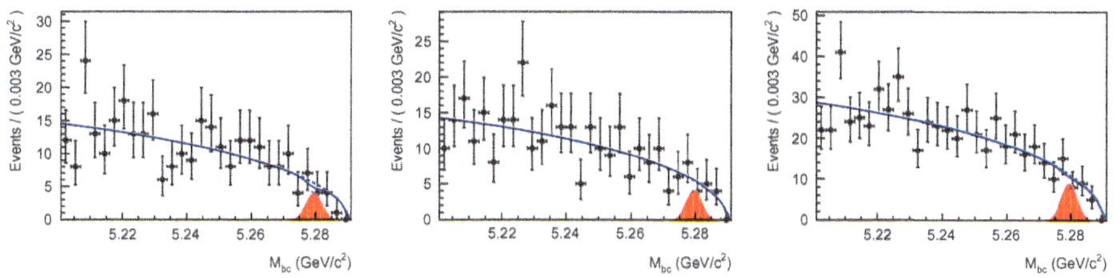

Fig. 3.3 The M_{bc} distributions for events that pass selection criteria for the decays $B^0 \to K^{*0}\mu^+e^-$ (left), $B^0 \to K^{*0}\mu^-e^+$ (middle), and also decays combined (right). The points with error bars are data, blue curve is the fit result, and the signal shape is depicted by the red shaded region with arbitrary normalization [37]

3.5 Search for Lepton-Flavor-Violating Decays $B^0 \to K^{*0}\mu^\pm e^\mp$

Over the recent years, measurements of the decays mediated by $b \to s\ell^+\ell^-$ quark-level transition hint for possible LFU violation, as discussed in Sect. 3.4. LFU is an important symmetry of the SM and its violation is usually accompanied by lepton flavor violation (LFV) [35, 36]. Very recently, Belle has reported a search for the LFV decays $B^0 \to K^{*0}\mu^\pm e^\mp$, where K^{*0} is reconstructed from $K^+\pi^-$ [37]. Backgrounds originating from $e^+e^- \to q\bar{q}$ $(q = u, d, c, s)$ continuum processes and other B decays are suppressed with two dedicated neural networks (NN). The $B^0 \to K^{*0}J/\psi$ decay is used as a control sample. Further, a set of vetoes are applied to suppress contributions from the decays $B^0 \to K^{*0}J/\psi(\to \ell^+\ell^-)$, in which one of the leptons is misidentified and swapped with the K^+ or π^-. Signal yields are obtained with an unbinned maximum-likelihood fit to the distributions of the kinematic variable, $M_{bc} = \sqrt{(E_{beam}/c^2)^2 - (p_B/c)^2}$, where E_{beam} is the beam energy and p_B is the momentum of the B candidate in the center-of-mass frame.

As shown in Fig. 3.3, there is no evidence for signal due to the LFV decay. Therefore, upper limits at 90% confidence level (CL) are set on $\mathcal{B}(B^0 \to K^{*0}\mu^+e^-) < 1.2 \times 10^{-7}$, $\mathcal{B}(B^0 \to K^{*0}\mu^-e^+) < 1.6 \times 10^{-7}$, and on both the decays combined $\mathcal{B}(B^0 \to K^{*0}\mu^\pm e^\mp) < 1.8 \times 10^{-7}$. These are the most stringent limits on these decays to date.

3.6 Search for $B^+ \to K^+\tau^+\tau^-$

The decay $B^+ \to K^+\tau^+\tau^-$ is mediated via $b \to s\ell^+\ell^-$ FCNC process involving the third-generation lepton family, which can provide additional sensitivity to NP [38]. The first search for $B^+ \to K^+\tau^+\tau^-$ is recently reported by the BaBar experiment [39]. In this study, the hadronic B-meson tagging method is used, where one of the two B mesons produced in $\Upsilon(4S) \to B^+B^-$ is reconstructed exclusively in many hadronic decay modes. The remaining tracks, clusters, and missing energy in the event is attributed to the signal B meson. Only leptonic decays of the τ are considered, which results in three signal decay topologies with a K^+, multiple missing neutrinos, and either e^+e^-, $\mu^+\mu^-$, or $e^+\mu^-$. The neutrinos account for the missing energy while any other neutral activity is discarded. Further, event shape variables are utilized to suppress continuum events. At this stage, the remaining backgrounds mostly arise from $B\bar{B}$ events, which are suppressed applying a criterion on the output of an NN formed with several input variables related to signal decay kinematics. No significant signal is observed and an upper limit on $\mathcal{B}(B^+ \to K^+\tau^+\tau^-) < 2.25 \times 10^{-3}$ is obtained at the 90% CL, whereas the SM prediction is in the range 1–2 $\times\ 10^{-7}$ [40, 41] .

3.7 Search for $B \to h\nu\bar{\nu}$

The $B \to h\nu\bar{\nu}$ decays, (where h refers to K^+, K_S^0, K^{*+}, K^{*0}, π^+, π^0, ρ^+, or ρ^0) are FCNC processes with a neutrino pair in the final state. These FCNC decays involve the Z boson alone, and hence are theoretically cleaner than $b \to s\ell^+\ell^-$ processes having a charged lepton pairs in the final state, where the photon also contributes [42].

Previously, these decays were searched for in Belle utilizing the hadronic tagging method [43] as well as in BaBar using both hadronic [44] and semi-leptonic tags [45]. The Belle analysis [46] is based on a more efficient semi-leptonic tagging method. In this analysis, the signal B daughter candidates are reconstructed through their subsequent decays: $K^{*0} \to K^+\pi^-$, $K^{*+} \to K^+\pi^0$ and $K_S^0\pi^+$, $\rho^+ \to \pi^+\pi^0$, $\rho^0 \to \pi^+\pi^-$, $K_S^0 \to \pi^+\pi^-$, and $\pi^0 \to \gamma\gamma$. Then, event shape variables are utilized to suppress the continuum background. Signal events are identified with the extra energy in the electromagnetic calorimeter, which is calculated by removing all the associated energy deposits due to tag and signal B mesons. The largest signal contribution is observed in the $B \to K^{*+}\nu\nu$ decay with a significance of 2.3σ. In absence of a statistically significant signal in any of the decay modes, upper limits on their BFs are set at the 90% CL. The result is summarized in Fig. 3.4 along with the expected values and previous measurements. These decays can be observed by Belle II with the uncertainties of similar size as that of current theoretical predictions [5].

3.8 Summary

Decays involving $b \to s$ quark-level transitions are forbidden at tree level in the SM, but can proceed via penguin loop or box diagrams in which various NP particles may also contribute. Belle reported the first evidence for isospin violation in the $B \to K^*\gamma$ decay [11]; also first measurement of the difference of CP asymmetries, between charged and neutral B meson is performed in the same analysis. In a sum-of-exclusive measurement of the decay $B \to X_s\gamma$ at Belle, a null isospin asymmetry is

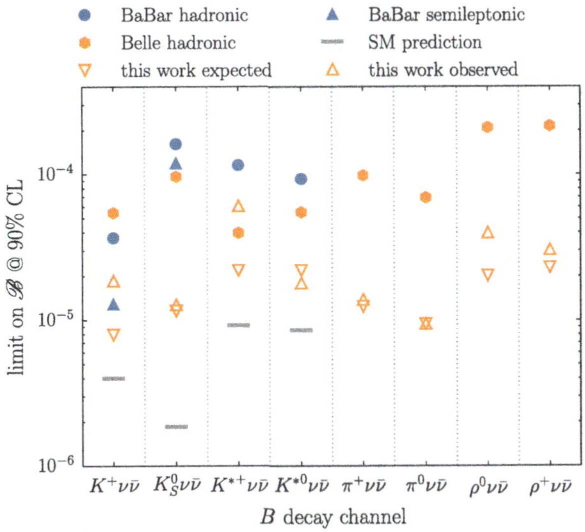

Fig. 3.4 Observed upper limits along with the expected values and previous measurement. SM predictions are also shown for the $K^{(*)}$ modes [46]

found, which can reduce theoretical uncertainty in the BF; similarly, the ΔA_{CP} value is found to be consistent with zero, helping constrain NP [10]. Belle's measurement of time-dependent CP violation parameters in $B^0 \to K_S^0\eta\gamma$ [23] is consistent with the null-asymmetry hypothesis within 2σ as well as SM predictions.

An angular analysis for the decay $B \to K^*\ell\ell$ is performed [32] for the first time in separate lepton-flavors; The results are consistent with both SM values and NP scenarios. As the measurement also hints at NP scenarios with possible LFU violation, this can eventually lead to LFV. The LFV decays $B^0 \to K^{*0}\mu^\pm e^\mp$ are searched at Belle. The most stringent upper limit on the BF of these LFV decays are obtained with no evidence for any signal event [37]. The decay $B^+ \to K^+\tau^+\tau^-$ is searched by BaBar for the first time and an upper limit on its BF is set at 90% CL [39]. Belle also reported a new search for the decay $B \to h\nu\nu$ [46] based on a more efficient semi-leptonic tagging method. The obtained upper limits are close to SM predictions for the $K^{(*)}$ modes and Belle II has brighter prospects to observe these decays.

References

1. The inclusion of charge conjugate decay modes is implied unless otherwise stated
2. M. Tanabashi et al. (Particle Data Group), Phys. Rev. D **98**, 030001 (2018)
3. Y. Amhis et al. (Heavy Flavor Averaging Group), Eur. Phys. J. C **77**, 895 (2017)
4. M. Misiak et al., Phys. Rev. Lett. **114**, 221801 (2015)
5. E. Kou et al., The Belle II Physics book. arXiv:1808.10567 [hep-ex]; submitted to Prog. Theor. Exp. Phys
6. S.J. Lee, M. Neubert, G. Paz, Phys. Rev. D **75**, 114005 (2007)
7. M. Benzke, S.J. Lee, M. Neubert, G. Paz, Phys. Rev. Lett. **106**, 141801 (2011)

8. R. Malm, M. Neubert, C. Schmell, JHEP **04**, 042 (2016)
9. M. Endo, T. Goto, T. Kitahara, S. Mishima, D. Ueda, K. Yamamoto, JHEP **04**, 019 (2018)
10. S. Watanuki et al. (Belle Collaboration), BELLE-CONF-1801. arXiv:1807.04236
11. T. Horiguchi et al. (Belle Collaboration), Phys. Rev. Lett. **119**, 191802 (2017)
12. M. Matsumori, A.I. Sanda, Y.-Y. Keum, Phys. Rev. D **72**, 014013 (2005)
13. D. Atwood, T. Gershon, M. Hazumi, A. Soni, Phys. Rev. D **71**, 076003 (2005)
14. Y. Ushiroda et al. (Belle Collaboration), Phys. Rev. D **74**, 111104 (2006)
15. B. Aubert et al. (BaBar Collaboration), Phys. Rev. D **78**, 071102 (2008)
16. B. Aubert et al. (BaBar Collaboration), Phys. Rev. D **79**, 011102 (2009)
17. J. Li et al. (Belle Collaboration), Phys. Rev. Lett. **101**, 251601 (2008)
18. P. del Amo Sanchez et al. (BaBar Collaboration), Phys. Rev. D **93**, 052013 (2016)
19. H. Sahoo et al. (Belle Collaboration), Phys. Rev. D **84**, 071101 (2011)
20. B. Grinstein, Y. Grossman, Z. Ligeti, D. Pirjol, Phys. Rev. D **71**, 011504 (2005)
21. M. Matsumori, A.I. Sanda, Phys. Rev. D **73**, 114022 (2006)
22. P. Ball, G.W. Jones, R. Zwicky, Phys. Rev. D **75**, 054004 (2007)
23. H. Nakano et al. (Belle Collaboration), Phys. Rev. D **97**, 092003 (2018)
24. W. Altmannshofer, D.M. Straub, Eur. Phys. J. C **75**, 382 (2015)
25. R. Aaij et al. (LHCb Collaboration), JHEP **02**, 104 (2016)
26. R. Aaij et al. (LHCb Collaboration), JHEP **09**, 179 (2015)
27. R. Aaij et al. (LHCb Collaboration), Phys. Rev. Lett. **113**, 151601 (2014)
28. S. Descotes-Genon, L. Hofer, J. Matias, J. Virto, JHEP **06**, 092 (2016)
29. S. Descotes-Genon, J. Matias, M. Ramon, J. Virto, JHEP **01**, 048 (2013)
30. S. Descotes-Genon, T. Hurth, J. Matias, J. Virto, JHEP **05**, 137 (2013)
31. B. Capdevila, S. Descotes-Genon, J. Matias, J. Virto, JHEP **10**, 075 (2016)
32. S. Wehle et al. (Belle Collaboration), Phys. Rev. Lett. **118**, 111801 (2017)
33. B. Capdevila, A. Crivellin, S. Descotes-Genon, J. Matias, J. Virto, JHEP **01**, 093 (2018)
34. W. Altmannshofer, P. Stangl, D.M. Straub, Phys. Rev. D **96**, 055008 (2017)
35. S.L. Glashow, D. Guadagnoli, K. Lane, Phys. Rev. Lett. **114**, 091801 (2015)
36. D. Guadagnoli, K. Lane, Phys. Lett. B **751**, 54 (2015)
37. S. Sandilya et al. (Belle Collaboration). arXiv:1807.03267 [to appear in Phys. Rev. D (RC)]
38. L. Calibbi, A. Crivellin, T. Ota, Phys. Rev. Lett. **115**, 181801 (2015)
39. J.P. Lees et al. (BaBar Collaboration), Phys. Rev. Lett. **118**, 031802 (2017)
40. J.L. Hewitt, Phys. Rev. D **53**, 4964 (1996)
41. C. Bouchard, G.P. Lepage, C. Monahan, H. Na, J. Shigemitsu, Phys. Rev. Lett. **111**, 162002 (2013)
42. A. Buras, J. Girrbach-Noe, C. Niehoff, D. Straub, JHEP **02**, 184 (2015)
43. O. Lutz et al. (Belle Collaboration), Phys. Rev. D **87**, 111103 (2013)
44. J.P. Lees et al. (BaBar Collaboration), Phys. Rev. D **87**, 112005 (2013)
45. J.P. Lees et al. (BaBar Collaboration), Phys. Rev. D **82**, 112002 (2010)
46. J. Grygier et al. (Belle Collaboration), Phys. Rev. D **96**, 091101(R) (2017)

Chapter 4
Charmless Hadronic B Decays and Direct CP Violation from e^+e^- B Factories

V. Gaur

Abstract We report a study of the decay $B^{\pm} \to K^+K^-\pi^{\pm}$ ($B^0 \to \pi^0\pi^0$) based on a data sample that contains 772×10^6 (752×10^6) $B\bar{B}$ events, and was collected at the $\Upsilon(4S)$ resonance with the Belle detector at the KEKB asymmetric energy e^+e^- collider. We then discuss about a Dalitz plot analysis of the decay $B^+ \to K_S^0\pi^+\pi^0$ performed using the full *BABAR* data set of $(470.9 \pm 2.8) \times 10^6$ $B\bar{B}$ events recorded at the $\Upsilon(4S)$ resonance. Finally, we present Belle preliminary results of the studies of $B^0 \to \eta\eta$, $B^0 \to K^-\pi^+K_S^0$, and $B^{\pm} \to K_S^0K_S^0h^{\pm}$ ($h = K, \pi$) based on data sets of 753×10^6, 772×10^6, and 772×10^6 $B\bar{B}$ events, respectively.

4.1 Introduction

Charmless hadronic B decays mainly proceed via the CKM-suppressed $b \to u$ tree and $b \to (s, d)g$ penguin diagrams, and have a branching fraction two to four orders of magnitude lower than the CKM-favored $b \to c$ transitions, allowing us to use them as a probe for physics beyond the standard model (SM). At Belle, B decay candidates are identified using two kinematic variables: the energy difference $\Delta E = \sum_i E_i - E_{\text{beam}}$ and beam-energy constrained mass $M_{\text{bc}} = \sqrt{E_{\text{beam}}^2 - \left|\sum_i \mathbf{p_i}\right|^2}$, where E_{beam} is the beam energy, and $\mathbf{p_i}$ and E_i are the center-of-mass (CM) momentum and energy of the ith daughter of the reconstructed B, respectively. At *BABAR*, B decay candidates are identified using similar kinematic variables: $\Delta E = E_B^* - \sqrt{s}/2$ and $m_{\text{ES}} = \sqrt{E_X^2 - \mathbf{p}_B^2}$ where $E_X = (s/2 + \mathbf{p}_{e^+e^-} \cdot \mathbf{p}_B)/E_{e^+e^-}$; E_B^* is the B energy in the CM frame, $\sqrt{s}$ is the CM energy, and $(E_{e^+e^-}, \mathbf{p}_{e^+e^-})$ and $(E_B, \mathbf{p}_B)$ are the four-momenta of the initial e^+e^- system and the B candidate, respectively, measured in the laboratory frame. Continuum $e^+e^- \to q\bar{q}$ ($q = u, d, s, c$) events are the primary

Throughout this paper, we use units in which $c = 1$.

V. Gaur (✉)
Virginia Polytechnic Institute and State University, Blacksburg, VA 24061, USA
e-mail: vipinhep@vt.edu

A. Giri and R. Mohanta (eds.), *16th Conference on Flavor Physics and CP Violation*, Springer Proceedings in Physics 234,
https://doi.org/10.1007/978-3-030-29622-3_4

source of background. To suppress this background, variables describing the event shape topology are combined into a multivariate classifier, such as a neural network, boosted decision tree, or Fisher discriminant. We use an unbinned extended maximum likelihood fit based on several discriminating variables; the fit usually includes signal, continuum, charm and charmless B background components.

4.2 $B^{\pm} \to K^{+}K^{-}\pi^{\pm}$

The $B^{\pm} \to K^{+}K^{-}\pi^{\pm}$ decays [1] proceed via the Cabibbo-suppressed $b \to u$ tree and $b \to d$ penguin transitions. Large CP asymmetries can occur due to interference between tree and loop diagrams with similar amplitudes. Final state interactions may be responsible for such CP violation [2]. We can also have contributions from beyond-the-SM particles in the loops. The previous best measurement of the branching fraction of $B^{\pm} \to K^{+}K^{-}\pi^{\pm}$ comes from *BABAR* [3] with $\mathcal{B}(B^{\pm} \to K^{+}K^{-}\pi^{\pm}) = [5.0 \pm 0.5(\text{stat}) \pm 0.5(\text{syst})] \times 10^{-6}$ while that of A_{CP} is from LHCb [4] with $A_{CP} = [-12.3 \pm 1.7(\text{stat}) \pm 1.2(\text{syst})]\%$. A structure is seen by both *BABAR* [3] and LHCb [4] at the low $K^{+}K^{-}$ invariant mass in $B^{\pm} \to K^{+}K^{-}\pi^{\pm}$; a large localized CP asymmetry is also observed in the same mass region.

In Belle's analysis, $B^{\pm}$ candidates are reconstructed by combining two oppositely charged kaons with a pion. Continuum background is suppressed with a neural network implemented using five event-shape variables. A tight requirement on the neural network output removes 99% of the continuum events while retaining 48% of the signal. A simultaneous $\Delta E - M_{\text{bc}}$ fit is performed in each bin of the $K^{+}K^{-}$ invariant mass with the signal, continuum background, generic B background (arising due to B decays mediated by the dominant $b \to c$ transition), and rare B background (from B decays via $b \to u, d, s$ transitions) as the fit components.

An excess of events is found in the $K^{+}K^{-}$ invariant mass below $1.5\,\text{GeV}/c^2$, confirming the earlier observations by *BABAR* [3] and LHCb [4]. Belle finds a strong evidence for large CP asymmetry of $[-90 \pm 17(\text{stat}) \pm 3(\text{syst})]\%$ with 4.8σ significance for the KK mass below $1.1\,\text{GeV}/c^2$. The overall branching fraction and CP asymmetry from Belle are $\mathcal{B}(B^{\pm} \to K^{+}K^{-}\pi^{\pm}) = [5.38 \pm 0.40(\text{stat}) \pm 0.35(\text{syst})] \times 10^{-6}$ and $A_{CP} = [-17.0 \pm 7.3(\text{stat}) \pm 1.7(\text{syst})]\%$.

4.3 $B^{0} \to \pi^{0}\pi^{0}$

The $B^{0} \to \pi^{0}\pi^{0}$ decay [5] proceeds via $b \to u$ tree and $b \to d$ penguin diagrams. We use branching fractions and time-integrated as well as time-dependent CP asymmetries to extract ϕ_2 [or α]. Branching fraction and CP asymmetry measurements of $B \to \pi\pi$ are sensitive to the angle ϕ_2 of the unitarity triangle (UT). Among three $B \to \pi\pi$ decays, the branching fraction and A_{CP} of $B^{0} \to \pi^{0}\pi^{0}$ are the least well determined. QCD predicts a branching fraction below 1×10^{-6} [6, 7]. The previously measured branching fractions are $[2.3^{+0.4}_{-0.5}(\text{stat})^{+0.2}_{-0.3}(\text{syst})] \times 10^{-6}$ and $[1.83 \pm 0.21(\text{stat}) \pm 0.13(\text{syst})] \times 10^{-6}$ by Belle [8] and *BABAR* [9] using a data sample of $275 \times 10^6 B\bar{B}$ and $467 \times 10^6 B\bar{B}$ events, respectively.

In Belle's new analysis, $B^0 \to \pi^0\pi^0$ candidates are reconstructed from the subsequent decay of each π^0 daughter to a pair of photons. For the continuum background suppression, a Fisher discriminant is constructed by combining the output of another such discriminant, formed out of modified Fox-Wolfram moments [10], with two event-shape variables. The resulting Fisher discriminant, T_c, lies in the range $(-1, +1)$. By discarding candidates with $T_c < -0.3$, 72% of the continuum background is rejected while retaining 98% of the signal. Among rare *B* decays, the major source of background comes from $B^+ \to \rho^+\pi^0$ where the charged pion from the subsequent $\rho^+ \to \pi^+\pi^0$ decay is lost. A simultaneous $\Delta E - M_{\text{bc}} - T_c$ fit is performed in each bin of a variable that distinguishes B^0 and $\bar{B}^0$ meson, with signal, continuum background, and backgrounds arising from $\rho\pi$ and other rare *B* decays as the fit components.

The overall branching fraction and *CP* asymmetry from Belle are $\mathcal{B}(B^0 \to \pi^0\pi^0) = [1.31 \pm 0.19(\text{stat}) \pm 0.19(\text{syst})] \times 10^{-6}$ with a significance of 6.4σ and $A_{CP} = [+14 \pm 36(\text{stat}) \pm 10(\text{syst})]\%$. These results are combined with the previous Belle results for $B^0 \to \pi^+\pi^-$ [11] and $B^+ \to \pi^+\pi^0$ [12] to constrain ϕ_2 via isospin relations [13]. Belle excludes this UT angle from the range 15.5°–75.0° at 95% confidence level.

4.4 $B^\pm \to K^0_S\pi^\pm\pi^0$

The $B^\pm \to K^0_S\pi^\pm\pi^0$ decays [14] proceed via $b \to u$ tree and $b \to s$ penguin diagrams. The interference between the two amplitudes can give rise to direct *CP* violation. We can constrain the UT angle ϕ_3 [or γ] using information from other $B \to K\pi\pi$ decays via an isospin analysis. QCD penguin contributions can be eliminated by constructing a linear combination of the weak decay amplitude for $B^+ \to (K^*\pi)^+$ to form a pure isospin $I = \frac{3}{2}$ state [15]:

$$\mathcal{A}_{\frac{3}{2}} = \mathcal{A}(K^{*0}\pi^+) + \sqrt{2}\mathcal{A}(K^{*+}\pi^0), \tag{4.1}$$

The phase $-\frac{1}{2}\text{Arg}\left(\bar{\mathcal{A}}_{\frac{3}{2}}/\mathcal{A}_{\frac{3}{2}}\right)$ is the angle ϕ_3 in the absence of the EWP contributions [16]. We can also probe effects similar to the $K\pi$ puzzle where QCD may be the cause of apparently anomalous *CP* violation.

$$\Delta A_{CP}(K\pi) = A_{CP}(K^+\pi^0) - A_{CP}(K^+\pi^-) \neq 0, \tag{4.2}$$

Previously, an upper limit on the branching fraction of $B^\pm \to K^0\pi^\pm\pi^0$ was set by CLEO [17], $\mathcal{B}(B^+ \to K^0\pi^+\pi^0) < 66 \times 10^{-6}$, at 90% confidence level using a data sample of 13.5 fb^{-1}.

In *BABAR*, $B^{\pm} \to K_S^0\pi^{\pm}\pi^0$ candidates are reconstructed by combining one K_S^0 candidate, formed from a pair of oppositely charged tracks, with a charged pion and a π^0 candidate. A boosted decision tree (*BDT*) based on four event-shape variables is implemented for continuum suppression. With a loose requirement on the *BDT* output ($BDT_{\rm out}$), 92% of the continuum background is rejected while retaining 70% of the signal. A simultaneous fit to $\Delta E - m_{\rm ES} - BDT_{\rm out}$ is performed over the Dalitz plot. The fit includes signal, continuum and B background as the components.

BABAR finds the first evidence for direct *CP* violation in $B^+ \to K^*(892)^+\pi^0$ with a significance of 3.4σ. Further, it reports the first measurement of the branching fraction and *CP* asymmetry for $B^+ \to K_0^*(1430)^+\pi^0$ with a significance of 5.4σ for the branching fraction. The results for various decay channels are listed in Table 4.1. The branching fractions and *CP* asymmetries of $B^+ \to K^*(892)^0\pi^+$, $B^+ \to K_0^*(1430)^0\pi^+$ and $B^+ \to K^*(892)^+\pi^0$ are then combined with previous *BABAR* measurements [18, 19], results of which are summarized in Table 4.2. Using the world-average value for direct *CP* violation in $B^0 \to K^*(892)^+\pi^-$ [20] and the final *BABAR* result for *CP* violation in $B^+ \to K^*(892)^+\pi^0$, ΔA_{CP} for $K^*\pi$ is found to be consistent with zero:

$$\Delta A_{CP}(K^*\pi) = A_{CP}(K^{*+}\pi^0) - A_{CP}(K^{*+}\pi^-) = -0.16 \pm 0.13\,. \quad (4.3)$$

Table 4.1 Branching fractions ($\mathscr{B}$) and *CP* asymmetries (A_{CP}) from a fit to the combined $B^{\pm}$ data sample. The first uncertainty is statistical, the second is systematic, and the third is due to the signal model

Decay channel	$\mathscr{B}$ (10^{-6})	A_{CP} (%)
$K^0\pi^+\pi^0$	$31.8 \pm 1.8 \pm 2.1^{+6.0}_{-0.0}$	$+7 \pm 5 \pm 3^{+2}_{-3}$
$K^*(892)^0\pi^+$	$10.1 \pm 1.7 \pm 1.0^{+0.2}_{-0.3}$	$-12 \pm 21 \pm 8^{+0}_{-11}$
$K^*(892)^+\pi^0$	$6.4 \pm 0.9 \pm 0.4^{+0.2}_{-0.3}$	$-52 \pm 14 \pm 4^{+4}_{-2}$
$K_0^*(1430)^0\pi^+$	$34.6 \pm 3.3 \pm 4.2^{+1.9}_{-1.8}$	$+14 \pm 10 \pm 4^{+13}_{-5}$
$K_0^*(1430)^+\pi^0$	$11.9 \pm 1.7 \pm 1.0^{+0.0}_{-1.3}$	$+26 \pm 12 \pm 8^{+12}_{-0}$
$\rho(770)^+K^0$	$6.5 \pm 1.1 \pm 0.8^{+0.0}_{-1.7}$	$+21 \pm 19 \pm 7^{+23}_{-19}$

Table 4.2 Combination of branching fractions ($\mathscr{B}$) and *CP* asymmetries (A_{CP}) of $B^+ \to K^*(892)^0\pi^+$, $B^+ \to K_0^*(1430)^0\pi^+$ and $B^+ \to K^*(892)^+\pi^0$ reported here with previous *BABAR* measurements [18, 19]. The first uncertainty is statistical and the second is systematic

Decay channel	$\mathscr{B}$ (10^{-6})	A_{CP} (%)
$K^*(892)^0\pi^+$	$10.5 \pm 0.6 \pm 0.9$	$+2.5 \pm 5 \pm 1.6$
$K^*(892)^+\pi^0$	$6.8 \pm 0.8 \pm 0.5$	$-39 \pm 12 \pm 3$
$K_0^*(1430)^0\pi^+$	$34.1 \pm 1.1 \pm 4.3$	$+4 \pm 3.3 \pm 3.3$

4.5 $B^0 \to \eta\eta$

The $B^0 \to \eta\eta$ decay [21] proceeds via $b \to u$ Cabibbo- and color-suppressed tree and $b \to d$ penguin diagrams. A precise measurement of the branching fraction of this channel would play an important role in improving the $|S_{c\bar{c}s} - S_{\eta' K_S}|$ bound based on the predictions of SU(3), where the *CP* violating parameters $S_{\eta' K_S} \sim \sin 2\phi_1$ and $S_{c\bar{c}s}$ are measured in a time-dependent analysis [22] of Cabibbo-favoured $b \to c\bar{c}s$ decays [23], respectively. $\mathcal{B}(B^0 \to \eta\eta)$ is expected to be in the range (0.1–1.0)$\times 10^{-6}$ based on the predictions of soft collinear effective theory [24], QCD factorization [25] and re-scattering effects [26]. The previous best upper limit on the branching fraction of $B^0 \to \eta\eta$ comes from *BABAR* [27], $\mathcal{B}(B^0 \to \eta\eta) < 1.0 \times 10^{-6}$ at 90% confidence level, using a data sample of 426 fb^{-1}.

In Belle, the $B^0 \to \eta\eta$ candidates are reconstructed from the sub-decay channels $\eta \to \gamma\gamma$ ($\eta_{\gamma\gamma}$) and $\eta \to \pi^+\pi^-\pi^0$ ($\eta_{3\pi}$). Photons that are not included in the final set of π^0 candidates are combined to form $\eta_{\gamma\gamma}$ candidates. For the continuum background suppression, a neural network is implemented based on four event-shape variables. With a loose requirement on the neural network output (C_{NB}), about 68, 65 and 59% of the continuum background is rejected while retaining 98, 98, and 97% of the $\eta_{\gamma\gamma}\eta_{\gamma\gamma}$, $\eta_{\gamma\gamma}\eta_{3\pi}$, and $\eta_{3\pi}\eta_{3\pi}$ signal, respectively. The remainder of the C_{NB} distribution strongly peaks near 1.0 for signal, making it difficult to model with an analytic function. However, its transformed variable has a Gaussian-like distribution.

$$C'_{NB} = \log\left[\frac{C_{NB} - C_{NB,\min}}{C_{NB,\max} - C_{NB}}\right], \tag{4.4}$$

A simultaneous $\Delta E - M_{\rm bc} - C'_{NB}$ fit is performed with the signal, continuum background, and *B* background as the fit components.

Belle finds $\mathcal{B}(B^0 \to \eta\eta) = [5.9^{+2.1}_{-1.8}(\text{stat}) \pm 1.4(\text{syst})] \times 10^{-7}$ with a significance of 3.3σ and places an upper limit on the bound $|S_{c\bar{c}s} - S_{\eta' K_S}| < 0.11$ at 90% confidence level.

4.6 $B^0 \to K^-\pi^+K^0_S$

The $B^0 \to K^{\pm}\pi^{\mp}K^0_S$ [28] decays proceeds via $b \to d$ penguin transition. $\mathcal{B}[B^0 \to (\bar{K}^{*0}K^0 + K^{*0}\bar{K}^0)]$ and $\mathcal{B}[B^0 \to K^{*\pm}K^{\mp}]$ are predicted to be in the range (0.2–2.0)$\times 10^{-6}$ and (0.2–1.0)$\times 10^{-7}$, respectively, based on the predictions of perturbative QCD [29], next-to-leading order QCD [30], QCD factorization [31, 32] and flavor SU(3) [33]. The previous best measurement of the branching fraction comes from *BABAR* [34], with $\mathcal{B}(B^0 \to K^{\pm}\pi^{\mp}K^0_S) = [3.2 \pm 0.5(\text{stat}) \pm 0.3(\text{syst})] \times 10^{-6}$, using a data sample of 424 fb^{-1}. An excess of events is observed at low the $K^0_S K^{\pm}$ invariant mass with no apparent contribution from an isospin partner of the $f_X(1500)$ decaying to $K^0_S K^{\pm}$; this is in contrast to a clear signal seen in $B^{\pm} \to K^+K^-\pi^{\pm}$ decays [1, 3, 4].

In Belle, $B^0 \to K^{\pm}\pi^{\mp}K_S^0$ candidates are reconstructed from a charged kaon, a charged pion, and one K_S^0 candidate reconstructed from a pair of oppositely charged tracks with pion mass assumed. For the continuum background suppression, a neural network is implemented based on five event-shape variables. With a loose requirement on the neural network output (C_{NN}), 93% of the continuum background is rejected while retaining 82% of of the signal. A simultaneous $\Delta E - M_{\rm bc} - C'_{NN}$ fit is performed with the signal, continuum background, generic B background, rare B background with $\Delta E < 0$, the same with $\Delta E > 0$, and remaining rare B background as the fit components. Here, C'_{NN} is the translated neural network output obtained using a similar translation function as shown in (4.4).

Belle finds $\mathcal{B}(B^0 \to K^{\pm}\pi^{\mp}K_S^0) = [3.60 \pm 0.33(\text{stat}) \pm 0.15(\text{syst})] \times 10^{-6}$ and a final-state asymmetry of $[-8.5 \pm 8.9(\text{stat}) \pm 0.2(\text{syst})]\%$.

4.7 $B^{\pm} \to K_S^0 K_S^0 h^{\pm}$ $(h = K, \pi)$

The $B^+ \to K_S^0 K_S^0 K^+ (B^+ \to K_S^0 K_S^0 \pi^+)$ decay [35] proceeds via the $b \to s$ $(b \to d)$ penguin transition. These are flavour changing neutral current processes, which are suppressed in the SM and hence provide a useful probe to search for physics beyond the SM. It is also possible to study the quasi-two-body resonances through a full amplitude analysis of the Dalitz plot and search for direct *CP* asymmetry. The decay $B^+ \to K_S^0 K_S^0 K^+$ has already been observed by Belle [36] with $\mathcal{B}(B^+ \to K_S^0 K_S^0 K^+) = [13.4 \pm 1.9(\text{stat}) \pm 1.5(\text{syst})] \times 10^{-6}$ using 78 fb^{-1} of data and studied later by *BABAR* [37] with $\mathcal{B}(B^+ \to K_S^0 K_S^0 K^+) = [10.6 \pm 0.5(\text{stat}) \pm 0.3(\text{syst})] \times 10^{-6}$ using 426 fb^{-1} of data. *BABAR* also reported an inclusive *CP* asymmetry of $[4^{+4}_{-5}(\text{stat}) \pm 2(\text{syst})]\%$ [37]. The decay $B^+ \to K_S^0 K_S^0 \pi^+$ has not yet been observed, with an upper limit available at 90% confidence level, $\mathcal{B}(B^+ \to K_S^0 K_S^0 \pi^+) < 5.1 \times 10^{-7}$, from *BABAR* [38].

In the new Belle analysis, $B^+ \to K_S^0 K_S^0 K^+$ $(B^+ \to K_S^0 K_S^0 \pi^+)$ candidates are formed by combining two K_S^0 candidates with a charged kaon (pion). In both decay channels, the K_S^0 candidate is reconstructed from a pair of oppositely charged tracks assumed to be pions. For the continuum background suppression, a neural network is implemented based on six event-shape variables. With a loose requirement on the neural network output (C_{NB}), 84% of the continuum background is rejected while retaining 91% of the signal. We translate C_{NB} to C'_{NB} using a transformation as shown in (4.4). A simultaneous $\Delta E - C'_{NB}$ fit is performed with the signal, continuum background, rare B peaking/feed-across background, and rare B combinatorial background (events that remain after removing signal and rare peaking channels) as the fit components. The feed-across background is mostly due to $K - \pi$ misidentification and contribute in the $M_{\rm bc}$ signal region having ΔE peak shifted from zero on the positive (negative) side for $B^+ \to K_S^0 K_S^0 K^+$ $(B^+ \to K_S^0 K_S^0 \pi^+)$. Figure 4.1 shows the ΔE and C'_{NB} projections of the fit to B^+ and B^- samples for $B^+ \to K_S^0 K_S^0 K^+$, and overall fit for $B^{\pm} \to K_S^0 K_S^0 \pi^{\pm}$.

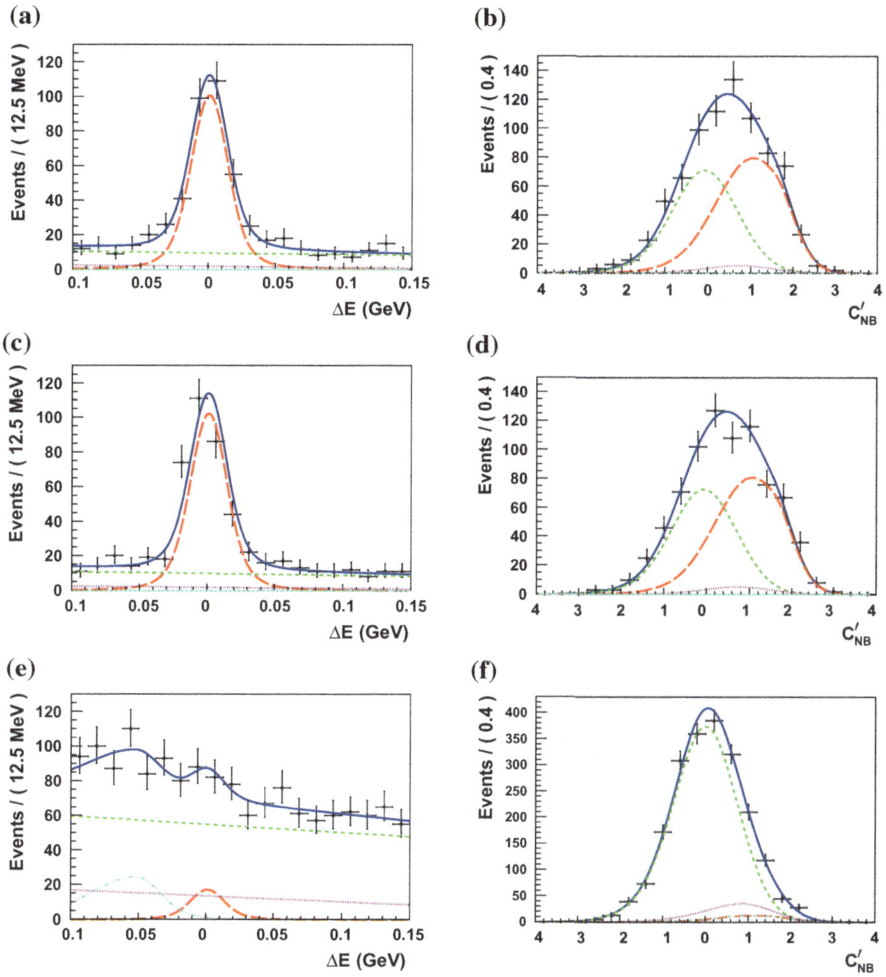

Fig. 4.1 The projections of ΔE (with $C'_{NB} > 0.0$) and C'_{NB} (with $|\Delta E| < 50\,\text{MeV}$) of the fit to B^+ (**a**, **b**) and B^- (**c**, **d**) samples for $B^\pm \to K_S^0 K_S^0 K^\pm$ and overall fit for $B^\pm \to K_S^0 K_S^0 \pi^\pm$ (**e**, **f**). Points with error bars are the data, solid (blue) curves are the total PDF, long dashed (red) curves are the signal component, dashed (green) curves are the continuum background, dotted (magenta) curves are the combinatorial $B\bar{B}$ background and dash-dotted (cyan) curves are the feed-across background

For $B^+ \to K_S^0 K_S^0 \pi^+$, Belle places an upper limit on its branching fraction of $\mathcal{B}(B^+ \to K_S^0 K_S^0 \pi^+) < 8.7 \times 10^{-7}$ at 90% confidence level. In case of $B^+ \to K_S^0 K_S^0 K^+$, the measured $\mathcal{B}(B^+ \to K_S^0 K_S^0 K^+) = [10.42 \pm 0.43(\text{stat}) \pm 0.22(\text{syst})] \times 10^{-6}$ and $A_{CP} = [+1.6 \pm 3.9(\text{stat}) \pm 0.9(\text{syst})]\%$.

4.8 Conclusion

Study of the aforementioned charmless hadronic *B* decays at Belle and *BABAR* have enabled the measurement of: new branching fraction upper limits, *CP* asymmetries and a new constrain on the CKM angle ϕ_2, providing a comprehensive comparison with theoretical models.

References

1. C.-L. Hsu et al. (Belle Collaboration), Phys. Rev. D **96**, 031101 (2017)
2. I. Bediaga, T. Frederico, O. Lourenco, Phys. Rev. D **89**, 094013 (2014)
3. B. Aubert et al. (*BABAR* Collaboration), Phys. Rev. Lett. **99**, 221801 (2007)
4. R. Aaij et al. (LHCb Collaboration), Phys. Rev. D **90**, 112004 (2014)
5. T. Julius et al. (Belle Collaboration), Phys. Rev. D **96**, 032007 (2017)
6. H. Li, S. Mishima, Phys. Rev. D **73**, 114014 (2006)
7. H. Li, S. Mishima, Phys. Rev. D **83**, 034023 (2011)
8. Y. Chao et al. (Belle Collaboration), Phys. Rev. Lett. **94**, 181803 (2005)
9. J.P. Lees et al. (*BABAR* Collaboration), Phys. Rev. D **87**, 052009 (2013)
10. S.H. Lee et al. (Belle Collaboration), Phys. Rev. Lett. **91**, 261801 (2003)
11. J. Dalseno et al. (Belle Collaboration), Phys. Rev. D **88**, 092003 (2013)
12. Y.-T. Duh et al. (Belle Collaboration), Phys. Rev. D **87**, 031103 (2013)
13. M. Gronau, D. London, Phys. Rev. Lett. **65**, 3381 (1990)
14. J.P. Lees et al. (*BABAR* Collaboration), Phys. Rev. D **96**, 072001 (2017)
15. M. Ciuchini, M. Pierini, L. Silvestrini, Phys. Rev. D **74**, 051301 (2006)
16. M. Gronau, D. Pirjol, A. Soni, J. Zupan, Phys. Rev. D **75**, 014002 (2007)
17. E. Eckhart et al. (CLEO Collaboration), Phys. Rev. Lett. **89**, 251801 (2002)
18. B. Aubert et al. (*BABAR* Collaboration), Phys. Rev. D **78**, 012004 (2008)
19. J.P. Lees et al. (*BABAR* Collaboration), Phys. Rev. D **84**, 092007 (2011)
20. Y. Amhis et al. (Heavy Flavor Averaging Group). arXiv:1207.1158v2 [hep-ex]
21. A. Abdesselam et al. (Belle Collaboration). arXiv:1609.03267v1 [hep-ex]
22. B. Aubert et al. (*BABAR* Collaboration), Phys. Rev. D **79**, 052003 (2009)
23. Y. Grossman, Z. Ligeti, Y. Nir, H. Quinn, Phys. Rev. D **68**, 015004 (2003)
24. A.R. Williamson, J. Zupan, Phys. Rev. D **74**, 014003 (2006)
25. H. Cheng, C. Chua, Phys. Rev. D **80**, 114008 (2009)
26. C. Chua, Nucl. Phys. A **844**, 260 (2010)
27. B. Aubert et al. (*BABAR* Collaboration), Phys. Rev. D **80**, 112002 (2009)
28. A. Abdesselam et al. (Belle Collaboration). arXiv:1807.06782v1 [hep-ex]
29. L. Guo, Q. Xu, Z. Xiao, Phys. Rev. D **75**, 014019 (2007)
30. A. Ali, Phys. Rev. D **58**, 094009 (1998)
31. D. Du, H. Gong, J. Sun, D. Yang, G. Zhu, Phys. Rev. D **65**, 094025 (2002)
32. M. Beneke, M. Neubert, Nucl. Phys. B **675**, 333 (2003)

33. C. Chiang, M. Gronau, Z. Luo, J.L. Rosner, D.A. Suprun, Phys. Rev. D **69**, 034001 (2004)
34. P.A. del Sanchez et al. (*BABAR* Collaboration), Phys. Rev. D **82**, 031101 (2010)
35. A.B. Kaliyar et al. (Belle Collaboration), Phys. Rev. D **99**, 031102 (2019)
36. A. Garmash et al. (Belle Collaboration), Phys. Rev. D **69**, 012001 (2004)
37. J.P. Lees et al. (*BABAR* Collaboration), Phys. Rev. D **85**, 112010 (2012)
38. B. Aubert et al. (*BABAR* Collaboration), Phys. Rev. D **79**, 051101 (2009)

Chapter 5
Recent Results with Charm Baryons

Shohei Nishida

Abstract Large data samples accumulated by BaBar, Belle, BESIII and LHCb experiments provide opportunities to perform studies of charmed baryons. The observation of excited Ω_c charm baryons by LHCb, and confirmation by Belle is reported. Another charmed-strange baryon $\Xi_c(2930)^0$ is observed by Belle. The decays of Ω_c and Λ_c into various final states are studied by the experiments, and the branching fractions are measured. In this proceedings, recent measurements related to charmed baryons are summarized.

5.1 Introduction

Charmed baryons (Λ_c, Σ_c, Ξ_c, Ω_c) consist of one heavy charm quark and two light u, d, s quarks, where large mass difference provides a natural way to classify these states using HQET. Doubly charmed baryons with two charm quark also exist, but are not well studied.

These charmed baryons can be studied with e^+e^- colliders and hadron colliders. They are produced by charm production process $e^+e^- \rightarrow c\bar{c}$ or prompt charm production at hadron colliders, and also by the decay of bottom mesons and baryons like B, B_s, Λ_b. BaBar and Belle, the B factory experiments at e^+e^- collider, has been performing many analyses on charmed baryons recently, though the operation of these experiments terminated several years ago. BESIII, e^+e^- experiment operated above $c\bar{c}$ threshold energy, provides unique data to study charmed baryons. LHCb, pp collider experiment targeting flavor physics, produces huge number of charm quarks, and obtains many new results on charm baryons.

Recent measurements on charmed baryons are summarized below.

S. Nishida (✉)
High Energy Accelator Research Organization (KEK) and Sokendai University, Tsukuba, Japan
e-mail: shohei.nishida@kek.jp

A. Giri and R. Mohanta (eds.), *16th Conference on Flavor Physics and CP Violation*, Springer Proceedings in Physics 234,
https://doi.org/10.1007/978-3-030-29622-3_5

5.2 Decay of Λ_c^+

BESIII accumulated data of 567 pb^{-1} at CM energy 4.599 GeV, which is 26 MeV above the threshold $e^+e^- \to \Lambda_c^+\overline{\Lambda}_c^-$. No other particle can be produced in addition to a pair of Λ_c^+ at this energy. If a $\overline{\Lambda}_c^-$ is reconstructed, it is gururanteed that the rest of particles come from a Λ_c^+. Hence, one can measure the absolute branching fraction of Λ_c^+ by tagging one $\overline{\Lambda}_c^-$.

Around 1.0×10^5 $\Lambda_c^+\overline{\Lambda}_c^-$ pairs are produced in this data set. Typically, BESIII reconstructs 12 Λ_c^+ modes and they obtain around 10^4 tagged events. With this technique, they measured the branching fractions $\mathcal{B}(\Lambda_c^+ \to \Lambda^0 e^+ \bar{\nu}_e) = (3.63 \pm 0.38 \pm 0.20)\%$ [1] and $\mathcal{B}(\Lambda_c^+ \to pK^-\pi^+) = (5.84 \pm 0.27 \pm 0.23)\%$ [2]. This technique is also useful to study modes with neutrons, and they observed $\Lambda_c^+ \to \Sigma^-\pi^+\pi^+\pi^0$ with $\Sigma^- \to n\pi^-$ for the first time [3].

Recently BESIII has performed a measurement of $\Lambda_c^+ \to \Xi^0 K^+$ and $\Lambda_c^+ \to \Xi(1530)^0 K^+$ [4]. In this analysis, one charged kaon is picked up after tagging $\overline{\Lambda}_c^-$, and the missing mass is measured. This kind of measurements is possible only at the clean e^+e^- environment. Figure 5.1 shows the missing mass distribution, where clear peaks of Ξ^0 and $\Xi(1530)^0$ are seen. They obtain $\mathcal{B}(\Lambda_c^+ \to \Xi^0 K^+) = (5.90 \pm 0.86 \pm 0.39) \times 10^{-3}$ and $\mathcal{B}(\Lambda_c^+ \to \Xi(1530)^0 K^+) = (5.02 \pm 0.99 \pm 0.31) \times 10^{-3}$. The results are more precise than the past measurements of the relative branching fractions.

BESIII has also measured the absolute branching fraction of the inclusive decay $\Lambda_c^+ \to \Lambda X$ [5]. In this analysis, $\overline{\Lambda}_c^-$ from $\bar{p}K_S^0$ and $\bar{p}K^+\pi^-$ is tagged and events with a Λ are selected. The measured branching fraction and CP asymmetry are $\mathcal{B}(\Lambda_c^+ \to \Lambda X) = (38.2^{+2.8}_{-2.2} \pm 0.8)\%$ and $A_{CP}(\Lambda_c^+ \to \Lambda X) = (2.1^{+7.0}_{-6.6} \pm 1.4)\%$. Considering the sum of exclusive modes of Λ_c^+ decay with a Λ is $(24.5 \pm 2.1)\%$ [6], there are still more unmeasured Λ_c^+ decay modes with a Λ in the final state.

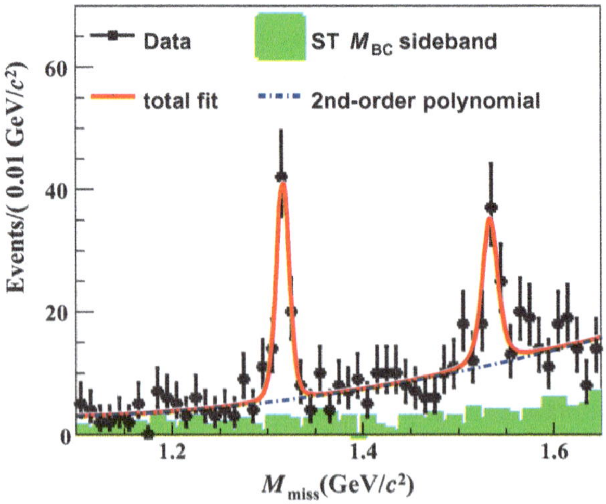

Fig. 5.1 Missing mass distribution in $\Lambda_c^+ \to \Xi^0 K^+$, $\Xi(1530)^0 K^+$ at BESIII

Belle has performed a measurement of $\Lambda_c^+ \to \Sigma\pi\pi$ using the full data (711 fb^{-1}) at $\Upsilon(4S)$ [7]. In this analysis, three decay modes $\Lambda_c^+ \to \Sigma^+\pi^+\pi^-$, $\Sigma^0\pi^+\pi^0$ and $\Sigma^+\pi^0\pi^0$ are studied, where Σ^+ and Σ^0 are reconstructed with $\Sigma^+ \to p\pi^0$ and $\Sigma^0 \to \Lambda^0\gamma$ ($\Lambda^0 \to p\pi^0$). $\Lambda_c^+ \to pK^-\pi^+$ is used as a normalization mode, and the branching ratios relative to this mode are measured.

Since the efficiencies of the signal and normalization modes are not uniform over the Dalitz plane, the analysis uses a model independent approach to divide the data into around 20 bins (7 bins for $\Sigma^+\pi^0\pi^0$) in the Dalitz distribution and to estimate the signal yield and the efficiency in each bin. Signal yields are extracted from the fit to Λ_c^+ mass distribution. Using $\mathcal{B}(\Lambda_c^+ \to pK^-\pi^+) = (6.35 \pm 0.33)\%$ [6], the branching fractions of $\Lambda_c^+ \to \Sigma^+\pi^+\pi^-$, $\Sigma^0\pi^+\pi^0$ and $\Sigma^+\pi^0\pi^0$ are calculated to be $(4.48 \pm 0.02 \pm 0.19 \pm 0.23)\%$, $(3.12 \pm 0.03 \pm 0.15 \pm 0.16)\%$ and $(1.26 \pm 0.04 \pm 0.11 \pm 0.07)\%$, respectively, where the third errors are from the uncertainty of the branching fraction of $\Lambda_c^+ \to pK^-\pi^+$. This is the first measurement of the branching ratio of $\Lambda_c^+ \to \Sigma^+\pi^0\pi^0$, and the results for other two modes are more precise than the past measurements.

LHCb has searched for the decay $\Lambda_c^+ \to p\mu^+\mu^-$ using 3.0 fb^{-1} data [8]. This decay is a flavor changing neutral current process, and the branching fraction is predicted to be at the order of 10^{-9} with the short distance effect. Considering the long distance effect it can be as large as 10^{-6}, but New Physics may further enhance the signal.

The analysis is performed in three different $M(\mu^+\mu^-)$ region. In the region at $M(\mu^+\mu^-) \sim M_\phi$, significant amount of the $\Lambda_c^+ \to p\phi$ decay is found, and this decay is treated as the normalization mode. In the region at $M(\mu^+\mu^-) \sim M_\omega$, they have observed the $\Lambda_c^+ \to p\omega$ signal and have measured the branching fraction to be $\mathcal{B}(\Lambda_c^+ \to p\omega) = (9.4 \pm 3.2 \pm 1.0 \pm 2.0) \times 10^{-4}$, where the third error is an error coming from relevant branching fractions. In the remaining region, which is for the non-resonant $\Lambda_c^+ \to p\mu^+\mu^-$ decay, no signal is observed. The upper limit is calculated to be $\mathcal{B}(\Lambda_c^+ \to p\mu^+\mu^-) < 7.7 \times 10^{-8}$ which is significantly improved from the previous limit 4.4×10^{-5} by BaBar [9].

5.3 Observation of $\Xi_c(2930)^0$

BaBar found some indication of excited Ξ_c^0 resonance around 2.93 GeV in the study of $B^- \to \Xi_c^0\overline{\Lambda}_c^- \to K^-\Lambda_c^+\overline{\Lambda}_c^-$ with 210 fb^{-1} data [10]. Recently Belle has studied the same decay mode $B^- \to K^-\Lambda_c^+\overline{\Lambda}_c^-$ with 711 fb^{-1} data [11]. This decay can also be used to study charmonium-like states, e.g. $Y(4660)$ through $B^- \to K^-Y(4660) \to K^-\Lambda_c^+\overline{\Lambda}_c^-$.

In the analysis by Belle, Λ_c^+ is reconstructed from $pK\pi^+$, pK_S^0, $\Lambda\pi^+$, $pK_S^0\pi^+\pi^-$ and $\Lambda\pi^+\pi^-\pi^+$. Figure 5.2 shows the $K^-\Lambda_c^+$ mass distribution from the data. The yields for each bin are obtained from the 2-dimensional fit to $M_{\rm bc}$ and $\Delta M_B \equiv M_B - m_B$, where m_B is the nominal mass of B meson. An excess is found around 2.93 GeV in the $M(K^-\Lambda_c^+)$ distribution. From the fit, the mass and width of this $\Xi_c(2930)^0$

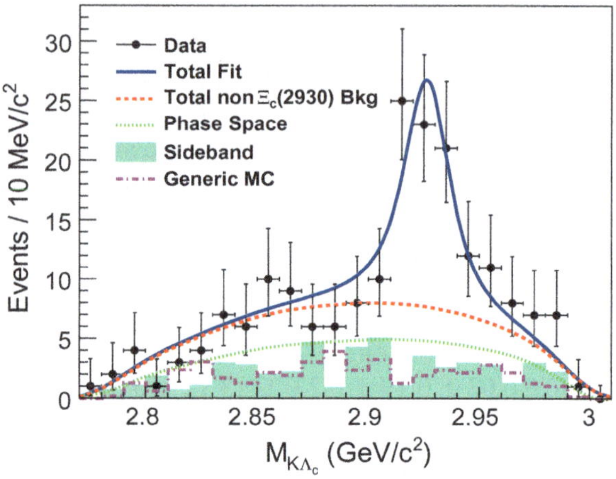

Fig. 5.2 $K^-\Lambda_c^+$ mass distribution in $B^- \to \Xi_c(2930)^0\overline{\Lambda}_c^-$ at Belle

resonance are calculated to be $2928.9 \pm 3.0^{+0.9}_{-12.0}$ and $19.5 \pm 8.4^{+5.9}_{-7.9}$ MeV. The significance of the $\Xi_c(2930)^0$ is 5.1σ. This is the first observation of this resonance. The product of the branching fraction $\mathcal{B}(B^- \to \Xi_c(2930)^0\overline{\Lambda}_c^-)\mathcal{B}(\Xi_c(2930)^0 \to K^-\Lambda_c^+)$ is obtained to be $(1.73 \pm 0.45 \pm 0.21) \times 10^{-4}$. On the other hand, no charmonium-like state is found in $M(\Lambda_c^+\overline{\Lambda}_c^-)$ distribution.

Belle has also searched for the charged $\Xi_c(2930)$ resonance in neutral B decays $B^0 \to K_S^0\Lambda_c^+\overline{\Lambda}_c^-$ [12], and has found the evidence of $\Xi_c(2930)^+$ with a significance of 3.9σ.

5.4 Hadronic Decay of Ω_c^0

The decay of Ω_c^0 is not well understood, and none of the absolute branching fractions are measured. CLEO [13] and BaBar [14] measured the branching ratios of $\Omega^-\pi^+\pi^0$, $\Omega^-\pi^+\pi^+\pi^-$, $\Xi^-K^-\pi^+\pi^+$ and $\Xi^0K^-\pi^+$ relative to $\Omega^-\pi^+$. FermiLab E687 experiment observed $\Omega_c^0 \to \Xi^+K^-K^-\pi+$ with more signal events than $\Omega^-\pi^+$ [15], but this mode is not confirmed by other experiments.

Belle has also studied hadronic Ω_c^0 decays using 980 fb^{-1} data [16]. In this analysis, Ω_c^0 is reconstructed with the decays mentioned above and, in addition, $\Omega_c^0 \to \Xi^-K^0\pi^+$, $\Xi^0K_S^0$, and $\Lambda^0K_S^0K_S^0$ using the sub-decay modes $\Xi^- \to \Lambda^0K^-$, $\Omega^- \to \Lambda^0\pi^-$, $\Xi^0 \to \Lambda\pi^0$, $\Xi^+ \to p\pi^0$ and $\Lambda \to p\pi^-$.

Figure 5.3 shows the invariant mass distribution of each sub-decay mode. Signal peaks of Ω_c^0 are seen in all the modes except $\Sigma^+K^-K^-\pi^+$. The intermediate resonant substructure is also examined. A contribution from the ρ is seen in the $\pi^+\pi^-$ mass distribution of $\Omega_c^0 \to \Omega^-\pi^+\pi^0$. Similarly, K^{*0} is found in $K^-\pi^+$ mass distribution in $\Omega_c^0 \to \Xi^-K^-\pi^+\pi^+$ and $\Omega_c^0 \to \Xi^0K^-\pi^+$, and $\Xi^0(1530)$ is found in $\Xi^-\pi^+$ mass distribution in $\Omega_c^0 \to \Xi^-K^-\pi^+\pi^+$. The obtained branching ratios with respect to

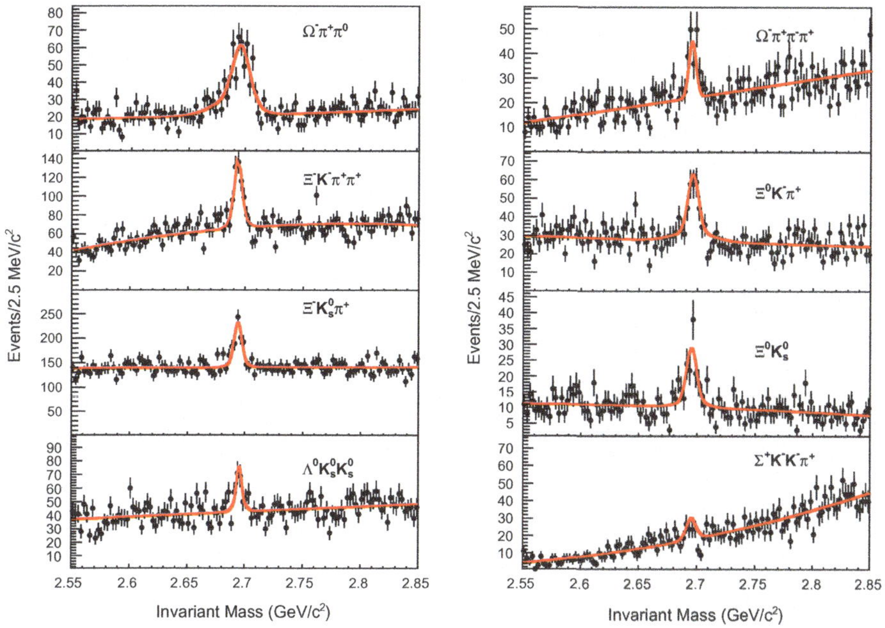

Fig. 5.3 Measurement of the hadronic Ω_c^0 decays at Belle. Invariant Ω_c^0 distributions are shown

$\Omega^-\pi^+$ are summarized in Table 5.1. This is the first measurement of the decays $\Omega_c^0 \to \Xi^-\bar{K}^0\pi^+$, $\Xi^0\bar{K}^0$ and $\Lambda\bar{K}^0\bar{K}^0$, and also the decays with intermediate resonances. Belle does not find $\Omega_c^0 \to \Sigma^+K^-K^-\pi^+$, although Fermilab E687 observed more events for this decay than for $\Omega_c^0 \to \Omega^-\pi^+$. They also find that the branching ratio $\mathcal{B}(\Omega_c^0 \to \Omega^0\pi^+\pi^+\pi-)/\mathcal{B}(\Omega_c^0 \to \Omega^0\pi^+)$ is much smaller than 1, while in other chamred baryon Y_c the ratio $\mathcal{B}(Y_c \to Y\pi^+\pi^+\pi^-)/\mathcal{B}(Y_c \to Y\pi^+)$ is larger than 1.

5.5 Excited Ω_c

The spectrum of excited Ω_c baryons is poorly known, and the only observed excited Ω_c was $\Omega_c(2770)^0$ decaying into $\Omega_c^0\gamma$ [17].

LHCb has searched for excited Ω_c through strong decay with $\Xi_c^+K^-$ final state using 3.3 fb^{-1} data [18], where Ξ_c^+ is reconstructed via $pK^-\pi^+$. Figure 5.4a shows the $\Xi_c^+K^-$ invariant mass distribution. Five narrow resonances $\Omega_c(3000)^0$, $\Omega_c(3050)^0$, $\Omega_c(3066)^0$, $\Omega_c(3090)^0$ and $\Omega_c(3199)^0$ with widths of 0–9 MeV are observed. A wide (∼60 MeV) resonance is also found at 3188 MeV.

Belle has also measured the excited Ω_c baryons decaying into $\Xi_c^+K^-$ using the full dataset of 980 fb^{-1} [19]. Ξ_c^+ is reconstructed from $\Xi^-\pi^+\pi^+$, $\Lambda^0K^-\pi^+\pi^+$, $\Xi^0\pi^+$, $\Xi^0\pi^+\pi^-\pi^+$, $\Sigma^+K^-\pi^+$, $\Lambda K_S^0\pi^+$ and $\Sigma^0K_S^0\pi^+$ in order to increase the statistics.

Table 5.1 Measured branching fraction of Ω_c^0 with respect to $\Omega^- K^+$ at Belle

Mode	Branching ratio with respect to $\Omega^-\pi^+$	Substructure	Previous measurement
$\Omega^-\pi^+$	1		
$\Omega^-\pi^+\pi^0$	$2.00 \pm 0.17 \pm 0.11$		$1.27 \pm 0.3 \pm 0.11$ [14]
$\Omega^-\rho^+$		>71%	
$\Omega^-\pi^+\pi^-\pi^+$	$0.32 \pm 0.05 \pm 0.02$		$0.28 \pm 0.09 \pm 0.01$ [14]
$\Xi^- K^-\pi^+\pi^+$	$0.68 \pm 0.07 \pm 0.03$		$0.46 \pm 0.13 \pm 0.03$ [14]
$\Xi^0(1530) K^-\pi^+$		$(33 \pm 9)\%$	
$\Xi^- \bar{K}^{*0}\pi^+$		$(55 \pm 16)\%$	
$\Xi^0 K^-\pi^+$	$1.20 \pm 0.16 \pm 0.08$		$4.0 \pm 2.5 \pm 0.4$ [13]
$\Xi^0 \bar{K}^{*0}$		$(57 \pm 10)\%$	
$\Xi^- \bar{K}^0\pi^+$	$2.12 \pm 0.24 \pm 0.14$		
$\Xi^0 \bar{K}^0$	$1.64 \pm 0.26 \pm 0.12$		
$\Lambda \bar{K}^0 \bar{K}^0$	$1.72 \pm 0.32 \pm 0.14$		
$\Sigma^+ K^- K^-\pi^+$	<0.32 (90% CL)		

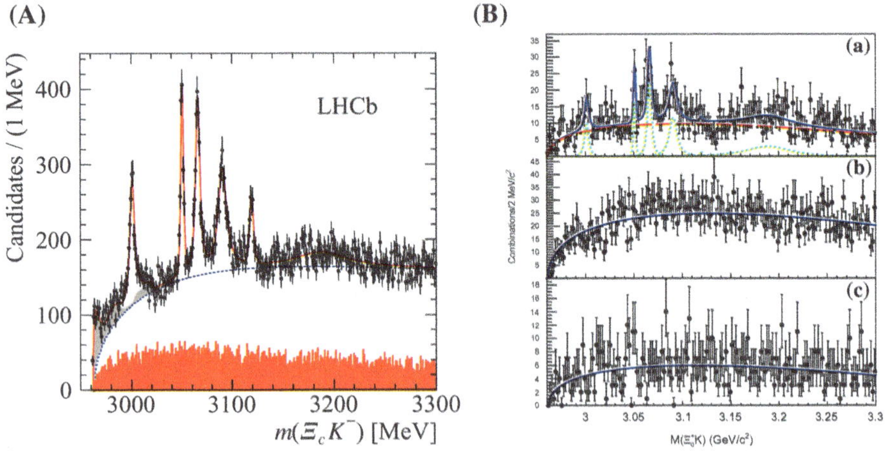

Fig. 5.4 **a** $\Xi_c^+ K^-$ mass distribution at LHCb. The solid (red) curve shows the fit result, and the dashed (blue) curve shows background component. The shaded (red) histogram shows the mass distribution in the Ξ_c sideband, and the the shaded (light gray) distributions indicate the feed-down from partially reconstructed excited Ω_c resonances. **b** The top plot shows $\Xi_c^+ K^-$ mass distribution at Belle. The solid (blue) curve shows the fit result. The middle and bottom plot show mass distribution for wrong sign $\Xi_c^+ K^+$ and Ξ_c^+ sideband, respectively

Figure 5.4b shows the $\Xi_c^+ K^-$ distribution and the fit result. Here the widths of the excited Ω_c baryons are fixed to the values obtained by LHCb. Belle has confirmed all the excited Ω_c baryons except $\Omega_c(3119)^0$.

5.6 Doubly Charmed Baryon Ξ_{cc}^{++}

The view on doubly charmed baryons is still unclear. The first observation of the doubly charmed baryon was done by SELEX experiment, where Ξ_{cc}^+ is observed in the $\Lambda_c^+ K^- \pi^+$ and $pD^+ K^-$ final states at the mass of 3518.7 ± 1.7 MeV with life time less than 33 fs [20]. However, this state was not confirmed by BaBar, Belle and LHCb [21]. Recently, LHCb observed another doubly charmed baryon Ξ_{cc}^{++} in the $\Lambda_c^+ K^- \pi^+ \pi^+$ final state at the mass of 3621.40 ± 0.78 MeV [22]. The two states, Ξ_{cc}^+ by SELEX and Ξ_{cc}^{++} by LHCb, have mass difference of around 100 MeV, which is too large to consider that they are in isodoublet.

LHCb has performed a new measurement of the lifetime of Ξ_{cc}^{++} using 1.7 fb^{-1} data [23]. They use $\Lambda_b^0 \to \Lambda_c^+ \pi^- \pi^+ \pi^-$ as a control mode, and the lifetime is obtained from the fit to the decay time distribution. They obtain the lifetime $\tau(\Xi_{cc}^{++}) = 0.256^{+0.024}_{-0.022} \pm 0.014$ ps for the first time. The lifetime of Ξ_{cc}^+ is predicted to be 3–4 times shorter than Ξ_{cc}^{++}, which leads to $\tau(\Xi_{cc}^+) \sim 60-90$ fs if we follow the prediction. Therefore, this result seems to point out an additional inconsistency between Ξ_{cc}^+ by SELEX and Ξ_{cc}^{++} by LHCb.

5.7 Production Rate of Baryons

The inclusive production cross sections of charmed baryons from e^+e^- are measured at Belle using 800 fb^{-1} data as shown in Fig. 5.5 [24]. The production rate has a relation of $(2J+1)\exp(-\alpha m)$, where J is the spin of the baryons, which is reproduced by relativistic string fragmentation model. Factor 3 difference is seen between Ξ_c and Λ_c in charmed baryons, while this tendency is not seen in hyperons. This indicates the di-quark structure of these charmed baryons.

5.8 Conclusion

In this proceedings, new experimental results on charmed baryons are reported. The spectrums and properties of charmed baryons are not well known yet. However, more analyses are going on using the data from e^+e^- experiments and LHCb, and Belle II starts taking data in near future. We expect more experimental results on charmed baryons.

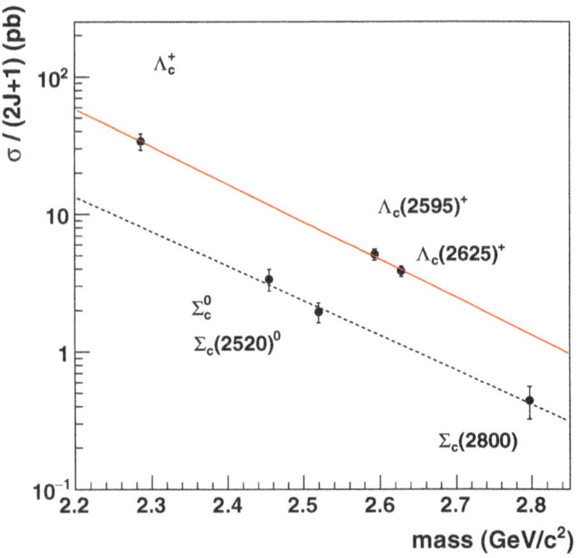

Fig. 5.5 Production cross section as a function of mass of charmed baryons. The solid and dashed lines show the fit results for Λ_c and Σ_c baryons

References

1. M. Ablikim et al. (BESIII Collaboration), Phys. Rev. Lett. **115**, 221805 (2015)
2. M. Ablikim et al. (BESIII Collaboration), Phys. Rev. Lett. **116**, 052001 (2016)
3. M. Ablikim et al. (BESIII Collaboration), Phys. Lett. B **772**, 388 (2017)
4. M. Ablikim et al. (BESIII Collaboration), Phys. Lett. B **783**, 200 (2018)
5. M. Ablikim et al. (BESIII Collaboration), Phys. Rev. Lett. **121**, 062003 (2018)
6. M. Tanabashi et al. (Particle Data Group), Phys. Rev. D **98**, 030001 (2018)
7. M. Berger et al. (Belle Collaboration). arXiv:1802.03421
8. R. Aaij et al. (LHCb Collaboration), Phys. Rev. D **97**, 091101 (2018)
9. J.P. Lees et al. (BaBar Collaboration), Phys. Rev. D **84**, 072006 (2011)
10. B. Aubert et al. (BaBar Collaboration), Phys. Rev. D **77**, 031101 (2008)
11. Y.B. Li et al. (Belle Collaboration), Eur. Phys. J. C **78**, 252 (2018)
12. Y.B. Li et al. (Belle Collaboration). arXiv:1806.09182
13. D. Cronin-Hennessy et al. (CLEO Collaboration), Phys. Rev. Lett. **86**, 3730 (2001)
14. B. Aubert et al. (BaBar Collaboration), Phys. Rev. Lett. **99**, 062001 (2007)
15. P.L. Frabetti et al. (E687 Collaboration), Phys. Lett. B **338**, 106 (1994)
16. J. Yelton et al. (Belle Collaboration), Phys. Rev. D **97**, 032001 (2018)
17. B. Aubert et al. (BaBar Collaboration), Phys. Rev. Lett. **86**, 3730 (2001); E. Solovieva et al. (Belle Collaboration), Phys. Lett. B **672**, 1 (2009)
18. R. Aaij et al. (LHCb Collaboration), Phys. Rev. Lett. **118**, 182001 (2017)
19. J. Yelton et al. (Belle Collaboration), Phys. Rev. D **97**, 051102 (2018)
20. M. Mattson et al. (SELEX Collaboration), Phys. Rev. Lett. **89**, 112001 (2002); A. Ocherashvili et al. (SELEX Collaboration), Phys. Lett. B **628**, 18 (2005)
21. B. Aubert et al. (BaBar Collaboration), Phys. Rev. D. **74**, 011103 (2006); R. Chistov et al. (Belle Collaboration), Phys. Rev. Lett. **97**, 162001 (2006); R. Aaij et al. (LHCb Collaboration), JHEP **12**, 090 (2013)
22. R. Aaij et al. (LHCb Collaboration), Phys. Rev. Lett. **119**, 112001 (2017)
23. R. Aaij et al. (LHCb Collaboration), Phys. Rev. Lett. **121**, 052002 (2018)
24. M. Niiyama et al. (Belle Collaboration), Phys. Rev. D **97**, 072005 (2018)

Chapter 6
Study of Onia and Heavy Quark QCD at the e^+e^- *B* Factories

J. Albert

Abstract We review five recent analyses (two from Belle and three from *BABAR*) in the areas of quarkonia, glueballs, and heavy quark quantum chromodynamics, three of which are new analyses (published just this year). They are as follows: (1) An observation and branching fraction measurement of the decay $\Upsilon(4S) \to \eta'\Upsilon(1S)$ by Belle [1]; (2) Measurement of the cross sections of $e^+e^- \to \eta$ + bottomonium measured near the $\Upsilon(5S)$, also by Belle [2]; (3) Measurement of several hadronic J/ψ decays and branching fractions by *BABAR* [3]; (4) A search for glueball-like mesons in $\Upsilon(1S) \to \gamma\pi^+\pi^-$ and $\Upsilon(1S) \to \gamma K^+K^-$ decays, also by *BABAR* [4]; and (5) A precision measurement of the mass difference between $D^{*+}(2010)$ and D^+ mesons, by *BABAR* [5].

6.1 Intro: Onia, Heavy Quark QCD, and the *B* Factories

Even though the initial primary goal of the e^+e^- *B* factories has been the study of *CP* violation in the weak interaction, a large fraction (including well over 100 papers in total) of the most cited publications from Belle and *BABAR* has been in the complementary area of the physics of the strong interaction, and particularly in heavy quark QCD and in quarkonia [6]. Multiple new mesons, and specifically several quarkonium states—including some unexpected ones with clearly exotic (non-pure-$q\bar{q}$) character [7]—have been discovered so far at both *B* factories, which has revolutionized the study of QCD and heavy mesons. Both *BABAR* and Belle continue to produce new results in the area of heavy quark QCD and quarkonia, and

J. Albert—(*BABAR* Collaboration member), representing the *BABAR* and Belle Collaborations.

J. Albert (✉)
University of Victoria, Victoria, BC V8W 3P6, Canada
e-mail: jalbert@uvic.ca

A. Giri and R. Mohanta (eds.), *16th Conference on Flavor Physics and CP Violation*, Springer Proceedings in Physics 234,
https://doi.org/10.1007/978-3-030-29622-3_6

so here we review a selection of five specific very recent (published in the past two years) results in onia and heavy quark QCD from Belle and *BABAR*, three of which have been published just this year.

6.2 Observation and Branching Fraction Measurement of $\Upsilon(4S) \to \eta'\Upsilon(1S)$ (Belle)

Heavy quark spin symmetry (HQSS), and in particular the QCD multipole expansion (QCDME) model that is based upon HQSS, predict the suppression of Υ decays to states containing an η, when compared with corresponding decays to states containing a $\pi^+\pi^-$ pair [8]. However, multiple results from both *BABAR* [9] and Belle [10] challenge that expectation. In particular, [9] measures the ratio of branching fractions $\frac{\mathscr{B}(\Upsilon(4S)\to\eta\Upsilon(1S))}{\mathscr{B}(\Upsilon(4S)\to\pi^+\pi^-\Upsilon(1S))}$ to be equal to 2.41 ± 0.42, which clearly does not constitute suppression. One suggestion [11] to address those results postulates a 4-quark component within the $\Upsilon(4S)$, and correspondingly provides a prediction of the previously-unmeasured ratio of branching fractions $R_{\eta'/\eta} \equiv \frac{\mathscr{B}(\Upsilon(4S)\to\eta'\Upsilon(1S))}{\mathscr{B}(\Upsilon(4S)\to\eta\Upsilon(1S))}$ to be within the range 0.2–0.6.

Belle measures this ratio $R_{\eta'/\eta}$ by measuring $\mathscr{B}(\Upsilon(4S) \to \eta'\Upsilon(1S))$ and comparing with the world-averaged value for $\mathscr{B}(\Upsilon(4S) \to \eta\Upsilon(1S))$. The $\Upsilon(1S)$ is reconstructed to $\mu^+\mu^-$, and the η' is reconstructed in its decays to $\rho^0\gamma$ and $\pi^+\pi^-\eta$, where $\eta \to \gamma\gamma$. Plots are shown in Fig. 6.1, and the branching fraction $\mathscr{B}(\Upsilon(4S) \to \eta'\Upsilon(1S)) = (3.43 \pm 0.88 \pm 0.21) \times 10^{-5}$ is measured, resulting in a measurement of $R_{\eta'/\eta} = 0.20 \pm 0.06$, at the low end of the "4-quark component" prediction [1, 11].

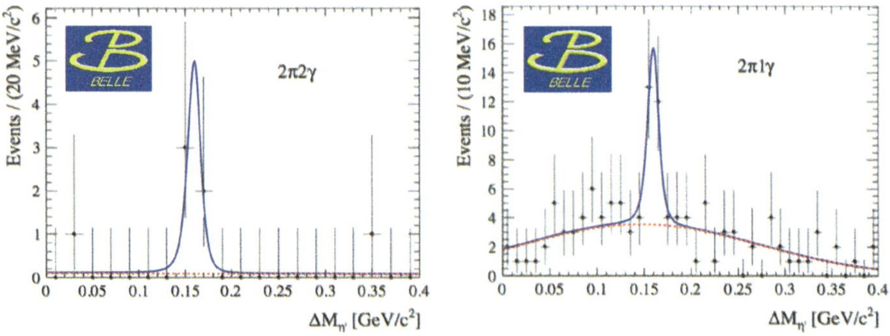

Fig. 6.1 Histograms of reconstructed $\Delta M_{\eta'} \equiv M(\Upsilon(4S)) - M(\Upsilon(1S)) - M(\eta')$ for (*left plot*) selected $\eta' \to \pi^+\pi^-\eta$ with $\eta \to \gamma\gamma$ events, and (*right plot*) selected $\eta' \to \rho^0\gamma$ with $\rho^0 \to \pi^+\pi^-$ events, from [1]. In both plots the fitted background is the red dotted curve, and total signal + background fit is the blue solid curve

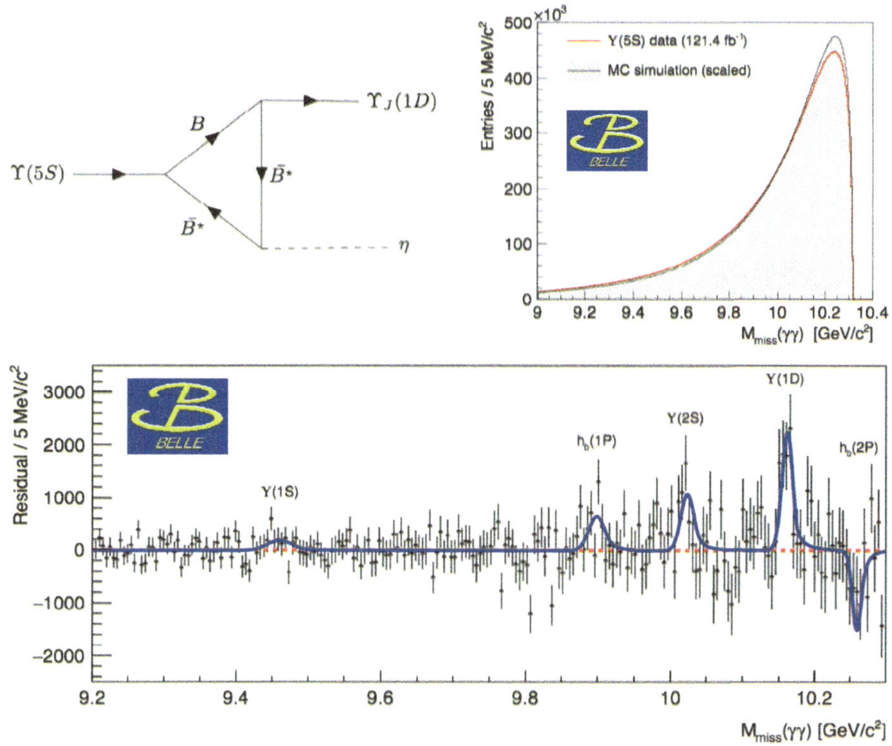

Fig. 6.2 From [2]: (*Upper left*) A virtual B meson loop contribution Feynman diagram to the $\Upsilon(5S) \rightarrow \eta+$ bottomonium process. The magnitude of such B meson loop contributions is difficult to predict theoretically, and thus measurements can be compared with various models of virtual B meson loop contributions to the process versus various models of contributions to the process from possible exotic (4-quark etc.) components of the $\Upsilon(5S)$ [12, 13]. (*Upper right*) Missing mass distribution $M_{\mathrm{miss}}(\gamma\gamma) \equiv \sqrt{(\langle \boldsymbol{p}_{e^+e^-} \rangle - \boldsymbol{p}_\eta)^2}$ for $\Upsilon(5S)$ data (red curve) and for MC simulation (black curve, scaled to match the total number of data events). (*Lower*) Background-subtracted missing mass distribution for selected events, together with the fitted and subtracted background (red dotted curve) and fitted total signal + background distribution (blue curve), showing the contributions from various bottomonium resonances

6.3 Measurement of the Cross Sections of $e^+e^- \rightarrow \eta+$ bottomonium (measured near the $\Upsilon(5S)$)(Belle)

In addition to the $\Upsilon(4S)$, exotic (4-quark etc.) components of other bottomonium resonances such as the $\Upsilon(5S)$ are also postulated [12]; and additionally the contribution of virtual B meson loops (as shown, e.g., in Fig. 6.2 upper left) to radiative $\Upsilon(5S)$ (and other radiative Υ) decays is theoretically uncertain [12, 13]. The two effects can be separated by precise measurement of cross sections of $e^+e^- \rightarrow \eta +$ bottomonium when measured at c.m. energies near higher Υ resonances.

Table 6.1 Table of significances (Σ), yields (N_{meas}), and measured visible cross sections (σ_v) for fitted bottomonium resonances, from [2] (Belle). The uncertainty on N_{meas} is statistical only, while fit-related systematic uncertainties are taken into account in the significance estimation. Upper limits on σ_v are at 90% confidence level

Process	Σ	N_{meas} [$\times 10^3$]	σ_v [pb]
$e^+e^- \to \eta\Upsilon(1S)$	1.5σ	1.7 ± 1.0	<0.34
$e^+e^- \to \eta h_b(1P)$	2.7σ	3.9 ± 1.5	<0.52
$e^+e^- \to \eta\Upsilon(2S)$	3.3σ	5.6 ± 1.6	$0.70 \pm 0.21 \pm 0.12$
$e^+e^- \to \eta\Upsilon(1D)$	5.3σ	9.3 ± 1.8	$1.14 \pm 0.22 \pm 0.15$
$e^+e^- \to \eta h_b(2P)$	–	-5.2 ± 3.6	<0.44

Belle measures these cross sections of $e^+e^- \to \eta +$ bottomonium near the $\Upsilon(5S)$ as shown in Fig. 6.2 and Table 6.1. The results appear consistent with models incorporating the effects of virtual B meson loops [12, 13].

6.4 Measurement of Several Hadronic J/ψ Decays and Branching Fractions (BABAR)

Nearly since the discovery of the J/ψ, its hadronic decays have provided a relatively clean laboratory for the study of QCD, for example in searches for glueballs and for light quark molecular states. Such studies naturally require very careful understanding, both experimentally and theoretically, of the various J/ψ decays. Although such studies have been ongoing at many detectors over the past 40 years, a number of J/ψ decays remained either unmeasured (e.g. $J/\psi \to K^0_S K^+\pi^-\pi^0$) or poorly measured (e.g. $J/\psi \to \pi^+\pi^-\pi^0$).

Initial-state radiation (ISR) events provide a clean source of many hadronic final states at B factories, including of course J/ψ-produced ones, as shown in Fig. 6.3. BABAR uses ISR-produced J/ψ decays to make the first measurement of $\mathcal{B}(J/\psi \to K^0_S K^+\pi^-\pi^0)$, finding it to be $(5.7 \pm 0.3 \pm 0.4) \times 10^{-3}$. That total branching fraction is, of course, composed of both resonant and non-resonant contributions. Figure 6.4 shows plots of both the total J/ψ peak from $K^0_S K^+\pi^-\pi^0$, and contributions from resonances that decay to that final state. BABAR measures the resonant branching fraction contributions to that final state to be $\mathcal{B}(J/\psi \to K^0_S K^+\rho^-) = (1.87 \pm 0.18 \pm 0.34) \times 10^{-3}$, $\mathcal{B}(J/\psi \to K^{*0} K^0_S \pi^0) = (0.43 \pm 0.10 \pm 0.27) \times 10^{-3}$, $\mathcal{B}(J/\psi \to K^{*0} K^-\pi^+) = (1.3 \pm 0.1 \pm 0.2) \times 10^{-3}$, $\mathcal{B}(J/\psi \to K^{*\pm} K^{\mp}\pi^0) = (1.0 \pm 0.1 \pm 0.3) \times 10^{-3}$, and $\mathcal{B}(J/\psi \to K^{*\pm} K^0_S \pi^{\mp}) = (0.8 \pm 0.1 \pm 0.3) \times 10^{-3}$ [3]. Additionally, in separate publications, BABAR also uses ISR-produced J/ψ to measure the branching fractions $\mathcal{B}(J/\psi \to K^0_S K^{\pm}\pi^{\mp}\eta) = (1.30 \pm 0.25 \pm 0.07) \times 10^{-3}$ [14] (which can be compared with the prior world average for that

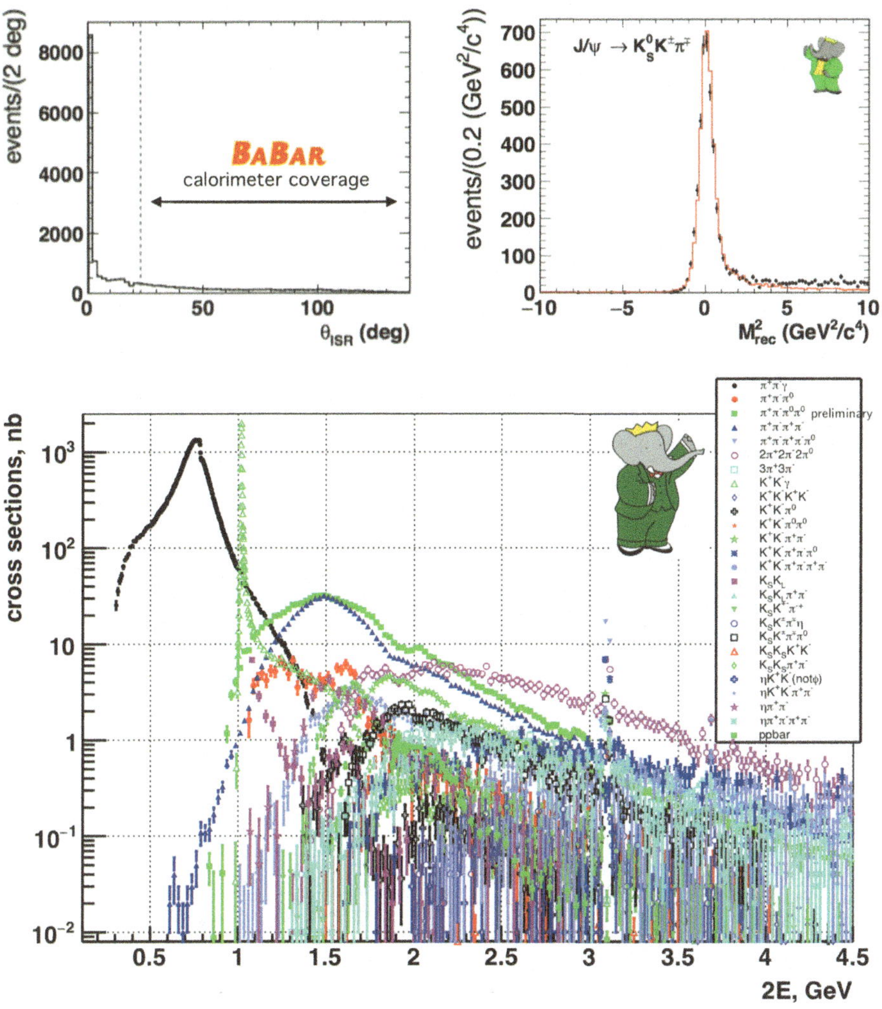

Fig. 6.3 From [3]: (*Upper left*) Distribution of the angle of the emitted initial-state radiation (ISR) photon in the laboratory frame (in degrees) away from the beamline for ISR events at *BABAR*. The distribution peaks near 0 degrees, which means that the ISR photon is not reconstructed in the calorimeter for the majority of ISR events. Nevertheless, such events can still be selected via the kinematics of the other final state particles from the interaction. (*Upper right*) Distribution of $M^2_{\rm rec} \equiv (\langle \mathbf{p}_{e^+e^-}\rangle - \mathbf{p}_{\rm hadrons})^2$ for selected ISR-produced $J/\psi \rightarrow K^0_S K^\pm \pi^\mp$ events. (*Lower*) Distribution of the invariant mass of the final-state hadronic system for selected ISR-produced events for many different hadronic final states at *BABAR*. Note that the J/ψ peak at 3.1 GeV can clearly be seen. ISR provides a very clean and high-statistics means of extracting quite pure samples of J/ψ and other charmonium and light quarkonium resonances at B factories

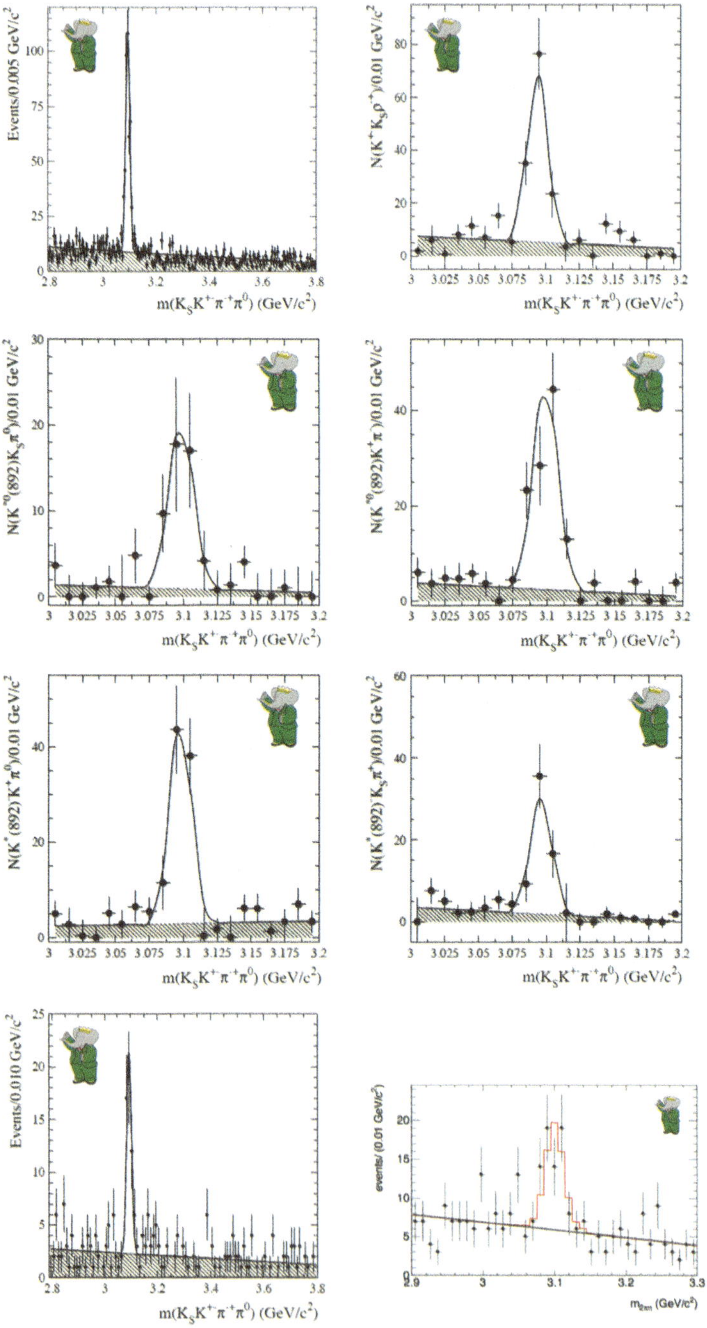
Events/0.005 GeV/c^2
m($K_S K^{+-}\pi^{-+}\pi^0$) (GeV/c^2)
N($K^+ K_S \rho^{-+}$)/0.01 GeV/c^2
N(K^{*0}(892)$K_S\pi^0$)/0.01 GeV/c^2
N(K^{*0}(892)$K^+\pi^-$)/0.01 GeV/c^2
N(K^*(892)$^- K^+\pi^0$)/0.01 GeV/c^2
N(K^*(892)$^- K_S\pi^+$)/0.01 GeV/c^2
Events/0.010 GeV/c^2
events/ (0.01 GeV/c^2)
m$_{2\pi\eta}$ (GeV/c^2)

◂**Fig. 6.4** From [3]: Each of the first 6 plots shows the invariant mass of the $K_S^0 K^\pm \pi^\mp \pi^0$ system in various ISR-produced $J/\psi \to K_S^0 K^\pm \pi^\mp \pi^0$ events at BABAR. The plot at (*top left*) shows all such events, without selection on resonant or non-resonant contributions to the $K_S^0 K^\pm \pi^\mp \pi^0$ final state. The plot at (*top right*) shows a selection on the contribution to that final state from $\rho^\mp \to \pi^\mp \pi^0$, i.e. $J/\psi \to K_S^0 K^\pm \rho^\mp$. The plot at (*upper middle left*) shows a selection on the contribution from $J/\psi \to K^{*0} K_S^0 \pi^0$. The plot at (*upper middle right*) shows a selection on the contribution from $J/\psi \to K^{*0} K^\pm \pi^\mp$. The plot at (*lower middle left*) shows a selection on the contribution from $J/\psi \to K^{*\pm} K^\mp \pi^0$. Finally, the plot at (*lower middle right*) shows a selection on the contribution from $J/\psi \to K^{*\pm} K_S^0 \pi^\mp$. The plot at (*bottom left*) shows the invariant mass of the $K_S^0 K^\pm \pi^\mp \eta$ system in all selected ISR-produced $J/\psi \to K_S^0 K^\pm \pi^\mp \eta$ events. The plot at (*bottom right*) shows the invariant mass of the $\pi^+\pi^-\eta$ system in all selected ISR-produced $J/\psi \to \pi^+\pi^-\eta$ events

mode of $(2.2 \pm 0.4) \times 10^{-3}$ [15]), and also the G-parity violating decay $\mathscr{B}(J/\psi \to \pi^+\pi^-\eta) = (4.2 \pm 0.8) \times 10^{-4}$ [16] (compared with the prior world average of $(4.0 \pm 1.7) \times 10^{-4}$ [15]).

6.5 Search for Glueball-Like Mesons in $\Upsilon(1S) \to \gamma\pi^+\pi^-$ and $\Upsilon(1S) \to \gamma K^+K^-$ Decays (BABAR)

The existence of gluonium (glueballs) remains a persistent prediction of lattice QCD, which predicts the lightest gluonium states to have quantum numbers $J^{PC} = 0^{++}$ and 2^{++} and to have mass below 2.5 GeV/c^2 [17]. One possible candidate for the $J^{PC} = 0^{++}$ glueball is the $f_0(1710)$, although early analyses assigned $J^{PC} = 2^{++}$ to the $f_0(1710)$ resonance. An unusually clean source of f mesons can be found in the $\pi^+\pi^-$ and K^+K^- spectra in $\Upsilon(1S)$ decays to $\gamma\pi^+\pi^-$ and to γK^+K^-. BABAR studies and fits these two spectra as shown in Fig. 6.5, and extracts yields and significances for the different resonances as shown in Table 6.2. The results are consistent with the $f_0(1710)$ glueball hypothesis found in [18].

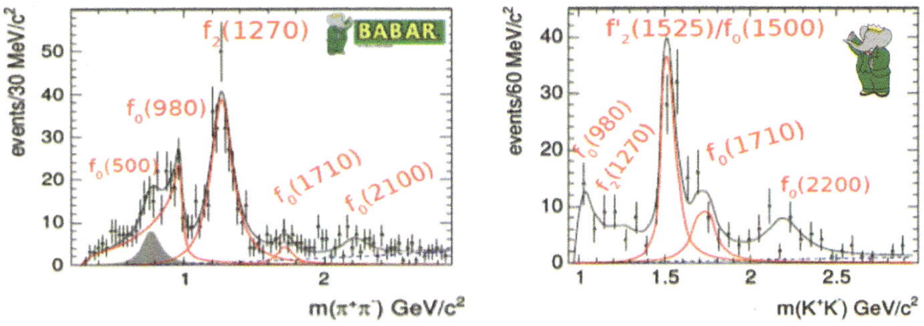

Fig. 6.5 Distributions of reconstructed invariant mass of the (*left plot*) $\pi^+\pi^-$ and (*right plot*) K^+K^- system in selected $\Upsilon(1S) \to \gamma\pi^+\pi^-$ and $\Upsilon(1S) \to \gamma K^+K^-$ events respectively, from [4]. Labels marking fitted resonance contributions are provided on both plots

Table 6.2 Table of yields and significances for fitted $\pi^+\pi^-$ and K^+K^- resonant (as well as $\pi^+\pi^-$ S-wave nonresonant) contributions to reconstructed $\Upsilon(1S) \to \gamma\pi^+\pi^-$ and $\Upsilon(1S) \to \gamma K^+K^-$ events within *BABAR* samples of $\Upsilon(2S) \to \Upsilon(1S)\pi^+\pi^-$ and $\Upsilon(3S) \to \Upsilon(1S)\pi^+\pi^-$ events, from [4]. Contributions from $f_2(1270)$ and from $f_0(1710)$ are highlighted in blue and red respectively. Systematic uncertainties are evaluated only for resonances for which branching fractions are computed

Resonances ($\pi^+\pi^-$)	Yield $\Upsilon(2S)$	Yield $\Upsilon(3S)$	Significance
S-wave	$133 \pm 16 \pm 13$	87 ± 13	12.8σ
$f_2(1270)$	$255 \pm 19 \pm 8$	$77 \pm 7 \pm 4$	15.9σ
$f_0(1710)$	$24 \pm 8 \pm 6$	$6 \pm 8 \pm 3$	2.5σ
$f_0(2100)$	33 ± 9	8 ± 15	
$\rho(770)^0$		54 ± 23	
Resonances (K^+K^-)	Yield $\Upsilon(2S)$ & $\Upsilon(3S)$		Significance
$f_0(980)$	47 ± 9		5.6σ
$f_J(1500)$	$77 \pm 10 \pm 10$		8.9σ
$f_0(1710)$	$36 \pm 9 \pm 6$		4.7σ
$f_2(1270)$	15 ± 8		
$f_0(2200)$	38 ± 8		

6.6 Precision Measurement of the Mass Difference Between $D^{*+}(2010)$ and D^+ Mesons (*BABAR*)

Both lattice QCD and chiral perturbation theory typically begin in the limit of SU(3) flavor symmetry and $m_b = m_c = \infty$, and then consider symmetry breaking (SB) due to $m_u \neq m_d \neq m_s \neq 0$, to finite m_b & m_c, and to EM interactions. This SB can be related to mass differences such as between the $D^{*+}(2010)$, D^+, and D^0 mesons [19]. Precise measurements of such differences constrain SB, leading to more precise predictions from the lattice and from chiral perturbation.

Using a fit to the reconstructed $D^{*+}(2010) - D^+$ mass difference spectrum as shown in Fig. 6.6, *BABAR* measures $\Delta M_+ \equiv m(D^{*+}(2010)) - m(D^+) = (140601.0 \pm 6.8 \pm 12.9)$ keV, over 5 times more precise than the previous world average for this quantity [5]. When combined with the previous *BABAR* result $\Delta M_0 \equiv m(D^{*+}(2010)) - m(D^0) = (145425.9 \pm 0.4 \pm 1.7)$ keV [20], one obtains $\Delta M_D \equiv m(D^+) - m(D^0) = (4824.9 \pm 6.8 \pm 12.9)$ keV, also over 5x more precise than previous world averages.

6.7 Conclusion

We have briefly summarized here five recent analyses from Belle and *BABAR* in the area of heavy quark QCD and quarkonia, a small fraction of the recent analyses in this area from the B factories (and an even smaller fraction of analyses in this

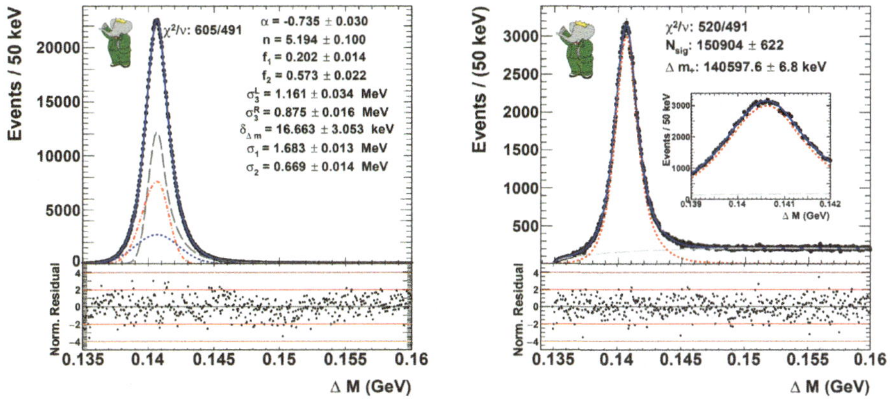

Fig. 6.6 Fits to the $\Delta M_{+} \equiv$ reconstructed $m(D^{+}\pi^{0})$ − reconstructed $m(D^{+})$ spectrum in (*left plot*) Monte Carlo simulated, and (*right plot*) data, events within selected $e^{+}e^{-} \rightarrow c\bar{c}$ events containing a $D^{*+}(2010)$, from [5]. The signal shapes in data and MC are fitted to a model composed of a sum of three Gaussian-like PDFs with a common mean: specifically a standard Gaussian + a Crystal Ball function [21] + a bifurcated Gaussian. Central values and statistical uncertainties of fitted parameters are shown on both plots

area from the B factories over the past 15 years). Heavy quark QCD and quarkonia analyses such as these have recently revealed QCD to be far more than the study of simple mesons and baryons, and the future promises much more excitement in what has now become a quest toward the understanding and predictability of exotic hadrons (in addition, of course, to their less-exotic relatives).

References

1. E. Guido et al. (Belle Collaboration), Phys. Rev. Lett. **121**, 062001 (2018), arXiv:1803.10303, https://journals.aps.org/prl/abstract/10.1103/PhysRevLett.121.062001
2. U. Tamponi et al. (Belle Collaboration), Eur. Phys. J. C **78**, 633 (2018), arXiv:1803.03225, https://epjc.epj.org/articles/epjc/abs/2018/08/10052_2018_Article_6086/10052_2018_Article_6086.html
3. J.P. Lees et al. (*BABAR* Collaboration), Phys. Rev. D **95**, 072007 (2017), arXiv:1702.01551, https://journals.aps.org/prd/abstract/10.1103/PhysRevD.95.072007
4. J.P. Lees et al. (*BABAR* Collaboration), Phys. Rev. D **97**, 112006 (2018), arXiv:1804.04044, https://journals.aps.org/prd/abstract/10.1103/PhysRevD.97.112006
5. J.P. Lees et al. (*BABAR* Collaboration), Phys. Rev. Lett. **119**, 202003 (2017), arXiv:1707.09328, https://journals.aps.org/prl/abstract/10.1103/PhysRevLett.119.202003
6. See, e.g., Chap. 18 (pp. 441–514) of https://link.springer.com/article/10.1140/epjc/s10052-014-3026-9 *The Physics of the B Factories*, ed. A.J. Bevan et al., Eur. Phys. J. C **74**, 3026 (2014), and references therein
7. Possibly the present best example of a clearly non-pure-$q\bar{q}$ neutral "quarkonium" state is the $Y(4260)$, as discussed in, e.g., pp. 7–8 of the http://pdg.lbl.gov/2018/reviews/rpp2018-rev-non-qqbar-mesons.pdf review on *Non-$q\bar{q}$ Mesons* by C. Amsler and C. Hanhart in M. Tanabashi et al. (Particle Data Group), Phys. Rev. D **98**, 030001 (2018), and references therein

8. See, e.g., (2006), arXiv:hep-ph/0601044, https://link.springer.com/article/10.100719, and references therein
9. B. Aubert et al. (*BABAR* Collaboration), Phys. Rev. D **78**, 112002 (2008), arXiv:0807.2014, https://journals.aps.org/prd/abstract/10.1103/PhysRevD.78.112002
10. U. Tamponi et al. (Belle collaboration), Phys. Rev. Lett. **115**, 142001 (2015), arXiv:1506.089014, https://journals.aps.org/prd/abstract/10.1103/PhysRevD.96.052005; E. Guido et al., (Belle Collaboration), Phys. Rev. D **96**, 052005 (2017), arXiv:1707.04973, https://journals.aps.org/prl/abstract/10.1103/PhysRevLett.115.142001
11. M.B. Voloshin, Mod. Phys. Lett. A **26**, 773 (2011), arXiv:1101.3542, https://www.worldscientific.com/doi/abs/10.1142/S0217732311035390
12. C. Meng, K.T. Chao, Phys. Rev. D **78**, 074001 (2008), arXiv:0806.3259, https://journals.aps.org/prd/abstract/10.1103/PhysRevD.78.074001
13. Y.A. Simonov, A.I. Veselov, Phys. Lett. B **673**, 211 (2009), arXiv:0810.0366, https://www.sciencedirect.com/science/article/pii/S0370269309001920
14. J.P. Lees et al. (*BABAR* Collaboration), Phys. Rev. D **95**, 092005 (2017), arXiv:1704.05009, https://journals.aps.org/prd/abstract/10.1103/PhysRevD.95.092005
15. C. Patrignani et al. (Particle Data Group), Chin. Phys. C **40**, 100001 (2016), and 2017 update, http://pdg.lbl.gov/2017/
16. J.P. Lees et al. (*BABAR* Collaboration), Phys. Rev. D **97**, 052007 (2018), arXiv:1801.02960, https://journals.aps.org/prd/abstract/10.1103/PhysRevD.97.052007
17. See, e.g., Y. Chen et al., Phys. Rev. D **73**, 014516 (2006), arXiv:hep-lat/0510074, https://journals.aps.org/prd/abstract/10.1103/PhysRevD.73.014516
18. R. Zhu, JHEP **1509**, 166 (2015), arXiv:1508.01445, https://link.springer.com/article/10.1007%2FJHEP09%282015%29166
19. J.L. Goity, C.P. Jayalath, Phys. Lett. B **650**, 22 (2007), arXiv:hep-ph/0701245, https://www.sciencedirect.com/science/article/pii/S0370269307005114
20. J.P. Lees et al., (*BABAR* Collaboration), Phys. Rev. Lett. **111**, 111801 (2013), arXiv:1304.5657, https://journals.aps.org/prl/abstract/10.1103/PhysRevLett.111.111801
21. The Crystal Ball function was first described in M.J. Oreglia, Ph.D. thesis, Stanford University Report No. SLAC-R-236 (1980), http://www-public.slac.stanford.edu/sciDoc/docMeta.aspx?slacPubNumber=slac-r-236; J.E. Gaiser, Ph.D. thesis, Stanford University Report No. SLAC-R-255 (1982), http://www-public.slac.stanford.edu/sciDoc/docMeta.aspx?slacPubNumber=slac-r-255; T. Skwarnicki, Ph.D. thesis, Cracow Institute of Nuclear Physics Report No. DESY-F31-86-02 (1986), http://www-library.desy.de/cgi-bin/showprep.pl?DESY-F31-86-02

Chapter 7
Measurement of Heavy Flavor Properties at CMS

Muhammad Alibordi

Abstract Measurements of lifetimes in the decay channels $B^0 \to J/\psi K^*(892)^0$, $B^0 \to J/\psi K^0_s$, $B^0_s \to J/\psi \pi^+\pi^-$, $B^0_s \to J/\psi\phi$, $\Lambda^0_b \to J/\psi \Lambda^0$ and $B^+_c \to J/\psi\pi^+$, and polarization of Λ_b in $\Lambda_b \to J/\psi\Lambda$ decays are presented. The data used are collected by the CMS experiment in proton–proton collisions at centre-of-mass energy $\sqrt{s} = 7$ and 8 TeV corresponding to an integrated luminosity of 5 fb^{-1} and 19.7 fb^{-1}, respectively. The measured lifetimes in μm are $c\tau_{B^0 \to J/\psi K^*(892)^0} = 453.0 \pm 1.6(\text{stat}) \pm 1.8(\text{syst})$, $c\tau_{B^0 \to J/\psi K^0_s} = 457.8 \pm 2.7(\text{stat}) \pm 2.8(\text{syst})$, $c\tau_{B^0_s \to J/\psi\phi} = 443.9 \pm 2.0(\text{stat}) \pm 1.5(\text{syst})$, $c\tau_{B^0_s \to J/\psi\pi^+\pi^-} = 502.7 \pm 10.2(\text{stat}) \pm 3.4(\text{syst})$, $c\tau_{\Lambda^0_b \to J/\psi\Lambda^0} = 442.9 \pm 8.2(\text{stat}) \pm 2.8(\text{syst})$ and $c\tau_{B^+_c \to J/\psi\pi^+} = 162.3 \pm 7.8(\text{stat}) \pm 4.2(\text{syst})$. The measured Λ^0_b polarization is $P = 0.00 \pm 0.06(\text{stat}) \pm 0.06(\text{syst})$ along with the asymmetry parameter $\alpha_1 = 0.14 \pm 0.14(\text{stat}) \pm 0.10(\text{syst})$.

7.1 Introduction

The quantum chromodynamics (QCD) based Heavy Quark Expansion (HQE) framework describes the decay rates for the decays of $B^0 \to J/\psi K^*$, $B^0 \to J/\psi K^0_s$, $B^0_s \to J/\psi\pi^+\pi^-$, $B^0_s \to J/\psi\phi$, $\Lambda^0_b \to J/\psi\Lambda^0$ and $B^+_c \to J/\psi\pi^+$, which can be a critical probe of the non-perturbative aspects of QCD. A detailed account of the theory of lifetimes can be found in [1]. While measuring the lifetimes of other b hadrons mentioned above we used similar methods, a different approach has been followed for the measurement of B^+_c lifetime. First the decay width difference between B^+_c and B^+ is measured and then from this difference the B^+_c lifetime is extracted by using the precisely known B^+ ($\to J/\psi K^+$) lifetime. The study of polarization measurement in the decays of $\Lambda_b \to J/\psi\Lambda$ provides an important avenue to strong interaction effects in hadronic decays. The main observables here are the polarization, P of Λ_b

Muhammad Alibordi—On behalf of the CMS Collaboration.

M. Alibordi (✉)
Indian Institute of Technology Madras, Chennai 600036, Tamil Nadu, India
e-mail: muhammad.muhammad.alibordi@cern.ch

A. Giri and R. Mohanta (eds.), *16th Conference on Flavor Physics and CP Violation*, Springer Proceedings in Physics 234,
https://doi.org/10.1007/978-3-030-29622-3_7

and the parity-violating asymmetry parameter in the decay $\Lambda \rightarrow p\pi^-$ [2–4]. The measurements of lifetime and polarization measurements are performed using data collected by the CMS experiment at the LHC. CMS comprises a series of coaxial cylindrical subdetectors. The innermost subdetector is silicon pixel and strip tracker. It is followed by an electromagnetic calorimeter, composed of lead tungstate scintillating crystals, surrounded by brass and scintillator sampling hadron calorimeter. All these subdetectors are placed inside a superconducting solenoid of inner diameter 6 m, which provides a 3.8 T magnetic field. Finally, the muon system, comprising drift tubes, resistive plate chambers and cathode strip chambers, is located outside of the solenoid [5, 6].

7.2 Event Reconstruction

For the lifetime measurements of $B^0 \rightarrow J/\psi K^*$, $B^0 \rightarrow J/\psi K_s^0$, $B_s^0 \rightarrow J/\psi\pi^+\pi^-$, $B_s^0 \rightarrow J/\psi\phi$, $\Lambda_b^0 \rightarrow J/\psi\Lambda^0$ and $B_c^+ \rightarrow J/\psi\pi^+$ decay channels, the J/ψ is reconstructed from two oppositely charged muons that are required to have transverse momentum $p_{\mathrm{T}} > 7.9$ GeV, pseudorapidity $|\eta| < 2.2$, a distance of closest approach to the event vertex < 0.5 cm, and a vertex fit χ^2 probability $>0.5\%$. The invariant mass is chosen to be within ± 5 times the experimental mass resolution around the nominal J/ψ mass [1]. Other final state candidates in different decay channels are reconstructed from charged tracks, originating from a common vertex as the J/ψ. Then the b hadrons are formed by combining the J/ψ with these tracks. For the polarization measurement of $\Lambda_b \rightarrow J/\psi\Lambda$, the J/$\psi$ is reconstructed as earlier whereas $\Lambda \rightarrow p\pi^-$ from a pair of oppositely charged tracks. The Λ_b candidates are formed by combining muon objects and tracks within the mass window $5.40 < m_{J/\psi\Lambda} < 5.84$ GeV.

7.3 Fit Model, Result and Systematics

7.3.1 Fit and Results

For the measurement of lifetimes in different decay channels, a simultaneous unbinned extended maximum-likelihood fit has been performed using b hadron mass, the proper decay length of b hadrons ct and decay length uncertainty σ_{ct}, as input parameters. The signal for the mass fit is modeled as one or two Gaussians. The background is described with a polynomial or an exponential depending upon decay channels. In the ct distribution the signal is modeled with an exponential convolved with the detector resolution whereas the background is described by a sum of exponentials convolved with the resolution. The σ_{ct} distributions are modeled with two gamma functions for signal and two exponentials convolved with a Gaussian function for background. The number of exponentials describing the background is different for different decay modes as determined from the data events in the mass sideband regions in each decay mode.

The fitted distributions of B^0, B^0_s and Λ_b at $\sqrt{s} = 8$ TeV are shown in Figs. 7.1, 7.2 and the lifetime results in μm corresponding to these channels are,

$$
\begin{aligned}
c\tau_{B^0\to J/\psi K^*(892)^0} &= 453.0 \pm 1.6(\text{stat}) \pm 1.8(\text{syst}) \\
c\tau_{B^0\to J/\psi K^0_s} &= 457.8 \pm 2.7(\text{stat}) \pm 2.8(\text{syst}) \\
c\tau_{B^0_s\to J/\psi\phi(1020)} &= 443.9 \pm 2.0(\text{stat}) \pm 1.5(\text{syst}) \\
c\tau_{B^0_s\to J/\psi\pi^+\pi^-} &= 502.7 \pm 10.2(\text{stat}) \pm 3.4(\text{syst}) \\
c\tau_{\Lambda^0_b\to J/\psi\Lambda^0} &= 442.9 \pm 8.2(\text{stat}) \pm 2.8(\text{syst})
\end{aligned}
\tag{7.1}
$$

The efficiency-corrected fitted ct distribution of the ratio of B_c^+ to B^+ gives decay width difference $\Delta\Gamma = 1.24 \pm 0.09$ ps^{-1} in Fig. 7.3. From the known lifetime of B^+ the extracted lifetime of B_c^+ is $c\tau_{B_c^+\to J/\psi\pi^+} = 162.3 \pm 7.8 \pm 4.2$ μm.

For the measurement of polarization, the Λ_b polarization P, and the angular decay parameters α_1, α_2, and γ_0 are determined. The results are obtained from an unbinned maximum-likelihood fit to the $J/\psi\,\Lambda$ invariant mass distribution and the three angular variables $\Theta_3 = (\cos\theta_\Lambda, \cos\theta_p, \cos\theta_\mu)$ are shown in Fig. 7.4. The extracted signal yields from both $J/\psi\,\Lambda$ and $J/\psi\,\bar{\Lambda}$ is 6000 and the measured values for different parameters are $P = 0.00 \pm 0.06 \pm 0.06$, $\alpha_1 = 0.14 \pm 0.14 \pm 0.10$, $\alpha_2 = -1.11 \pm 0.04 \pm 0.05$ and $\gamma_0 = -0.27 \pm 0.08 \pm 0.11$. The values of helicity amplitudes are,

$$
\begin{aligned}
|T_{++}|^2 &= 0.05 \pm 0.04(\text{stat}) \pm 0.04(\text{syst}) \\
|T_{+0}|^2 &= -0.10 \pm 0.04(\text{stat}) \pm 0.04(\text{syst}) \\
|T_{-0}|^2 &= 0.51 \pm 0.03(\text{stat}) \pm 0.04(\text{syst}) \\
|T_{--}|^2 &= 0.52 \pm 0.04(\text{stat}) \pm 0.04(\text{syst})
\end{aligned}
\tag{7.2}
$$

7.3.2 Systematic Uncertainties

For lifetime measurements, common systematic uncertainties are statistical uncertainty in the Monte Carlo samples, modeling of the mass distribution shape, and also channel specific uncertainties depending upon selection, reconstruction and fitting. Monte Carlo statistical uncertainty contributes the most among all uncertainties; in $B^0 \to J/\psi K^0_s$ decays it is 2.4 μm and in $\Lambda^0_b \to J/\psi\Lambda^0$ decays it has the value of

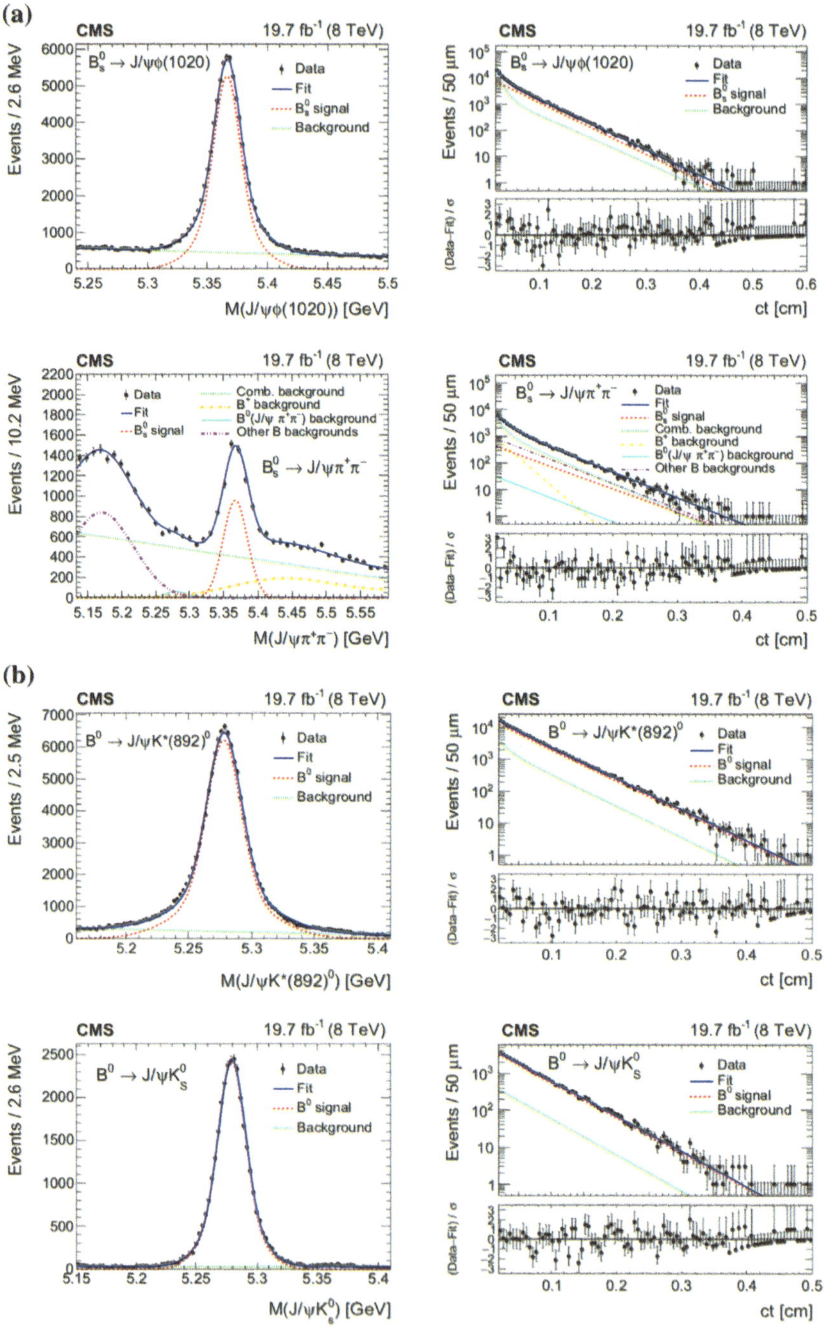

Fig. 7.1 Mass and ct distributions for B^0 and B_s^0, points with error bars and data, and the curves are projections of the fit to various event components as shown in the plots [1]

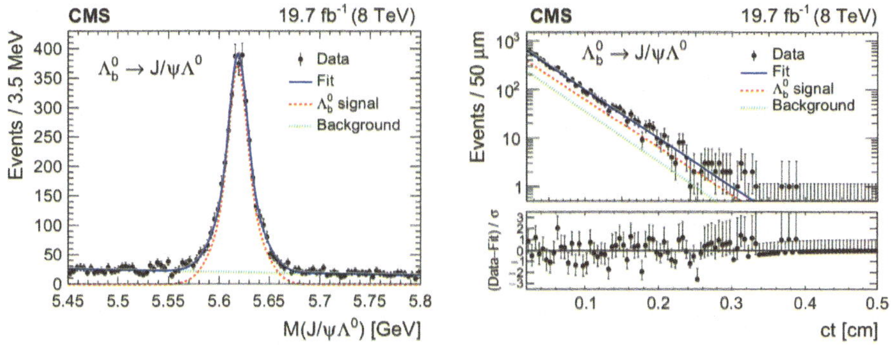

Fig. 7.2 Mass and ct distribution of Λ_b in the decay channel of $\Lambda_B^0 \to J/\psi \Lambda^0$, points with error bars and data, and the curves are projections of the fit to various event components as shown in the plots [1]

2.3 μm. Uncertainties arising for ct range regulation and S-wave contamination is absent in all decay channels except $B_s^0 \to J/\psi\phi$. Modulation in ct accuracy contributes almost at same scale ~1.3–1.4 μm in all cases. The measured values for polarization (P), asymmetry parameter (α_1), longitudinal polarization (α_2) and its related parameter (γ_0) in $\Lambda_b^0 \to J/\psi \Lambda^0$ decays are reconstruction wise biased by 5.7, 9.8, 2.0 and 9.1, respectively, up to the order 10^{-2}. The γ_0 is systematically uncertain by 5.0×10^{-2} due to the angular distribution used for the background fit. The angular efficiency contributes to the systematic uncertainty in α_2 measurement at 3.0×10^{-2}. There are other sources of uncertainties in the polarization measurement such as the weighting procedure for asymmetry parameter α_Λ.

7.4 Summary

The measurement of lifetimes for b hadrons in the channels $B^0 \to J/\psi K^*$, $B^0 \to J/\psi K_s^0$, $B_s^0 \to J/\psi\pi^+\pi^-$, $B_s^0 \to J/\psi\phi$, $\Lambda_b^0 \to J/\psi \Lambda^0$ and $B_c^+ \to J/\psi\pi^+$, and the polarization in the decays of $\Lambda_b^0 \to J/\psi \Lambda^0$ are briefly described. The measurement of Λ_b^0 lifetime is compatible with its measured value from other experiments in the same channel. The B_c^+ lifetime is in agreement with LHCb [7] and more precise than CDF [8] and D0 [9]. In the polarization measurement α_2 is compatible with -1, which implies that a Λ of a positive-helicity state from Λ_b is suppressed. The measured value of α_1 is also consistent with LHCb [10] and ATLAS [11].

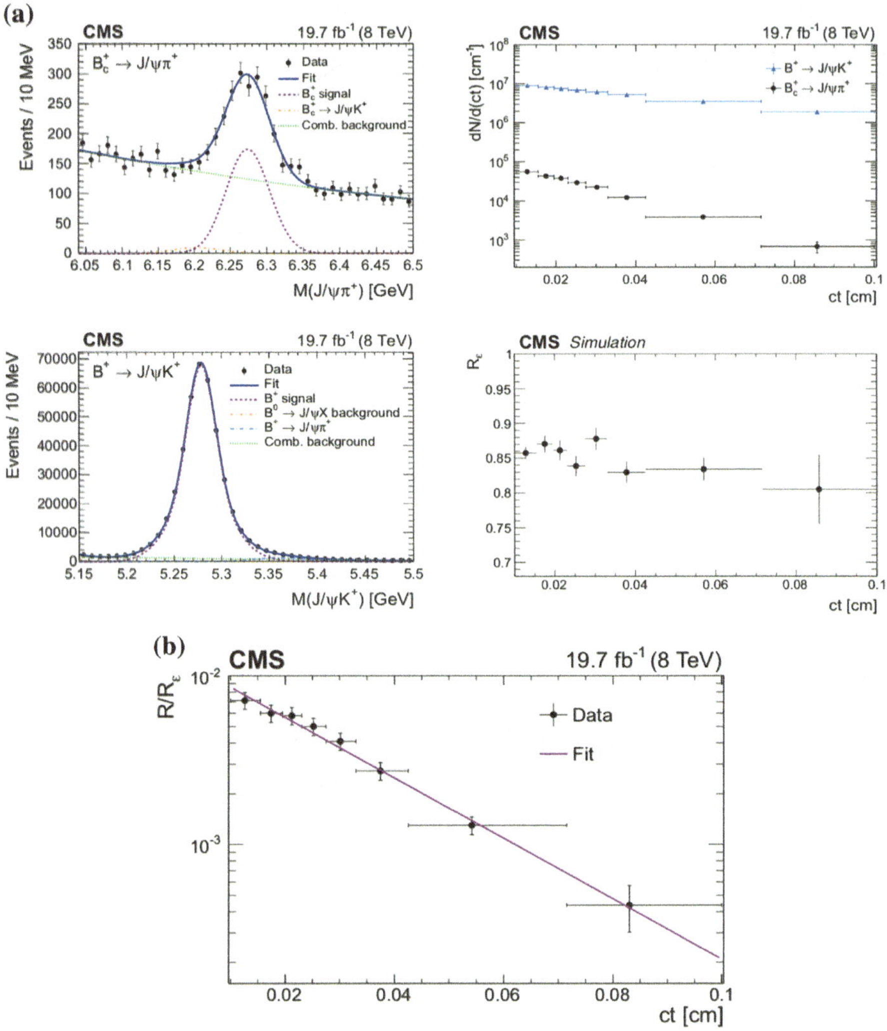

Fig. 7.3 Mass and efficiency distributions for B_c^+ and B^+ (**a**) and ratio of the B_c^+ to B^+ efficiency corrected ct distribution (**b**) [1]

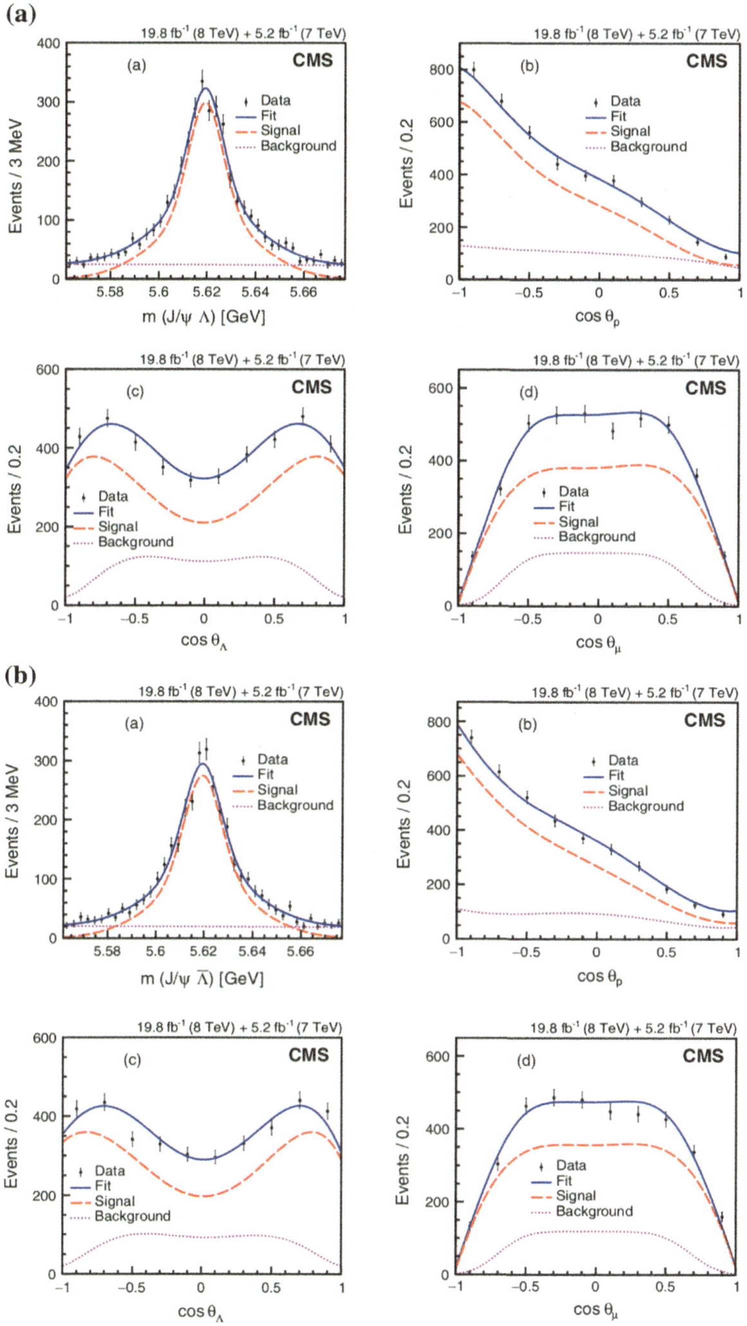

Fig. 7.4 Fitted distribution of mass and cosines of the three angles ($\cos\theta_\Lambda$, $\cos\theta_p$, $\cos\theta_\mu$) for both Λ_b (**a**) and $\bar{\Lambda}_b$ (**b**) are shown [2]

References

1. C.M.S. Collaboration, Measurement of b hadron lifetimes in pp collisions at $\sqrt{s} = 8$ TeV. Eur. Phys. J. C. **78**, 457 (2018). https://doi.org/10.1140/epjc/s10052-018-5929-3
2. C.M.S. Collaboration, Measurement of the Λ_b polarization and angular parameters in $\Lambda_b \rightarrow J/\psi \Lambda$ decays from pp collisions at $\sqrt{s}$ =7 and 8 TeV. Phys. Rev. D **97**, 072010 (2018). https://doi.org/10.1103/PhysRevD.97.072010
3. G. Hiller et.al., Photon polarization from helicity suppression in radiative decays of polarized Λ_b to spin-3/2 baryons. Phys. Lett. B **649**, 152–158 (2007). https://doi.org/10.1016/j.physletb.2007.03.056
4. M. Kramer, Angular correlations in $Ab \rightarrow A + V$: polarization measurements, HQET and CP violation. Nucl. Phys. B (Proc. Suppl.) **50**, 125–129 (1996), PII:S0920-5632(96)00378-7
5. The Compact Muon Solenoid Phase II Upgrade, Technical Proposal, CERN-LHCC-2015-010, LHCC-P-008, CMS-TDR-15-02, June 2015 (2015)
6. CMS Collaboration, The CMS experiment at the CERN LHC, 2008 JINST 3 S08004. https://doi.org/10.1088/1748-0221/3/08/s08004
7. LHCb Collaboration, Measurement of the lifetime of the B_c^+ meson using the $B_c^+ \rightarrow J/\psi\pi^+$ decay mode. Phys. Lett. B **742**, 29 (2015). https://doi.org/10.1016/j.physletb.2015.01.010. arXiv:1411.6899
8. C.D.F. Collaboration, Measurement of the B_c^+ meson lifetime using the decay mode $B_c^+ \rightarrow J/\psi e^{\pm}\nu_e$. Phys. Rev. Lett. **97**, 012002 (2006). https://doi.org/10.1103/PhysRevLett.97.012002. arXiv:hep-ex/0603027
9. D0 Collaboration, Measurement of the lifetime of the $B_c^{\pm}$ meson in the semileptonic decay channel. Phys. Rev. Lett. **102**, 092001 (2009). https://doi.org/10.1103/PhysRevLett.102.092001. arXiv:0805.2614
10. LHCb Collaboration, Measurements of the $\Lambda_b \rightarrow J/\psi \Lambda$ decay amplitudes and the Λ_b polarisation in p-p collisions at $\sqrt{s}$= 7TeV. Phys. Lett. B **724**, 27 (2013)
11. ATLAS Collaboration, Measurement of the parity-violating asymmetry parameter α_b and the helicity amplitudes for the decay $\Lambda_b \rightarrow J/\psi \Lambda$ with the ATLAS detector. Phys. Rev. D **89**, 092009 (2014)

Chapter 8
Prospects of CMS in B Physics in Phase 2

Rajarshi Bhattacharya

Abstract The Standard Model (SM) of Particle Physics precisely predicts the cross-sections for the extremely rare Flavour Changing Neutral Current (FCNC) processes $B_{d/s} \rightarrow \mu\mu$. Although, these processes are suppressed in the SM, beyond Standard Model (BSM) contributions can significantly enhance their decay rates. Therefore, a precise measurement of branching fractions and the $B_s \rightarrow \mu\mu$ lifetime is of prime concern for experiments like CMS in the quest for new physics. The SM also precisely predicts the CP violating phase ϕ_s in the $B_s \rightarrow \phi\phi \rightarrow$4 kaons decay process. Any significant deviation from the SM prediction for this rare FCNC process would provide a clear signature for BSM. During the High Luminosity (HL) period of the LHC, also known as LHC Phase-2, CMS will continue to look for $B_{d/s} \rightarrow \mu\mu$ and will also study $B_s \rightarrow \phi\phi \rightarrow$ 4 kaons. The latter will be possible for the first time in CMS due to inclusion of the tracking information in the Level-1 trigger. In this article, we shall discuss the CMS Phase-2 Tracker, Level 1 Tracking and present results obtained from Phase-2 simulations for $B_{d/s} \rightarrow \mu\mu$ and $B_s \rightarrow \phi\phi \rightarrow$ 4 kaons.

8.1 High Luminosity LHC

The LHC will undergo a major upgrade during the third Long Shutdown (LS3) period resulting in a ten-fold increase in instantaneous luminosity (5–7 $\times$ 10^{-34} $\text{fb}^{-1}\text{s}^{-1}$) compared to the current value [1]. Over a ten year period beginning in 2026, the LHC is expected to deliver a total integrated luminosity of ≥ 3 ab^{-1} at $\sqrt{s} = 14$ TeV to CMS [2]. This increase is equivalent to ten times the data collected during LHC Runs 1–3 [1].

During the HL-LHC period, the increase in luminosity, will cause the number of interactions per bunch crossing (known as pileup or PU) to increase significantly with

Rajarshi Bhattacharya—On behalf of the CMS Collaboration.

R. Bhattacharya (✉)
Saha Institute of Nuclear Physics, HBNI, Kolkata 700064, India
e-mail: rajarshi.bhattacharya@cern.ch

A. Giri and R. Mohanta (eds.), *16th Conference on Flavor Physics and CP Violation*, Springer Proceedings in Physics 234,
https://doi.org/10.1007/978-3-030-29622-3_8

the average pileup or <PU> foreseen to have a value between 140 and 200, which is five times the current value. Without an detector upgrade the increased pileup will severely impact physics prospects especially for B physics during Phase-2.

To retain the full physics potential during HL-LHC, the CMS detector will undergo a major upgrade, known as Phase-2 upgrade [3]. The present tracking detector will be replaced with a new improved detector. Tracking information will also be made available at Level-1 (L1) trigger for the first time in CMS [4], which would play a crucial role in the physics studies we are going to discuss.

8.2 CMS Phase-2 Tracker and L1 Tracking

A layout of the CMS Phase-2 tracker [4] is shown in Fig. 8.1a. The Phase-2 tracker will be extended from $|\eta| \approx 2.4$ to 4.0. It will consist of 4 pixel layers in the innermost barrel and 12 endcap discs on each side of the barrel region. Surrounding the pixel, the outer tracker will have 6 barrel layers and 5 endcap discs on either sides. The unit of the outer tracker is known as "p_T module", where each module will consist of two closely placed silicon sensors, separated by a small distance (between 1 and 4 mm). A p_T module can correlate hits in its two layers to create a "stub". By construction, this

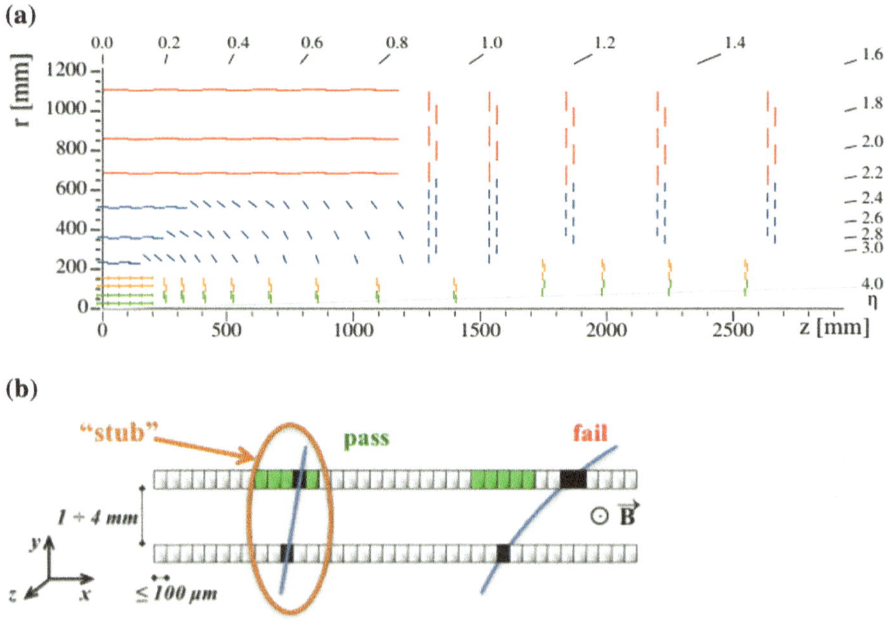

Fig. 8.1 **a** Layout of the CMS Phase-2 tracker layout and **b** concept of the "p_T module" [4]

module can accept "stubs" with $p_T \geq 2$ GeV. L1 tracks are formed by connecting stubs in different layers. The new tracker will have less material, better radiation hardness, and improved p_T and vertex resolution.

8.3 $B_{d/s} \to \mu\mu$

$B_{d/s} \to \mu\mu$ is an FCNC process that proceeds through electroweak penguin or box diagrams and hence is suppressed in the Standard Model (SM). It is further suppressed due to helicity suppression. The SM predicts the branching fraction for these two processes to be [5]

- $\mathrm{B}(B_s \to \mu\mu) \approx (3.57 \pm 0.16) \times 10^{-9}$
- $\mathrm{B}(B_d \to \mu\mu) \approx (1.02 \pm 0.06) \times 10^{-10}$

Several extensions of the SM, such as supersymmetry [6–8] and models with a non-standard Higgs sector [9], may enhance or suppress the branching fractions. Any deviation of the measured branching fractions from the SM values would imply presence of new physics. Therefore, precise measurements of the two branching fractions are an important goal of various experiments in their pursuit for new physics. So far, the measured branching fraction values from various experiments are found to be in agreement with the SM prediction albeit with large uncertainties.

The expected improvements in the mass measurement of B_d and B_s from the CMS Phase-2 upgrade is discussed in this section. B candidates are formed from two oppositely charged muon candidates having $p_T > 4$ GeV if $|\eta| < 1.4$ or $p_T > 2$ GeV if $|\eta| > 1.4$. A Gaussian fit is performed on the reconstructed di-muon invariant mass to find the peak position and resolution of the distribution. In Fig. 8.2a an overlay of the mass distributions for B_s and B_d in Phase-2 is shown, where the pseudorapidity of the most forward muon candidate is restricted to $|\eta_f| < 1.4$. An improvement in the mass resolution due to improvement in the p_T resolution of tracks for Phase-2 leads to a better separation between the B_d and B_s mass peaks compared to the current CMS data. In Fig. 8.2b, the B_s mass resolution is shown as a function of $|\eta_f|$ both for Run 2 (current LHC running) and Phase-2. The mass resolution has a ≈40% gain for $|\eta_f| < 1$ in Phase-2 compared to Run-2.

The mass peak separation with respect to $|\eta_f|$ is shown in Fig. 8.3a. For Phase-2, a 25% improvement in mass peak separation can be achieved in the region $|\eta_f| < 1.4$, which reduces the overlap between B_s and B_d signals as well as helps rejecting some rare semi-leptonic decays.

The leakage variable, defined as the ratio of the number of B_d over B_s events in the B_d invariant mass range, is a measure of overlap between the B_s and B_d events and is plotted with respect to $|\eta_f|$ for both Run-2 and Phase-2 in Fig. 8.3b. From the plot, it can be seen that Phase-2 performs better in the region $|\eta_f| < 0.8$. High leakage in the high $|\eta_f|$ region does not allow a clear separation between the $B_d \to \mu\mu$ and $B_s \to \mu\mu$ events.

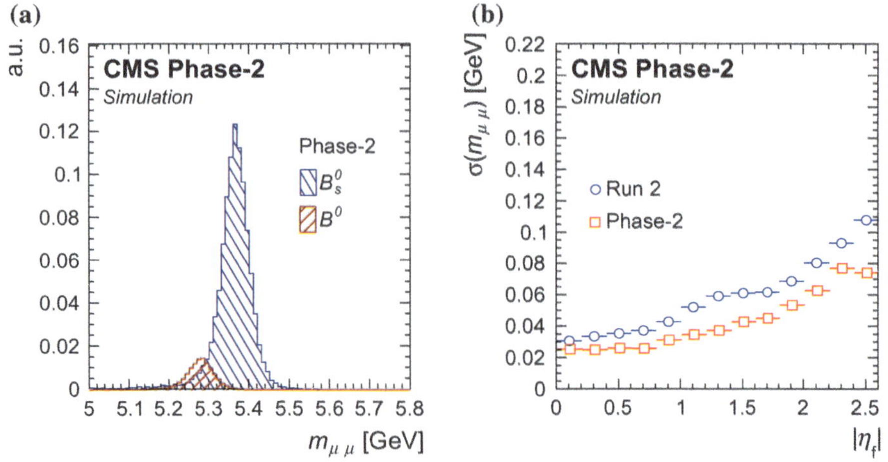

Fig. 8.2 **a** Dimuon mass distributions for B_s (blue dashed area) and B_d (red dashed area) in Phase-2 for pseudorapidity of the most forward muon candidate, $|\eta_f| < 1.4$ [4]. The distribution for $B_s \rightarrow \mu\mu$ is normalized to unit area, whereas the distribution for $B_d \rightarrow \mu\mu$ is normalized to SM predicted value. **b** B_s mass resolution with respect to pseudorapidity of the most forward muon candidate $|\eta_f|$ [4]. The blue circles are for Run-2 (current LHC running) and red squares are for Phase-2

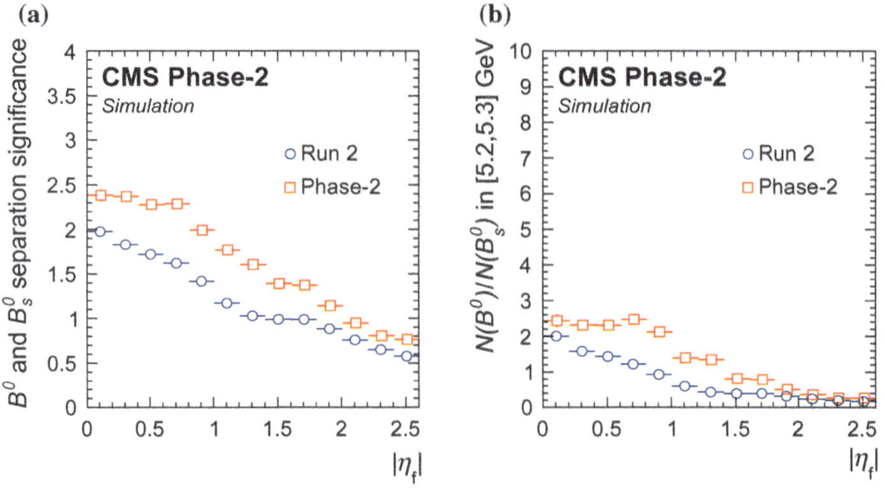

Fig. 8.3 **a** Mass peak separation between the B_s and B_d mass distribution with respect to pseudorapidity of the most forward muon candidate $|\eta_f|$ [4]. **b** Leakage of the B_s events in the B_d invariant mass region with respect to $|\eta_f|$ [4]. The blue circles represent Run-2 (current LHC running) where the red squares represent same for Phase-2

At the end of Phase-2 running, we expect to observe $B_d \rightarrow \mu\mu$ events with 5.6–8.0σ confidence level and measure the branching fractions of $B_d \rightarrow \mu\mu$ and $B_s \rightarrow \mu\mu$ and their ratios more precisely [10].

8.4 $B_s \rightarrow \phi\phi \rightarrow$4 kaons

$B_s \rightarrow \phi\phi \rightarrow$4 kaons is an FCNC process, which is forbidden at tree level in the SM and proceeds through $b \rightarrow s\bar{s}s$ penguin loop. This process could provide a new insight to the CP violating phase in the B_s system and may receive contributions from heavy particles that are beyond direct reach of the LHC.

This is also a rare process with a fully hadronic final state. The world average of its branching fraction B($B_s \rightarrow \phi\phi$) is $\approx (1.84 \pm 0.18) \times 10^{-5}$ [11]. Triggering on such low-p_T, fully hadronic final states is impossible at CMS without tracking at L1. The final state kaons have very low p_T, close to the threshold of the L1 tracking. So, it is usually difficult to reconstruct all the tracks corresponding to the four final-state kaons, and this is the main source of inefficiency in this analysis.

This analysis tries to estimate CMS sensitivity about triggering the $B_s \rightarrow \phi\phi \rightarrow$4 kaons events with L1 tracking. First, ϕ candidates are reconstructed from two oppositely charged tracks coming from the same primary vertex, with the kaon mass assigned to each track. Then the B_s candidates are reconstructed from pairs of ϕ candidates in an event coming from the same primary vertex. Events having at least one B_s candidate, compatible with the nominal B_s mass [11], are selected.

To study signal efficiency, $B_s \rightarrow \phi\phi \rightarrow$4 kaons events were generated and simulated through the Phase-2 CMS detector with an average pileup of 140 or 200. An offline study is performed to check whether there is any further loss of efficiency at that level. To study the event rate at L1 trigger, minimum bias events with <PU> = 70, 140 and 200 are used.

Three different selection baselines are studied to optimize the signal efficiency and event rate. For the extreme pileup scenario, <PU> = 200, a signal efficiency of 30% can be achieved at a rate ~15 kHz at L1. Figure 8.4a shows the invariant mass distribution of B_s candidates while Fig. 8.4b shows the rate and efficiencies for three pileup scenarios.

8.5 Conclusion

With a total $3ab^{-1}$ data of Phase-2, CMS expects significant improvement in the mass resolution for $B_{d/s} \rightarrow \mu\mu$ which in turn would yield a better separation of $B_s \rightarrow \mu\mu$ and $B_d \rightarrow \mu\mu$ mass peaks. We expect to observe $B_d \rightarrow \mu\mu$ events with 5σ and make more precise measurement of B($B_{d/s} \rightarrow \mu\mu$) and their ratio. Measurement in the very forward region will be challenging due to high leakage between B_s and B_d events.

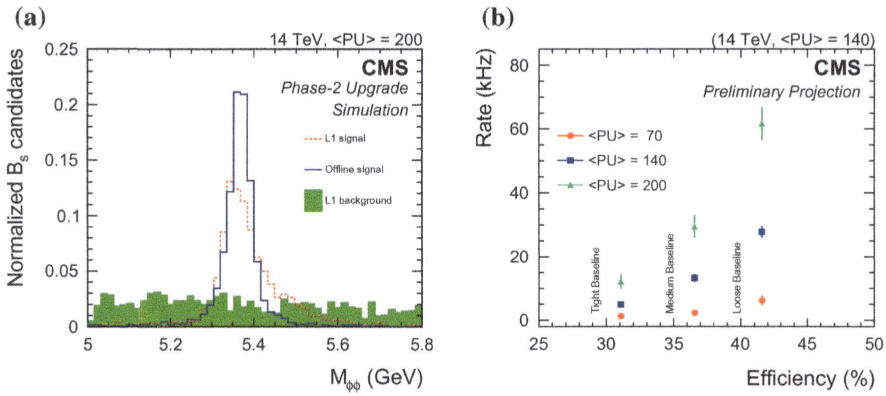

Fig. 8.4 **a** Invariant mass of ϕ pair candidates [4]. The dashed red line represents L1 signal, whereas the blue solid line denotes the offline signal. The green filled area represents L1 background. All the plots are normalized to unit area. **b** Rate versus efficiency for different selection baselines and different pileup scenarios [12]. Uncertainties are statistical only

Study of the $B_s \to \phi\phi \to$4 kaons process with L1 tracking showcases the power of L1 tracking in CMS. The signal is distinctly visible even in the extreme pileup scenario, i.e. , <PU> = 200. Further work continues to optimize the signal event rate and efficiency.

References

1. G. Apollinari et al., High-Luminosity Large Hadron Collider (HL-LHC): Preliminary Design Report. CERN-2015-005. https://doi.org/10.5170/CERN-2015-005
2. CMS Collaboration, The CMS experiment at the CERN LHC. JINST **3**, S08004 (2008). https://doi.org/10.1088/1748-0221/3/08/s08004
3. CMS Collaboration, Technical Proposal for the Phase-II Upgrade of the CMS Detector. CERN-LHCC-2015-010, LHCC-P-008, CMS-TDR-15-02 (2015)
4. CMS Collaboration, The Phase-2 Upgrade of the CMS Tracker. CERN-LHCC-2017-009, CMS-TDR-014 (2017)
5. C. Bobeth et al., $B_{s,d} \to l^+l^-$ in the standard model with reduced theoritical uncertainty. Phys. Rev. Lett. **112**, 101801 (2014). https://doi.org/10.1103/PhysRevLett.112.101801. arxiv:1311.0903
6. S.R. Choudhury, A.S. Cornell, N. Gaur, G.C. Joshi, Signatures of new physics in dileptonic B decays. Int. J. Mod. Phys. A **21**, 2617 (2006). https://doi.org/10.1142/S0217751X06029491. arXiv:hep-ph/0504193
7. J.K. Parry, Lepton flavor violating Higgs boson decays, $\tau \to \mu\gamma$ and $B_s^0 \to \mu^+\mu^-$ in the constrained MSSM+NR with large tanβ. Nucl. Phys. B **760**, 38 (2007). https://doi.org/10.1016/j.nuclphysb.2006.10.011. arXiv:hep-ph/0510305
8. J.R. Ellis, J.S. Lee, A. Pilaftsis, B-meson observables in the maximally CP-violating MSSM with minimal flavour violation. Phys. Rev. D **76**, 115011 (2007). https://doi.org/10.1103/PhysRevD.76.115011. arXiv:0708.2079

9. J.R. Ellis, K.A. Olive, Y. Santoso, V.C. Spanos, On $B_s^0 \to \mu^+\mu^-$ and cold dark matter scattering in the MSSM with non-universal Higgs masses. JHEP **05**, 063 (2006). https://doi.org/10.1088/1126-6708/2006/05/063. arXiv:hep-ph/06031306
10. CMS Collaboration, B Physics Analyses for the Phase-II Upgrade Technical Proposal. CMS-PAS-FTR-14-015. http://cds.cern.ch/record/2036007?ln=en
11. M. Tanabashi et al. (Particle Data Group), The review of particle physics. Phys. Rev. D **98**, 030001 (2018)
12. CMS Collaboration, ECFA 2016: prospects for selected standard model measurements with the CMS experiment at the high-luminosity LHC. CMS-PAS-FTR-16-006. http://cds.cern.ch/record/2262606?ln=en

Chapter 9
Model Independent New Physics Analysis in $\Lambda_b \to \Lambda(\to N\pi)\ell^+\ell^-$ Decay

Diganta Das

Abstract We study the rare four body $\Lambda_b \to \Lambda(\to N\pi)\ell^+\ell^-$ decay in beyond the Standard Model. We take a model-independent approach and consider a set of dimension-six four-fermion operators which includes, in addition to the Standard Model operators, new (axial-)vector operators, and (pseudo-)scalar operators. Working in transversity basis we present a full angular distribution of $\Lambda_b \to \Lambda(\to N\pi)\ell^+\ell^-$ decay where the leptons are massive and the Λ_b is unpolarized.

9.1 Introduction

The $b \to s\ell^+\ell^-$ flavor changing neutral current transitions are known for their sensitivity to physics in and beyond the Standard Model (SM), also known as new physics (NP). In the past decades, $b \to s\ell^+\ell^-$ mediated $B \to K^*\ell^+\ell^-$ decay have been the topic of intense theoretical and experimental investigations as its full angular distribution gives access to a multitude of related observables that can uniquely test the SM [1]. Recently, the LHCb collaboration has presented measurements of a set of $\Lambda_b \to \Lambda\mu^+\mu^-$ observables [2, 3]. This is the first semi-leptonic baryonic decay studied at the LHCb whose branching ratio was first measured by the CDF [4]. The full angular distribution of $\Lambda_b \to \Lambda(\to N\pi)\ell^+\ell^-$ also give access to a plethora of observables [5] making it at par with the $B \to K^*\ell^+\ell^-$ decay. If the polarization of Λ_b is also considered, the number of observables become more than three fold compared to the unpolarized case [6]. Therefore, $\Lambda_b \to \Lambda(\to N\pi)\ell^+\ell^-$ decay can be a unique test ground for physics in and beyond the SM.

Talk presented at the FPCP 2018, Hyderabad.

Diganta Das—Speaker.

D. Das (✉)
Department of Physics and Astrophysics, University of Delhi, Delhi 110007, India
e-mail: diganta99@gmail.com

A. Giri and R. Mohanta (eds.), *16th Conference on Flavor Physics and CP Violation*, Springer Proceedings in Physics 234,
https://doi.org/10.1007/978-3-030-29622-3_9

In this paper we present a model independent NP analysis of $\Lambda_b \to \Lambda(\to N\pi)\ell^+\ell^-$ decay where we keep the lepton masses finite but consider the Λ_b to be unpolarized. We extend the SM operator basis by including new (axial-)vector (VA) and (pseudo-)scalar (SP) operators. Tensor operators has been neglected for simplicity as they give rise to a large number of terms in the angular coefficients due to the presence of lepton masses. We work in transversity basis and present a full angular distribution of the $\Lambda_b \to \Lambda(\to N\pi)\ell^+\ell^-$ decay. From the angular distribution a number of observables are studied in the presence of the NP couplings. For our numerical analysis, $\Lambda_b \to \Lambda$ form factors at low dilepton invariant mass squared, or large recoil, are taken from the calculations in the light cone sum-rules (LCSR) [7]. At large dilepton invariant mass squared, or low recoil, the form factors are taken from lattice QCD calculations [8].

The paper is organized as follows. The effective Hamiltonian is described in Sect. 9.2 and the decay amplitudes are derived in Sect. 9.3. In Sect. 9.4 we describe the angular distribution followed by numerical analysis in Sect. 9.5. We conclude in Sect. 9.6.

9.2 Effective Hamiltonian

In the SM the $b \to s\ell^+\ell^-$ transition is described in terms of hadronic operators $\mathcal{O}_{1-6,8}$, radiative operator $\mathcal{O}_7$, and semi-leptonic operators $\mathbb{O}_{9,10}$. We extend the SM operator basis by including new VA and SP operators so that the effective Lagrangian reads [9]

$$\mathcal{H}^{\text{eff}} = -\frac{4G_F}{\sqrt{2}} V_{tb}V_{ts}^* \frac{\alpha_e}{4\pi}\Big(\sum_i \mathcal{C}_i\mathcal{O}_i + \sum_j \mathcal{C}'_j\mathcal{O}'_j\Big), \tag{9.1}$$

where $i = 7, 9, 10, V, A, S, P$ and $j = V, A, S, P$. The operators $\mathcal{O}^{(\prime)}$ read

$$\begin{aligned}
&O_7 = \frac{m_b}{e}\big[\bar{s}\sigma^{\mu\nu}P_R b\big]F_{\mu\nu}, \quad \mathcal{O}_9 = \big[\bar{s}\gamma^\mu P_L b\big]\big[\ell\gamma_\mu\ell\big], \quad \mathcal{O}_{10} = \big[\bar{s}\gamma^\mu P_L b\big]\big[\ell\gamma_\mu\gamma_5\ell\big],\\
&\mathcal{O}_V^{(\prime)} = \big[\bar{s}\gamma^\mu P_{L(R)} b\big]\big[\ell\gamma_\mu\ell\big], \quad \mathcal{O}_A^{(\prime)} = \big[\bar{s}\gamma^\mu P_{L(R)} b\big]\big[\ell\gamma_\mu\gamma_5\ell\big],\\
&\mathcal{O}_S^{(\prime)} = \big[\bar{s}P_R(L) b\big]\big[\ell\ell\big], \quad \mathcal{O}_P^{(\prime)} = \big[\bar{s}P_R(L) b\big]\big[\ell\gamma_5\ell\big].
\end{aligned} \tag{9.2}$$

The details of the SM Wilson coefficients can be found in [9]. In the SM $\mathcal{C}^{(\prime)}_{V,A}$, $\mathcal{C}^{(\prime)}_{S,P}$ vanish. In this Hamiltonian we assume only the factorizable corrections to the hadronic operators $\mathcal{O}_{1-6,8}$. These corrections can be absorbed in to the Wilson coefficients $\mathcal{C}^{\text{eff}}_{7,9}$ corresponding to the operators $\mathcal{O}_{7,9}$. The no-factorizable corrections which are ignored for simplicity are expected to play a significant role at large recoil [10, 11]. In (9.2) G_F is the Fermi constant, $\alpha_e = e^2/4\pi$ is the fine structure constant, $P_{L,R} = (1 \mp \gamma)/2$ are the chiral projectors, $V_{tb}V_{ts}^*$ are the CKM elements, and we have defined $\sigma_{\mu\nu} = i[\gamma_\mu, \gamma_\nu]/2$. The b-quark mass appearing in $\mathcal{O}_7$ is the running mass in the modified minimal subtraction ($\overline{\text{MS}}$) scheme.

9.3 Decay Amplitudes

The $\Lambda_b \to \Lambda(\to N\pi)\ell^+\ell^-$ decay proceeds in two steps,

$$\Lambda_b(p, s_p) \to \Lambda(k, s_k)\ell^+(q_1)\ell^-(q_2), \quad \Lambda(k, s_k) \to N(k_1, s_N)\pi(k_2). \tag{9.3}$$

Here p, k, k_1, k_2, q_1 and q_2 are the momenta of Λ_b, Λ, N, π and the positively and negatively charged leptons in their respective rest frame. The projections of the baryonic spins on to the z-axis in their respective rest frames are denoted by $s_{p,k,N}$. Momentum conservation implies $k = k_1 + k_2$ and the four momentum of the dilepton pair is defined as $q^\mu = q_1^\mu + q_2^\mu$.

The four-body decay can be completely described in terms of four kinematic variables, dilepton invariant mass squared q^2 and three angles θ_ℓ, θ_Λ and ϕ. The angle θ_ℓ is defined as the angle made by ℓ^- with the $+z$-direction in the dilepton rest frame. The angle θ_Λ is made by N with respect to the $+z$-direction in the $N\pi$ rest frame, and the angle ϕ is defined as the angle between the $\ell^+\ell^-$ and $N\pi$ decay planes.

The $\Lambda_b \to \Lambda$ hadronic form factors corresponding to different operators are conveniently defined in terms of ten q^2 dependent helicity form factors $f^V_{t,0,\perp}$, $f^A_{t,0,\perp}$, $f^T_{0,\perp}$, $f^{T5}_{0,\perp}$ [12]. Following the parametrization of the $\Lambda_b \to \Lambda$ hadronic matrix elements, the transversity amplitudes corresponding to the operator basis (9.2) were derived in [9]. For VA type operators there are ten amplitudes which read

$$A^{L,(R)}_{\perp_1} = -\sqrt{2}N\left(f^V_\perp\sqrt{2s_-}\mathcal{C}^{L,(R)}_{\text{VA}+} + \frac{2m_b}{q^2}f^T_\perp(m_{\Lambda_b} + m_\Lambda)\sqrt{2s_-}\mathcal{C}^{\text{eff}}_7\right), \tag{9.4}$$

$$A^{L,(R)}_{\parallel_1} = \sqrt{2}N\left(f^A_\perp\sqrt{2s_+}\mathcal{C}^{L,(R)}_{\text{VA}-} + \frac{2m_b}{q^2}f^{T5}_\perp(m_{\Lambda_b} - m_\Lambda)\sqrt{2s_+}\mathcal{C}^{\text{eff}}_7\right), \tag{9.5}$$

$$A^{L,(R)}_{\perp_0} = \sqrt{2}N\left(f^V_0(m_{\Lambda_b} + m_\Lambda)\sqrt{\frac{s_-}{q^2}}\mathcal{C}^{L,(R)}_{\text{VA}+} + \frac{2m_b}{q^2}f^T_0\sqrt{q^2 s_-}\mathcal{C}^{\text{eff}}_7\right), \tag{9.6}$$

$$A^{L,(R)}_{\parallel_0} = -\sqrt{2}N\left(f^A_0(m_{\Lambda_b} - m_\Lambda)\sqrt{\frac{s_+}{q^2}}\mathcal{C}^{L,(R)}_{\text{VA}-} + \frac{2m_b}{q^2}f^{T5}_0\sqrt{q^2 s_+}\mathcal{C}^{\text{eff}}_7\right), \tag{9.7}$$

$$A_{\perp t} = -2\sqrt{2}Nf^V_t(m_{\Lambda_b} - m_\Lambda)\sqrt{\frac{s_+}{q^2}}(\mathcal{C}_{10} + \mathcal{C}_A + \mathcal{C}'_A), \tag{9.8}$$

$$A_{\parallel t} = 2\sqrt{2}Nf^A_t(m_{\Lambda_b} + m_\Lambda)\sqrt{\frac{s_-}{q^2}}(\mathcal{C}_{10} + \mathcal{C}_A - \mathcal{C}'_A). \tag{9.9}$$

Here

$$\mathcal{C}^{L(R)}_{\text{VA},+} = (\mathcal{C}^{\text{eff}}_9 \mp \mathcal{C}_{10}) + (\mathcal{C}_V \mp \mathcal{C}_A) + (\mathcal{C}'_V \mp \mathcal{C}'_A), \tag{9.10}$$

$$\mathcal{C}^{L(R)}_{\text{VA},-} = (\mathcal{C}^{\text{eff}}_9 \mp \mathcal{C}_{10}) + (\mathcal{C}_V \mp \mathcal{C}_A) - (\mathcal{C}'_V \mp \mathcal{C}'_A), \tag{9.11}$$

and

$$s_{\pm} = (m_{\Lambda_b} \pm m_{\Lambda})^2 - q^2. \tag{9.12}$$

The normalization constant is given as

$$N(q^2) = G_F V_{tb} V_{ts}^* \alpha_e \sqrt{\tau_{\Lambda_b} \frac{q^2 \sqrt{\lambda(m_{\Lambda_b}^2, m_{\Lambda}^2, q^2)}}{3.2^{11} m_{\Lambda_b}^3 \pi^5} \beta_\ell}, \quad \beta_\ell = \sqrt{1 - \frac{4m_\ell^2}{q^2}}. \tag{9.13}$$

Corresponding to the SP operators there are four transversity amplitudes

$$A_{S\perp} = 2\sqrt{2} N f_t^V \frac{m_{\Lambda_b} - m_{\Lambda}}{m_b} \sqrt{s_+}(C_S + C_S'), \tag{9.14}$$

$$A_{S\parallel} = -2\sqrt{2} N f_t^A \frac{m_{\Lambda_b} + m_{\Lambda}}{m_b} \sqrt{s_-}(C_S - C_S'), \tag{9.15}$$

$$A_{P\perp} = -2\sqrt{2} N f_t^V \frac{m_{\Lambda_b} - m_{\Lambda}}{m_b} \sqrt{s_+}(C_P + C_P'), \tag{9.16}$$

$$A_{P\parallel} = 2\sqrt{2} N f_t^A \frac{m_{\Lambda_b} + m_{\Lambda}}{m_b} \sqrt{s_-}(C_P - C_P'). \tag{9.17}$$

The subsequent decay of $\Lambda \to N\pi$ is governed by $\Delta S = 1$ effective Hamiltonian and the decay amplitudes can be written in terms of two form factors ξ and ω [5]. In the decay distribution, the only relevant quantity that enters is the parity violating parameter that is defined in terms ξ and ω as

$$\alpha_\Lambda = \frac{-2\mathrm{Re}(\xi\omega)}{\sqrt{\frac{r_-}{r_+}}|\xi|^2 + \sqrt{\frac{r_+}{r_-}}|\omega|^2}, \quad \text{where} \quad r_{\pm} = (m_\Lambda \pm m_N)^2 - m_\pi^2. \tag{9.18}$$

9.4 Angular Distribution

The four-fold distribution of $\Lambda_b \to \Lambda(\to N\pi)\ell^+\ell^-$ can be written as

$$\frac{d^4\mathcal{B}}{dq^2 d\cos\theta_\ell d\cos\theta_\Lambda d\phi} = \frac{3}{8\pi} K(q^2, \cos\theta_\ell, \cos\theta_\Lambda, \phi), \tag{9.19}$$

where $K(q^2, \cos\theta_\ell, \cos\theta_\Lambda)$ reads in terms of angular coefficients and trigonometric functions as

$$\begin{aligned} K(q^2, \cos\theta_\ell, \cos\theta_\Lambda) = {} & (K_{1ss}\sin^2\theta_\ell + K_{1cc}\cos^2\theta_\ell + K_{1c}\cos\theta_\ell) \\ & + (K_{2ss}\sin^2\theta_\ell + K_{2cc}\cos^2\theta_\ell + K_{2c}\cos\theta_\ell)\cos\theta_\Lambda \\ & + (K_{3sc}\sin\theta_\ell\cos\theta_\ell + K_{3s}\sin\theta_\ell)\sin\theta_\Lambda\sin\phi \\ & + (K_{4sc}\sin\theta_\ell\cos\theta_\ell + K_{3s}\sin\theta_\ell)\sin\theta_\Lambda\cos\phi. \end{aligned} \tag{9.20}$$

Each of the angular coefficients can be decomposed as [9]

$$K_{\{\cdots\}} = \mathcal{K}_{\{\cdots\}} + \frac{m_\ell}{\sqrt{q^2}}\mathcal{K}'_{\{\cdots\}} + \frac{m_\ell^2}{q^2}\mathcal{K}''_{\{\cdots\}}, \tag{9.21}$$

where $\{\cdots\}$ stands for the suffixes $1ss$, $1cc$, $1c$, $2ss$, $2cc$, $2c$, $3sc$, $3s$, $4sc$, $4s$. The detailed expressions of $\mathcal{K}$, $\mathcal{K}'$ and $\mathcal{K}''$ can be found in [9]. The angular distribution (9.19) allows us to construct a set observables through weighted integrals over the angles that can be measured in the experiments. A set of such observables were discussed in [9, 13].

9.5 Numerical Analysis

In this section we carry out the numerical analysis of few $\Lambda_b \to \Lambda\mu^+\mu^-$ and $\Lambda_b \to \Lambda(\to p\pi)\tau^+\tau^-$ observables. Detailed discussions can be found in [9] and [13]. For this analysis, the form factors at large recoil, or low q^2, is taken from calculations in light-cone sum rules [7]. At low recoil, or large q^2, the form factors are taken from lattice QCD calculations [8]. The value of the parity violating parameter for the pair $N\pi \equiv p\pi$ is given by $\alpha_\Lambda = 0.642 \pm 0.013$. The values of other inputs are collected in the appendix of [9].

We begin by showing in Fig. 9.1 the SM predictions of differential branching ratio of $\Lambda_b \to \Lambda\mu^+\mu^-$. Here the bands correspond to the uncertainties coming from form factors and other inputs. The crosses indicate data from LHCb [2]. In Fig. 9.1 we also show our predictions for two benchmark NP scenarios with only VA coupling and with only SP couplings. For the VA coupling we chose values that are consistent with the current global fits to $b \to s\mu^+\mu^-$ data [14]. The benchmark values of the SP couplings are consistent with the exclusive $B_s \to \mu^+\mu^-$ data. Detailed discussions for other $\Lambda_b \to \Lambda\mu^+\mu^-$ observables can be found in [13]. In Fig. 9.2 we show the SM predictions for $\Lambda_b \to \Lambda(\to p\pi)\tau^+\tau^-$ differential branching ratio and leptonic forward-backward asymmetry $A^\tau_{\rm FB}$. Here also, the bands correspond to the uncertainties coming from the form factors and different inputs. Detailed discussions of $\Lambda_b \to \Lambda(\to p\pi)\tau^+\tau^-$ observables in the SM and NP can be found in [9].

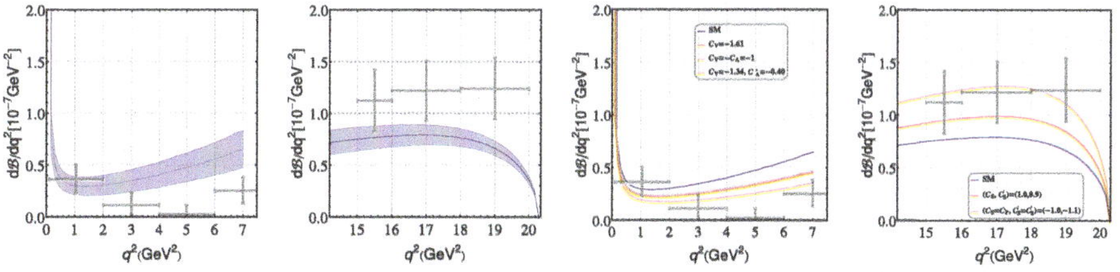

Fig. 9.1 The differential branching ratio of $\Lambda_b \to \Lambda\mu^+\mu^-$ in the SM and NP

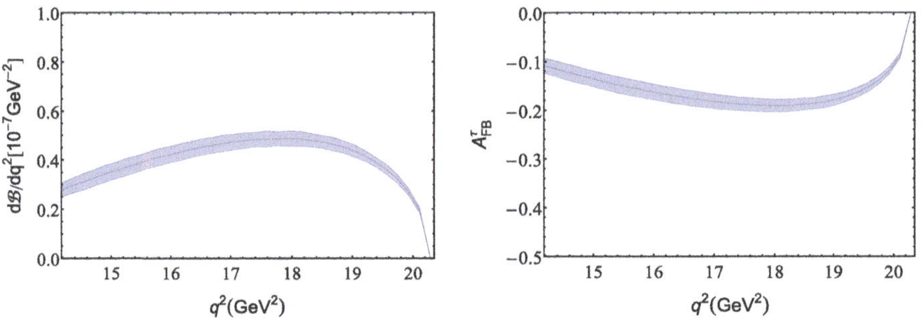

Fig. 9.2 The SM predictions of $\Lambda_b \to \Lambda(\to p\pi)\tau^+\tau^-$ observables

9.6 Summary and Conclusion

In this paper we have presented a full angular distribution of four-body rare decay $\Lambda_b \to \Lambda(\to N\pi)\ell^+\ell^-$. The operator basis include, in addition to the SM operator basis, new (axial-)vector and (pseudo-)scalar operators. In this analysis, the masses of the leptons are taken to be finite and the Λ_b is taken to be unpolarized. Working in transversity basis we have expressed the angular coefficients in terms of transversity amplitudes. In the numerical section we have studied the $\Lambda_b \to \Lambda(\to N\pi)\ell^+\ell^-$ decay for $\ell = \mu$ and $\ell = \tau$.

Acknowledgements I would like to thank the organizers of FPCP 2018 at Hyderabad for a wonderful conference. The work is supported by the DST, Govt. of India under INSPIRE Faculty Fellowship.

References

1. F. Kruger, L.M. Sehgal, N. Sinha, R. Sinha, Phys. Rev. D **61**, 114028 (2000), Erratum: [Phys. Rev. D **63**, 019901 (2001)] [hep-ph/9907386]
2. R. Aaij et al. (LHCb Collaboration), JHEP **1506**, 115 (2015). arXiv:1503.07138 [hep-ex]
3. R. Aaij et al. (LHCb Collaboration). arXiv:1808.00264 [hep-ex]
4. T. Aaltonen et al., CDF Collaboration. Phys. Rev. Lett. **107**, 201802 (2011). arXiv:1107.3753 [hep-ex]
5. P. Böer, T. Feldmann, D. van Dyk, JHEP **1501**, 155 (2015). arXiv:1410.2115 [hep-ph]
6. T. Blake, M. Kreps, JHEP **1711**, 138 (2017). arXiv:1710.00746 [hep-ph]
7. Y.M. Wang, Y.L. Shen, JHEP **1602**, 179 (2016). arXiv:1511.09036 [hep-ph]
8. W. Detmold, S. Meinel, Phys. Rev. D **93**(7), 074501 (2016). arXiv:1602.01399 [hep-lat]
9. D. Das, JHEP **1807**, 063 (2018). arXiv:1804.08527 [hep-ph]
10. M. Beneke, T. Feldmann, D. Seidel, Nucl. Phys. B **612**, 25 (2001) [hep-ph/0106067]
11. M. Beneke, T. Feldmann, D. Seidel, Eur. Phys. J. C **41**, 173 (2005) [hep-ph/0412400]
12. T. Feldmann, M.W.Y. Yip, Phys. Rev. D **85**, 014035 (2012). Erratum: Phys. Rev. D **86**, 079901 (2012). arXiv:1111.1844 [hep-ph]
13. D. Das, Eur. Phys. J. C **78**(3), 230 (2018). arXiv:1802.09404 [hep-ph]
14. W. Altmannshofer, C. Niehoff, P. Stangl, D.M. Straub, Eur. Phys. J. C **77**(6), 377 (2017). arXiv:1703.09189 [hep-ph]

Chapter 10
Open Heavy-Flavour Measurements in Small Systems with ALICE at the LHC

P. Dhankher

Abstract Heavy quarks (charm and beauty) are produced at the initial stages of hadronic collisions in hard scattering processes. In pp collisions, their production provides a test bed for perturbative QCD calculations and a crucial baseline for p–Pb and Pb–Pb collisions. Furthermore, differential measurements, for example their production as a function of charged-particle multiplicity, may provide insight into the role of Multiple Parton Interactions (MPI) and the interplay between hard and soft mechanism for particle production. In the p–Pb collision system, the formation and kinematic properties of heavy-flavour hadrons can be influenced at all stages by Cold Nuclear Matter effects and concurrent MPI. In addition, collective effects may be present. In this contribution, recent open heavy-flavour results from pp collisions at $\sqrt{s_{\rm NN}} = 5.02$, 7, 8 and 13 TeV and p–Pb collisions at $\sqrt{s_{\rm NN}} = 5.02$ and 8.16 TeV, with ALICE at the LHC are presented. The results include the production cross-section, nuclear modification factor and multiplicity dependence studies of the production of D mesons, charmed baryons and heavy-flavour hadron decay's electrons at mid-rapidity and muons at forward rapidity. The results will also be compared with theoretical model predictions.

10.1 Introduction

A large ion collider experiment (ALICE) at the Large Hadron Collider (LHC) is a dedicated heavy-ion experiment designed to study physics of strongly interacting matter at high temperature and extreme densities, where quarks and gluons are deconfined to a new state of matter known as Quark Gluon Plasma (QGP). Heavy quarks (charm and beauty), due to their large masses ($m_{\rm q} \gg \lambda_{\rm QCD}$), are produced at the initial stages of hadronic collisions in hard scattering processes. Therefore, they

P. Dhankher—For the ALICE collaboration.

P. Dhankher (✉)
Indian Institute of Technology Bombay, Mumbai 400076, Maharashtra, India
e-mail: preeti.dhankher@cern.ch

A. Giri and R. Mohanta (eds.), *16th Conference on Flavor Physics and CP Violation*, Springer Proceedings in Physics 234,
https://doi.org/10.1007/978-3-030-29622-3_10

are an effective tool to probe the hot dense QCD matter (QGP). Their production as a function of multiplicity reveals the importance of MPI, interplay between the hard and soft mechanism for particle production, as well as the connection between open and hidden production of heavy-flavour [1].

10.2 Heavy-Flavour Reconstruction with ALICE

With the excellent particle identification capabilities of the ALICE apparatus, the presence of heavy-flavour hadrons is either detected directly via the reconstruction of hadronic decays of D-mesons (D^0, D^+, D^{*+} and D^+) and Λ_c^+ baryon at mid-rapidity, or indirectly by finding a single electron at mid-rapidity, or muon produced at forward and backward rapidity via a semi-leptonic decay channel.

The charged tracks are reconstructed using the Time Projection chamber (TPC) following the Inner Tracking System (ITS) in radial direction. The particle identification is done using the ionisation energy loss dE/dx in the TPC, aided by either the Time-of-Flight detector (TOF) at low p_T region or by the Electromagnetic Calorimeter (EMCal) at high p_T region. The charged-particle multiplicities in ALICE in the forward and backward rapidity is determined using V0C and V0A detectors which are arrays of 32 scintillating counters, installed on either side of the ALICE interaction point. At mid-rapidity, SPD tracklet is used which is reconstructed by connecting the hits in both Silicon Pixel Detector (SPD) layers with origin at the vertex. A more detailed description of the ALICE detector can be found in [2].

10.3 Results

In this section, the recent results of heavy-flavour production cross-section, nuclear modification factor RpPb and their multiplicity dependence studies in pp and p–Pb collisions at various center-of-mass energies are shown.

10.3.1 Heavy-Flavour Production in pp and p–Pb Collisions

The differential cross sections of prompt D-mesons (D^0, D^+, D^{*+} and D_s^+), heavy-flavour decay electrons (HFE), electrons from beauty-hadron decays and heavy-flavour decay muons (HFM) measured in pp collisions at $\sqrt{s_{NN}}$ = 2.76 TeV [3], 7 TeV [4, 5], 8 TeV and 13 TeV. The p_T differential cross section of D^0-meson measured upto $p_T = 35$ GeV$/c$ at mid-rapidity is consistent with the FONLL pQCD calculations [6] within uncertainties as shown in Fig. 10.1. The p_T differential cross section of electrons from heavy-flavour decays measured by ALICE and ATLAS [7] in pp collisions at $\sqrt{s_{NN}} = 7$ TeV in different rapidity intervals is shown in Fig. 10.2.

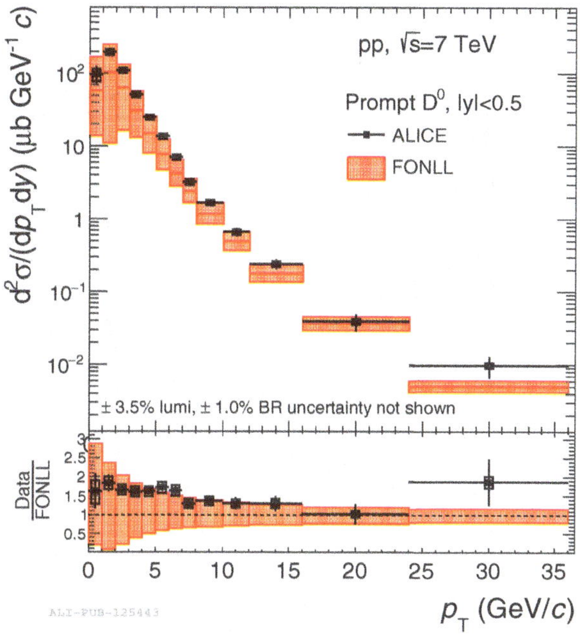

Fig. 10.1 D^0 differential cross section in pp collisions at $\sqrt{s_{NN}} = 7$ TeV compared with FONLL pQCD calculations

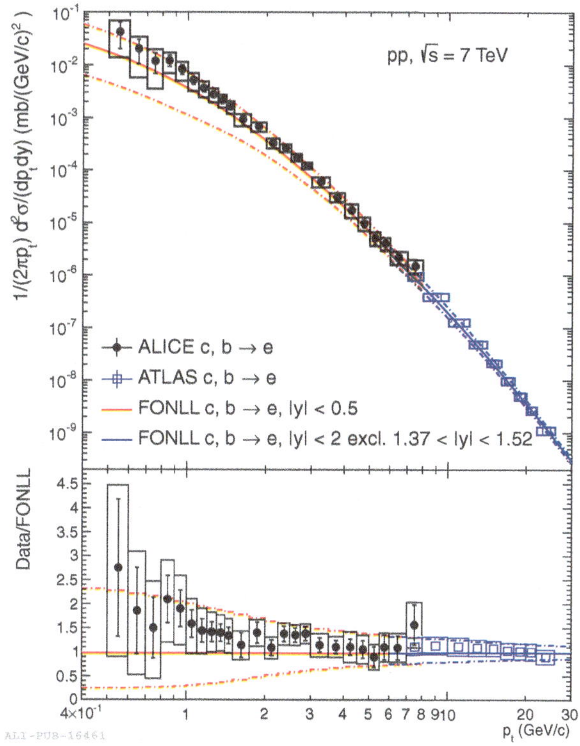

Fig. 10.2 Heavy-flavour decay electron p_T differential cross section measured by ALICE and ATLAS [7] in pp collisions at $\sqrt{s_{NN}} = 7$ TeV in different rapidity intervals compared with FONLL pQCD calculations

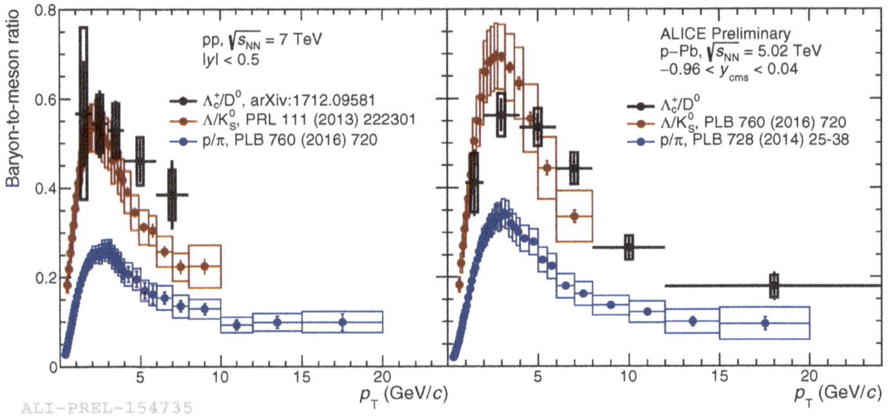

Fig. 10.3 Λ_c^+/D^0 production ratio in p–Pb collisions at $\sqrt{s_{NN}}$ = 5.02 TeV and pp collisions at $\sqrt{s_{NN}}$ = 7 TeV, compared with Λ_c/K^0_s and p/π ratios in the same systems

To investigate charm baryon hadronisation mechanisms, the measurement of the charmed baryon-to-meson ratio (Λ_c^+/D^0) is carried out by the ALICE [8]. Figure 10.3 shows the Λ_c^+/D^0 ratio in pp collisions at $\sqrt{s_{NN}}$ = 7 TeV and in p–Pb collision at $\sqrt{s_{NN}}$ = 5.02 TeV and the comparison with light-flavour (Λ_c/K^0_s and p/π) in the same systems. The trend observed is similar to the baryon-to-meson ratio in light-flavours.

10.3.2 Nuclear Matter Effects and v_2 in p–Pb Collisions

The measurement of heavy-flavour production in p–Pb collisions provides access to the cold nuclear matter (CNM) effects. Figure 10.4 shows the nuclear modification factor of prompt D-meson compared with models while in Fig. 10.5, R_{pPb} of strange D-meson compared with non-strange D-meson in p–Pb collisions at $\sqrt{s_{NN}}$ = 5.02 TeV. The measurement of R_{pPb} of prompt D-meson are consistent with no nuclear modification in the whole momentum range. The R_{pPb} of strange is similar to that of non-strange D-mesons.

Figure 10.6 shows the ratio of most central collisions to the most peripheral collisions known as Q_{cp} measurement for average D-meson compared with charged particles in p–Pb collisions at $\sqrt{s_{NN}}$ = 5.02 TeV. The two measurement are in agreement with each other. As shown in Fig. 10.7, a positive elliptic flow coefficient v_2, obtained from the decomposition of correlation distribution of HFE, is observed in $1.5 < p_T < 4$ GeV/c with a significance of $\sim 5\sigma$ and comparable with charged particles in mid-rapidity for p–Pb collisions at $\sqrt{s_{NN}}$ = 5.02 TeV.

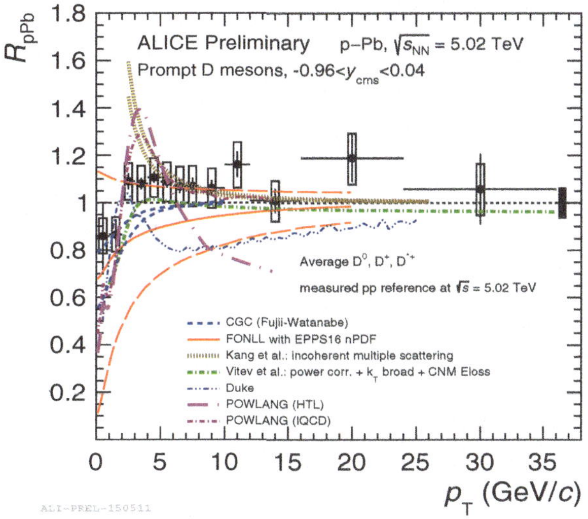

Fig. 10.4 Nuclear modification factor R_{pPb} of prompt D mesons in p–Pb collisions at $\sqrt{s_{NN}}$ = 5.02 TeV at central rapidity

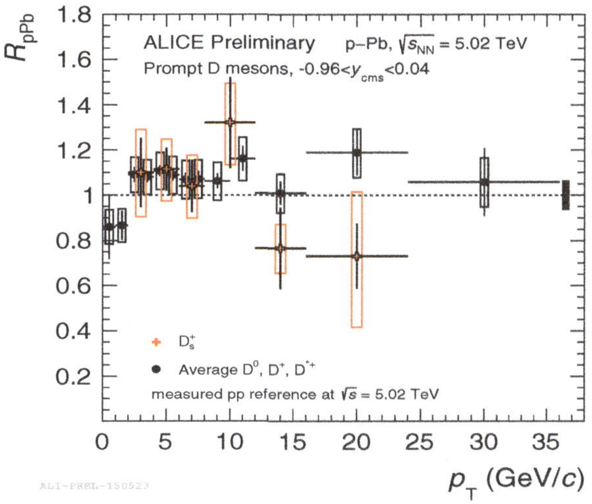

Fig. 10.5 Nuclear modification factor R_{pPb} of strange D mesons in p–Pb collisions at $\sqrt{s_{NN}}$ = 5.02 TeV at central rapidity compared with non-strange D meson

Figure 10.8 shows the nuclear modification factor of muons in the forward (2.03 $< y <$ 3.53) as well as in backward ($-4.46 < y <$ 2.96) rapidity ranges [9]. The modification in the shadowing region(forward) is consistent with unity, while some enhancement is indicated at lower p_T values in the anti-shadowing region (backward).

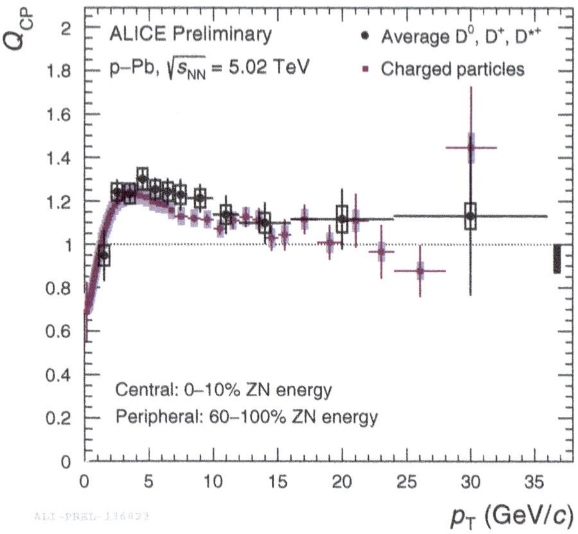

Fig. 10.6 Q_{cp} of average D-meson compared with charged particles in p–Pb collisions at $\sqrt{s_{NN}}$ = 5.02 TeV

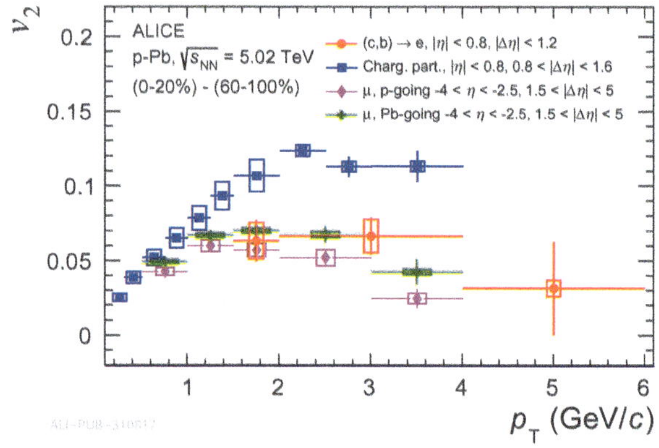

Fig. 10.7 HFE v_2 as a function of p_T compared to the v_2 of unidentified charged particles and inclusive muons

10.3.3 Multiplicity Dependence in Heavy-Flavour Production

The self-normalised yield (ratio of the yield in a particular multiplicity interval normalised to the multiplicity integrated yield) of average D-meson in pp collisions at $\sqrt{s_{NN}}$ = 7 TeV shows stronger than linear dependence on self-normalised charged-particle multiplicity [10]. Figure 10.9 shows the comparison of D-meson and J/ψ measurement, which are consistent with each other. This suggests that the observed trend is dependent on the hard QCD process, not on hadronization mechanism. A similar trend is seen for HFM and HFE. Figure 10.10 shows the comparison of HFM measurements with the D-meson in pp collisions.

The left panel in Fig. 10.11 shows the dependence of the relative yield of HFE on relative charged-particle multiplicity in pp collision at $\sqrt{s_{NN}}$ = 13 TeV in p_T

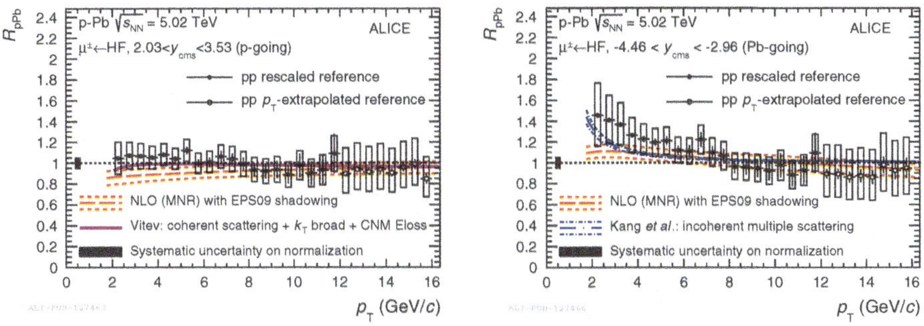

Fig. 10.8 Nuclear modification factor $R_{\rm pPb}$ of HFM in the forward $(2.03 < y < 3.53)$ and backward $(-4.46 < y < 2.96)$ rapidity ranges in p–Pb collisions at $\sqrt{s_{\rm NN}}$ = 5.02 TeV

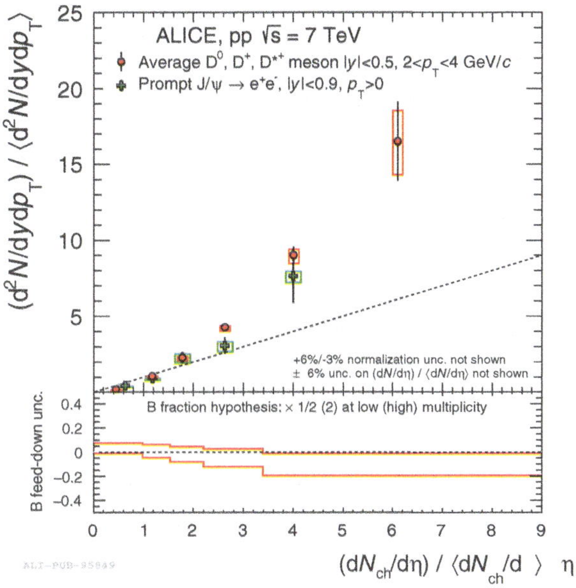

Fig. 10.9 Self-normalised yield of average D-meson compared with J/ψ as a function of multiplicity in pp collisions at $\sqrt{s_{\rm NN}}$ = 7 TeV at mid-rapidity

intervals. The high $p_{\rm T}$ measurement shows a steeper increase. The results are in good agreement with the PYTHIA8.2 predictions which incorporates MPI. The right panel in Fig. 10.11 shows the comparison of HFE measurement in p–Pb collisions at $\sqrt{s_{\rm NN}}$ = 8.16 TeV with the D-meson in p–Pb collisions at $\sqrt{s_{\rm NN}}$ = 5.02 TeV. The data points for the D-meson measurement are shifted to have a better visibility. The results show good agreement with each other.

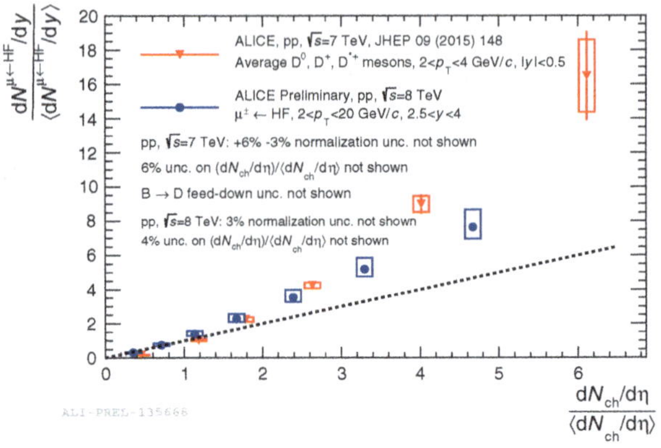

Fig. 10.10 Self-normalised yield of HFM as a function of multiplicity in the forward rapidity region in pp collisions at $\sqrt{s_{\mathrm{NN}}} = 8$ TeV compared with self-normalised yield of average D-meson as a function of multiplicity at mid-rapidity in pp collisions at $\sqrt{s_{\mathrm{NN}}} = 7$ TeV

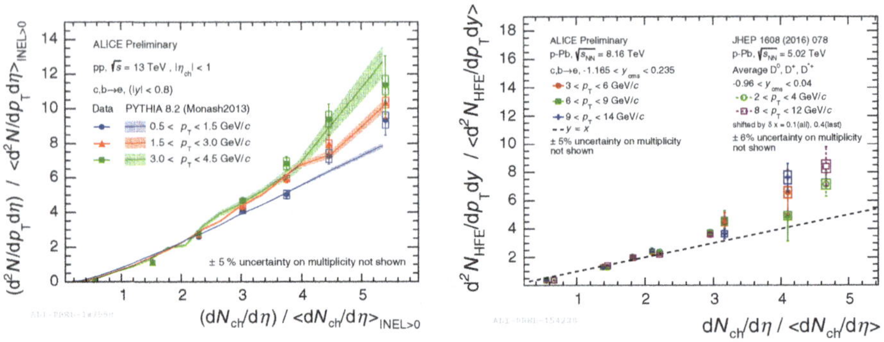

Fig. 10.11 Self-normalised yield of HFE as a function of multiplicity for different p_{T} ranges in pp collisions at $\sqrt{s_{\mathrm{NN}}} = 13$ TeV compared with PYTHIA8.2 predictions on left, in p–Pb collisions at $\sqrt{s_{\mathrm{NN}}} = 8.16$ TeV compared with D-meson in p–Pb collisions at $\sqrt{s_{\mathrm{NN}}} = 5.02$ TeV on right

10.4 Summary and Outlook

Overview of selected heavy-flavour results from pp collisions at $\sqrt{s_{\mathrm{NN}}} = 5.02$, 7, 8 and 13 TeV and p–Pb collisions at $\sqrt{s_{\mathrm{NN}}} = 5.02$ and 8.16 TeV with ALICE at the LHC has been presented. FONLL pQCD calculations, describe the pp data within uncertainties. The nuclear modification measurements are consistent with no modification in the whole momentum range. The effect of anti-shadowing can be seen from the production of heavy-flavour decay muons in the backward direction at low p_{T}. The self-normalised yield of open, hidden charm [10], HFM and HFE show stronger than linear dependence on self-normalised charged-particle multiplicity and this trend can be qualitatively explained by models with MPI.

The detector upgrades [11] during the long shutdown 2 (including the ITS, TPC and muon forward tracker) will considerably increase the precision of measurements. The improved impact parameter resolution together with the increased luminosity of the LHC will improve the significance of upcoming measurements by a factor of about ten.

References

1. L. Frankfurt, M. Strikman, C. Weiss, Phys. Rev. D **83**, 054012 (2011)
2. K. Aamodt et al. (ALICE Collaboration), J. Instrum. **3**, S08002 (2008)
3. B. Abelev et al. (ALICE Collaboration), J. High Energy. Phys. **1207**, 191 (2012); Phys. Rev. D **91**, 012001 (2015); Phys. Rev. Lett. **109**, 112301 (2012); Phys. Lett. B **738**, 97 (2014)
4. J. Adam et al. (ALICE Collaboration), Phys. Rev. C **94**, 054908 (2016)
5. B. Abelev et al. (ALICE Collaboration), J. High Energy. Phys. **1201**, 128 (2012); Phys. Lett. B **718**, 279 (2012); Phys. Rev. D **86**, 112007 (2012); Phys. Lett. B **708**, 265 (2012); Phys. Lett. B **721**, 13 (2013)
6. M. Cacciari et al., J. High Energy. Phys. **1210**, 137 (2012)
7. G. Aad et al. (ATLAS Collaboration), Phys. Lett. B **707**, 438 (2012)
8. S. Acharya et al. (ALICE Collaboration), J. High Energy. Phys. **1804**, 108 (2018)
9. S. Acharya et al. (ALICE Collaboration), Phys. Lett. B **770**, 459 (2017)
10. J. Adam et al. (ALICE Collaboration), J. High Energy. Phys. **1509**, 148 (2015)
11. B. Abelev et al. (ALICE Collaboration), J. Phys. G **41**, 087001 (2014)

Chapter 11
Performance of the Belle II SVD

K. Lalwani, H. Aihara, T. Aziz, S. Bacher, S. Bahinipati, E. Barberio, T. Baroncelli, T. Baroncelli, A. K. Basith, G. Batignani, A. Bauer, P. K. Behera, V. Bertacchi, S. Bettarini, B. Bhuyan, T. Bilka, F. Bosi, L. Bosisio, A. Bozek, F. Buchsteiner, G. Caria, G. Casarosa, M. Ceccanti, D. Červenkov, T. Czank, N. Dash, M. De Nuccio, Z. Doležal, F. Forti, M. Friedl, B. Gobbo, J. A. M. Grimaldo, K. Hara, T. Higuchi, C. Irmler, A. Ishikawa, H. B. Jeon, C. Joo, M. Kaleta, J. Kandra, N. Kambara, K. H. Kang, P. Kodyš, T. Kohriki, S. Koike, I. Komarov, M. Kumar, R. Kumar, W. Kun, P. Kvasnička, C. La Licata, L. Lanceri, J. Y. Lee, S. C. Lee, J. Libby, T. Lueck, P. Mammini, A. Martini, S. N. Mayekar, G. B. Mohanty, T. Morii, K. R. Nakamura, Z. Natkaniec, Y. Onuki, W. Ostrowicz, A. Paladino, E. Paoloni, H. Park, K. Prasanth, A. Profeti, I. Rashevskaya, K. K. Rao, G. Rizzo, P. K. Resmi, M. Rozanska, D. Sahoo, J. Sasaki, N. Sato, S. Schultschik, C. Schwanda, J. Stypula, J. Suzuki, S. Tanaka, H. Tanigawa, G. N. Taylor, R. Thalmeier, T. Tsuboyama, P. Urquijo, L. Vitale, S. Watanuki, M. Watanabe, I. J. Watson, J. Webb, J. Wiechczynski, S. Williams, H. Yin and L. Zani

Abstract The Belle II at SuperKEKB will accumulate e^+e^- collision data at an unprecedented instantaneous luminosity of 8×10^{35} cm^{-2} s^{-1}, about 40 times larger than its predecessor (Belle). Such a dramatic increase in luminosity will result in

---

S. Koike and M. Watanabe—Deceased authors.

E. Barberio · T. Baroncelli · T. Baroncelli · G. Caria · G. N.Taylor · P. Urquijo
J. Webb · S. Williams
School of Physics, University of Melbourne, Melbourne, VIC 3010, Australia

A. Bauer · F. Buchsteiner · M. Friedl · C. Irmler · S. Schultschik · C. Schwanda
R. Thalmeier · H. Yin
Institute of High Energy Physics, Austrian Academy of Sciences, 1050 Vienna, Austria

T. Bilka · D. Červenkov · Z. Doležal · J. Kandra · P. Kodyš · P. Kvasnička
Faculty of Mathematics and Physics, Charles University, 121 16 Prague, Czech Republic

S. Bahinipati · N. Dash
Indian Institute of Technology Bhubaneswar, Satya Nagar, India

A. K. Basith · P. K. Behera · J. Libby · P. K. Resmi
Indian Institute of Technology Madras, Chennai 600036, India

B. Bhuyan
Indian Institute of Technology Guwahati, Assam 781039, India

K. Lalwani (✉) · M. Kumar
Malaviya National Institute of Technology Jaipur, Jaipur 302017, India
e-mail: kavita.phy@mnit.ac.in

A. Giri and R. Mohanta (eds.), *16th Conference on Flavor Physics and CP Violation*, Springer Proceedings in Physics 234,
https://doi.org/10.1007/978-3-030-29622-3_11

a harsh background environment in Belle II. In this paper, we present the detailed performance studies of the Belle II silicon vertex detector with test beam data and Phase II data.

11.1 Introduction

The silicon vertex detector (SVD) is one of the important sub-detectors of the Belle II experiment [1] at SuperKEKB [2]. It consists of four layers of double-sided silicon strip detectors (DSSDs) arranged in concentric cylinders around the Belle II interaction point. In conjunction with the DEPFET-based pixel detector (PXD), the SVD plays a key role in the precise measurement of the decay vertex position as well as the reconstruction of low-momentum tracks. The excellent performance of the Belle II SVD will enable the measurements of CP asymmetry in the B-meson system with higher precision. To achieve the physics goals, reconstruction of tracks with a high efficiency and good resolution is needed. Belle II will collect e^+e^- collision data at an instantaneous luminosity of $8 \times 10^{35}\ \text{cm}^{-2}\ \text{s}^{-1}$, which is about 40 times larger than its previous experiment (Belle). At such a high luminosity, Belle II will face harsh background environment. To validate the SVD performance in such high rate and challenging background conditions, systematic studies are required with offline reconstruction software.

R. Kumar
Punjab Agricultural University, Ludhiana 141004, India

T. Aziz · S. N. Mayekar · G. B. Mohanty · K. Prasanth · K. K. Rao · D. Sahoo
Tata Institute of Fundamental Research, Mumbai 400005, India

G. Batignani · V. Bertacchi · S. Bettarini · G. Casarosa · M. De Nuccio · F. Forti
T. Lueck · A. Martini · E. Paoloni · G. Rizzo · L. Zani
Dipartimento di Fisica, Università di Pisa, 56127, Pisa, Italy

G. Batignani · V. Bertacchi · S. Bettarini · F. Bosi · G. Casarosa · M. Ceccanti · M. De Nuccio
F. Forti · T. Lueck · P. Mammini · A. Martini · E. Paoloni · A. Profeti · G. Rizzo · L. Zani
INFN Sezione di Pisa, 56127, Pisa, Italy

L. Bosisio · C. La Licata · L. Lanceri · L. Vitale
Dipartimento di Fisica, Università di Trieste, 34127, Trieste, Italy

L. Bosisio · B. Gobbo · I. Komarov · C. La Licata · L. Lanceri · I. Rashevskaya · L. Vitale
INFN Sezione di Trieste, 34127, Trieste, Italy

T. Higuchi · C. Joo · T. Morii · A. Paladino
Kavli Institute for the Physics and Mathematics of the Universe (WPI),
University of Tokyo, Kashiwa 277-8583, Japan

M. Watanabe
Department of Physics, Niigata University, Niigata 950-2181, Japan

11.2 Overview of the SVD

The innermost sub-detector of Belle II is the Vertex Detector (VXD), which comprises of two layers of PXD and four layers of SVD. The VXD is responsible for the reconstruction of the vertices of the decaying particles, which helps in the study of time-dependent CP violation. As the Belle II SVD [3] is shifted outward to a maximum radius of 140 mm, it enhances the fiducial coverage provided by VXD. Special characteristics of the SVD are to reconstruct the decays of neutral particle like K_S^0 as well as to track low momentum particles like the pions produced in the decay of D^* mesons that do not reach to the drift chamber. Furthermore, the SVD can provide very precise timing hit information and use of the ionization energy loss for better hit selection and discrimination. The SVD along with the PXD provides an improved impact parameter resolution as compared to Belle. The SVD consists of DSSDs and their arrangement is shown in Fig. 11.1. Strips on the two sides of SVD provide the x and y coordinates of the hits (z coordinate is defined by the sensor position). The two orthogonal sides of the strips are known as U (parallel to the beam direction, P side with a small pitch of 50–75 μm) and V (perpendicular to the beam direction, N side with a large pitch 160–240 μm). The SVD sensors come in the different dimensions: the innermost layer 3 has small rectangular sensor of thickness 320 μm, while the outer three layers are composed of two types of sensors: large rectangular sensors (thickness 320 μm) for the barrel part and slanted sensors (thickness 300 μm) with a trapezoidal shape, which improves precision for the forward boosted particles while extending of polar angle coverage.

T. Czank · A. Ishikawa · S. Watanuki
Department of Physics, Tohoku University, Sendai 980-8578, Japan

H. Aihara · J. A. M. Grimaldo · W. Kun · Y. Onuki · J. Sasaki · H. Tanigawa · I. J. Watson
Department of Physics, University of Tokyo, Tokyo 113-0033, Japan

K. Hara · N. Kambara · T. Kohriki · S. Koike · K. R. Nakamura · N. Sato · J. Suzuki
S. Tanaka · T. Tsuboyama
High Energy Accelerator Research Organization (KEK), Tsukuba 305-0801, Japan

H. B. Jeon · K. H. Kang · S. C. Lee · H. Park
Department of Physics, Kyungpook National University, Daegu 702-701, Korea

J. Y. Lee
Department of Physics and Astronomy, Seoul National University, Seoul 151-742, Korea

S. Bacher · A. Bozek · M. Kaleta · Z. Natkaniec · W. Ostrowicz · M. Rozanska · J. Stypula
J. Wiechczynski
The Henryk Niewodniczanski Institute of Nuclear Physics, 31-342 Krakow, Poland

I. Rashevskaya
TIFPA - INFN, 38123, Trento, Italy

M. Watanabe
Nippon Dental University, Niigata 951-8580, Japan

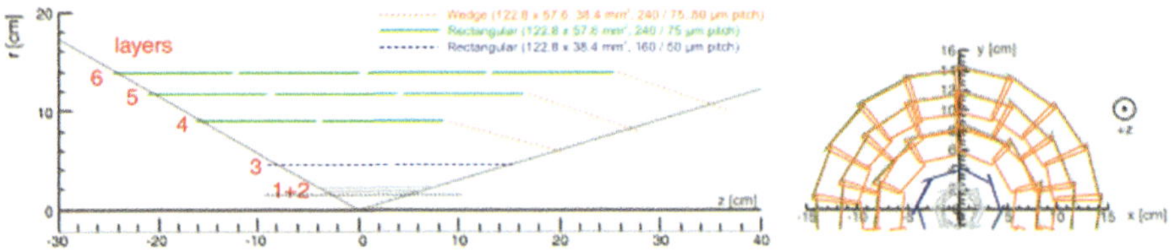

Fig. 11.1 A cross sectional view of VXD in the r-z (left) and x-y plane (right, top view) [3]

11.3 Performance Studies of the SVD

The performance of SVD is studied using the test beam data, taken at DESY, Hamburg on February 2017. The corresponding set-up relies on 4 GeV electron beam being incident on a detector system comprising of 2 PXD layers and 4 SVD layers in the presence of 1T magnetic field. In the SVD sensors, with an intermediate floating strips between two readout strips, the charge seen on the floating electrode is coupled to the readout strip through capacitive charge division, via a network shown schematically in Fig. 11.3 (left). The charge loss due to the presence of additional capacitive coupling between the floating strip and the backside of the sensor is significant in the sensors with large pitch, as the backside capacitance increases linearly with the implant pitch and becomes comparable to the interstrip capacitance (C_{int}). This effect is modeled in simulations and the resulting sensitivity of the cluster charge to C_{int} can be exploited in following two ways: (a) measure the actual C_{int} of SVD

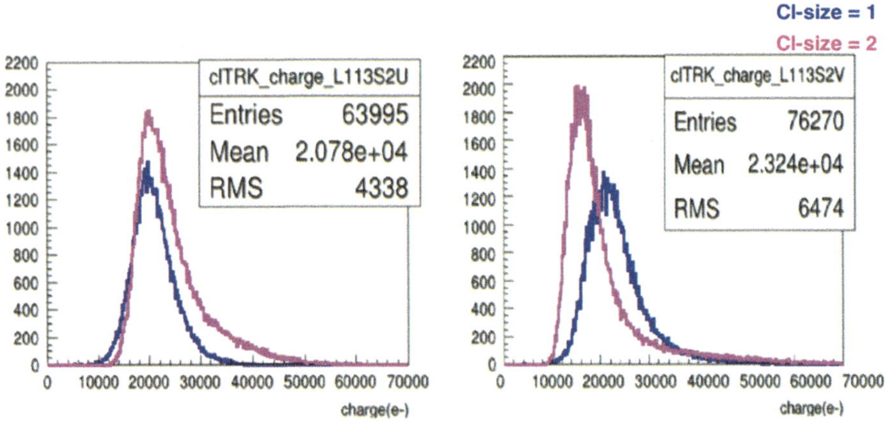

Fig. 11.2 Cluster charge for layer 3 (sensor 2): U side (left) and V side (right) with test beam data

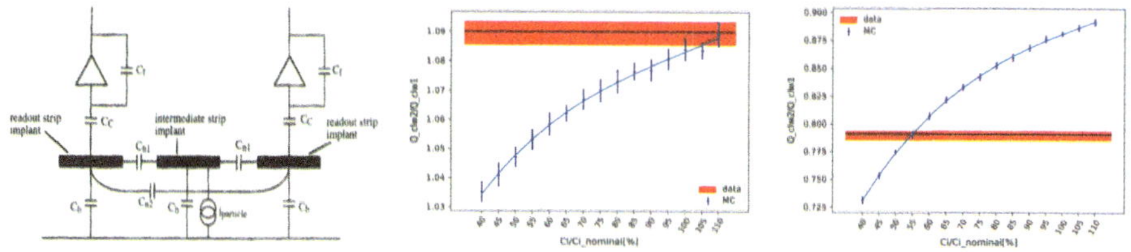

Fig. 11.3 Schematic layout of a detector making use of capacitive charge division (left) [4]; Measurement of C_{int} for layer 3 (sensor 2): U side (center) and V side (right)

sensors; (b) tune this important parameter for the full SVD simulation. Therefore, this measurement is carried out by comparing the test beam data and full SVD simulation. Figure 11.2 shows the cluster charge for layer 3 (sensor 2) for both U (left) and V side (right). It can be seen that the V side, with large pitch, has lower cluster charge for cluster size 2, due to the above mentioned charge loss effect. For the U side, with small pitch, this effect is not present and charge for cluster size 2 is comparable or even a bit higher. Since complete charge reconstruction is instead possible for cluster size 1, if the second strip of cluster is below threshold. Further, the effect of C_{int} on the cluster charge is studied with the simulated data as a function of C_{int}, used to model charge division among the floating and readout strip, and compared with the results of test beam analysis to measure C_{int} for each sensor/side of the SVD. Figure 11.3 shows results from the measurement of C_{int}, where cluster charge ratio $R = \frac{Q_{\text{clw}_2}}{Q_{\text{clw}_1}}$ is plotted as a function of C_{int} for layer 3, U (center) and V side (right). Here Q_{clw_1} and Q_{clw_2} are the cluster charge measured for strip with cluster size 1 and 2, respectively. In addition, the performance of the SVD is also studied using the first collision data with Phase-II operation of Belle II. The set-up for data collection under Phase-II operation consists of a subset of the VXD comprising six layers with a single ladder per layer, installed in spring 2018. The cluster signal-to-noise ratio (S/N) with tracks of different momentum and inclination is calculated for both U and V side as shown in Fig. 11.4 (left). The S/N ratio is measured to be greater than 20 for N-side, which performs better than the P-side especially for 1-strip cluster. This is expected due to the longer P side strips resulting in a larger capacitive load to the charge preamplifier used for the readout and hence higher noise on P-side. Further, signal hit time is also measured with first data as shown in Fig. 11.4 (right), where the RMS of signal hit time corresponding to a bunch crossing is found to be in the order of 3 ns, which is nicely matching with the design expectation. This study will be useful to remove the off-time hits as well as to improve the tracking performance of the SVD.

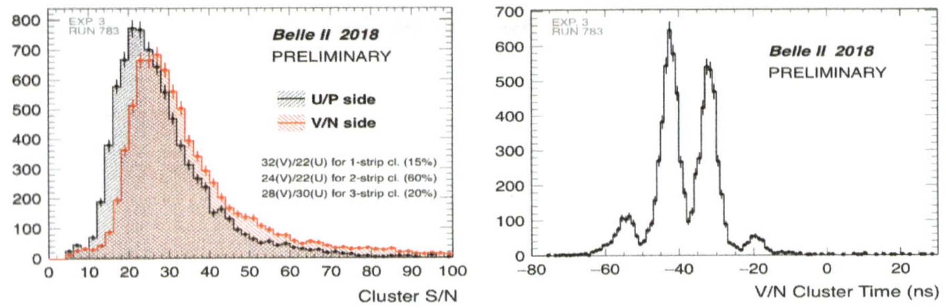

Fig. 11.4 S/N ratio (left), Signal Hit time (right) with first data (Phase-II)

11.4 Summary

The performance of the Belle II SVD is validated using test beam data as well as first collision data (Phase II). The effect of C_{int} on the cluster charge is studied in the test beam data and compared with simulations to measure C_{int} for each SVD sensor/side. The S/N ratio with the Phase-II data are in agreement with design expectations. It is further demonstrated that the SVD is clearly able to distinguish bunch crossings of 16 ns apart.

References

1. T. Abe et al., Belle II Technical Design Report. arXiv:1011.0352
2. B. Wang, The Belle II Experiment and SuperKEKB Upgrade. arXiv:1511.09434
3. K. Adamczyk et al., The silicon vertex detector of the Belle II experiment. Nucl. Instr. Methods. Phys. Res. Sect. A **824**, 406–410 (2016)
4. M. Krammer, H. Pernegger, Nucl. Instr. Methods Phys. Res. Sect. A **397**, 232–242 (1997)

Chapter 12
Predictions of Angular Observables for $\bar{B}_s \to K^*\ell\ell$ and $\bar{B} \to \rho\ell\ell$ in Standard Model

Bharti Kindra and Namit Mahajan

Abstract Exclusive semileptonic decays based on $b \to s$ transitions have been attracting a lot of attention as some angular observables deviate significantly from the Standard Model (SM) predictions in specific q^2 bins. B meson decays induced by other Flavor Changing Neutral Current (FCNC), $b \to d$, can also offer a probe to new physics with an additional sensitivity to the weak phase in Cabibbo-Kobayashi-Masakawa (CKM) matrix. We provide predictions for angular observables for $b \to d$ semileptonic transitions, namely $\bar{B}_s \to K^*\ell^+\ell^-$, $\bar{B}^0 \to \rho^0\ell^+\ell^-$, and their CP-conjugated modes including various non-factorizable corrections.

12.1 Introduction

Experimental evidence of new physics has been found in the channels involving FCNC $b \to s\ell^+\ell^-$ and charged current $b \to c\ell\nu$. However, the $b \to d$ counterpart of the weak decay, i.e. $b \to d\ell^+\ell^-$, has not caught much attention perhaps because of low branching ratio. The weak phases incorporate CKM matrix elements $\xi_q^i = V_{qi}^* V_{qb}$, where $q \in \{u, c, t\}$ and $i \in \{s, d\}$. For $b \to s\ell\ell$ transition, $\xi_{c,t}^s \sim \lambda^2$ and $\xi_u^s \sim \lambda^4$ where $\lambda = 0.22$. Since $u\bar{u}$ contribution introduces CKM phase which is negligible for $b \to s\ell\ell$, CP violating quantities are very small in SM. On the other hand, since $\xi_u^d \sim \xi_c^d \sim \xi_t^d \sim \lambda^4$ for $b \to d\ell\ell$, the B decays mediated through this transition allow for large CP violating quantities. Also, leading order contribution in this case is smaller than the leading contribution in $b \to s\ell\ell$ which makes it more sensitive to new particles and interactions. In this work, we focus on two such decay channels, $B_S \to \bar{K}^*\ell^+\ell^-$ and $B \to \rho\ell^+\ell^-$ [1].

B. Kindra (✉) · N. Mahajan
Physical Research Laboratory, Ahmedabad, India
e-mail: bharti@prl.res.in

B. Kindra
Indian Institute of Technology, Gandhinagar, India

A. Giri and R. Mohanta (eds.), *16th Conference on Flavor Physics and CP Violation*, Springer Proceedings in Physics 234,
https://doi.org/10.1007/978-3-030-29622-3_12

12.2 Decay Amplitude

We follow the effective Hamiltonian approach as used in [2] to write the Hamiltonian and decay amplitude. Th amplitude is written as a product of short-distance contributions through Wilson coefficients and long-distance contribution which is further expressed in terms of form factors,

$$\mathcal{M} = \frac{G_F\alpha}{\sqrt{2}\pi} V_{tb}V_{td}^{*}\Big\{\Big[\langle V|\bar{d}\gamma^{\mu}(C_9^{\rm eff}P_L)b|P\rangle - \frac{2m_b}{q^2}\langle V|\bar{d}\, i\,\sigma^{\mu\nu}q_{\nu}(C_7^{\rm eff}P_R)b|P\rangle\Big](\bar{\ell}\gamma_{\mu}\ell)$$
$$+ \langle V|\bar{d}\gamma^{\mu}(C_{10}^{\rm eff}P_L)b|P\rangle(\bar{\ell}\gamma_{\mu}\gamma_5\ell) - 16\pi^2\frac{\bar{\ell}\gamma^{\mu}\ell}{q^2}\mathcal{H}_{\mu}^{\text{non-fac}}\Big\}. \quad (12.1)$$

Wilson coefficients ($C_i's$) are computed upto next-to-next-to leading order (NNLO) [3] and form factors are computed using the method of Light Cone Sum Rules (LCSR) and QCD lattice calculation [4]. $\mathcal{H}_{\mu}^{\text{non-fac}}$ represents the non-factorizable contribution of non-local hadronic matrix element. This results from four quark and chromomagnetic operators combined with virtual photon emission which then decays to lepton pair through electromagnetic interaction. These corrections are given in terms of hard-scattering kernels ($\mathcal{T}_a^q s$), where $a \in \{\perp, \parallel\}$ and $q \in \{u, c\}$, which are convoluted with B(B_S)-meson and $\rho(\bar{K}^*)$ distribution amplitudes. The non-factorizable corrections included here are spectator scattering $\mathcal{T}_a^{q,\text{spec}}$, weak annihilation $\mathcal{T}_a^{q,\text{WA}}$, and soft gluon emission $\Delta C_9^{q,\text{soft}}$. These corrections have been computed in [5–7] except charm loop corrections corresponding to up quark in th eloop. For present work, we are assuming that its contribution is less than 10% of C_9: $\Delta C_{9,u}^{\text{soft}} = ae^{i\theta}$; $|a| \in \{0, 0.5\}$, $\theta \in \{0, \pi\}$.

These corrections are then added to transversity amplitudes in the following way:

$$A_{\perp L,R}(q^2) = \sqrt{2\lambda}\, N\left[2\frac{m_b}{q^2}(C_7^{\rm eff}T_1(q^2) + \Delta T_{\perp}) + (C_9^{\rm eff} \mp C_{10} + \Delta C_9^1(q^2))\frac{V(q^2)}{M_B + M_V}\right] \quad (12.2)$$

$$A_{\parallel L,R}(q^2) = -\sqrt{2}N(M_B^2 - M_V^2)\Big[2\frac{m_b}{q^2}(C_7^{\rm eff}T_2(q^2) + 2\frac{E(q^2)}{M_B}\Delta T_{\perp})$$
$$+ (C_9^{\rm eff} \mp C_{10} + \Delta C_9^2(q^2))\frac{A_1(q^2)}{M_B - M_V}\Big] \quad (12.3)$$

$$A_{0L,R}(q^2) = -\frac{N}{2M_V\sqrt{q^2}}\Big[2m_b\big((M_B^2 + 3M_V^2 - q^2)(C_7^{\rm eff}T_2(q^2)\big)$$
$$- \frac{\lambda}{M_B^2 - M_v^2}(C_7^{\rm eff}T_3(q^2) + \Delta T_{\parallel})) + (C_9^{\rm eff} \mp C_{10} + \Delta C_9^3)$$
$$\big((M_B^2 + M_V^2 - q^2)(M_B + M_V)A_1(q^2) - \frac{\lambda}{M_B + M_V}A_2(q^2)\big)\Big] \quad (12.4)$$

$$A_t(q^2) = \frac{N}{\sqrt{s}}\sqrt{\lambda}2\, C_{10}\, A_0(q^2) \quad (12.5)$$

where,

$$\Delta T_{\perp} = \frac{\pi^2}{N_c}\frac{f_P f_{V,\perp}}{M_B}\frac{\alpha_s C_F}{4\pi}\int \frac{d\omega}{\omega}\Phi_{P,-}(\omega)\int_0^1 du\ \Phi_{V,\perp}(u)\Big(T_{\perp}^{c,\text{spec}} + \frac{\xi_u}{\xi_t}(T_{\perp}^{u,\text{spec}})\Big) \tag{12.6}$$

$$\Delta T_{\parallel} = \frac{\pi^2}{N_c}\frac{f_P f_{V,\parallel}}{M_B}\frac{M_V}{E}\sum_{\pm}\int \frac{d\omega}{\omega}\Phi_{P}(\omega)\int_0^1 du\ \Phi_{V,\parallel}(u)\Big[T_{\parallel}^{c,WA} + \frac{\xi_u}{\xi_t}T_{\parallel}^{u,WA}$$
$$\frac{\alpha_s C_F}{4\pi}(T_{\parallel}^{c,\text{spec}} + \frac{\xi_u}{\xi_t}T_{\parallel}^{,\text{spec}})\Big] \tag{12.7}$$

$$\Delta C_9^i = \Delta C_{9,c}^{i,soft} + \Delta C_{9,u}^{i,soft} \tag{12.8}$$

12.3 Observables

The angular decay distribution of $B \to V(\to M_1 M_2)\ell^+\ell^-$ is given in terms of angular functions $(I_i(q^2, \theta_V, \theta_l, \phi)$, the value of which can be obtained by integrating data over specific values of the parameters. We consider an optimized set of observables constricted choosing specific combinations of these angular functions. The observables considered here are:

– Form Factor Dependent observables.

$$\frac{d\Gamma}{dq^2} = \frac{1}{4}(3I_1^c + 6I_1^s - I_2^c - 2I_2^s) \quad A_{FB}(q^2) = \frac{-3I_6^s}{3I_1^c + 6I_1^s - I_2^c - 2I_2^s}$$
$$F_L(q^2) = \frac{3I_1^c - I_2^c}{3I_1^c + 6I_1^s - I_2^c - 2I_2^s} \tag{12.9}$$

– Form Factor independent observables.

$$P_1 = \frac{I_3}{2I_2^s},\ P_2 = \beta_l\frac{I_6^s}{8I_2^s},\ P_3 = -\frac{I_9}{4I_2^s},\ P_4' = \frac{I_4}{\sqrt{-I_2^c I_2^s}}$$
$$P_5' = \frac{I_5}{2\sqrt{-I_2^c I_2^s}},\ P_6' = -\frac{I_7}{2\sqrt{-I_2^c I_2^s}},\ P_8' = -\frac{I_8}{2\sqrt{-I_2^c I_2^s}} \tag{12.10}$$

– Lepton Flavor Universality violating observables.

$$R_{K^*}^{B_S} = \frac{\left[\mathcal{BR}(B_s \to \bar{K}^*\mu^+\mu^-)\right]_{q^2\in\{q_1^2,q_2^2\}}}{\left[\mathcal{BR}(B_s \to \bar{K}^*e^+e^-\right]_{q^2\in\{q_1^2,q_2^2\}}} \tag{12.11}$$

These observables are valid for $B_S \to \bar{K}^*\ell\ell$. For $B \to \rho\ell\ell$, angular functions are replaced with time-dependent angular functions, since the final state in this case is self conjugate [1].

12.4 Results

The binned values for the decay modes in study are listed in Tables 12.1 and 12.2, where the first uncertainty is due to form factors and scond uncertainty is due to soft-gluon emission from up quark. Moreover, the full branching ratio for $B_S \to \bar{K}^*\ell\ell$ is $(3.356 \pm 0.814) \times 10^{-8}$ which is consistent with the recent measurement [8].

Table 12.1 Observables for $\bar{B}_s \to K^*\mu^+\mu^-$ and $B_s \to \bar{K}^*\mu^+\mu^-$ using form factors based on LCSR and QCD lattice calculation

Observable	$\bar{B}_s \to K^*\mu^+\mu^-$		$B_s \to \bar{K}^*\mu^+\mu^-$	
	[0.1−1] GeV^2	[1−6] GeV^2	[0.1−1] GeV^2	[1−6] GeV^2
P_1	$0.017 \pm 0.132 \pm 0.001$	$-0.096 \pm 0.128 \pm 0.005$	$0.015 \pm 0.135 \pm 0.001$	$-0.087 \pm 0.118 \pm 0.005$
P_2	$0.122 \pm 0.013 \pm 0.001$	$0.026 \pm 0.081 \pm 0.036$	$0.114 \pm 0.012 \pm 0.001$	$0.054 \pm 0.081 \pm 0.034$
P_3	$0.001 \pm 0.003 \pm 0.0$	$0.004 \pm 0.009 \pm 0.002$	$0.001 \pm 0.006 \pm 0.0$	$0.004 \pm 0.009 \pm 0.002$
P_4'	$-0.704 \pm 0.063 \pm 0.009$	$0.543 \pm 0.167 \pm 0.014$	$-0.736 \pm 0.064 \pm 0.008$	$0.453 \pm 0.176 \pm 0.016$
P_5'	$0.437 \pm 0.044 \pm 0.016$	$-0.422 \pm 0.124 \pm 0.046$	$0.445 \pm 0.045 \pm 0.016$	$-0.377 \pm 0.130 \pm 0.047$
P_6'	$-0.091 \pm 0.005 \pm 0.016$	$-0.087 \pm 0.010 \pm 0.002$	$-0.048 \pm 0.004 \pm 0.001$	$-0.064 \pm 0.004 \pm 0.002$
P_8'	$0.027 \pm 0.007 \pm 0.016$	$0.042 \pm 0.010 \pm 0.017$	$0.048 \pm 0.009 \pm 0.016$	$0.036 \pm 0.008 \pm 0.019$
$R_{K^*}^{B_s}$	$0.945 \pm 0.008 \pm 0.001$	$0.998 \pm 0.004 \pm 0.0$	$0.944 \pm 0.007 \pm 0.001$	$0.998 \pm 0.004 \pm 0.0$
$BR \times 10^9$	$4.439 \pm 0.648 \pm 0.086$	$8.251 \pm 1.872 \pm 0.357$	$5.082 \pm 0.699 \pm 0.101$	$8.763 \pm 1.959 \pm 0.375$
A_{FB}	$-0.048 \pm 0.008 \pm 0.001$	$0.001 \pm 0.021 \pm 0.009$	$-0.047 \pm 0.007 \pm 0.001$	$-0.012 \pm 0.020 \pm 0.009$
F_L	$0.576 \pm 0.066 \pm 0.014$	$0.872 \pm 0.035 \pm 0.007$	$0.553 \pm 0.065 \pm 0.014$	$0.862 \pm 0.035 \pm 0.007$

Table 12.2 Binned values of observables for the process $B \to \rho\mu^+\mu^-$ and $\bar{B} \to \rho\mu^+\mu^-$ for tagged events to be measured at Belle. Form factors are based on LCSR form factors

	$B \to \rho\mu^+\mu^-$		$\bar{B} \to \rho\mu^+\mu^-$	
	[0.1–1] GeV^2	[1–6] GeV^2	[0.1–1] GeV^2	[1–6] GeV^2
$\langle P_1 \rangle$	$0.009 \pm 0.177 \pm 0.001$	$-0.065 \pm 0.116 \pm 0.003$	$0.010 \pm 0.175 \pm 0.001$	$-0.069 \pm 0.120 \pm 0.003$
$\langle P_2 \rangle$	$0.082 \pm 0.0 \pm 0.001$	$0.021 \pm 0.056 \pm 0.023$	$0.076 \pm 0.008 \pm 0.0$	$-0.042 \pm 0.050 \pm 0.024$
$\langle P_3 \rangle$	$0 \pm 0.005 \pm 0.0$	$0.001 \pm 0.005 \pm 0.002$	$0.001 \pm 0.001 \pm 0.0$	$0.002 \pm 0.005 \pm 0.002$
$\langle P'_4 \rangle$	$-0.724 \pm 0.081 \pm 0.047$	$0.508 \pm 0.161 \pm 0.029$	$-0.703 \pm 0.080 \pm 0.046$	$0.569 \pm 0.154 \pm 0.017$
$\langle P'_5 \rangle$	$0.276 \pm 0.004 \pm 0.027$	$-0.270 \pm 0.083 \pm 0.085$	$0.246 \pm 0.003 \pm 0.030$	$-0.321 \pm 0.074 \pm 0.098$
$\langle P'_6 \rangle$	$-0.043 \pm 0.003 \pm 0.001$	$-0.061 \pm 0.004 \pm 0.002$	$-0.075 \pm 0.005 \pm 0.001$	$-0.073 \pm 0.010 \pm 0.002$
$\langle P'_8 \rangle$	$0.025 \pm 0.005 \pm 0.016$	$0.025 \pm 0.005 \pm 0.018$	$0.031 \pm 0.005 \pm 0.007$	$0.030 \pm 0.006 \pm 0.017$
$\langle R_\rho \rangle$	$0.936 \pm 0.008 \pm 0.001$	$0.997 \pm 0.003 \pm 0.0$	$0.950 \pm 0.167 \pm 0.002$	$1.064 \pm 0.392 \pm 0.0$
$\langle BR \rangle \times 10^9$	$5.233 \pm 0.711 \pm 0.080$	$8.714 \pm 1.668 \pm 0.366$	$4.736 \pm 0.656 \pm 0.077$	$8.414 \pm 1.649 \pm 0.365$
$\langle A_{FB} \rangle$	$-0.038 \pm 0.005 \pm 0.001$	$-0.007 \pm 0.019 \pm 0.007$	$-0.034 \pm 0.005 \pm 0.001$	$0.014 \pm 0.022 \pm 0.006$
$\langle F_L \rangle$	$0.495 \pm 0.067 \pm 0.014$	$0.813 \pm 0.037 \pm 0.007$	$0.514 \pm 0.072 \pm 0.014$	$0.838 \pm 0.046 \pm 0.006$

References

1. B. Kindra, N. Mahajan. arXiv:1803.05876 [hep-ph]
2. W. Altmannshofer, P. Ball, A. Bharucha, A.J. Buras, D.M. Straub, M. Wick, JHEP **0901**, 019 (2009). https://doi.org/10.1088/1126-6708/2009/01/019. arXiv:0811.1214 [hep-ph]
3. H.M. Asatrian, K. Bieri, C. Greub, M. Walker, Phys. Rev. D **69**, 074007 (2004). https://doi.org/10.1103/PhysRevD.69.074007 [hep-ph/0312063]
4. A. Bharucha, D.M. Straub, R. Zwicky, JHEP **1608**, 098 (2016). https://doi.org/10.1007/JHEP08(2016). arXiv:1503.05534 [hep-ph]
5. M. Beneke, T. Feldmann, D. Seidel, Nucl. Phys. B **612**, 25 (2001). https://doi.org/10.1016/S0550-3213(01)00366-2 [hep-ph/0106067]
6. M. Beneke, T. Feldmann, D. Seidel, Eur. Phys. J. C **41**, 173 (2005). https://doi.org/10.1140/epjc/s2005-02181-5 [hep-ph/0412400]
7. A. Khodjamirian, T. Mannel, A.A. Pivovarov, Y.-M. Wang, JHEP **1009**, 089 (2010). https://doi.org/10.1007/JHEP09(2010). arXiv:1006.4945 [hep-ph]
8. R. Aaij et al. (LHCb Collaboration), JHEP **1807**, 020 (2018). https://doi.org/10.1007/JHEP07(2018)020. arXiv:1804.07167 [hep-ex]

Chapter 13
Radiative B Decays at LHCb

Brij Kishor Jashal and Arantza Oyanguren Campos

Abstract Rare radiative B decays are sensitive probes of New Physics through the study of branching fractions, CP asymmetries and other observables related to the photon polarization. The LHCb experiment has performed several measurements with radiative B decays. These results provide constraints on predictions of models beyond the Standard Model, and are at present key to understanding the nature of flavor physics.

13.1 Introduction

Rare radiative decays are sensitive probes to new physics (NP) particles since they proceed via Flavour Changing Neutral Currents (FCNC), which are forbidden at tree level in the Standard Model (SM). New particles can enter in the loops and modify the helicity structure of the photon.

Photons in $b \to s\gamma$ transitions are predicted to be left-handed in the SM since the recoiling s quark that couples to a W boson is left-handed. However, some new physics models, particularly Left-Right Symmetric models predict an anomalous component of polarized photons, primarily due to the exchange of a heavy fermion in the penguin loop [1].

The $b \to s\gamma$ transition can be expressed in terms of Wilson coefficients: $C_7^{'}$ in $'\mathrm{H}_{eff} = \frac{\mathrm{G_f}}{\sqrt{2}} V_{ts}^* V_{tb} \big(C_7 O_7 + C_7' O_7'\big) + h.c.$ where it is related to left-(right)-handed polarized photons [2].

There are two ways to access the photon polarization in radiative b-hadron decays.

Poster presented in the FPCP 2018 conference, Hyderabad, India.
B. K. Jashal and A. Oyanguren Campos—On behalf of LHCb collaboration.

B. K. Jashal (✉) · A. Oyanguren Campos
Instituto de Física Corpuscular (IFIC—University of Valencia and CSIC), València, Spain
e-mail: Brij@cern.ch

B. K. Jashal
Tata Institute of Fundamental Research, Mumbai, India

A. Giri and R. Mohanta (eds.), *16th Conference on Flavor Physics and CP Violation*, Springer Proceedings in Physics 234,
https://doi.org/10.1007/978-3-030-29622-3_13

1. Time dependent analyses using the phenomenon of interference in mixing and decays of neutral B mesons, such as $B_s \rightarrow \phi\gamma$, $B_d \rightarrow \rho\gamma$ and $B_d \rightarrow \omega\gamma$ where the observables are time dependent decay widths and CP asymmetries.
2. Angular analyses of
   - Radiative B_s decays with three final-state hadrons such as $B^+ \rightarrow K^-\pi^+\pi^+\gamma$
   - decays of b-baryons ($\Lambda_b \rightarrow \Lambda\gamma$ ...)
   - decays with an electron pair in the final state $\gamma \rightarrow e^-e^+$

Here we discuss some measurements performed by LHCb: the time dependent analysis of $B_s \rightarrow \phi\gamma$ [3] events and the angular analysis of $B^+ \rightarrow K^-\pi^+\pi^+\gamma$ decays [4]. We also present ongoing studies related to decays of radiative b-baryons [5].

13.2 The LHCb Experiment

The LHCb experiment at the Large Hadron Collider (LHC) at CERN [6, 7] is one of the main experiments designed to study decays of *b* and *c* hadrons. The single-arm forward spectrometer design of LHCb detector is optimal for heavy $q\bar{q}$ pair production since at higher energies this process predominantly occurs in the forward or backward direction. The sub-systems of the detector include Vertex Locator, Tracking System, RICH detectors, Calorimeter System (ECAL, HCAL) and Muon system. Rare radiative decays are difficult to reconstruct in the pp collisions. LHCb provides decay time resolution for radiative decays of 50 fs, B meson mass resolution of about 90 MeV/c^2 and photon resolution (ECAL) $\sigma/E \sim 10\%/\sqrt{E} \oplus 1\%$.

13.3 Photon Polarization Measurements

In LHCb, there are three analyses of FCNC radiative decays where the photon polarization is accessible.[1]

13.3.1 Analysis of $B^+ \rightarrow K^+\pi^-\pi^+\gamma$ Decays [4]

In this analysis, information about the photon polarization is obtained by studying the angular distribution of the photon with respect to the normal to the plane defined by the momenta of the three final-state hadrons in their centre-of-mass frame. The photon polarization parameter, λ_γ, is extracted by studying $B \rightarrow K_J\,\gamma$ decays where K_J (J^P) is a resonant state decaying to $K\pi\pi$. Measuring the up-down asymmetry, A_{UD},

[1] An additional measurement sensitive to C_7' is the study of $B^0 \rightarrow K^{*0}e^+e^-$ decays at low four-momentum transfer [8].

defined as $A_{ud} = \frac{N(K\pi\pi\gamma)_{\cos\theta>0} - N(K\pi\pi\gamma)_{\cos\theta<0}}{N(K\pi\pi\gamma)_{\cos\theta>0} + N(K\pi\pi\gamma)_{\cos\theta<0}}$ gives access to λ_γ. The data sample corresponds to an integrated luminosity of 3 fb^{-1} collected in pp collisions at 7 and 8 TeV centre-of-mass energies. By reconstructing the invariant mass of the system composed of one kaon and two-opposite charged pions, plus an energetic photon, a total of 13876 ± 153 signal events were observed after background subtraction. This analysis presents the first observation of a photon polarization different from zero at a significance of 5.2σ. This result cannot be translated directly into a measurement of the photon polarization without precise knowledge of the resonances that populate the Kππ system.

13.3.2 Time Dependent Analysis of $B_s^o \to \phi\gamma$ Decays [3]

LHCb has performed the first measurement of one of the mixing-induced and CP violating parameters of the $B_s^o \to \phi\gamma$ decay $\left(\mathcal{A}^{\Delta}_{\phi\gamma}, S_{\phi\gamma}, C_{\phi\gamma}\right)$ appearing in the time-dependent decay rate $\Gamma_{B_s^o \to \phi\gamma}(t) \propto e^{-\Gamma\,\mathrm{dt}}\left[\cosh\left(\frac{\Delta\Gamma_s t}{2}\right) - \mathcal{A}^{\Delta}\left(\frac{\Delta\Gamma_s t}{2}\right)\right]$ [9, 10]. The polarization parameter $\mathcal{A}^{\Delta}$ gives access to the ratio of right over left handed photon polarization amplitudes in $b \to s\gamma$ transitions. The φ meson is reconstructed in two kaons of opposite sign. In this analysis $B^0 \to K^*\gamma$ (with $K^{*0} \to K^+\pi^-$) is used as the control channel. A 3 fb^{-1} dataset collected at centre-of-mass energies of 7 and 8 TeV is used. A total of about 4000 $B_s^0 \to \phi\gamma$ decays are reconstructed, and a measured value of $\mathcal{A}^{\Delta} = -0.98^{+0.46+0.23}_{-0.52-0.20}$ is obtained. The result is compatible with the SM expectation of $\mathcal{A}^{\Delta} = -0.047^{+0.029}_{-0.025}$, within two standard deviations as explained in Fig. 13.1.

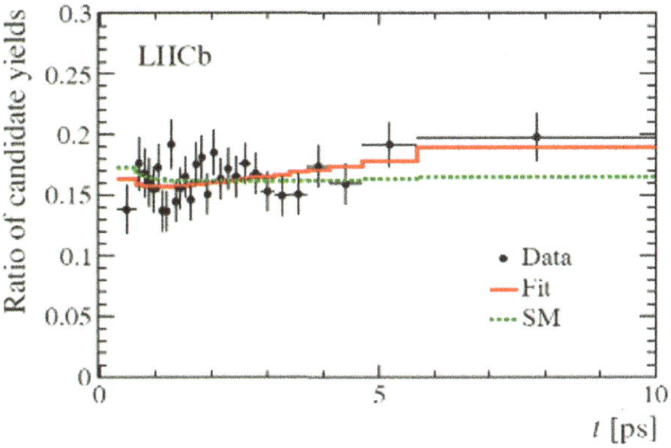

Fig. 13.1 Decay-time dependence of the ratio of the yields of $B_s^o \to \phi\gamma$ and $B^0 \to K^*\gamma$, with the fit overlaid along with the expected distribution for the central value of the SM prediction. Figure extracted from [3]

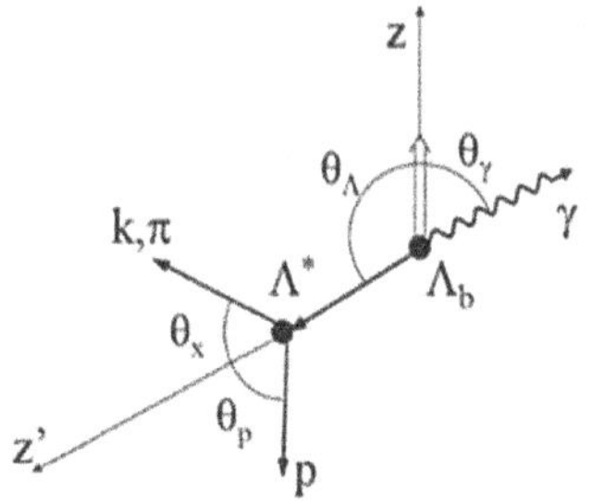

Fig. 13.2 Schematic view of the decay

13.3.3 Decays of b-Baryons (Λ_b, Ξ_b, Ω_b)

In the *b*-baryon sector, there are at least three decays where the photon polarization is accessible through the study of the angular distribution of the decay products: $\Lambda_b \rightarrow \Lambda\gamma$, $\Xi_b^- \rightarrow \Xi^-\gamma$ and $\Omega_b^- \rightarrow \Omega^-\gamma$.

The $\Lambda_b \rightarrow \Lambda\gamma$ decay mode is predicted to have the largest relative branching fraction. It has never been observed and an upper limit on the branching fraction is obtained by CDF [11].

In the $\Lambda_b \rightarrow \Lambda\gamma$ decay mode, where Λ is reconstructed from a final-state proton (p) and a pion (π) (as shown in Fig. 13.2), the photon polarization can be accessed via a measurement of angle θ_p. However, due to the long lifetime of Λ, the majority of its decays occur outside the VELO which makes it very challenging to reconstruct the $\Lambda_b \rightarrow \Lambda\gamma$ decay.

There are several approaches to reconstruct the signal candidates and access the photon polarization information, but all approaches requires major improvements in the existing trigger strategies. Among others, one such strategy, which could significantly improve the efficacy of this analysis, would be the inclusion of tracks, which are in the downstream of the magnet. The improved trigger available after LHCb upgrade and the larger statistics dataset collected in Runs 3 and 4 will enable us to exploit this channel up to its full potential.

13.4 New Physics Constraints

Considering together all $b \rightarrow s\gamma$ analyses sensitive to $C7$ and $C7'$, the results are compatible at present with the SM predictions, Constraints on C7 and C7′ obtained from the Flavio framework [12] can be seen in Fig. 13.3.

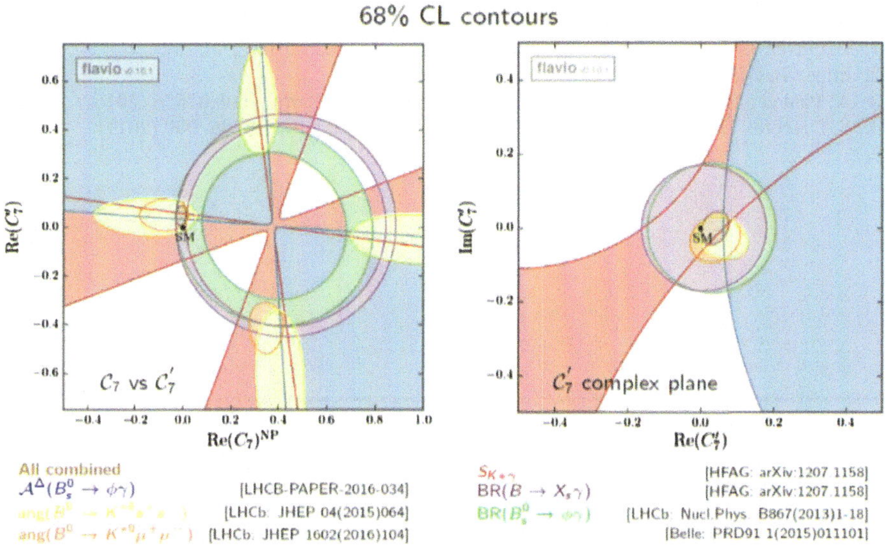

Fig. 13.3 Constraints on C7 and C7' obtained from the Flavio framework [12–14]

13.5 Conclusion

The measurement of photon polarization in rare radiative B decays is an observable that is very sensitive to possible NP contributions and LHCb is performing the measurements using several channels and different observables. These measurements are important to constrain the Wilson coefficient $\mathbf{C_7}'$ and key to disentangle different new physics scenarios. Despite the γ reconstruction in pp collisions being very challenging, LHCb has been able to provide precise results in several decay channels. The new physics constraints have helped to provide more stringent constraints on NP models. It is expected that the results obtained from the LHCb Run 2 dataset will further test the limits of the Standard Model and provide even better new physics constraints.

References

1. D. Atwood, M. Gronau, A. Soni, PRL **79**(97), 185
2. Y. Grossman et al., JHEP **06**, 029 (2000)
3. R. Aaij et al., Phys. Rev. Lett. **118**, 021801 (2017)
4. R. Aaij et al. *PRL **112**, 161801 (2014)
5. M. Gronau, D. Pirjol, PRD **66**(02), 054008
6. LHCb: Int. J. Mod. Phys. **30** 1530022 (2015)
7. LHCb: JINST **3**, S08005 (2008)
8. LHCb: arXiv:1501.03038v2
9. F. Yu, E. Kou, C. Lü, JHEP **12**, 102 (2013)

10. F. Munheim et al., Phys. Lett. B **664**, 174–179 (2008)
11. L. Oliver, J.-C. Raynal, R. Sinha, Phys. Rev. D **82**, 117502 (2010)
12. https://flav-io.github.io/
13. A. Paul et al., Constraints on NP from radiative *B* decays, arXiv:1608.02556 (2016)
14. S. Descotes-Genon et al., Exploring NP in the C7-C7′ plane. JHEP **06**, 099 (2011)

Chapter 14
Spectroscopy of Heavy-Light Flavor B_c Mesons in a Non-relativistic Scheme

Smruti Patel, Tanvi Bhavsar and P. C. Vinodkumar

14.1 Introduction

The recently discovered charm-bottom heavy quarkonium completes the well investigated charmonium and bottomonium families and offers a new perspective in the study of the nonrelativistic dynamics of the strong interactions. The discovery of the Bc meson by the Collider Detector at Fermilab (CDF) Collaboration [1] in $p\bar{p}$ collisions at $\sqrt{s} = 1.8$ TeV has demonstrated the possibility of the experimental study of this system and has stimulated considerable interest in B_c spectroscopy [2–4]. The statistics, however, is expected to increase significantly in future experiments at Tevatron and LHC greatly improving the accuracy of the data. Note that only the pseudoscalar (spin singlet) state has been observed so far while the vector (spin triplet) meson B_c^* is still to be discovered. This distinguishes $c\bar{b}$ quarkonium from the $b\bar{b}$ system, where it is the pseudoscalar η_b meson, which asks for experimental detection. Both in mass and in size, the mesons with beauty and charm are intermediate between the $c\bar{c}$ and $b\bar{b}$ states. Therefore, an analysis of B_c could shed new light on the balance between the perturbative and non-perturbative effects and further check whether a perturbative analysis provides a reliable starting point for them.

14.2 Phenomenology

There are many methods to estimate the mass of a hadron, among which phenomenological potential model is a fairly reliable one, specially for heavy hadrons. For the description of the quarkonium bound states, we adopt the phenomenological potential of the form which is expressed in terms of a vector (Coulomb) plus a scalar (confining) part given by

S. Patel (✉) · T. Bhavsar · P. C. Vinodkumar
Government Science College, Veer Narmad South Gujarat University, Songadh, Gujarat, India
e-mail: fizixsmriti31@gmail.com

A. Giri and R. Mohanta (eds.), *16th Conference on Flavor Physics and CP Violation*, Springer Proceedings in Physics 234,
https://doi.org/10.1007/978-3-030-29622-3_14

$$V(r) = V_V + V_S = \frac{-4\alpha_s}{3r} + \frac{Ar^2}{(1+4Br)^{\frac{1}{2}}} - V_{nS} \tag{14.1}$$

Here, A = 0.55 GeV^3, B = 1.0 GeV and V_{nS} is a state dependent constant potential. Similar type of potential is used by [5, 6] for the study of quarkonia in the light and heavy sector. The parameters α_s and A are same as in [5]. The optimized spectroscopic parameters thus correspond to the fitted quark masses, the potential strength V_{nS} and the corresponding radial wave functions. The quark masses are m_c = 1.317 GeV, m_b = 4.5 GeV while the state dependent constant potential V_{nS} is fixed as 0.12 GeV. And its state dependence is given by the recursion relation,

$$V_{nS} = V_{(n-1)} + 0.1(n+3) \tag{14.2}$$

We now present the details of our calculations by using the potential with $n = 1$ of (14.1). Different degenerate $n^{2S+1}L_J$ low-lying states of $c\bar{b}$ mesons are calculated by including spin dependent part of the usual one gluon exchange potential [3, 7]. The Schrödinger equation with the potential given by (14.1) is numerically solved using the Mathematica notebook of the Runge-Kutta method [8] to obtain the energy eigenvalues and the corresponding wave functions.

14.3 Decay Rates of Heavy Quarkonia

Apart from the masses of the low lying mesonic states, the correct predictions of the decay rates are important features of any successful model. The spectroscopic parameters obtained including the predicted masses and the resultant radial wave functions are being used here to compute the decay constants and decay rates.

In the non-relativistic limit the decay constant is proportional to the wave function at the origin and is given by van Royen-Weisskopf formula [12]

$$f_{V/P} = \sqrt{\frac{12}{M_{V/P}}}\Psi_{V/P}(0). \tag{14.3}$$

Here $\Psi_{V/P}(0)$ is wave function for vector and pseudoscalar state at the origin respectively. And M_V and M_P are the masses of the vector and pseudoscalar meson respectively. We compute the radiative leptonic decay width using the following relation [11, 13]

$$\Gamma(B_c \to \gamma\ell\bar{\nu}) = G_F^2 \frac{|V_{cb}|^2}{8\pi^2} f_{B_c}^2 m_{B_c}^2 \frac{m_l^2}{m_{B_c}^2}\left(1 - \frac{m_l^2}{m_{B_c}^2}\right)^2 \tag{14.4}$$

where G_F is the Fermi coupling constant and $|Vcb|$ is the CKM matrix element [13].

There have been a number of recent studies on processes involving strong decays, radiative decays and leptonic decays of vector mesons. Such studies offer a direct probe of hadron structure and help in revealing some aspects of the underlying quark-gluon dynamics that are complementary to what is learnt from pseudo scalar mesons. It can be a crucial test for potential model. These processes can be calculated with perturbative quantum electrodynamics (QED) with corrections from the strong interaction.

14.4 Results and Discussions

In the present paper, we have studied the spectroscopy of heavy light meson B_c by employing the non-relativistic potential model. We have computed the masses of ground state and radial excitations of B_c meson. The calculated masses are compared with other theoretical models and with available experimental data which are displayed in Table 14.1. Overall we obtain a good t to the spectrum. For B_c mesons, pseudoscalar states are experimentally observed for ground and first radially excited state only and we found our results closer to the present experimental value. Apart from this, we have also computed vector and pseudoscalar decay constants as well as radiative leptonic decay width. The calculated values of the pseudoscalar and vector decay constants of the B_c meson in our model are displayed in Table 14.2. In the absence of precise experimental measurements for radiative leptonic decay width we compared our results with other theoretical model predictions and is in good agreement with the predictions from other theoretical models as shown in the Table 14.3. In conclusion we have successfully studied the properties of the B_c meson within the current theoretical framework. Future experimental studies of the yet unobserved states and determination of their J^P will be significant helpful for the test and validation of the model.

Table 14.1 Mass spectra of B_c meson systems (in GeV)

State	Present	Exp. [9]	[10]	[4]	[11]
1^3S_1	6.339	–	6.333	6.333	6.321
1^1S_0	6.275	6.275	6.264	6.272	6.272
2^3S_1	6.884	–	6.899	6.882	6.900
2^1S_0	6.844	6.842	6.856	6.842	6.864
3^3S_1	7.261	–	7.28	7.258	7.338
3^1S_0	7.228	–	7.244	7.226	7.306
4^3S_1	7.594	–	–	7.609	7.714
4^1S_0	7.565	–	–	7.585	7.684

Table 14.2 Vector decay constant (f_V) and pseudoscalar decay constant (f_P) of B_c meson systems (in GeV)

State	f_V	[11]	[2]	[7]	f_P	[2]	[7]	[11]	[14]
1S	0.489	0.434	0.500	0.460 + 0.060	0.530	0.500	0.460 + 0.060	0.432	0.517
2S	0.457	0.356	–	–	0.495	–	–	0.355	–
3S	0.448	0.326	–	–	0.487	–	–	0.325	–
4S	0.443	0.308	–	–	0.482	–	–	0.307	–

Table 14.3 Radiative leptonic decay width of B_c meson systems (in GeV)

State	Present	[11]	[13]
1S	1.46×10^{-17}	1.59×10^{-17}	6.44×10^{-17}

Acknowledgements The work is part of a research project No. SERB/F/8749/2015-16 funded by DST-SERB, India.

References

1. CDF Collaboration, F. Abe et al., Phys. Rev. Lett. **81**, 2432 (1998), OPAL Collaboration, K. Ackerstaff et al., Phys. Lett. B **420**, 157 (1998)
2. E.J. Eichten, C. Quigg, Phys. Rev. D **49**, 5845 (1994). arXiv:hep-ph/9402210
3. S. Godfrey, N. Isgur, Phys. Rev. D **32**, 189 (1985)
4. D. Ebert, R.N. Faustov, V.O. Galkin, Eur. Phys. J. C **71**, 1825 (2011)
5. S. Patel, P.C. Vinodkumar, S. Bhatnagar, Chin. Phys. C **40**, 102 (2016)
6. S. Bhatnagar, J. Mahecha, Y. Mengesha, Phys. Rev D **90**, 014034 (2014)
7. S.S. Gershtein, V.V. Kiselev, A.K. Likhoded, A.V. Tkabladze, Phys. Rev. D **51**, 3613 (1995)
8. W. Lucha, F. Shoberl, Int. J. Mod. Phys. C **10** (1999). arXiv:hep-ph/9811453
9. C. Patrignani et al., Particle data group. Chin. Phys. C **40**, 100001 (2016)
10. V.O. Galkin, A. Yu, Mishurov, R.N. Faustov, Yad. Fiz. **55**, 2175 (1992)
11. N.R. Soni et al. (2017). arxiv:1707.07144
12. R. Van Royen, V.F. Weisskopf, Nuovo Cimento **50**, 617 (1967)
13. C. Chao-Hsi, J.P. Cheng, C.D. Lu. arXiv:hep-ph/9712325v2
14. L.P. Fulcher, Phys. Rev. D **60**, 074006 (1999). arXiv:hep-ph/9806444

Chapter 15
Search of Rare Annihilation Decay at Belle

Rajeev Kumar

Abstract The decay $B^0 \to J/\psi\gamma$ is investigated using $B\bar{B}$ events collected at the $\Upsilon(4S)$ resonance with the Belle detector at the KEKB asymmetric-energy e^+e^- collider. Event selection is carried out by using standard criteria, and a reconstruction efficiency of $(33.9 \pm 0.2)\%$ is obtained. The background is estimated using large inclusive J/ψ Monte Carlo sample and is mainly dominated by the decays $B^0 \to J/\psi\pi^0$, $B^0 \to J/\psi\eta$ and $B^0 \to J/\psi K_L$. If there is any enhancement in the decay rate of $B^0 \to J/\psi\gamma$, it will be sensitive to charged right-handed currents $(V + A)$ in addition of $(V - A)$, and this admixture will give a significant contribution to new physics.

15.1 Introduction

The decays of B mesons to final states containing charmonium, in general, constitute a sensitive laboratory for the study of electroweak interactions as well as quantum chromodynamics (QCD). A particular approach to predict the branching fractions for these type of decays is to factorize the decays into a short-distance contribution, which can be computed with perturbative QCD, and a long-distance contribution for which non-perturbative approach can be used. The degree to which factorization assumption holds good leads to large theoretical uncertainties. Thereofre experimental measurements are crucial to find out the most appropriate theoretical approach.

Within the framework of Standard Model (SM), the decay $B^0 \to J/\psi\gamma$[1] proceeds through a W boson exchange diagram as shown in Fig. 15.1, where one quark radiates

[1]The charge conjigation is applied throughout.

Rajeev Kumar—Belle Collaboration.

R. Kumar (✉)
Department of Mathematics, Statistics & Physics, Punjab Agri University,
Ludhiana 141004, India
e-mail: rajeevsharma@pau.edu

A. Giri and R. Mohanta (eds.), *16th Conference on Flavor Physics and CP Violation*, Springer Proceedings in Physics 234,
https://doi.org/10.1007/978-3-030-29622-3_15

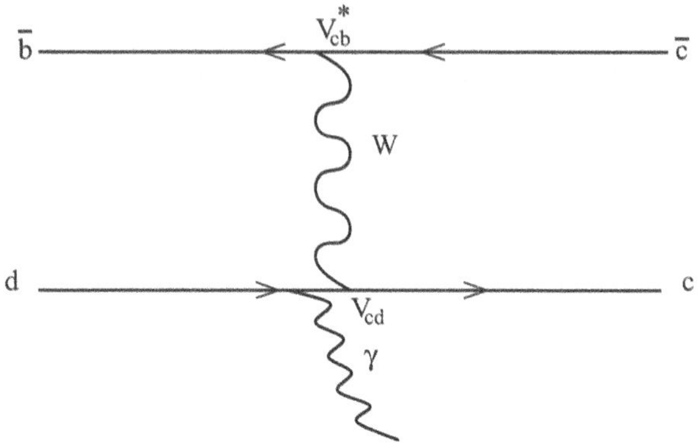

Fig. 15.1 Feynman diagram of the leading order contribution to the decay $B^0 \to J/\psi\gamma$ in the Standard Model. Radiation of the photon from the other quark is suppressed by a factor of Λ_{QCD}/m_b [12]

a photon. This is a very rare decay with a typical branching fraction of 5.398×10^{-8} predicted in naive factrorization [10] and 2.435×10^{-9} in the framework of QCD factorization [12]. This decay is sensitive to New Physics, for example right handed currents [12] or non-spectator charm in the B^0 meson [6]. An upper limit of $1.6(1.5) \times 10^{-6}$ on the branching fraction was set at 90% confidence level by the BaBar (LHCb) collaboration [1, 5].

The analysis reported in these proceedings is based on data collected at the $\Upsilon(4S)$ resonance by the Belle detector, produced by the asymmetric-energy e^+e^- KEKB collider. The details about Belle detector and KEKB accelerator are described in detail else where [3, 8].

15.2 Event Selection

Events with B-meson candidates are first filtered by applying general hadronic event selection criteria. These include a requirement on charged tracks (at least three of them should originate from an event vertex consistent with the interaction point); a requirement on the reconstructed center-of-mass (CM) energy ($E^{CM} > 0.2\sqrt{s}$, where $\sqrt{s}$ is the total CM energy); a requirement on the longitudinal (along beam direction) component of the reconstructed CM momentum ($|p_z^{CM}| < 0.5\sqrt{s}/c$) and the total electromagnetic calorimeter (ECL) energy ($0.1\sqrt{s} < E_{ECL}^{CM} < 0.8\sqrt{s}$) with at least two energy clusters. To suppress $e^+e^- \to q\bar{q}(q = u, d, s, c)$ continuum background, we require that the ratio of the second and zeroth order Fox-Wolfram moments [7] be less than 0.5. To remove charged particle tracks that either are poorly measured or do not come from the interaction region, we require them to originate within 1.5 cm in the radial direction and 5 cm along the beam direction with respect to the interaction point.

15.3 Selection of J/ψ and γ

A correct identification of leptons is vital for the reconstruction of J/ψ. The J/ψ mesons are reconstructed in its decay channel $J/\psi \to \ell^+\ell^-$, where ℓ is a muon or an electron. A charged track is considered to be a muon if its muon likelihood ratio $\mathcal{R}(\mu)$ is greater than 0.1, where $\mathcal{R}(\mu) = \mathcal{L}_\mu/(\mathcal{L}_\mu + \mathcal{L}_\pi + \mathcal{L}_K)$. The likelihoods $\mathcal{L}_\mu$, $\mathcal{L}_\pi$, and $\mathcal{L}_K$ for a track to be identified muon, pion, and kaon, respectively, are calculated using the method explained in [2]. The electrons are identified by combining dE/dx from the drift chamber with, E/p and shower shape in the ECL. The four-momenta of soft photons within 50 mrad of the e^+ or e^- direction are included in the invariant mass calculation of J/ψ to recover the di-electron events that are radiating a photon. The invariant mass windows are -0.07 (-0.14) GeV/c^2 $\leq M_{\ell^+\ell^-} - m_{J/\psi} \leq 0.040$ GeV/c^2 to select J/ψ candidates in the $\mu^+\mu^-(e^+e^-)$ channels where $m_{J/\psi}$ denotes the world average of J/ψ mass [11].

Photon candidates are identified as ECL clusters that are not associated with any charged track. To select good photon candidates, we require that the energy deposited in the 3×3 to 5×5 ECL cells to be greater than 0.95. To suppress photons arising from $\pi^0 \to \gamma\gamma$ and $\eta \to \gamma\gamma$ decays, we require that the probability of a candidate photon being associated with another photon to form a π^0 or an η meson to be less than 0.25.

15.4 B-meson Reconstruction

The B-meson is reconstructed by combining a J/ψ with a selected photon candidate. The energy difference, $\Delta E \equiv E_B^* - E_{\text{beam}}^*$ and the beam-constrained mass $M_{\text{bc}} \equiv \sqrt{E_{\text{beam}}^{*2} - p_B^{*2}}$, are used to separate signal from background, where E_{beam}^* is the run dependent beam energy, and E_B^* and p_B^* are the reconstructed energy and momentum of the B-meson candidates in the CM frame. We retain candidates in the region 5.24 GeV/c^2 $\leq M_{\text{bc}} \leq 5.29$ GeV/c^2 and -0.35 GeV $\leq \Delta E \leq 0.15$ GeV. A correctly reconstructed $B^0 \to J/\psi\gamma$ decay should have $\Delta E = 0$ and M_{bc} equals to mass of the B-meson [11]. Therefore, we demand that $B^0 \to J/\psi\gamma$ candidates be within a signal region in the ΔE versus M_{bc} plane defined by $5.27 < M_{bc} < 5.29$ GeV/c^2 and $-0.2 < \Delta E < 0.1$ GeV. About 0.1 million signal Monte Carlo (MC) events are generated using EvtGen [9] and the response of Belle detector is siumlated with the GEANT4 software package [4]. To determine the reconstruction efficiency, a three-dimensional unbinned maximum likelihood fit is performed to ΔE, M_{bc} and $\cos(\theta)$ of J/ψ as shown in Fig. 15.2 and obtained efficiency is $33.9 \pm 0.2\%$. The signal ΔE shape is modeled by a Crystal Ball function, M_{bc} by a sum of two Gaussian functions and the helicity angle of J/ψ by $\mathrm{a} - \mathrm{b}\sin^2(\theta)$. Various sources of background in this analysis include contributions from other B decays and from continuum production of $e^+e^- \to q\bar{q}$, where $q = u, d, s, c$.

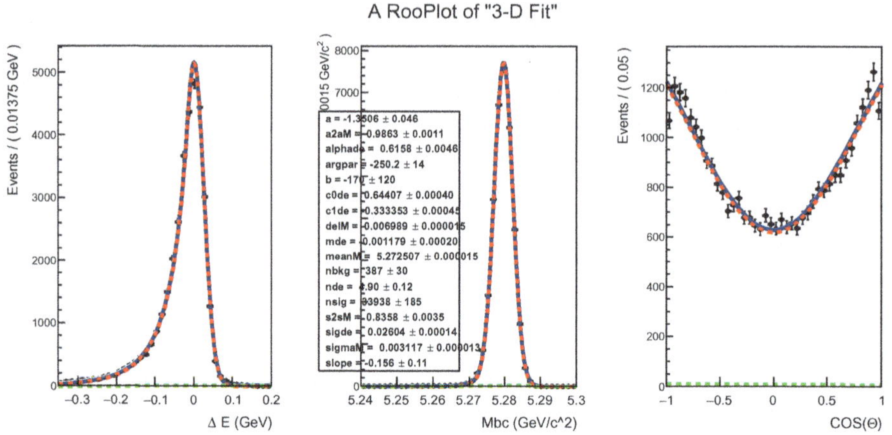

Fig. 15.2 Fitted ΔE, $M_{\rm bc}$ and helicity angle ($\cos\theta$) of J/ψ from signal MC sample

The simulated background sample contains both of these backgrounds. The studies of on-resonance simulated continuum background and off-resonance real data indicate that the fraction of continuum events satisfying our selection criteria is negligible. The background from hadronic decays of B mesons is estimated using a large MC sample having size 100 times that of data. The small fraction of this background in the signal region is primarily due to $B^0 \to J/\psi\pi^0(\eta)$ decays in which a photon from $\pi^0(\eta) \to \gamma\gamma$ is misidentified as the signal photon of $B^0 \to J/\psi\gamma$. Also, the background from $B^0 \to J/\psi K_L^0$ decays has a peaking structure in the signal ΔE region due to $K_L^0 \to 3\pi^0$ decays in the electromagnetic calorimeter for which the resulting showers overlap and are incorrectly interpreted as a shower from a single photon. A three-dimensional fit is performed to model the shape of different background and to calculate the contribution of each component in the signal region as shown in Fig. 15.3. The estimated background yields are 52 events for $B^0 \to J/\psi\pi^0(\eta)$ and 36 events for $B^0 \to J/\psi K_L^0(K_S^0)$ mode, resulting in a total background yield of 88 ± 11 events. The remaining components have a flat structure in the signal region. The next step of this analysis is to check the stability of fitter using toymc and gsim test and then to look for the signal in real data sample. On summarizing, the Belle has started the analysis of decay $B^0 \to J/\psi\gamma$ using 711 fb^{-1} data sample. Using this large data sample, we are expecting a first evidence or better upper limit from the previous measurements by BaBar and LHCb collaborations.

We are highly greatful to the KEKB group for excellent operation of the accelerator, the KEK cryogenics group for efficient solenoid operations, and the KEK computer group and the NII for valuable computing and SINET3 network support. We acknowledge support from MEXT and JSPS (Japan); ARC and DEST (Australia); NSFC (China); DST (India); MOEHRD, KOSEF, and KRF (Korea); KBN (Poland); MES and RFAAE (Russia); ARRS (Slovenia); SNSF (Switzerland); NSC and MOE (Taiwan); and DOE (USA).

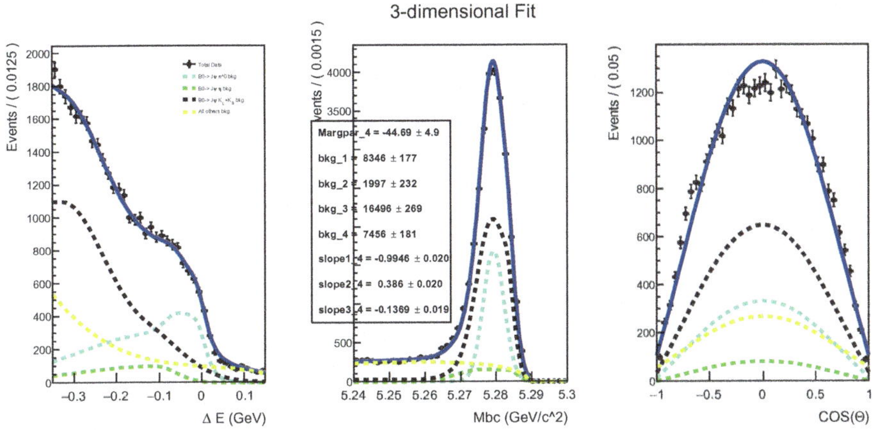

Fig. 15.3 The fitted ΔE, M_{bc} and helicity angle ($\cos\theta$) from inclusive J/ψ MC sample

References

1. R. Aaij, C. Beteta et al., Search for rare decays $b^0 \rightarrow j/\psi\gamma$ and $b_s^0 \rightarrow j/\psi\gamma$. Phys. Rev. D **92**, 112,002 (2015. https://doi.org/10.1103/PhysRevD.92.112002
2. A. Abashian, K. Abe et al., Muon identification in the belle experiment at kekb. Nucl. Instrum. Meth. A **491**, 69–82 (2002)
3. A. Abashian et al., The belle detector. Nucl. Instrum. Meth. A **479**, 117–132 (2002). https://doi.org/10.1016/S0168-9002(01)02013-7
4. S. Agostinelli, J. Allison et al., Geant4-a simulation toolkit. Nucl. Instrum. Meth. A **506**, 250–303 (2003). https://doi.org/10.1016/S0168-9002(03)01368-8
5. B. Aubert, R. Barate et al., Search for the decay $b^0 \rightarrow j/\psi\gamma$. Phys. Rev. D **70**, 091,104 (2004). https://doi.org/10.1103/PhysRevD.70.091104
6. S. Brodsky, S. Gardner, Evading the CKM hierarchy: intrinsic charm in b decays. Phys. Rev. D **65**, 054,016 (2002). https://doi.org/10.1103/PhysRevD.65.054016
7. G. Fox, S. Wolfram, Observable for the analysis of event shapes in e^+e^- annihilation and other processes. Phys. Rev. Lett. **41**, 1581 (1978). https://doi.org/10.1103/PhysRevLett.41.1581
8. S. Kurokawa, E. Kikutani, Overview of the kekb accelerators. Nucl. Instrum. Meth. A **499**, 1–7 (2003). https://doi.org/10.1016/S0168-9002(02)01771-0
9. D. Lange, The evtgen particle decay simulation package. Nucl. Instrum. Meth. A **462**, 152–155 (2001). https://doi.org/10.1016/S0168-9002(01)00089-4
10. M. Neubert, B. Stech, Non-leptonic weak decays of b mesons, in *Heavy Flavours II*, vol. 15, ed. by A. Buras, M. Linder (World Scientific Publishing, Singapore, 1988), pp. 294–344
11. C. Partignani et al., The review of particle physics. Chin. Phys. C **40**, 100,001 (2016). https://doi.org/10.1088/1674-1137/40/10/100001
12. Y. Yang, L. Gongru, R. Wang, The rare annihilation decays $\bar{b}_{s,d}^0 \rightarrow j/\psi\gamma$. Eur. Phys. J. C **34**, 291–296 (2004). https://doi.org/10.1140/epjc/s2004-01711-y

Chapter 16
Decay $D_s^+ \to K^{(*)0}\ell^+\nu_\ell$ in Covariant Quark Model

N. R. Soni and J. N. Pandya

Abstract Semileptonic decay widths of D_s^+ mesons ($D_s^+ \to K^{(*)0}\ell^+\nu_\ell$) are presented. The required transition form factors are computed in the entire physical range of momentum transfer in the framework of covariant quark model (CQM). We compute the branching fractions incorporating the form factors and obtain the ratio of the partial decay width $\Gamma(D_s^+ \to K^0\mu^+\nu_\mu)/\Gamma(D_s^+ \to K^0e^+\nu_e) = 0.98$ which is in close resemblance with isospin symmetry.

16.1 Introduction

The semileptonic decays of charmed mesons provide the best tool in exploring the key aspects of heavy quark decays and heavy-light meson spectroscopy. The semileptonic decays are important to study because they provide insight into the quark flavor mixing and hence the elements of Cabibbo Kobayashi Maskawa (CKM) matrix.

In this paper, we compute the semileptonic branching fractions of charmed mesons in the frame work of covariant quark model (CQM) developed by Efimov and Ivanov [1–4]. The CQM is the effective field theoretical approach for the hadronic interactions with their constituents. The exclusive semileptonic branching fraction of D_s^+ mesons are reported in [5] using the CLEO-c data from e^+e^- annihilations at Cornell Electron Storage Ring. These decays are also studied in light front quark model (LFQM) [6] and heavy meson chiral theory (HMχT) [7].

N. R. Soni · J. N. Pandya (✉)
Applied Physics Department, Faculty of Technology and Engineering,
The Maharaja Sayajirao University of Baroda, Vadodara 390001, Gujarat, India
e-mail: jnpandya-apphy@msubaroda.ac.in

N. R. Soni
e-mail: nrsoni-apphy@msubaroda.ac.in

A. Giri and R. Mohanta (eds.), *16th Conference on Flavor Physics and CP Violation*, Springer Proceedings in Physics 234,
https://doi.org/10.1007/978-3-030-29622-3_16

16.2 Form Factors and Branching Fractions

For the channel $D_s^+ \to K^{(*)0}\ell^+\nu_\ell$, the matrix element can be written as

$$\mathcal{M}(D_s^+ \to K^{(*)0}\ell^+\nu_\ell) = \frac{G_F}{\sqrt{2}} V_{cd} \langle K^{(*)0}|\bar{s}O^\mu c|D_s^+\rangle \ell^+ O^\mu \nu_\ell \tag{16.1}$$

where G_F and O^μ is the fermi coupling constant and weak Dirac matrix respectively. The invariant form factors for D_s^+ decays into K^0 and $K^*(892)^0$ can be written as

$$\mathcal{M}^\mu_{D_s^+ \to K^0} = P^\mu F_+(q^2) + q^\mu F_-(q^2) \tag{16.2}$$

and

$$\begin{aligned}\mathcal{M}^\mu_{D_s^+ \to K^{*0}} = \frac{1}{m_1 + m_2} \varepsilon^\dagger_\nu \{ -g^{\mu\nu} Pq A_0(q^2) + P^\mu P^\nu A_+(q^2) \\ + q^\mu P^\nu A_-(q^2) + i\epsilon^{\mu\nu\alpha\beta} P_\alpha q_\beta V(q^2)\}.\end{aligned} \tag{16.3}$$

where $P = p_1 + p_2$ and $q = p_1 - p_2$ with p_1, p_2 to be the momenta of D_s^+ and $K^{(*)0}$ mesons respectively. The form factors $F_\pm$, V, A_0, $A_\pm$ are computed in the framework of CQM in the whole physical range of momentum transfer. The invariant form factors are expressed in terms of helicity form factors in the following way:

For $D_s^+ \to K^0$ channel:

$$\begin{aligned} H_t &= \frac{1}{\sqrt{q^2}}(PqF_+ + q^2F_-), \\ H_\pm &= 0 \quad \text{and} \quad H_0 = \frac{2m_1|\mathbf{p_2}|}{\sqrt{q^2}} F_+ \end{aligned} \tag{16.4}$$

For $D_s^+ \to K^{(*)0}$ channel:

$$\begin{aligned} H_t &= \frac{1}{m_1+m_2}\frac{m_1|\mathbf{p_2}|}{m_2\sqrt{q^2}}\left((m_1^2 - m_2^2)(A_+ - A_-) + q^2A_-\right) \\ H_\pm &= \frac{1}{m_1+m_2}(-(m_1^2 - m_2^2)A_0 \pm 2m_1|\mathbf{p_2}|V) \\ H_0 &= \frac{1}{m_1+m_2}\frac{1}{2m_2\sqrt{q^2}}(-(m_1^2 - m_2^2)(m_1^2 - m_2^2 - q^2)A_0 + 4m_1^2|\mathbf{p_2}|^2A_+). \end{aligned} \tag{16.5}$$

where $|\mathbf{p_2}| = \lambda^{1/2}(m^2_{D_s^+}, m^2_{K^{(*)0}}, q^2)/2m_{D_s^+}$ is the momentum of the $K^{(*)0}$ meson in the rest frame of D_s^+ meson. Also $m_1 = m_{D_s^+}$ and $m_2 = m_{K^{(*)0}}$. The form factors are used in computing the semileptonic branching fractions. The relations reads [8]

$$\frac{d\Gamma(D_s^+ \to K^{(0)*}\ell^+\nu_\ell)}{dq^2} = \frac{G_F^2|V_{cd}|^2|\mathbf{p_2}|q^2v^2}{12(2\pi)^3m_{D_s^+}^2}((1+\delta_\ell)(\mathcal{H}_U+\mathcal{H}_L)+3\delta_\ell\mathcal{H}_S) \tag{16.6}$$

where δ_ℓ is the helicity flip factor, v is the velocity type parameter and $\mathcal{H}$'s are the bilinear combination of the helicity structure functions. The detailed description of the helicity amplitudes are given in [9–11].

16.3 Results

The necessary numerical computational techniques and the model parameters are given in the recent paper [12, 13] and references therein. The other model independent parameters such as meson masses, life-time of D_s^+ meson, Fermi coupling constant and CKM matrices are taken from PDG [14]. The form factors (16.2) and (16.3) are computed in the entire physical range of momentum transfer $0 \le q^2 \le q^2_{max} = (m^2_{D_s^+} - m^2_{K^{(*)0}})$ using the *Mathematica* and shown in the Fig. 16.1 as a function of square of momentum transfer.

The semileptonic differential branching are computed using (16.6) and presented in Fig. 16.2. The branching fractions are computed by integrating the differential branching fractions numerically. We obtain the following results:

$$\mathcal{B}(D_s^+ \to K^0e^+\nu_e) = 0.21\%, \quad \mathcal{B}(D_s^+ \to K^0\mu^+\nu_\mu) = 0.20\%,$$
$$\mathcal{B}(D_s^+ \to K^*(892)^0e^+\nu_e) = 0.19\%, \quad \mathcal{B}(D_s^+ \to K^*(892)^0\mu^+\nu_\mu) = 0.18\%$$

The experimental results are only available for the electron channel from the CLEO-c data [5]. Our results are in accordance with the CLEO-c data and also match with the results from LFQM data [6]. We also obtain the ratio $\Gamma(D_s^+ \to K^0\mu^+\nu_\mu)/\Gamma(D_s^+ \to K^0e^+\nu_e) = 0.98$ confirming the isospin symmetry of the kaons. The deviation from

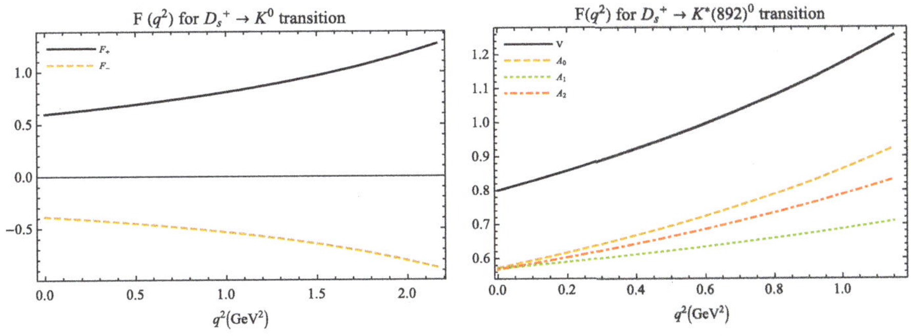

Fig. 16.1 Form factors

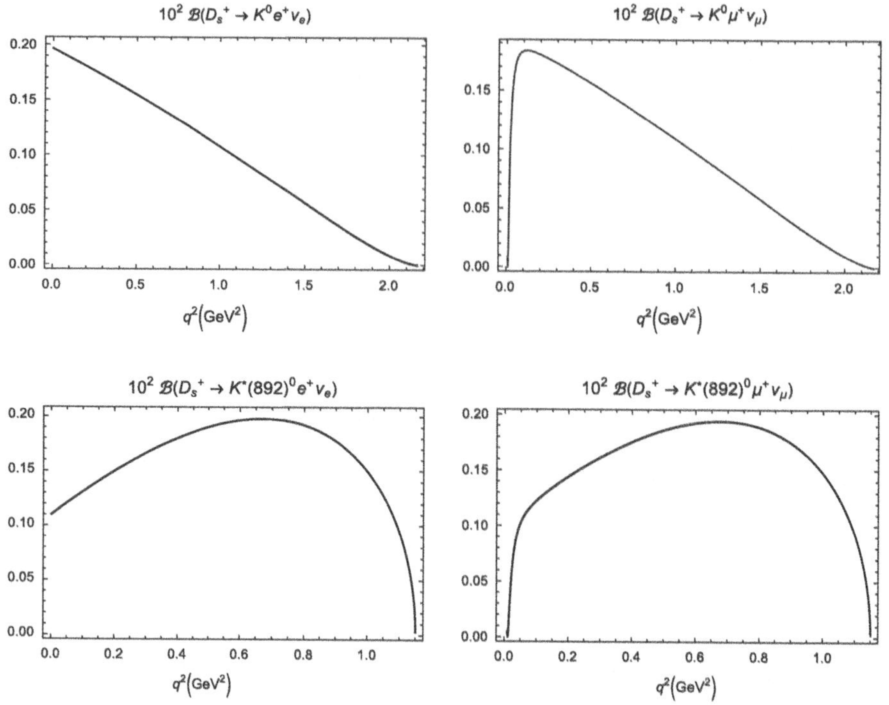

Fig. 16.2 Branching fractions

the Standard Model might be due to inherent uncertainties in the numerical computations. The muonic channels are yet to be confirmed from the experiments.

Acknowledgements JNP acknowledges the financial support from the University Grants Commission of India under Major Research Project F.No.42-775/2013(SR).

References

1. G.V. Efimov, M.A. Ivanov, Int. J. Mod. Phys. A **4**, 2031 (1989)
2. G.V. Efimov, M.A. Ivanov, *The Quark Confinement Model of Hadrons* (IOP, Bristol, 1993)
3. A. Faessler, T. Gutsche, M.A. Ivanov, J.G. Körner, V.E. Lyubovitskij, Eur. Phys. J. Direct C **4**, 1 (2002)
4. T. Branz, A. Faessler, T. Gutsche, M.A. Ivanov, J.G. Körner, V.E. Lyubovitskij, Phys. Rev. D **81**, 034010 (2010)
5. J. Hietala, D. Cronin-Hennessy, T. Pedlar, I. Ship, Phys. Rev. D **92**, 012009 (2015)
6. H.-Y. Cheng, X.-W. Kang, Eur. Phys. J. C **77**, 587 (2017)
7. S. Fajfer, J.F. Kamenik, Phys. Rev. D **72**, 034029 (2005)
8. M.A. Ivanov, J.G. Körner, C.T. Tran, Phys. Rev. D **94**, 094028 (2016)
9. M.A. Ivanov, J.G. Körner, C.T. Tran, Phys. Rev. D **92**, 114022 (2015)

10. T. Gutsche, M.A. Ivanov, J.G.Körner, V.E. Lyubovitskij, P. Santorelli, N. Habyl, Phys. Rev. D **91**, 074001 (2015) [Erratum: Phys. Rev. D **91**, 119907 (2015)]
11. M.A. Ivanov, J.G. Körner, C.T. Tran, Phys. Rev. D **95**, 036021 (2017)
12. N.R. Soni, J.N. Pandya, Phys. Rev. D **96**, 016017 (2017)
13. N.R. Soni, M.A. Ivnov, J.G. Korner, J.N. Pandya, P. Santorelli, C.T. Tran. Phys. Rev. D **98**, 114031 (2018)
14. M. Tanabashi et al., (Particle Data Group). Phys. Rev. D **98**, 030001 (2018)

Chapter 17
Lepton Mass Effects and Angular Observables in $\Lambda_b \to \Lambda(\to P\pi)\ell^+\ell^-$

Ria Sain

Abstract The flavor changing rare decay $B \to K^*(\to K\pi)\ell^+\ell^-$ is one of the most studied modes due to its sensitivity to physics beyond the standard model and several discrepancies have come to light among the plethora of observables that are measured. In this work we revisit the analogous baryonic decay mode $\Lambda_b \to \Lambda(\to p\pi)\ell^+\ell^-$ and we present a complete set of 10 angular observables that can be measured using this decay mode. Our calculations are done retaining the finite lepton mass so that the signal of lepton non-universality observed in $B \to K^*\ell^+\ell^-$ can be corroborated by the corresponding baryonic decay mode. We show that due to the parity violating nature of the subsequent $\Lambda \to p\pi$ decay there exists at least one angular asymmetry that is non-vanishing in the large recoil limit unlike the case in $B \to K^*\ell^+\ell^-$ decay mode, making it particularly sensitive to new physics that violates lepton flavor universality.

17.1 Introduction

Recent experimental outcomes from LHCb show some significant discrepancies in several observables from semileptonic $b \to s$ decay channels. For $B \to K^{(*)}\ell^+\ell^-$ decay, the measurements of R_K shows 2.6σ shift from Standard Model (SM) for $q^2 \in [1:6]\,\text{GeV}^2$ and in R_{K^*}, it shows 2.1σ and 2.4σ for $q^2 \in [0.045:1.1]\,\text{GeV}^2$ and $[1.1:6]\,\text{GeV}^2$ respectively [1–3]. The $\Lambda_b \to \Lambda(\to p\pi)\ell^+\ell^-$ shares the same semileptonic quark level process. So, it is necessary to have a detailed study for this decay substantiate the idea of lepton non-universality in FCNC processes. It is

R. Sain (✉)
The Institute of Mathematical Sciences, Taramani, Chennai 600113, India
e-mail: riasain@imsc.res.in

Homi Bhabha National Institute Training School Complex,
Anushakti Nagar, Mumbai 400085, India

A. Giri and R. Mohanta (eds.), *16th Conference on Flavor Physics and CP Violation*, Springer Proceedings in Physics 234,
https://doi.org/10.1007/978-3-030-29622-3_17

obvious that an observable R_Λ can be proposed in the same spirit as $R_{K^{(*)}}$ for the corresponding baryonic decay $\Lambda_b \to \Lambda\ell^+\ell^-$ as,

$$R_\Lambda = \frac{\displaystyle\int_{q^2_{\min}}^{q^2_{\max}} \frac{\mathrm{d}\Gamma(\Lambda_b \to \Lambda\mu^+\mu^-)}{\mathrm{d}q^2}\mathrm{d}q^2}{\displaystyle\int_{q^2_{\min}}^{q^2_{\max}} \frac{\mathrm{d}\Gamma(\Lambda_b \to \Lambda e^+e^-)}{\mathrm{d}q^2}\mathrm{d}q^2} \tag{17.1}$$

One should expect the lepton mass effect to play a significant role on R_Λ in the low-q^2 region, just as $R_{K^{(*)}}$. The discrepancy between the SM expectation and the experimentally observed value of $R_{K^{(*)}}$ was largest in the low-q^2 region. In order to disentangle the new physics contribution that may manifest as lepton non-universality, one must take into account the effect of finite leptons mass in SM calculation of R_Λ. We, therefore, derive the expression for R_Λ without any approximation. Also, we show the mass dependent expressions for the angular asymmetries and their ratios. We show that due to the parity violating nature of the $\Lambda \to p\pi$ decay, hadronic angular asymmetries are non-vanishing in the large recoil or low-q^2 limit unlike the case in $B \to K^*(\to K\pi)\ell^+\ell^-$. This follows since ratios of hadronic forward-backward asymmetry for $\ell = \mu$ and $\ell = e$, in the low-q^2 region, is a ratio of two finite quantities for $\Lambda_b \to \Lambda(\to p^+\pi^-)\ell^+\ell^-$ decay. This observable for the $\Lambda_b \to \Lambda(\to p\pi^-)\ell^+\ell^-$ decay makes it particularly sensitive to new physics that violates lepton flavor universality.

17.2 Full Angular Distribution for Decay of $\Lambda_b \to \Lambda(\to p^+\pi^-)\ell^+\ell^-$

The full angular distribution for $\Lambda_b \to \Lambda(\to p\pi^-)(\to \ell^+\ell^-)$ [4–6] can be described fully by four independent kinematic variables: the dilepton invariant mass squared q^2, the polar angles θ_l, θ_Λ and the azimuthal angle ϕ defined by the decay products in their respective centre-of-mass (CM) frames. The angular co-efficients $I_1 \cdots I_{10}$ are derived in terms of the transversality amplitudes and are presented in details in Table II in [7].

We have derived the angular distribution without any approximations. In particular, we retain the finite lepton mass effects and the two time like amplitudes. These contributions play a significant role in estimating accurately the size of lepton non-universality that may show up in the mode within SM. Among the 10 angular coefficients three forward-backward observables A^l_{FB}, A^h_{FB} and A^{hl}_{FB} are the most important. We have plotted these in Fig. 17.1. We construct the ratios of A^{μ}_{FB}, $A^h_{FB,\mu}$, $A^{h\mu}_{FB}$ to the corresponding quantities for the electron i.e. A^e_{FB}, $A^h_{FB,e}$, A^{he}_{FB} which are $R_{A^l_{FB}}$,$R_{A^h_{FB}}$,$R_{A^{hl}_{FB}}$ similar to R_Λ and the plots are shown in Fig. 17.2.

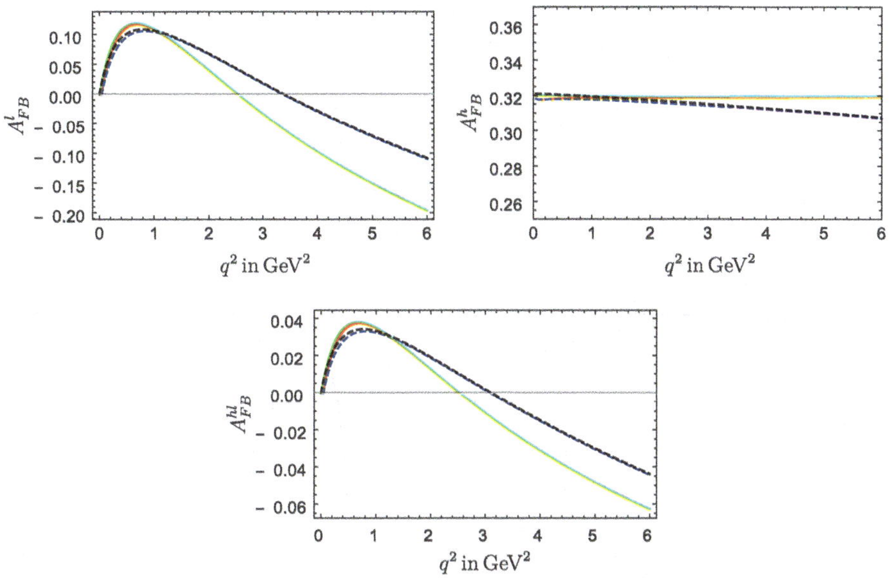

Fig. 17.1 The q^2-dependence of the forward-backward asymmetries A^l_{FB}, A^h_{FB}, A^{hl}_{FB} for electron and muon are presented. For muons, the q^2-dependence of the asymmetries are given by the solid red line, obtained using covariant quark model form-factors and the blue dotted line is for SCET form factors. For electrons, the solid green line represents the q^2-dependence using covariant quark model form factors whereas the black dotted line represents the q^2-dependence due to SCET form factors

Table 17.1 The binned values of the observables R_Λ, $R_{A^l_{FB}}$, $R_{A^h_{FB}}$. Only $R_{A^h_{FB}}$ for $q^2 \sim$ 1–6 GeV^2 does not show a Gaussian behavior as it is peaked towards unity

Binned ratios	Bin 1 $q^2 \sim$ 0.045–1 GeV^2	Bin 2 $q^2 \sim$ 1–6 GeV^2
R_Λ	0.907 ± 0.003	0.9885 ± 0.0002
$R_{A^l_{FB}}$	0.9469 ± 0.0007	0.998 ± 0.196
$R_{A^h_{FB}}$	0.993 ± 0.001	>0.9973 (0.999% C.L.)

17.3 Numerical Estimation

We numerically estimate R_Λ, and the ratios of the asymmetries $R_{A^l_{FB}}$, $R_{A^h_{FB}}$ (Table 17.1). In order to test the dependence of these observables on different form factor parametrizations ,we have used both the covariant quark model [4] and soft-collinear effective theory (SCET) [9–11] to calculate all the observables. We find that all above mentioned ratios are remarkably insensitive to the choice of factors as can be seen from Figs. 17.1 and 17.2.

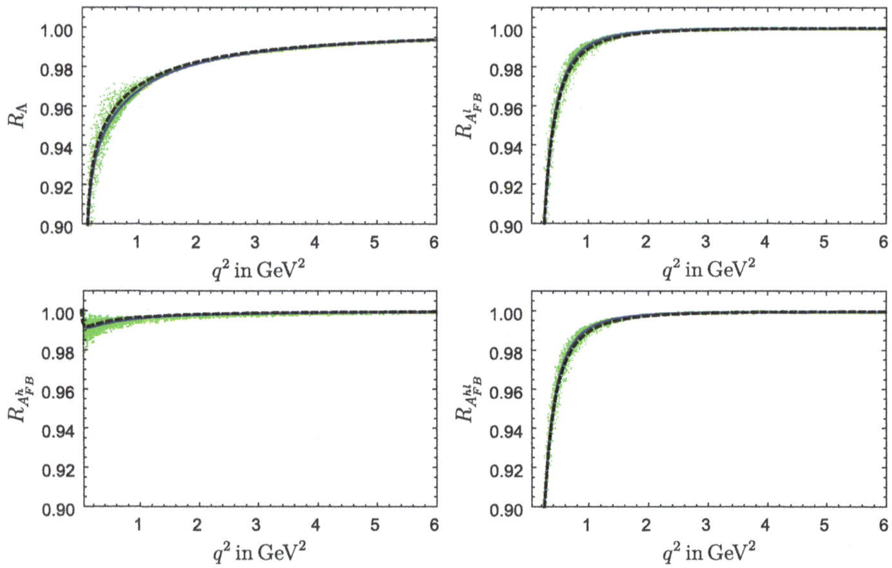

Fig. 17.2 The q^2 dependence for R_Λ and the ratios of forward-backward asymmetries $R_{A^l_{FB}}$, $R_{A^h_{FB}}$ and $R_{A^{hl}_{FB}}$. The blue solid line is found using the values of form factors in covariant quark model given in. This line almost coincides with the estimate using SCET form factors, represented as the black dotted line. Note that, while the two asymmetries $A^h_{FB,\mu}$ and $A^h_{FB,e}$ are form factor dependent the ratio $R_{A^h_{FB}}$ is independent of the choice of form-factors (CQM or SCET). The green band represents the possible values of the ratios obtained by randomly generating 10^4 points corresponding to $\pm 30\%$ error in each of the covariant quark model form factor estimates.

Acknowledgements We would like to thank Prof. Rahul Sinha and Shibasis Roy for collaborating in the paper [7] and for giving valuable suggestions for the presentation.

References

1. R. Aaij et al., LHCb collaboration. JHEP **1708**, 055 (2017)
2. S. Bifani (on behalf of the LHCb Collab.), CERN seminar on April 18, 2017
3. R. Aaij et al., [LHCb Collab.], Phys. Rev. Lett. **113**, 151601 (2014)
4. T. Gutsche, M.A. Ivanov, J.G. Körner, V.E. Lyubovitskij, P. Santorelli, Phys. Rev. D **87**, 074031 (2013). arXiv:1301.3737 [hep-ph]
5. P. Böer, T. Feldmann, D. van Dyk, JHEP **1501**, 155 (2015). arXiv:1410.2115 [hep-ph]
6. T. Blake, M. Kreps, JHEP **1711**, 138 (2017). https://doi.org/10.1007/JHEP11(2017). arXiv:1710.00746 [hep-ph]
7. S. Roy, R. Sain and R. Sinha, Phys. Rev. D **96**, 11, 116005 (2017). https://doi.org/10.1103/PhysRevD.96.116005 arXiv:1710.01335 [hep-ph]
8. Y.M. Wang, Y.L. Shen, C.D. Lu, Phys. Rev. D **80**, 074012 (2009). arXiv:0907.4008 [hep-ph]

9. W. Wang, Phys. Lett. B **708**, 119 (2012). arXiv:1112.0237 [hep-ph]
10. T. Feldmann and M. W. Y. Yip, Phys. Rev. D **85**, 014035 (2012) Erratum. Phys. Rev. D **86**, 079901 (2012). arXiv:1111.1844 [hep-ph]
11. T. Mannel, Y.M. Wang, JHEP **1112**, 067 (2011). arXiv:1111.1849 [hep-ph]

Chapter 18
Simulation Study of $B \to K\ell\ell(')$ at Belle

S. Choudhury, S. Sandilya, K. Trabelsi and A. Giri

Abstract We report the simulation study of R_K for B^+B^- and $B^0\bar{B}^0$ at Belle. The R_K (J/ψ) is measured for Belle full data sample 772×10^6 $B\bar{B}$ pairs and is consistant with unity. The R_K measurement for $B \to K\ell\ell$ is an important probe to search for New Physics (NP). In lot of NP models LFU violation comes together with LFV. We are also searching for LFV in $B \to K\mu e$ and $B \to Ke\mu$.

18.1 Introduction

The FCNC (Flavor Changing Neutral Current) $B \to K\mu\mu$ and $B \to Kee$ involve the quark transition from $b \to s$ and are forbidded in tree level in SM. These type of reactions have penguin and box contributions. Observable R_K, which is the ratio of B.R from $B \to K\mu\mu$ to $B \to Kee$ is LFU ratio. This was measured for LHCb [1] and it shows a devation of 2.6 σ for a bin of $1 < q^2 < 6$ GeV^2/c^4, which is the most precise measurement till date. Belle [2] also has measured R_K for whole q^2 region for a data sample of 657×10^6 $B\bar{B}$ pair and result is consistant with unity.

18.2 Belle Detector

The Belle detector is a large-solid-angle magnetic spectrometer that consists of a silicon vertex detector (SVD), a 50-layer central drift chamber (CDC), an array of aerogel threshold Cherenkov counters (ACC), a barrel-like arrangement of time-of-

S. Choudhury (✉) · A. Giri
IIT Hyderabad, Hyderabad, India
e-mail: ph16resch11007@iith.ac.in

K. Trabelsi
KEK, Tsukuba, Japan

S. Sandilya
University of Cincinnati, Cincinnati, Ohio, USA

A. Giri and R. Mohanta (eds.), *16th Conference on Flavor Physics and CP Violation*, Springer Proceedings in Physics 234,
https://doi.org/10.1007/978-3-030-29622-3_18

flight scintillation counters (TOF), and an electromagnetic calorimeter comprised of CsI(TI) crystals (ECL). All the sub-detectors are located inside a superconducting solenoid coil that provides a 1.5 T magnetic field. An iron flux-return placed outside of the coil is instrumented to detect K_L^0 mesons and muons. The z-axis is alligned with the direction opposite the e^+ beam. The detector is described in detail elsewhere [3].

18.3 MC Simulation

The selection is optimized with MC simulation sample. For, signal MC we have generated 1 million signal events using BTOSLLBALL decay model [4] for LFU modes and PHSP model for LFV modes with the package EvtGen [5] and the detector simulation is done with GEANT3 [6]. The generic MC which have same luminosity as that of Belle data sample contain charged, mixed and charm and uds samples. The charm and uds combinedly called as continuum MC and charged-mixed as $B\bar{B}$. The inclusive J/ψ sample which has 100 times more luminosity than Belle data sample is used to investigate peaking background from J/ψ. The rare MC, which has 50 times more data contain real signal events. We have scaled these luminosity to Belle luminosity for our use.

18.4 Event Selection

We reconstruct $B \to K\ell\ell(')$ by combining kaon (charged or neutral) with two oppositely signed leptons. Here, ℓ can be either electron or muon. The impact parameter cut on the charged particle tracks are applied, along the z-axis $|dz| < 4$ cm and in radial direction $|dr| < 1$ cm. Kaon are selected from likelihood ratio $L = \frac{L_K}{L_K + L_\pi}$ where L_K and L_π are relative likelihoods of kaon and pion respectively. For our selection $L_{K/\pi} > 0.6$ which corresponds to efficiency of >92% and fake rate of <10%. Similarly, electron and muon are selected with $L_e > 0.9$ and $L_\mu > 0.9$ and this correspond to efficiency of >92% (90%) and fake rate of <0.3% (<1.4%) for electron and muon respectively. The energy emitted by high energy electrons are recovered by cone of 50 m rad along the initial momentum direction of the track. K_S^0 are selected using Belle specific package nisKsFinder. The kinematic variables which differentiate signal from background are beam energy constraint mass (M_{bc}) and energy difference (ΔE).

$M_{bc} = \sqrt{(E_{\text{beam}}/c^2)^2 - p_{\text{B}}^2}$

$\Delta E = E_{\text{B}} - E_{\text{beam}}$

where, E_{B} and p_{B} are energy and momentum of B candidate respectively. E_{beam} is the beam energy. Events are selected within the range of $5.2 > M_{bc} < 5.29$ GeV/c^2 and $-0.13 < \Delta E < 0.25$ GeV. K_S, $K^\pm$, $\mu^\pm$ and $e^\pm$ are combined to form B candidate.

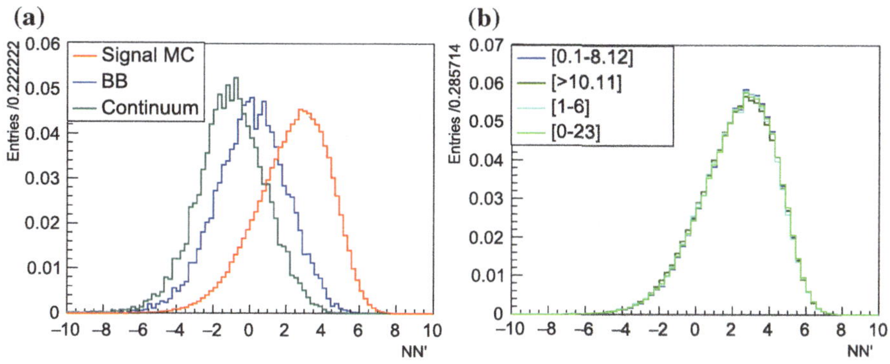

Fig. 18.1 **a** Translated $\mathcal{O}_{\mathcal{NN}}$ distribution. Red color represent signal MC, deep green and blue color represents continuum and $B\bar{B}$ backgrounds respectively. **b** The $\mathcal{O}_{\mathcal{NN}'}$ shape for different q^2 region

18.5 Background Rejection

After reconstructing the B candidate, we found out that the two main sources of background are continuum events ($e^+e^- \rightarrow q\bar{q}$ and $B\bar{B}$). The $B\bar{B}$ background can be separated into 3 important parts.
(a) Same side semileptonic decay.
(b) Both side semileptonic decay.
(c) Fake leptons.
We have discover some event shape variables and vertex quality variables which effectively separate signal from background. B We have performed NN [7] with equal number of signal and background events, where the background events contain continuum as well as $B\bar{B}$ events and they are scaled according to their luminosity. The NN output ($\mathcal{O}_{\mathcal{NN}}$) we obtain is translated to $\mathcal{O}_{\mathcal{NN}'}$ using a log function, which is shown below:
$\mathcal{O}_{\mathcal{NN}'} = \log(\frac{\mathcal{O}_{\mathcal{NN}} - \mathcal{O}_{\mathcal{NN}_{\min}}}{\mathcal{O}_{\mathcal{NN}_{\max}} - \mathcal{O}_{\mathcal{NN}}})$
Here, $\mathcal{O}_{\mathcal{NN}_{\min}}$ is the minimum NN cut and is choosen to be -0.6. $\mathcal{O}_{\mathcal{NN}_{\max}}$ is the maximum NN cut and is mostly found from signal MC. This minimum cut reduce around 75% background with a 5–6% signal efficiency loss. The $\mathcal{O}_{\mathcal{NN}'}$ has similar distribution for different q^2 region in except for the first bin $q^2 < 1$ GeV^2/c^4. The translated NN and NN$'$ distribution in different q^2 region is shown in Fig. 18.1.

18.6 Signal Yield Extraction

We perform a 3D fit of M_{bc}, ΔE and $\mathcal{O}_{\mathcal{NN}'}$. The range are $5.2 < M_{bc} < 5.29$ GeV/c^4, $-0.15 < \Delta E < 0.25$ GeV and $-10 < \mathcal{O}_{\mathcal{NN}'} < 8$. There are peaking backgrounds for the modes containing $\mu\mu$ and these backgrounds are vetoed by applying invariant

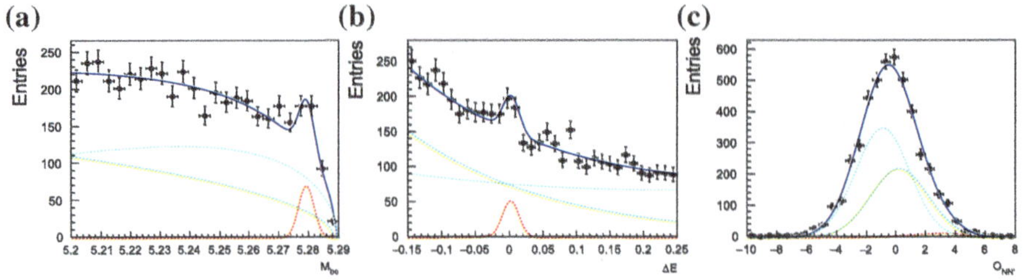

Fig. 18.2 3D fit results for the whole q^2 region in case of $B^+ \to K^+\mu^+\mu^-$ mode

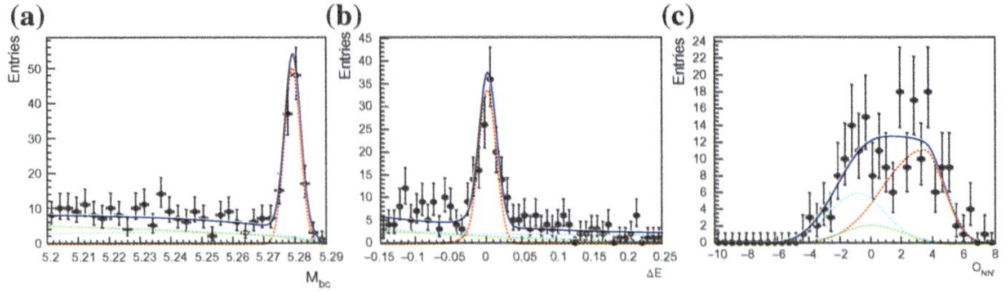

Fig. 18.3 Projection plots in the signal region for whole q^2 in case of $B^+ \to K^+\mu^+\mu^-$

mass cut using pion mass hypothesis. The 3D fit results are shown in Fig. 18.2 for $B^\pm \to K^\pm\mu^\pm\mu^\mp$. The signal enhanced projection is shown in Fig. 18.3. The three dimension is divided into signal and background parts. The part with $M_{bc} > 5.27$ GeV/c^2 and $-0.025 < \Delta E < 0.025$ GeV and $\mathscr{O}_{\mathcal{N}\mathcal{N}'} > 1.\triangledown$ are chosen as signal region.

18.7 Results

The BR for $B \to KJ/\psi$ and $B \to K\ell\ell$ are summarized in Tables 18.1, 18.2 , respectively.

Table 18.1 Branching ratio calculation for $B \to KJ/\psi$ modes. These are our control sample

Mode	N_{sig} (Data)	B (10^{-5}) (Data)	$\mathscr{PDG}$ B (10^{-5})
$B^+ \to K^+J/\psi(\to \mu^+\mu^-)$	17096 ± 131	6.04 ± 0.05	(6.12 ± 0.03)
$B^+ \to K^+J/\psi(\to e^+e^-)$	17850 ± 134	6.06 ± 0.04	(6.13 ± 0.03)
$B^0 \to K_S^0J/\psi(\to \mu^+\mu^-)$	5413 ± 74	2.68 ± 0.04	(2.59 ± 0.04)
$B^0 \to K_S^0J/\psi(\to e^+e^-)$	5173 ± 72	2.62 ± 0.04	(2.59 ± 0.04)

The R_K uncertainty for the whole q^2 and $1 < q^2 < 6$ bin are given in Table 18.3.

Table 18.2 Branching ratio estimation for $B \to K\ell\ell$ modes for the whole q^2 region

Mode	N_{sig} (MC)	B (10^{-7}) (MC)
$B^+ \to K^+\mu^+\mu^-$	140 ± 14	4.5 ± 0.5
$B^+ \to K^+e^+e^-$	138 ± 13	4.5 ± 0.5
$B^0 \to K_S^0\mu^+\mu^-$	37 ± 7	1.8 ± 0.4
$B^0 \to K_S^0e^+e^-$	42 ± 7	2.1 ± 0.4

Table 18.3 Expected uncertainty in R_K for $B \to K\ell\ell$

	Whole q^2	$1 < q^2 < 6$
R_K Uncertainty	0.13	0.23

18.8 Conclusion

The $R_K(J/\psi)$ is consistent with unity within the uncertainty for data and are $R_K(J/\psi) = 0.99 \pm 0.01$ (1.02 ± 0.02) for charged B case (neutral B case). The R_K uncertainty has decreased by 15% for whole q^2 bin and by 10% for bin of $1 < q^2 < 6$ $\mathrm{GeV}^2/\mathrm{c}^4$ in 3D fit as compared to 1D fit for $B \to K\ell\ell$ model.

Acknowledgements We thank the KEKB group for excellent operation of the accelerator; the KEK cryogenics group for efficient solenoid operations; and the KEK computer group the NII, and PNNL/EMSL for valuable computing and SINET5 network support. We acknowledge support from MEXT, JSPS and Nagoya's TLPRC (Japan); ARC (Australia); FWF (Austria); NSFC and CCEPP (China), MSMT (Czechia); CZF, DFG, EXC153, and VS (Germany); DST (India); INFN (Italy); MOE, MSIP, NRF, RSRI, FLRFAS project and GSDC of KISTI (Korea); MNiSW and NCN (Poland); MES and RFAAE (Russia); ARRS (Slovenia); IKERBASQUE and MINECO (Spain); SNSF (Switzerland); MOE and MOST (Taiwan); and DOE and NSF (USA).

References

1. LHCb Collaboration, Angular analysis of the $B^0 \to K^{*0}\mu^{\pm}\mu^{\mp}$ decay using $3\,fb^{-1}$ of integrated luminosity. arXiv:1512.04442
2. Belle Collaboration, *Lepton-flavor-dependent* angular analysis of $B \to K^{*0}\ell^{\pm}\ell^{\mp}$. arXiv:1612.05014
3. Belle Collaboration, A. Abashian et al., Nucl. Instr. and Meth. **A479**, 117 (2002)

4. A. Ali, Patricia Ball, L.T. Handoko, G. Hiller, Phys. Rev. D **61**, 074024 (2000)
5. D.J. Lange, Nucl. Instrum. Methods Phys. Res. Sect. A **462**, 152 (2001)
6. R. Brun et al., GEANT, CERN Report No. DD/EE/84-1 (1984)
7. M. Feindt et al., A hierarchical neurobayes-based algorithm for full reconstruction of *b* mesons at *b* factories

Part II
Charm, Kaon and Heavy Leptons

Chapter 19
Search for $K^+ \to \pi^+\nu\overline{\nu}$: First NA62 Results

Jurgen Engelfried

Abstract NA62 has searched for the $K^+ \to \pi^+\nu\overline{\nu}$ decay using a new kaon decay in-flight technique. One candidate event, compatible with the Standard Model prediction, has been observed from a sample of 1.2×10^{11} decays. Assuming that the event

---

The speakers acknowledges support from CONACyT to attend this conference.
For the NA62 Collaboration.

The NA62 Collaboration: R. Aliberti, F. Ambrosino, R. Ammendola, B. Angelucci, A. Antonelli, G. Anzivino, R. Arcidiacono, M. Barbanera, A. Biagioni, L. Bician, C. Biino, A. Bizzeti, T. Blazek, B. Bloch-Devaux, V. Bonaiuto, M. Boretto, M. Bragadireanu, D. Britton, F. Brizioli, M.B. Brunetti, D. Bryman, F. Bucci, T. Capussela, A. Ceccucci, P. Cenci, V. Cerny, C. Cerri, B. Checcucci, A. Conovaloff, P. Cooper, E. Cortina Gil, M. Corvino, F. Costantini, A. Cotta Ramusino, D. Coward, G. D'Agostini, J. Dainton, P. Dalpiaz, H. Danielsson, N. De Simone, D. Di Filippo, L. Di Lella, N. Doble, B. Dobrich, F. Duval, V. Duk, J. Engelfried, T. Enik, N. Estrada-Tristan, V. Falaleev, R. Fantechi, V. Fascianelli, L. Federici, S. Fedotov, A. Filippi, M. Fiorini, J. Fry, J. Fu, A. Fucci, L. Fulton, E. Gamberini, L. Gatignon, G. Georgiev, S. Ghinescu, A. Gianoli, M. Giorgi, S. Giudici, F. Gonnella, E. Goudzovski, C. Graham, R. Guida, E. Gushchin, F. Hahn, H. Heath, T. Husek, O. Hutanu, D. Hutchcroft, L. Iacobuzio, E. Iacopini, E. Imbergamo, B. Jenninger, K. Kampf, V. Kekelidze, S. Kholodenko, G. Khoriauli, A. Khotyantsev, A. Kleimenova, A. Korotkova, M. Koval, V. Kozhuharov, Z. Kucerova, Y. Kudenko, J. Kunze, V. Kurochka, V.Kurshetsov, G. Lanfranchi, G. Lamanna, G. Latino, P. Laycock, C. Lazzeroni, M. Lenti, G. Lehmann Miotto, E. Leonardi, P. Lichard, L. Litov, R. Lollini, D. Lomidze, A. Lonardo, P. Lubrano, M. Lupi, N. Lurkin, D. Madigozhin, I. Mannelli, G. Mannocchi, A. Mapelli, F. Marchetto, R. Marchevski, S. Martellotti, P. Massarotti, K. Massri, E. Maurice, M. Medvedeva, A. Mefodev, E. Menichetti, E. Migliore, E. Minucci, M. Mirra, M. Misheva, N. Molokanova, M. Moulson, S. Movchan, M. Napolitano, I. Neri, F. Newson, A. Norton, M. Noy, T. Numao, V. Obraztsov, A. Ostankov, S. Padolski, R. Page, V. Palladino, C. Parkinson, E. Pedreschi, M. Pepe, M. Perrin-Terrin, L. Peruzzo, P. Petrov, F. Petrucci, R. Piandani, M. Piccini, J. Pinzino, I. Polenkevich, L. Pontisso, Yu. Potrebenikov, D. Protopopescu, M. Raggi, A. Romano, P. Rubin, G. Ruggiero, V. Ryjov, A. Salamon, C. Santoni, G. Saracino, F. Sargeni, V. Semenov, A. Sergi, A. Shaikhiev, S. Shkarovskiy, D. Soldi, V. Sougonyaev, M. Sozzi, T. Spadaro, F. Spinella, A. Sturgess, J. Swallow, S. Trilov, P. Valente, B. Velghe, S. Venditti, P. Vicini, R. Volpe, M. Vormstein, H. Wahl, R. Wanke, B. Wrona, O. Yushchenko, M. Zamkovsky, A. Zinchenko.

J. Engelfried (✉)
Instituto de Física, Universidad Autónoma de San Luis Potosí, San Luis Potosi, Mexico
e-mail: Jurgen.Engelfried@cern.ch
URL: http://www.ifisica.uaslp.mx/-jurgen

A. Giri and R. Mohanta (eds.), *16th Conference on Flavor Physics and CP Violation*, Springer Proceedings in Physics 234,
https://doi.org/10.1007/978-3-030-29622-3_19

is background, an upper limit of 1.4×10^{-9} (95% CL) has been placed. Prospects for further improvements of the measurement are given.

19.1 Introduction

Among the many B and K decays sensitive to high energy scales, $K^+ \to \pi^+\nu\overline{\nu}$ stands out for the precision of the Standard Model (SM) prediction and its strong CKM [1, 2] and GIM [3] suppressions. The SM prediction [4–6] has been constantly updated and the latest update reads [7]:

$$\mathcal{B}(K^+ \to \pi^+\nu\overline{\nu})_{\rm SM} = (8.39 \pm 0.30) \times 10^{11}\left(\frac{|V_{cb}|}{0.0407}\right)^{2.8}\left(\frac{\gamma}{73.2^\circ}\right)^{0.74} \tag{19.1}$$

From the above equation one can see that, apart from the uncertainty in the CKM parameters (which can be reduced in the future), the irreducible theoretical error is only 3.6%. Taking typical values for V_{cb} and γ, the numerical prediction is[1]

$$\mathcal{B}(K^+ \to \pi^+\nu\overline{\nu})_{\rm SM} = (0.84 \pm 0.10) \times 10^{-10}. \tag{19.2}$$

This number is used in this paper to determine the expected number of signal events. Experimentally the decay has been studied with decays at rest with E787/E949 [9] at Brookhaven measuring

$$\mathcal{B}(K^+ \to \pi^+\nu\overline{\nu})_{\rm exp} = (17.3^{+11.5}_{-10.5}) \times 10^{-11}. \tag{19.3}$$

The difference between the precision of the theoretical prediction and the large experimental error motivates the construction of a new experiment using the decay-in-flight technique [10].

Extension to the SM to which $K^+ \to \pi^+\nu\overline{\nu}$ is sensitive include: Randall-Sundrum models with protective custodial symmetry [11]; MSSM analyses [12, 13]; simplified Z and Z0 [14]; littlest Higgs with T-Parity [15]; LFU violation models [16].

Constraints from existing experiments which include kaon mixing, correlations with other K and B decays and limits from direct searches leave a significant window of opportunity to be explored. Here we report the first results of NA62 based on a sample of about 1.2×10^{11} decays collected in 2016 at the CERN SPS.

[1]Taking the latest PDG [8] values gives $\mathcal{B}(K^+ \to \pi^+\nu\overline{\nu})_{\rm SM} = (0.93 \pm 0.07) \times 10^{-10}$.

19.2 The NA62 Experiment

A cross section of the NA62 experiment is shown in Fig. 19.1. Details about the beam and the detector can be found in [10]. The in-flight technique provides large acceptance and good suppression of the backgrounds because the interactions between the kaon decay products and the detector material are minimized. Protons from the CERN SPS are slowly extracted and hit a 40 cm long Be target. A secondary positively charged hadron beam of central momentum equal to 75 GeV/c and momentum bite of 1%, and a nominal intensity of up to 750×10^6 particles per second, is selected and transported 102 m downstream of the target where it enters an evacuated decay region. The momentum and direction of each beam particle is tracked by a Si pixel tracker (GTK) and the kaons (about 6% of the beam) are tagged by a differential Cherenkov counter (KTAG). The evacuated decay region is surrounded by stations of lead glass (LAV) to veto kaon decays with photons in the final state. In the forward region photons are detected by a liquid krypton calorimeter (LKr) and by other electromagnetic calorimeters covering smaller angles (IRC and SAC) with respect to the beam direction. The coverage of the acceptance is completed by a lateral muon detector (MUV0) to veto pion decaying into muons at large angles and by a hadronic sampling calorimeter (HASC) to veto charged particles from kaon decays at small angles. The direction and momentum of the π^+ originating from the kaon decay is measured by a magnetic spectrometer made of four stations of drift tubes (Straw) housed in the vacuum tank at the end of the decay region and by a dipole magnet placed between the second and the third Straw station. The nature of the charged particle is determined by particle identification based on a ring imaging Cherenkov counter (RICH), electromagnetic (LKr) and hadronic (MUV1 and MUV2) calorimeters. Events with large deposits of energy in the LKr or hits in the muon veto (MUV3) are rejected by the Level 0 trigger which requires a tight time

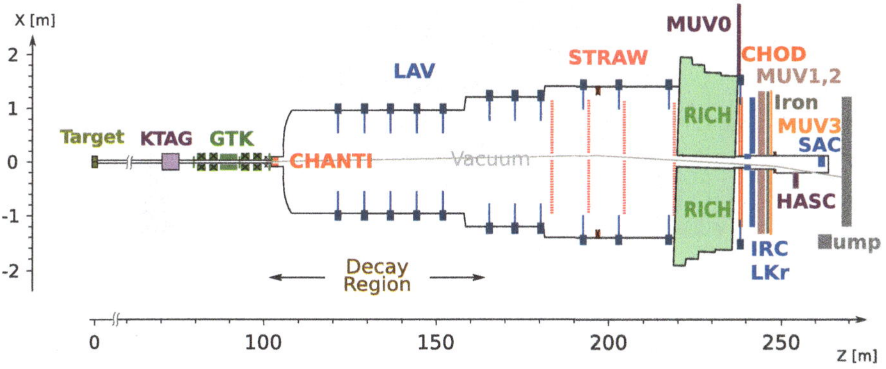

Fig. 19.1 Horizontal cross section of the NA62 detector. The RICH is slightly off-axis with respect to the other detectors in order to take into account the kick of the dipole magnet on the undecayed beam particles

coincidence between a plastic scintillator hodoscope (CHOD)and the RICH. Events with hits in the CHANTI anti-counter placed downstream of the GTK are rejected in order to veto inelastic interactions occurring in the last station of the beam tracker. All detector have excellent time resolution, some <100 ps, and are aligned during the offline processing to better than 10 ps.

Even though the experiment is designed for the measurement of the $K^+ \to \pi^+\nu\overline{\nu}$ branching ratio, a long list of other kaon and pion physics topics can be addressed with the acquired dataset. A first result on the search for heavy neutral leptons in K^+ decays was presented [17] at this conference.

19.3 Analysis Strategy

Events with more than one Straw track are kept only if no combination of tracks forms a vertex. In Fig. 19.2 the squared missing mass, $m^2_{\text{miss}} = (p_K - p_\pi)^2$, is shown for the events passing the preliminary selection. Events with extra clusters in the LKr are rejected if they are found more than 10 cm away from the impact position of

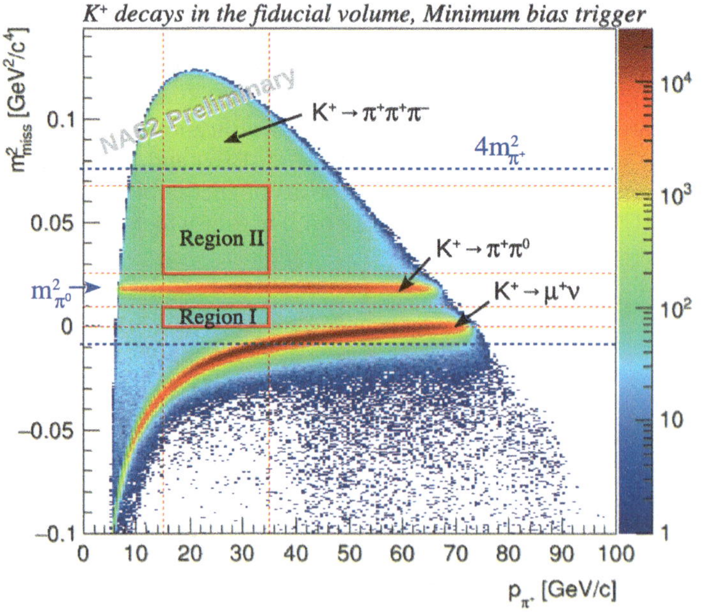

Fig. 19.2 The scatter plot of typical kaon decays plotting the missing mass squared $m^2_{\text{miss}} = (p_K - p_\pi)^2$ as a function of the Straw track momentum under the hypothesis that the beam particle is a kaon and the track reconstructed in the Straw is a π^+. Notice the deformation of the two body kinematics for $K^+ \to \mu^+\nu$ decays as a function of the pion momentum: this is the effect of having assigned the wrong particle hypothesis to the Straw track. Also indicated are the two signal regions

the π^+ on the LKr. Events with hits in either the LAV, SAV, IRC or HASC in time with the π^+ are rejected. Muon candidates are rejected combining the information from the calorimeters with the RICH. Signal events are characterized by a three body kinematics, while the frequent $K^+ \to \mu^+\nu$ and $K^+ \to \pi^+\pi^0$ decay events have a distinctive two body distribution. For the time being, simple cut and count signal regions are defined using the squared missing mass recoiling against the Straw track assumed to be a π^+.

This allows one to confine the search to areas which are not swamped by two body decays. Events with Straw track momentum larger than 35 GeV/c are rejected to better suppress the $K^+ \to \pi^+\pi^0$ decays: in this case the π^0 is geometrically bound to deposit at least 40 GeV of energy in the calorimeters where such a large amount of energy can hardly be missed. Events with Straw track momentum less than 15 GeV/c are rejected because the pion is below or too close to the Cherenkov threshold of the RICH.

The single event sensitivity (SES) is found to be SES $= (3.15 \pm 0.01_{\text{stat}} \pm 0.24_{\text{syst}}) \times 10^{-10}$. The largest uncertainty is due to the estimation of the loss due to random veto which will be improved in the future. The signal acceptance was found to be 4%, of which about 1% is in Region I (suitably defined between the $K^+ \to \mu^+\nu$ and $K^+ \to \pi^+\pi^0$ peaks) and 3% is in Region II (defined between the $K^+\pi^+\pi^0$ peak and the area populated by $K^+ \to \pi^+\pi^+\pi^-$ decays). The number of expected signal and background events are given in Table 19.1.

The inspection of the signal regions revealed one candidate event in Region II, as shown in Fig. 19.3. The ring of the decay track as seen by the RICH is shown in Fig. 19.4: it is beautifully consistent to be a pion. Assuming that the event is background, a 95% CL upper limit has been placed:

$$\mathcal{B}(K^+ \to \pi^+\nu\overline{\nu}) < 14 \times 10^{-10}. \tag{19.4}$$

For comparison, if the candidate is taken to be signal, the corresponding branching ratio (68% CL) reads:

$$\mathcal{B}(K^+ \to \pi^+\nu\overline{\nu}) = 28^{+44}_{-23} \times 10^{-11}. \tag{19.5}$$

Table 19.1 Summary of the signal and background expectations for the 2016 $K^+ \to \pi^+\nu\overline{\nu}$ data

Process	Expected events in the signal regions
$K^+ \to \pi^+\nu\overline{\nu}$	$0.267 \pm 0.001_{\text{stat}} \pm 0.02_{\text{syst}} \pm 0.032_{\text{ext}}$
$K^+ \to \pi^+\pi^0(\gamma)$ IB	$0.064 \pm 0.007_{\text{stat}} \pm 0.006_{\text{syst}}$
$K^+ \to \mu^+\nu(\gamma)$ IB	$0.020 \pm 0.003_{\text{stat}} \pm 0.003_{\text{syst}}$
$K^+ \to \pi^+\pi^-e^+\nu$	$0.018^{+0.024}_{0.017}\vert_{\text{stat}} \pm 0{:}009_{\text{syst}}$
$K^+ \to \pi^+\pi^+\pi^-$	$0.002 \pm 0.001_{\text{stat}} \pm 0.002_{\text{syst}}$
Upstream background	$0.050^{+0/090}_{0.030}\vert_{\text{stat}}$
Total background	$0.15 \pm 0{:}09_{\text{stat}} \pm 0.01_{\text{syst}}$

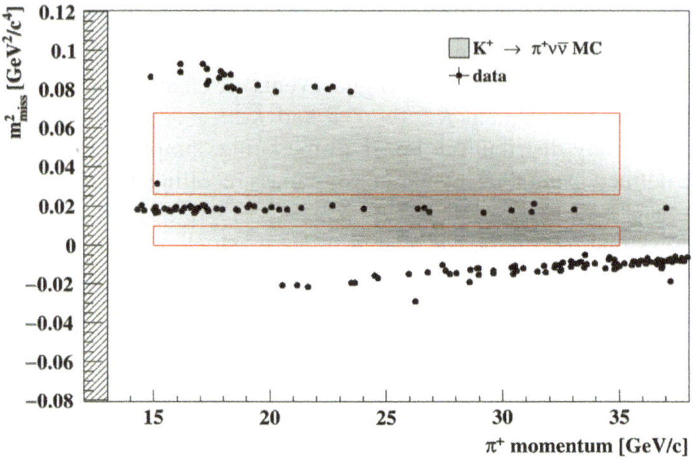

Fig. 19.3 The data passing the analysis. The inspection of the signal boxes revealed one candidate event in Region II

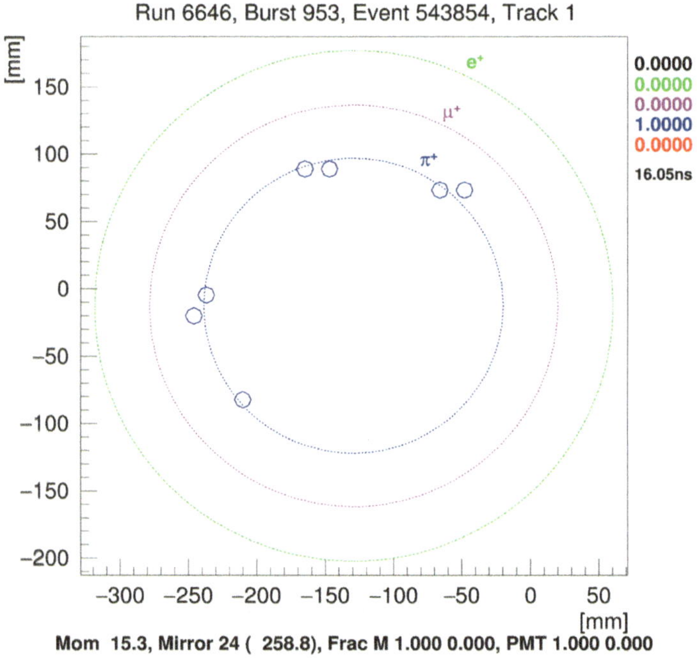

Fig. 19.4 Display of the RICH hits for the candidate event. The rings indicate the different particle hypotheses, the numbers on the right are the normalized likelihoods for the different hypotheses; the average time of the hits is 16.05 ns within the relevant readout window for this trigger, the positive track of 15.3 GeV/c passed thru mirror 24 and all the Cherenkov light is fully contained on the mirror and the photomultiplier arrays. The candidate fits very nicely the π^+ hypothesis

This result is not yet competitive with those obtained with the decay at rest but it shows that the new in-flight technique works. NA62 has already 20 times more statistics on tape and the analysis of the data is ongoing. Data taking is underway and will continue until the CERN Long Shutdown 2 (LS2) when the accelerator complex stops for two years of maintenance. By the end of 2018 NA62 should have accumulated about 20 SM signal events. NA62 is seeking approval to continue data taking after LS2 in order to complete the measurement with a precision of about 10%.

References

1. N. Cabibbo, Phys. Rev. Lett. **10**, 531 (1963). https://doi.org/10.1103/PhysRevLett.10.531
2. M. Kobayashi, T. Maskawa, Prog. Theor. Phys. **49**, 652 (1973). https://doi.org/10.1143/PTP.49.652
3. S.L. Glashow, J. Iliopoulos, L. Maiani, Phys. Rev. D **2**, 1285 (1970). https://doi.org/10.1103/PhysRevD.2.1285
4. G. Buchalla, A.J. Buras, Nucl. Phys. B **400**, 225 (1993). https://doi.org/10.1016/0550-3213(93)90405-E
5. G. Buchalla, A.J. Buras, Nucl. Phys. B **412**, 106 (1994). https://doi.org/10.1016/0550-3213(94)90496-0 [hep-ph/9308272]
6. M. Misiak, J. Urban, Phys. Lett. B **451**, 161 (1999). https://doi.org/10.1016/S0370-2693(99)00150-1 [hep-ph/9901278]
7. A.J. Buras, D. Buttazzo, J. Girrbach-Noe, R. Knegjens, JHEP **1511**, 033 (2015). https://doi.org/10.1007/JHEP11(2015)033. [arXiv:1503.02693] [hep-ph]
8. M. Tanabashi et al., Particle data group. Phys. Rev. D **98**, 030001 (2018)
9. A.V. Artamonov et al. (BNL-E949 Collaboration), Phys. Rev. D **79**, 092004 (2009). https://doi.org/10.1103/PhysRevD.79.092004. [arXiv:0903.0030] [hep-ex]
10. E. Cortina Gil et al. (NA62 Collaboration), JINST **12**(05), P05025 (2017). https://doi.org/10.1088/1748-0221/12/05/P05025. [arXiv:1703.08501] [physics.ins-det]
11. M. Blanke, A.J. Buras, B. Duling, K. Gemmler, S. Gori, JHEP **0903**, 108 (2009). https://doi.org/10.1088/1126-6708/2009/03/108. [arXiv:0812.3803] [hep-ph]
12. T. Blažek, P. Maták, Int. J. Mod. Phys. A **29**(27), 1450162 (2014). https://doi.org/10.1142/S0217751X14501620 [arXiv:1410.0055] [hep-ph]
13. G. Isidori, F. Mescia, P. Paradisi, C. Smith, S. Trine, JHEP **P0608**, 064 (2006). https://doi.org/10.1088/1126-6708/2006/08/064 [hep-ph/0604074]
14. A.J. Buras, D. Buttazzo, R. Knegjens, JHEP **1511**, 166 (2015). https://doi.org/10.1007/JHEP11(2015)166.[arXiv:1507.08672] [hep-ph]
15. M. Blanke, A.J. Buras, S. Recksiegel, Eur. Phys. J. C **76**(4), 182 (2016). https://doi.org/10.1140/epjc/s10052-016-4019-7. [arXiv:1507.06316] [hep-ph]
16. M. Bordone, D. Buttazzo, G. Isidori, J. Monnard, Eur. Phys. J. C **77**(9), 618 (2017). https://doi.org/10.1140/epjc/s10052-017-5202-1. [arXiv:1705.10729] [hep-ph]
17. N. Estrada-Tristán (NA62 Collaboration), These proceedings

Chapter 20
Latest Results in Kaon Physics on the KLOE Data and Status of Analysis of the KLOE-2 Data

Eryk Czerwiński

Abstract The KLOE-2 Collaboration concluded its data-taking at the DAΦNE ϕ-factory of the Laboratori Nazionali di Frascati of INFN, collecting 5.5 fb^{-1} of data at the center of mass energy corresponding to the mass of the $\phi(1020)$. This sample together with the 2.5 fb^{-1} collected by the KLOE Collaboration until 2006 constitutes the largest sample in the world collected at the peak of the ϕ resonance. With the kaons produced in the decay $\phi \rightarrow K_S K_L$ it is possible to perform very precise test of the discrete symmetries T, CP, and CPT, and moreover the entanglement of the neutral kaons can be exploited to perform unique tests of quantum mechanics. Also large samples of η mesons are produced in ϕ radiative decays, allowing to study the rare decays of this meson. KLOE-2 has also installed two tagging devices for electrons and positrons scattered in $\gamma\gamma$ processes, with the main goal to study the single π^0 production. Search of new vector bosons in the hidden dark sector is also performed with the KLOE/KLOE-2 data. An information about the latest results with KLOE data on CPT test with charge asymmetry of K_S semileptonic decays and a preliminary result on the CPT test in transitions is given.

20.1 DAΦNE Collider and KLOE-2 Detector

The DAΦNE electron-positron collider at the INFN National Laboratory of Frascati (LNF) consists of two separate storage rings with the center-of-mass energy of the colliding beams set to the mass of the $\phi(1020)$. One of the DAΦNE interaction points is a location of the KLOE-2 detection system: the 4 m diameter large cylindrical drift chamber [1] surrounded by an electromagnetic calorimeter [2] immersed in 0.5 T magnetic field. In addition the following subdetectors are used: the Inner Tracker [3] for improvement of resolution on the vertex position and acceptance for tracks with

Eryk Czerwiński—On behalf of KLOE-2 collaboration.

E. Czerwiński (✉)
Faculty of Physics, Astronomy and Applied Computer Science,
Institute of Physics, Jagiellonian University, Kraków, Poland
e-mail: eryk.czerwinski@uj.edu.pl

A. Giri and R. Mohanta (eds.), *16th Conference on Flavor Physics and CP Violation*, Springer Proceedings in Physics 234,
https://doi.org/10.1007/978-3-030-29622-3_20

low transverse momentum; two pairs of small angle tagging devices to detect low (Low Energy Tagger—LET [4]) and high (High Energy Tagger—HET [5]) energy e^+e^- originated from $e^+e^- \to e^+e^- X$ reactions; crystal calorimeters (CCALT) to cover the low polar angle region to increase acceptance for very forward electrons and photons down to 8° [6]; and a tile calorimeter (QCALT) used for the detection of photons coming from K_L decays in the drift chamber [7]. High performance of drift chamber for momentum and vertex reconstruction ($\sigma_{p_\perp}/p_\perp < 0.4\%$ for $\theta > 45°$; ~150 μm in transverse plane) and excellent time and energy resolution of the calorimeter ($\sigma_t = 57\,\text{ps}/\sqrt{E(\text{GeV})} \oplus 100\,\text{ps}; \sigma_E/E = 5.7\%/\sqrt{E(\text{GeV})}$) ensure high quality of collected data.

20.2 Charge Asymmetry of K_S

In general the charge asymmetry for semileptonic decays of neutral kaons (A_S and A_L, for short- and long-lived kaon, respectively) can be defined with their semileptonic decay widths:

$$\begin{aligned} A_{S,L} &= \frac{\Gamma(K_{S,L} \to \pi^- e^+ \nu) - \Gamma(K_{S,L} \to \pi^+ e^- \bar{\nu})}{\Gamma(K_{S,L} \to \pi^- e^+ \nu) + \Gamma(K_{S,L} \to \pi^+ e^- \bar{\nu})} \\ &= 2\left[Re\left(\epsilon_{S,L}\right) - Re(y) \pm Re(x_-)\right] \end{aligned} \quad (20.1)$$

to the first order in parameters ϵ_S, ϵ_L. These parameters can be expressed in terms of the CP and CPT violation parameters ϵ_K and δ_K:

$$\epsilon_{L/S} = \epsilon_K \mp \delta_K . \quad (20.2)$$

The $K \to \pi e \nu$ decays can be described with the following decay amplitudes:

$$\begin{aligned} \langle \pi^- e^+ \nu | \, H_{weak} \, | K^0 \rangle &= \mathcal{A}_+ , \\ \langle \pi^+ e^- \bar{\nu} | \, H_{weak} \, | \bar{K^0} \rangle &= \bar{\mathcal{A}}_- , \\ \langle \pi^+ e^- \bar{\nu} | \, H_{weak} \, | K^0 \rangle &= \mathcal{A}_- , \\ \langle \pi^- e^+ \nu | \, H_{weak} \, | \bar{K^0} \rangle &= \bar{\mathcal{A}}_+ , \end{aligned} \quad (20.3)$$

where H_{weak} stands for weak Hamiltonian. Sum and difference of the A_S and A_L allow to search for the CPT symmetry violation, either in the decay amplitudes through the parameter y or in the mass matrix through the parameter δ_K:

$$\begin{aligned} A_S + A_L &= 4Re(\epsilon) - 4Re\,(y) , \\ A_S - A_L &= 4Re(\delta_K) + 4Re\,(x_-) . \end{aligned} \quad (20.4)$$

where:

$$x = \frac{\bar{\mathcal{A}}_+}{\mathcal{A}_+}, \quad \bar{x} = \left(\frac{\mathcal{A}_-}{\bar{\mathcal{A}}_-}\right)^*, \quad y = \frac{\bar{\mathcal{A}}_-^* - \mathcal{A}_+}{\bar{\mathcal{A}}_-^* + \mathcal{A}_+},$$
$$x_\pm = \frac{x \pm \bar{x}^*}{2} = \frac{1}{2}\left[\frac{\bar{\mathcal{A}}_+}{\mathcal{A}_+} \pm \left(\frac{\mathcal{A}_-}{\bar{\mathcal{A}}_-}\right)^*\right]. \tag{20.5}$$

The charge asymmetry was measured by the KTeV experiment for long-lived kaon [8] and by the KLOE—for the short-lived one [9]:

$$A_L = (3.322 \pm 0.058_{stat} \pm 0.047_{syst}) \times 10^{-3},$$
$$A_S = (1.5 \pm 9.6_{stat} \pm 2.9_{syst}) \times 10^{-3}. \tag{20.6}$$

Recently the KLOE data set of 1.7 fb^{-1} was analyzed for more precise A_S measurement [10]. In this analysis a K_L interaction in the calorimeter is considered as a tag for K_S decay. Semileptonic decays are selected based on determination of particle time of flight from the decay point to the calorimeter. In the final event sample a certain amount of background is allowed to perform a MC-based event counting. The best separation between the signal and background components is obtained with the variable:

$$M^2(e) = \left[E_{K_S} - E(\pi) - E_\nu\right]^2 - p^2(e), \tag{20.7}$$

where $E_\nu = |\mathbf{p}_{K_S} - \mathbf{p}(e) - \mathbf{p}(\pi)|$.

The numbers of semileptonic decay events obtained from the fit are 34579 ± 251 for $K_S \to \pi^- e^+ \nu$ and and 36874 ± 255 for $K_S \to \pi^+ e^- \nu$, resulting in the following value of the charge asymmetry:

$$A_S = (-4.9 \pm 5.7_{stat} \pm 2.6_{syst}) \times 10^{-3}, \tag{20.8}$$

which is consistent with the previous determination on an independent data sample [9] and improving the statistical accuracy by almost a factor of two.

Combining both measurements taking into account the correlations of the systematical uncertainties one obtains:

$$A_S = (-3.8 \pm 5.0_{stat} \pm 2.6_{syst}) \times 10^{-3}. \tag{20.9}$$

The above result together with the KTeV result on A_L [8] yields for the sum and difference of asymmetries:

$$(A_S - A_L)/4 = Re(\delta_K) + Re(x_-) = (-1.8 \pm 1.4) \times 10^{-3}, \tag{20.10}$$
$$(A_S + A_L)/4 = Re(\epsilon_K) - Re(y) = (-0.1 \pm 1.4) \times 10^{-3}. \tag{20.11}$$

Using $Re(\delta_K) = (2.5 \pm 2.3) \times 10^{-4}$ [11] and $Re(\epsilon_K) = (1.596 \pm 0.013) \times 10^{-3}$ [12] the CPT violating parameters $Re(x_-)$ and $Re(y)$ can be extracted:

$$Re(x_-) = (-2.0 \pm 1.4) \times 10^{-3}, \tag{20.12}$$

$$Re(y) = (1.7 \pm 1.4) \times 10^{-3}. \tag{20.13}$$

Values of these parameters are consistent with CPT invariance and improve by almost a factor of two the previous results [9]. The obtained $Re(x_-)$ and $Re(y)$ values are the most precise results to date and agree with the world average value [11]. Improvements are expected in the future with the analysis of the additional $\sim$5.5 fb^{-1} collected by KLOE-2 [13] with a statistical uncertainty at the level of 3×10^{-3}.

20.2.1 CPT and T Symmetry Test in Transition

A direct test of CPT and T symmetry is possible at the KLOE-2 detector with neutral kaons [14]. These states can be defined as states with definite flavour or CP:

$$\mathcal{S}\,|\mathrm{K}^0\rangle = +1\,|\mathrm{K}^0\rangle\,, \mathcal{S}\,|\bar{\mathrm{K}}^0\rangle = -1\,|\bar{\mathrm{K}}^0\rangle\,, \tag{20.14}$$

$$\begin{aligned} |\mathrm{K}_+\rangle &= \frac{1}{\sqrt{2}}\left[|\mathrm{K}^0\rangle + |\bar{\mathrm{K}}^0\rangle\right] \text{ with CP} = +1, \\ |\mathrm{K}_-\rangle &= \frac{1}{\sqrt{2}}\left[|\mathrm{K}^0\rangle - |\bar{\mathrm{K}}^0\rangle\right] \text{ with CP} = -1, \end{aligned} \tag{20.15}$$

and identified through observation of their decay products. The state of one kaon is identified at the moment of decay of the other kaon. Identification of its state at the moment of its decay after time Δt allows to observe a transition between the strangeness and CP-definite states. For each of the possible transitions a time-dependent ratio of probabilities can be defined as an observable of the T symmetry test. Two of those ratios can be measured at KLOE-2:

$$\begin{aligned} R_2(\Delta t) &= \frac{P[\mathrm{K}^0(0) \to \mathrm{K}_-(\Delta t)]}{P[\mathrm{K}_-(0) \to \mathrm{K}^0(\Delta t)]} \sim \frac{\mathrm{I}(\ell^-, 3\pi^0; \Delta t)}{\mathrm{I}(\pi\pi, \ell^+; \Delta t)}, \\ R_4(\Delta t) &= \frac{P[\bar{\mathrm{K}}^0(0) \to \mathrm{K}_-(\Delta t)]}{P[\mathrm{K}_-(0) \to \bar{\mathrm{K}}^0(\Delta t)]} \sim \frac{\mathrm{I}(\ell^+, 3\pi^0; \Delta t)}{\mathrm{I}(\pi\pi, \ell^-; \Delta t)}, \end{aligned} \tag{20.16}$$

with a sensitivity at the level of 10^{-3}. For $\Delta t >> \tau_S$ their deviation from unity should be proportional to a real part of δ_K which is a T violating parameter of the neutral kaon system [14].

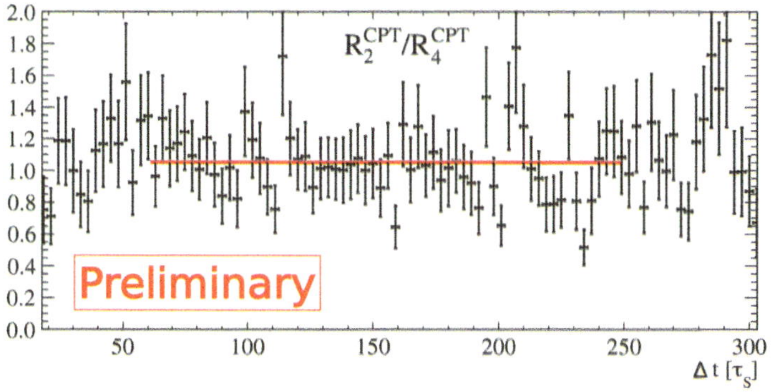

Fig. 20.1 Distribution of the CPT sensitive double ratio of neutral kaon double decay rates presented as a function of difference between entangled kaon decay times (Δt). The statistical uncertainty of a fit of a constant (red line) for $\Delta t >> \tau_S$ region is 0.011 with the KLOE data-set

Ratios sensitive to CPT violation are the following [15]:

$$R_{2,CPT}(\Delta t) = \frac{P[\mathrm{K}^0(0) \to \mathrm{K}_-(\Delta t)]}{P[\mathrm{K}_-(0) \to \mathrm{K}^0(\Delta t)]},$$
$$R_{4,CPT}(\Delta t) = \frac{P[\bar{\mathrm{K}}^0(0) \to \mathrm{K}_-(\Delta t)]}{P[\mathrm{K}_-(0) \to \bar{\mathrm{K}}^0(\Delta t)]}, \tag{20.17}$$

leading to a double ratio as CPT violation observable in the $\Delta t \gg \tau_S$ limit equal to $1 - 8\Re\delta_K - 8\Re x_-$.

Such double ratio constitutes a CPT violation observable which was never measured so far. Currently the analysis of KLOE data sample of 1.7 fb^{-1} integrated luminosity is ongoing and a preliminary distribution of the ratio of ratios defined in 20.17 is presented at Fig. 20.1.

The analysis to be performed at KLOE-2 data-set will allow for 10^{-3} statistical accuracy test of T and CPT.

20.3 Summary

In the recent years a long list of physics results from neutral kaon was provided by the KLOE experiment: measurement of V_{us}, study of CP and CPT discrete symmetries, search for decoherence of entangled kaons. Presently the KLOE-2 detector collected 5.5 fb^{-1} of integrated luminosity data sample for broader and more precise results. The ongoing analyses and newly collected data promises significant progress to be made in the field of discrete symmetries and quantum mechanics test as well as low-energy structure of mesons [13].

Acknowledgements We warmly thank our former KLOE colleagues for the access to the data collected during the KLOE data taking campaign. We thank the DAΦNE team for their efforts in maintaining low background running conditions and their collaboration during all data taking. We want to thank our technical staff: G. F. Fortugno and F. Sborzacchi for their dedication in ensuring efficient operation of the KLOE computing facilities; M. Anelli for his continuous attention to the gas system and detector safety; A. Balla, M. Gatta, G. Corradi and G. Papalino for electronics maintenance; C. Piscitelli for his help during major maintenance periods. This work was supported in part by the Polish National Science Centre through the Grants No. 2013/11/B/ST2/04245, 2014/14/E/ST2/00262, 2014/12/S/ST2/00459, 2016/21/N/ST2/01727, 2016/23/N/ST2/01293, 2017/26/M/ST2/00697.

References

1. M. Adinolfi et al., Nucl. Instr. Meth. A **488**, 51 (2002)
2. M. Adinolfi et al., Nucl. Instr. Meth. A **482**, 364 (2002)
3. A. Balla et al., Acta Phys. Pol. B Proc. Suppl. **6**(4), 1053 (2013)
4. D. Babusci et al., Nucl. Instr. Meth. A **617**, 81 (2010)
5. F. Archilli et al., Nucl. Instr. Meth. A **617**, 266 (2010)
6. M. Cordelli et al., Nucl. Instr. Meth. A **718**, 81 (2013)
7. A. Balla et al., Nucl. Instr. Meth. A **718**, 95 (2013)
8. A. Alavi-Harati et al., Phys. Rev. Lett. **88**, 181601 (2002)
9. F. Ambrosino et al., Phys. Lett. B **636**, 173 (2006)
10. A. Anastasi et al., JHEP **1809**, 021 (2018)
11. C. Patrignani et al. (Particle Data Group), Chin. Phys. C **40**(10), 100001 (2016)
12. F. Ambrosino et al. (KLOE Collaboration), JHEP **0612**, 011 (2006)
13. G. Amelino-Camelia et al., Eur. Phys. J. C **68**, 619 (2010)
14. J. Bernabeu, A. Di Domenico, P. Villanueva-Perez, Nucl. Phys. B **868**, 102 (2013)
15. J. Bernabeu, A. Di Domenico, P. Villanueva-Perez, JHEP **1510**, 139 (2015)

Chapter 21
COMET/Mu2e/MEG

Satoshi Mihara

Abstract Charged Lepton Flavor Violating (cLFV) processes are thought to be powerful tools to investigate the physics beyond the Standard Model (SM) as all of them are extremely suppressed in the framework of SM even if we take the neutrino oscillation into account. There are intensive efforts ongoing to find signals of cLFV processes at high-power proton accelerator facilities in Japan, the US, and Switzerland. We report these experimental efforts and their status in this presentation.

21.1 Introduction

The muon has been playing an important role while the Standard Model (SM) was constructed; non-observation of the $\mu \rightarrow e\gamma$ process, and non-observation of electron production in reactions of neutrinos, produced as decay products of muons, provided the idea of lepton flavor, namely the generation. Once the SM was established, motivation of muon rare process searches violating the lepton flavor conservation has been turned to be a way to search for the new physics beyond the SM. This is simply because new physics models such as the Grand Unification Theory (GUT) with Super-Symmetric (SUSY) extension naively predict existence of charged Lepton Flavor Violating (cLFV) processes with branching ratios (or reaction rate) just below the current limits, while SM contributions to such processes are extremely suppressed even if we take the neutrino oscillation into account [1]. Thus once any cLFV process is observed in an experiment, it is a clear evidence of new physics and its branching ratio provides us precious information of the energy scale of new physics. However as such contribution of new physics to cLFV process(es) is through a loop or intermediate particle exchange as illustrated in Fig. 21.1, it would not be possible to clarify the exact particle contributing to the process(es). Nevertheless such new particle(s) can contribute to each of these processes differently, we expect

S. Mihara (✉)
KEK, J-PARC Center, and SOKENDAI, 1-1 Oho,
Tsukuba, Ibaraki 305-0801, Japan
e-mail: satoshi.mihara@kek.jp

A. Giri and R. Mohanta (eds.), *16th Conference on Flavor Physics and CP Violation*, Springer Proceedings in Physics 234,
https://doi.org/10.1007/978-3-030-29622-3_21

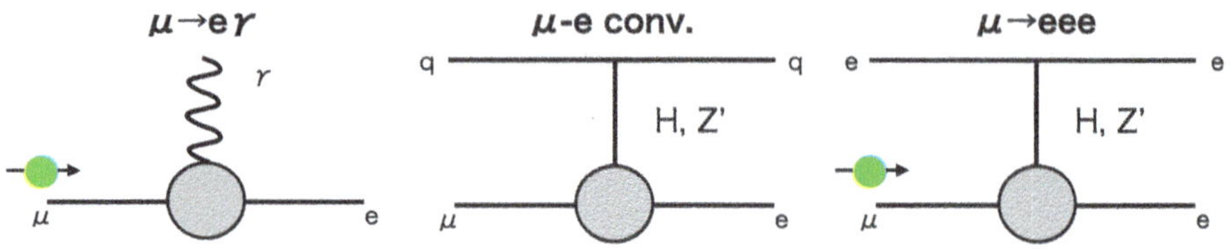

Fig. 21.1 Charged Lepton Flavor Violating processes of muon. (left) $\mu \rightarrow e\gamma$, (center) $\mu - e$ conversion, and (right) $\mu \rightarrow eee$. The shaded circles indicate contributions of new physics. Photon exchange diagram would also contribute in $\mu - e$ conversion and $\mu \rightarrow eee$ processes

that we can extract valuable information of new physics by investigating all processes with similar sensitivities and comparing them. As for $\mu \rightarrow e\gamma$ and $\mu \rightarrow eee$ processes, it should also be noted that we can derive information of the contributing physics by measuring angular distributions of final state particles with respect to the muon polarization. As for the $\mu - e$ conversion process, we can address possible contribution of the new force by measuring the target mass dependence of the signal rate [2].

21.2 Muon cLFV Experiments

High intensity muon beam is imperative to conduct searches of muon cLFV experiments as the upper limits achieved so far are all below the level of 10^{-12}. Muons are to be produced as decay products of pions produced in proton-nucleus interactions. Thus all of muon cLFV experiments being conducted and/or planned are carried out at high-power proton accelerator facilities. There are three major facilities where high-power proton beam is available at Japan Proton Accelerator Research Complex (J-PARC) in Japan, Fermi National Accelerator Laboratory (FNAL) in the U.S., and Paul Scherrer Institute (PSI) in Switzerland. Two types of muon beam are used depending on requirement by the experiment; one is a pulsed muon beam and the other is a continuous muon beam. The $\mu - e$ conversion search experiment needs a pulsed muon beam whose pulse time structure is as long as the muon life time in muonic atoms of about 1 μs. This is because only one particle (electron) should appear in the final state and can be separated from beam induced prompt background by requiring the signal to appear in a delayed timing. On the other hand $\mu \rightarrow e\gamma$ and $\mu \rightarrow eee$ search experiments require a continuous muon beam as there are more than one particle in the final state and accidental overlaps of background events can be well distinguished by requiring time coincidence of final state particles. The PSI cyclotron has a proton beam time structure of 19.7 ns and thus muons produced as decay products of pions whose life time is 26 ns are provided in a continuous mode, which is suitable for coincidence type experiments as explained above. In this presen-

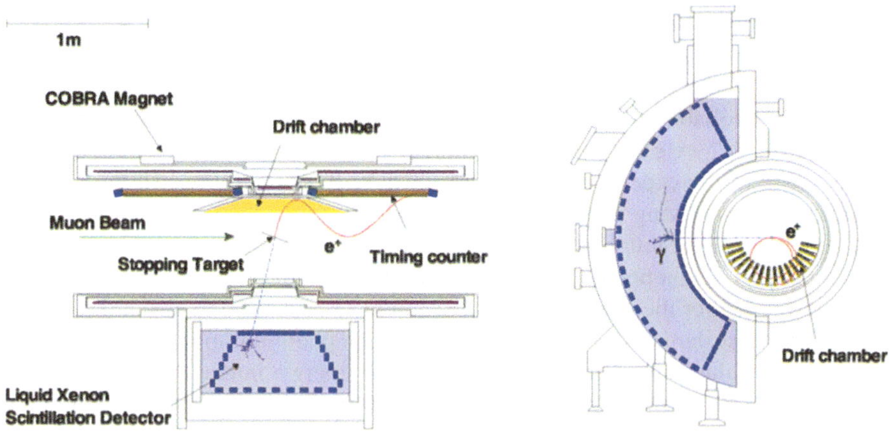

Fig. 21.2 Schematic view of the MEG experiment at PSI

tation we report the status and prospects of MEG and MEG II experiments searching for the $\mu \to e\gamma$ decay, and COMET and Mu2e experiments searching for the $\mu - e$ conversion process.

21.2.1 MEG and MEG II

The MEG experiment was designed and conducted at PSI to search for the cLFV muon process, $\mu^+ \to e^+\gamma$ with a better sensitivity than the previous experiment, MEGA, conducted at Los Alamos National Laboratory (LANL) [3]. A schematic view of the MEG experiment setup is shown in Fig. 21.2; the muon stopping target of 205 μm thick polyethylene and polyester layered structure is located at the center of a positron spectrometer. The positron spectrometer is composed of low-mass segmented drift chamber modules (0.002 X_0) to measure the positron momentum, and timing counters made of plastic scintillator bars with photomultiplier readout to measure the positron timing. The spectrometer is installed in a solenoidal magnet with a gradient magnetic field, 1.27 T at the center to 0.47 T at either end of the magnet, to sweep out efficiently low-momentum positrons produced in muon Michel decays. A liquid-xenon photon detector is located beside the spectrometer magnet to measure the energy, position, and timing of gammas from the muon stopping target. The thickness of the central part of the spectrometer magnet is made to be thin (0.197 X_0) to allow high transmission efficiency of gammas to the detector. Details of the MEG setup is summarized in [4] and references therein.

The MEG experiment carried out physics data acquisition in the period of 2009–2013 and accumulated the statistic of 7.5×10^{14} muon stops on the target. No significant excess of events has been observed in the signal region so far, and thus a new upper limit of the branching ratio of the decay $Br(\mu^+ \rightarrow e^+\gamma) < 4.2 \times 10^{-13}$ at 90% confidence level is set [5]. This is the most stringent result of the muon cLFV search at the moment.

The MEG collaboration has started detector upgrade to improve the sensitivity to search for $\mu^+ \rightarrow e^+\gamma$ just after completing the 1st phase of data acquisition. The upgrade of the experiment, called MEG II, will achieve about 10 times better sensitivity by improving detector resolutions almost twice in every component and by using twice larger muon beam rate of 7×10^7Hz.

In the liquid-xenon photon detector in MEG 2-in. diameter photomultipliers specially developed for the use in the liquid xenon have been used to detect the scintillation light. Distribution of the scintillation light on photomultipliers on the surface of the active volume provided information of the energy, interaction position, and interaction timing of incoming gammas. In the MEG II experiment photomultiplier arrangement is modified on the lateral sides to enlarge the active volume and realize better detection efficiency of the scintillation light, and SiPMs with sufficient sensitivity to the ultra-violet component of the liquid-xenon scintillation light are employed on the front surface instead of photomultipliers. Better position resolution is expected thanks to the smaller dimension of the SiPM and higher transmission efficiency of gammas penetrating the front wall is anticipated owing to thinner structure of the SiPM.

The positron spectrometer is also upgraded in MEG II. The positron tracker used in MEG was composed of 16 segmented drift chambers, and is replaced in MEG II by a single volume Cylindrical Drift Chamber (CDC) with stereo wire readout. The CDC is designed to achieve high transparency and low multiple Coulomb scattering for the signal positron. The positron timing counter located at both ends of the spectrometer magnet is replaced by pixelated timing counters composed of small tile plastic scintillators with SiPM readout. The size of tiles and their configuration are optimized to achieve the best performance of timing measurement. A new detector component is also implemented to identify the radiative muon decay with a low-momentum positron in forward region; a gamma in such event deposits relatively large energy in the liquid-xenon detector while the associated position could not be identified in the MEG setup. The detector, called Radiative Decay Counter (RDC), composed of LYSO crystals and plastic scintillators with SiPM readout is located at the downstream end of the spectrometer magnet.

The MEG II experiment intend to achieve a sensitivity of 6×10^{-14} in 3 years data acquisition period. All detector components are installed and engineering running is starting as of late 2018. A schematic view of the MEG II detector is presented in Fig. 21.3.

21.2.2 COMET

Search for the $\mu - e$ conversion process in muonic atoms is a complementary approach to the $\mu \rightarrow e\gamma$ search in muon cLFV physics. Muonic atoms are formed

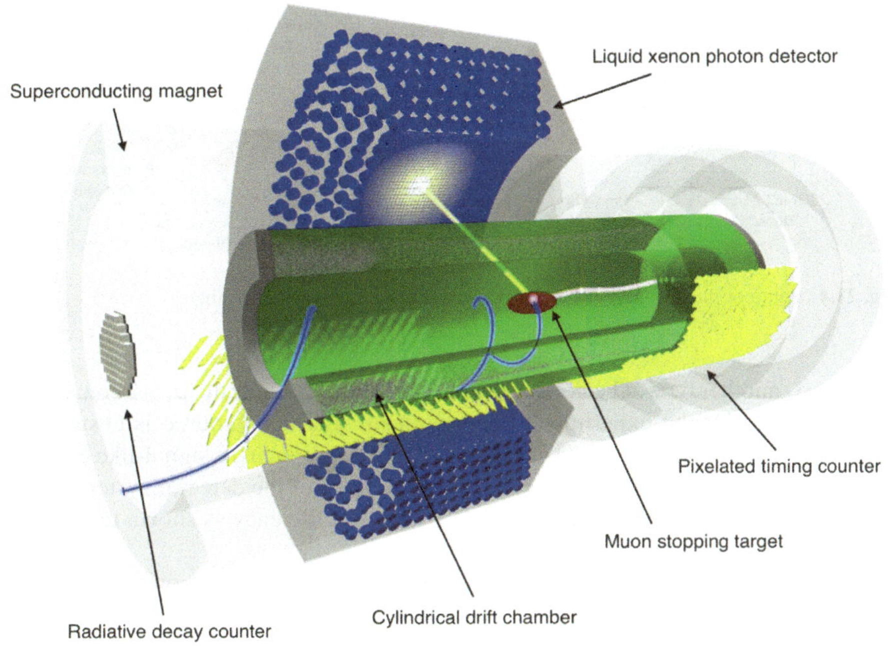

Fig. 21.3 Schematic view of the MEG II experiment at PSI

by stopping μ^- in a material; after being trapped by a nucleus, the muon goes down to the 1s state with emitting X-ray and then is captured by the nucleus ($\mu^- + (A, Z) \rightarrow \nu_\mu + (A, Z-1)$) or decays in orbit ($\mu^- \rightarrow e^- \nu_\mu \bar{\nu}_\mu$). In either case there are neutrino(s) in the final state, conserving the lepton flavor number. On the other hand the $\mu - e$ conversion process does not have any neutrino in the final state, violating the lepton flavor number conservation, and thus is a clear evidence of new physics once it is observed. The electron in the final state will carry energy as large as the muon mass minus binding energy of muon in the muonic atom, and recoil energy of the remaining atom. For the case of muonic aluminum, the $\mu - e$ conversion electron should have energy of 104.97 MeV.

The muon life time in a muonic aluminum is 860 ns, which is shorter than that in free space because of the capture process. The life time is utilized to distinguish the signal from pion induced background; π^0 produced in a charge exchange process decays to two gammas, and then one of them produces an energetic electron in a pair-creation process, which is potentially misidentified as the signal. The process occurs in a prompt timing of pion arrival at the muon stopping target and thus can be removed by requesting the signal to be observed in a delayed timing. The COMET experiment uses the pulsed proton beam of 8 GeV provided at J-PARC. This is to suppress anti-proton production, which potentially induces background when it arrives at the target even with very low energy. It is known that the anti-proton production cross section by proton interaction on a nucleus rapidly decreases below 8 GeV.

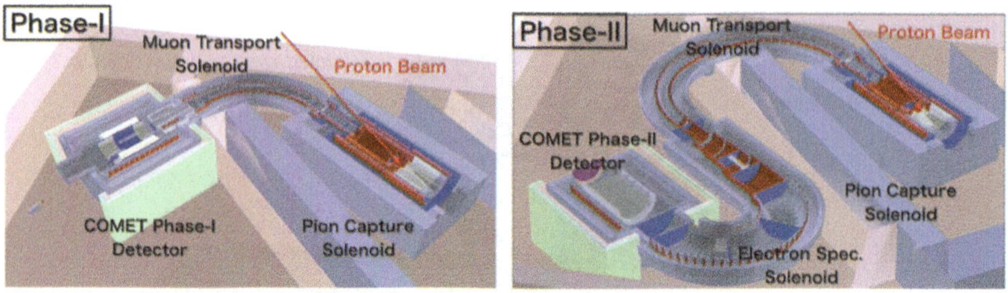

Fig. 21.4 Setup of the COMET experiment Phase I (left) and Phase II (right)

There is another type of background source when we conduct the $\mu - e$ conversion search, caused by the experimental environment. The main source is cosmic-ray muons, arriving at any area of the detector system and producing a signal-like electron in the signal timing window. Therefore the major part of the detector system has to be covered with cosmic-ray veto counters with high efficiency as shown in green in Fig. 21.4.

The COMET experiment intends to achieve the $\mu - e$ conversion search sensitivity better than 10^{-16} using a pulsed proton beam provided at J-PARC. The proton beam pulse-to-pulse timing is optimized to 1.2 μs by filling every two acceleration buckets in the J-PARC Main Ring (MR). The proton beam is extracted after beam acceleration to 8 GeV in the slow-extraction mode with applying acceleration RF voltage during extraction period to keep the pulse structure. The COMET experiment uses a unique system to produce and transport high-intensity pulsed muon beam. A pion production target made of graphite with 1.5 interaction length is located inside a solenoidal magnet with a gradient magnetic field, 5 T at the target location and 3 T at the matching section to the downstream side, to realize efficient collection of pions. Pions, decaying to muon in flight, are transported to the experiment area by a curved solenoid with a magnetic field of 3 T. The curved solenoid, compared to the conventional beam line element combination like dipole and quadrupole magnets, has larger momentum acceptance, transporting higher intensity muon beam. It should also be noted that particle charge can be selected by locating a collimator system in an appropriate location in the curved solenoid [1].

The COMET experiment intends to achieve the final sensitivity in two steps; an intermediate sensitivity better than 10^{-14} is aimed at in Phase I and the final sensitivity better than 10^{-16} in Phase II. Schematic views of the detector system of Phase I and II of the COMET experiment are shown in Fig. 21.4.

In COMET Phase II, the muon transport solenoid magnet is extended as shown in Fig. 21.4 (right) to realize better beam momentum and charge selections, allowing to increase the primary proton beam power above 50 kW. The spectrometer system to identify the signal electron is located after another 180-degree curved solenoid to reduce the detector hit rate by selecting particle momentum from the muon stopping target efficiently.

Facility construction for the COMET experiment is in progress both in the new beam line to transport protons from MR to the experiment area, and the solenoid magnet system including the cryogenic facility to supply liquid helium for its operation. In parallel to facility construction, beam acceleration and extraction test dedicated to the COMET beam operation was carried out in early 2018. Protons were injected to every two J-PARC MR acceleration buckets, in total four out of nine, to realize the required proton pulse time structure. Then they were accelerated to 8 GeV and extracted to the experiment hall, injecting the existing primary target currently used in particle and nuclear physics experiments. Secondary particles generated by proton-nucleus interactions at the target were measured by hodoscope counters to evaluate the proton pulse time structure. The result indicates that pulsing of the proton beam is good enough to suppress the beam related background in the COMET experiment although it was found that there is small amount of proton leakage from a bucket to in-between buckets only in the first 100 μs during beam extraction. This will be further investigated in future studies.

The new beam line to transport protons from MR to the experiment hall will be ready by summer 2020, followed by beam element installation to deliver the proton beam to the COMET experiment area. Construction of the solenoid magnet system will be completed in 2022. Delivery of the proton beam to generate the pulsed muon beam for the COMET experiment is expected after that. The main physics detector CDC is ready and conditioning is in progress. After installing the detector solenoid magnet in the experiment area, a long-term cosmic-ray test is planned to understand possible contribution of cosmic ray to the background even before the whole facility is ready to start the experiment.

21.2.3 Mu2e

The Mu2e experiment is another $\mu - e$ conversion search with a target sensitivity better than 10^{-16} at FNAL. The Mu2e uses the proton beam provided by the FNAL accelerator chain; the proton beam is optimized for $\mu - e$ conversion search as in the case of the COMET experiment. The Mu2e experiment realizes necessary beam pulsing by RF re-bunching in the Delivery Ring and extract it to the experiment area in the slow extraction mode.

A pion production primary target is located in a pion collection solenoid magnet with a gradient magnetic field, 4.6 T at one end and 2.5 T at the other end, realizing collection efficiency as high as possible. The pions and their decay products, muons, produced in the backward direction are transported to a muon stopping target through a two curved solenoid magnets with opposite curvatures forming an S-shape transport solenoid magnet as shown in Fig. 21.5. A collimator is located at the center of the S-shape, and remove positive particles and higher momentum particles.

The Mu2e detector to identify the $\mu - e$ conversion signal electron is composed of two detectors; a straw-tube tracking detector and CsI calorimeter located in the large solenoid magnet containing the muon stopping target. The straw-tube tracker

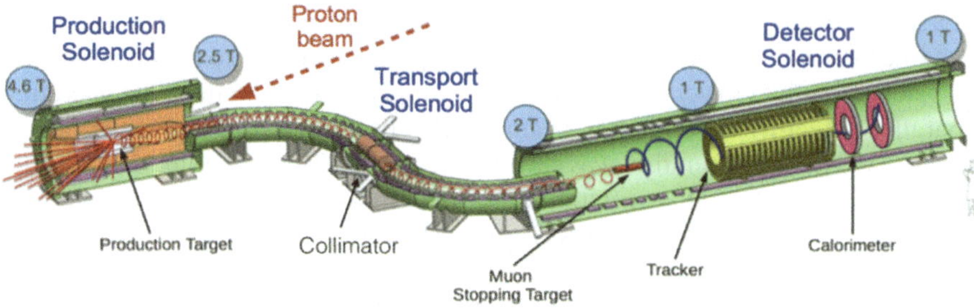

Fig. 21.5 Experimental setup of the Mu2e experiment

measures precisely the momentum of electrons from the muon stopping target made of aluminum, and the CsI calorimeter measures the electron energy. Facility construction and detector preparation is in progress to start the engineering run in 2022, followed by physics data acquisition for three years to achieve the target sensitivity better than 10^{-16}. The setup of the Mu2e experiment is shown in Fig. 21.5.

21.3 Summary and Outlook

Search for muon cLFV is a unique tool to investigate the physics beyond the Standard Model in a complementary way to high-energy frontier experiments. This is simply because muon cLFV processes are strictly forbidden in the Standard Model even if we take the neutrino oscillation into account while there is no fundamental rule to suppress cLFV processes completely in models beyond the SM. Further development of high-power proton accelerator facilities will certainly boost realization of future muon cLFV experiments.

Acknowledgements The author would like to thank the MEG & MEG II, COMET, and Mu2e collaborations for providing materials presented in this article.

References

1. S. Mihara, J. Miller, P. Paradisi, G. Piredda, Ann. Rev. Nucl. Part. Sci. **63**, 531 (2013). https://doi.org/10.1146/annurev-nucl-102912-144530
2. V. Cirigliano et al., Phys. Rev. D **80**, 013002. https://doi.org/10.1103/PhysRevD.80.013002
3. M. Ahmed et al., Phys. Rev. D **65**, 112002. https://doi.org/10.1103/PhysRevD.65.112002
4. J. Adam et al., Euro. Phys. J. C **73**, 2365. https://doi.org/10.1140/epjc/s10052-013-2365-2
5. A.M. Baldini et al., Euro. Phys. J. C **76**, 434. https://doi.org/10.1140/epjc/s10052-016-4271-x

Chapter 22
Heavy Flavour Spectroscopy and Exotic States at LHC

Nairit Sur

Abstract The proton proton collision data collected at high centre of mass energies (7, 8, and 13 TeV) in the Large Hadron Collider provide an excellent environment for precision spectroscopy studies of beauty and charm hadrons. The general purpose experiments, ATLAS and CMS, and the forward-spectrometer experiment LHCb have investigated many interesting aspects of hadron spectroscopy over the years. Some of the latest results on spectroscopy of conventional and exotic hadrons are reviewed.

22.1 Introduction

In the last fifteen years, experimental evidence of new multi-quark bound states has been collected that does not fit the expectations for the conventional qqq baryons or the $q\bar{q}$ mesons. The exact nature of many of these new states still remains a puzzle. Even when confirmed by more than one experiment, most of the quantum numbers of these states have not yet been experimentally determined. Spectroscopic studies of heavy quark states provide deeper understanding of the physics and dynamics of quarks at the hadron level and valuable insight into the various QCD inspired phenomenological models. Proton-proton collisions at high centre of mass energies at the LHC ensure abundant production of $b\bar{b}$ and $c\bar{c}$ pairs which provide great opportunities for performing heavy flavour spectroscopy studies.

On behalf of ATLAS, CMS and LHCb Collaborations.

N. Sur (✉)
Tata Institute of Fundamental Research, 1, Homi Bhabha Road, Colaba, Mumbai 400005, India
e-mail: nairit.sur@cern.ch

A. Giri and R. Mohanta (eds.), *16th Conference on Flavor Physics and CP Violation*, Springer Proceedings in Physics 234,
https://doi.org/10.1007/978-3-030-29622-3_22

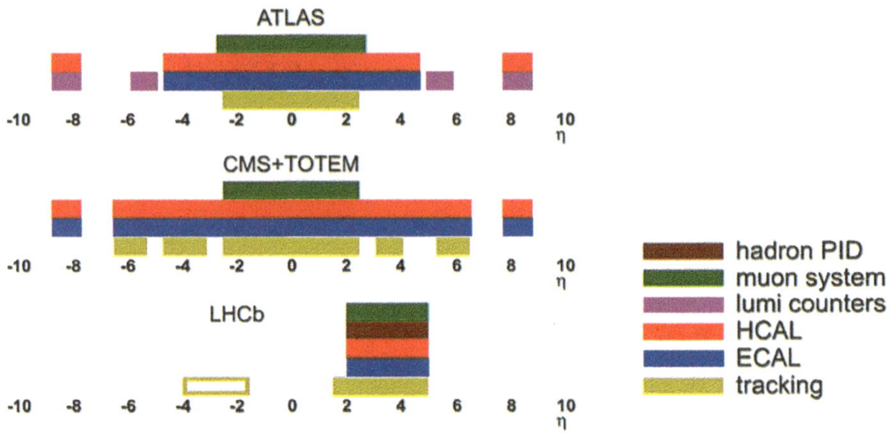

Fig. 22.1 Detector acceptance of the LHC experiments over the pseudorapidity η range

22.1.1 *LHC and the Detectors*

The detectors at the LHC are complementary to each other in the study of heavy flavour spectroscopy by covering different acceptance and kinematic ranges as seen in Fig. 22.1. The general purpose detectors ATLAS [1] and CMS [2] have excellent tracking and muon identification in the central region combined with a robust trigger system to collect data at high luminosities and high pile-up scenarios. LHCb [3] on the other hand is optimized for strongly forward peaked heavy quark production. It has excellent track momentum and vertex resolution, calorimetry and particle identification over the whole acceptance region.

22.2 X(5568)

22.2.1 *X(5568) at the TeVatron*

On Feb 2016, the DØ collaboration reported evidence for a narrow structure with mass $M = 5567.8 \pm 2.9^{+0.9}_{-1.9}$ MeV and width $\Gamma = 21.9 \pm 6.4^{+5}_{-2.5}$ MeV from a possible decay $X(5568)^{\pm} \rightarrow B_S^0\pi^{\pm}$, where $B_S^0 \rightarrow J/\psi\phi$, $J/\psi \rightarrow \mu^+\mu^-$ and $\phi \rightarrow K^+K^-$ [4] (Fig. 22.2). This is a potential tetraquark candidate with all four different quark flavours—b, u, d, s. A similar structure was also seen in the semileptonic decay channel where $B_S^0 \rightarrow \mu^{\mp}D_S^{\pm}X$ and $D_S^{\pm} \rightarrow \phi\pi^{\pm}$ [5]. The high rate of B_S^0 production from X(5568) is rather puzzling. The ratio of the yield of the new state X(5568) to the yield of the B_S^0 meson is given by $\rho_x^{D0} \equiv \frac{\sigma(p\bar{p} \rightarrow X + \text{anything}) \times \mathcal{B}(X \rightarrow B_S^0\pi)}{\sigma(p\bar{p} \rightarrow B_S^0 + \text{anything})} = (8.6 \pm 1.9 \pm 1.4)\%$.

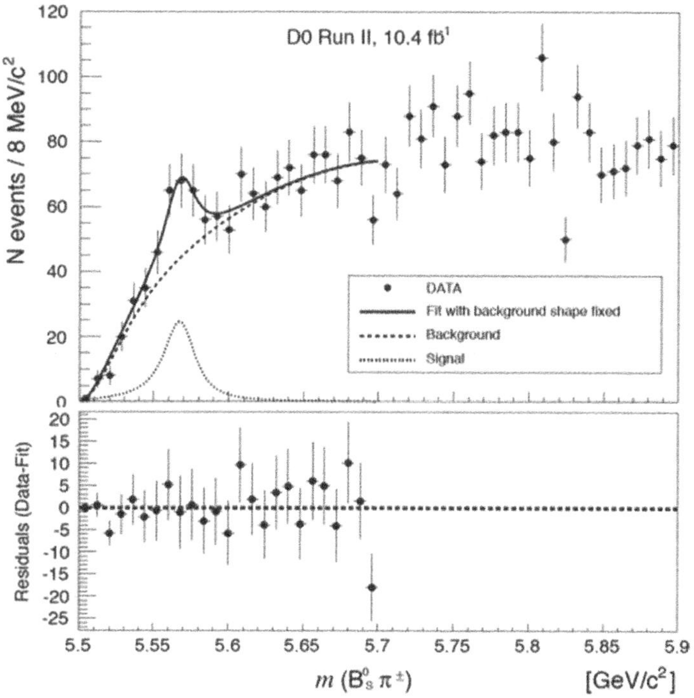

Fig. 22.2 The $B_s^0\pi^\pm$ mass distribution together with the background distribution and the fit results at DØ [4]

22.2.2 *X(5568) at the LHC*

In LHCb The combined $B_S^0 \to D_S\pi$ and $B_S^0 \to J/\psi\phi$ sample in the Run 1 dataset (3 fb^{-1}) is much larger than the DØ sample. Assuming the same production rate, LHCb should observe a huge signal. However, no significant deviation from background is observed in Fig. 22.3.

In ATLAS and CMS For every case of different selection criteria, background parameterization, fit range and method of data description, the yield for X(5568) remains consistent with zero. Stringent limits have been set to ρ_X as seen in Fig. 22.4. In conclusion, ATLAS, CMS, and LHCb do not confirm the presence of the X(5568) state.

22.3 $\chi_{b1}(3P) - \chi_{b2}(3P)$ Mass Split in CMS

Analyzing the full LHC Run 2 dataset (13 TeV, 80 fb^{-1}), CMS has observed for the first time the split in the $\chi_{b1}(3P) - \chi_{b2}(3P)$ doublet and measured the masses of the

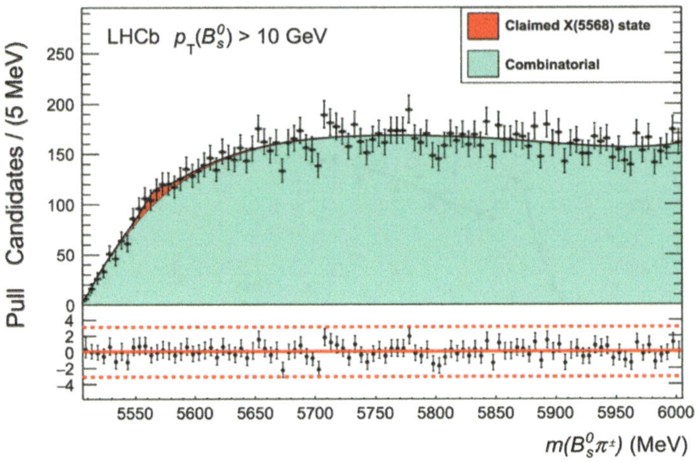

Fig. 22.3 Results of the fit to the $B_S^0\pi^\pm$ mass distribution for candidates (both $B_S^0 \to D_S\pi$ and $B_S^0 \to J/\psi\phi$ modes combined). The component for the claimed X(5568) state is included in the fit but is not significant [6]

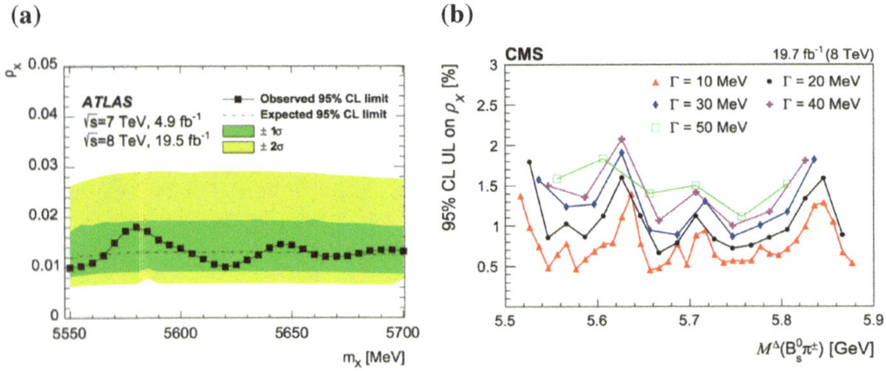

Fig. 22.4 Upper limits on ρ_X at 95% CL at different masses of a hypothetical resonant state $X(5568)^\pm \to B_S^0\pi^\pm$, **a** limits from ATLAS [7], **b** limits from CMS [8]

two states [9]. $\chi_b(3P)$ is reconstructed in the $\Upsilon(3S) + \gamma$ mode where the low energy γ is detected through $\gamma \to e^+e^-$ conversion inside the silicon tracker. The photon energy scale is calibrated using high yield $\chi_{c1} \to J/\psi + \gamma$ processes and tested with $\chi_b(1P, 2P)$ states for high accuracy mass measurements. This observation significantly constrains theoretical predictions, which predict mass splits in a wide range [−2, 18] GeV. This is the first time that the $J = 1$ and $J = 2$ states are well resolved and their masses are individually measured (Fig. 22.5).

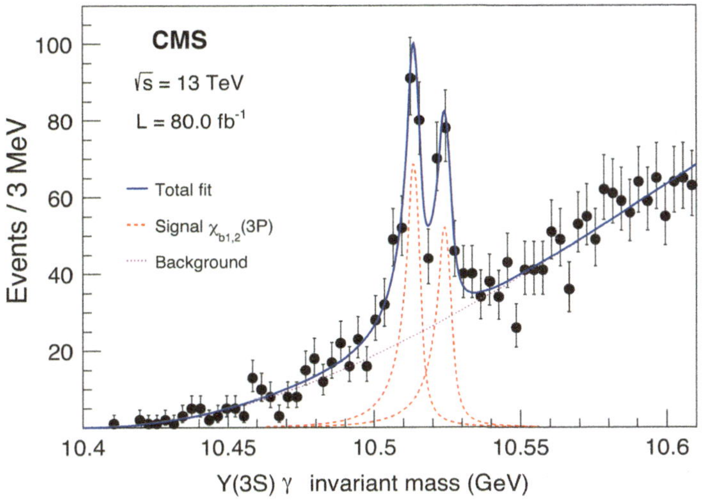

Fig. 22.5 Invariant mass distribution of the $\chi_{bJ}(3P) \rightarrow \Upsilon(3S)\gamma$ [9]; $\chi_{b1}(3P)$ mass M1 = 10513.42 $\pm$ 0.41(stat) $\pm$ 0.18(syst) MeV; $\chi_{b2}(3P)$ mass M2 = 10524.02 $\pm$ 0.57(stat) $\pm$ 0.18(syst) MeV; Mass difference ΔM = 10.6 $\pm$ 0.64(stat) $\pm$ 0.17(syst) MeV

22.4 Beauty-full Tetraquarks at LHCb

It is interesting to notice that none of the (30+) exotic hadrons observed since 2003 contain more than two heavy quarks. Recent studies predict that $b\bar{b}b\bar{b}$ tetraquarks below the $\eta_b\eta_b$ threshold (18.8 GeV) can decay to $\Upsilon(nS)\mu^+\mu^-$ [10]. The corresponding cross section times branching fraction is expected to be ~4 fb and the width ~1.2 MeV. LHCb looked at 6.3 fb^{-1} of 7, 8 and 13 TeV data where $\Upsilon(nS)$ was reconstructed through $\mu^+\mu^-$ decays. But no significant excess was observed in the $M(X_{b\bar{b}b\bar{b}})$ mass range [16, 26] GeV [11]. In the limits on $\sigma \times \mathcal{B}$ in the range [17.5, 20] GeV, the largest deviation from background of 2.5σ (local) is seen at $M(X_{b\bar{b}b\bar{b}})$ = 19.35 GeV which is above the $\eta_b\eta_b$ and $\Upsilon(1S)\Upsilon(1S)$ thresholds (Fig. 22.6). Improved sensitivity for this state will be obtained with data collected during future running periods of the LHC using an upgraded LHCb detector.

22.5 A New $\mathrm{B}^*_{\mathrm{S}2}$ Decay at CMS

CMS studied the excited $\mathrm{B_S}$ states $\mathrm{B_{S1}}(5830)^0$ and $\mathrm{B}^*_{\mathrm{S}2}(5840)^0$ in the B^+K^- and $B^0K^0_S$ decay modes [12] (Fig. 22.7). The study reports the first observation of $B^*_{S2} \rightarrow B^0K^0_s$ and $B_{S1} \rightarrow B^{0*}K^0_s$. A few highlights of this study include the measurement of the B^*_{S2} natural width $\Gamma_{B^*_{S2}} = 1.52 \pm 0.34$(stat) $\pm$ 0.3(syst) MeV and the first

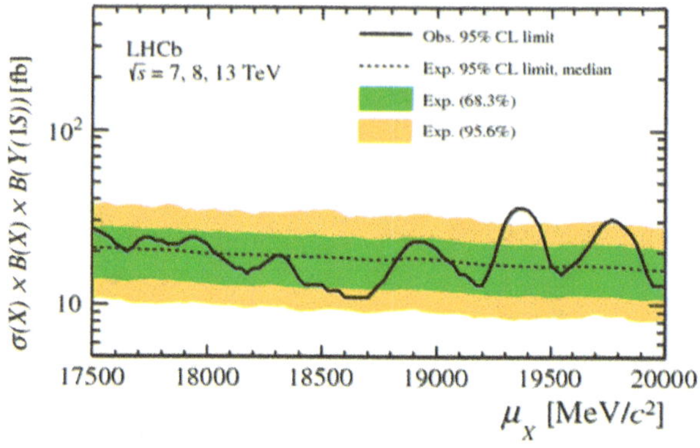

Fig. 22.6 The 95% CL upper limits on $S \equiv \sigma(pp \to X) \times \mathcal{B}(X \to \Upsilon(1S)\mu^+\mu^-) \times \mathcal{B}(X \to \Upsilon(1S)\mu^+\mu^-)$ as functions of the X mass hypothesis [11]

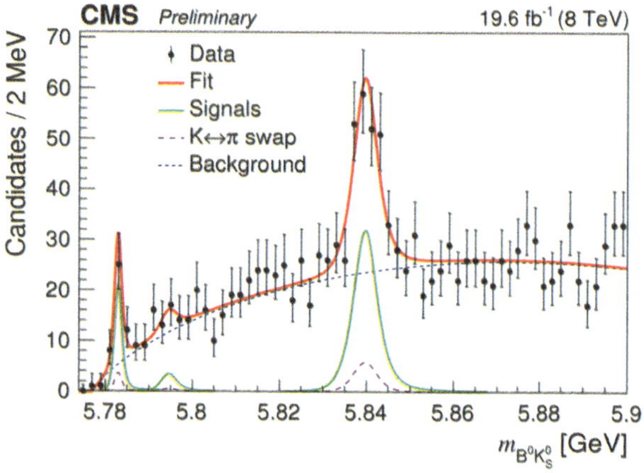

Fig. 22.7 Invariant mass distributions of $B^0K^0_S$ candidates with the results of the fit overlaid. The three peaks correspond to (from the left) $B_{s1} \to B^{0*}K^0_s$, $B_{s2} \to B^{0*}K^0_s$ and $B^*_{s2} \to B^0K^0_s$ respectively [12]

measurement of the mass difference between neutral and charged excited mesons, $m_{B^{*0}} - m_{B^{*+}} = 0.91 \pm 0.24\text{(stat)} \pm 0.09\text{(syst)} \pm 0.02\text{(PDG) MeV}$.

22.6 A New Ξ_b^- Resonance At LHCb

LHCb observed a peak in both $\Lambda^b_0 K^-$ and $\Xi^0_b \pi^-$ spectra with 4.5 fb^{-1} of proton-proton collision data at $\sqrt{s} = 7$, 8 and 13 TeV (Fig. 22.8). This corresponds to a new resonance $\Xi_b(6627)^-$ [13]. Fully reconstructed candidates from its hadronic decay mode $\Xi_b(6627)^- \to \Lambda^b_0 K^-$, $\Lambda^b_0 \to \Lambda^+_c \pi^-$ have been used for mass measurements while the semileptonic decay modes have been used to determine its production. The mass and natural width are measured to be $m_{\Xi_b}(6627)^- = 6226.9 \pm 2.0\text{(stat)} \pm 0.3\text{(sys)} \pm 0.2\text{(PDG) MeV}$ and $\Gamma_{\Xi_b}(6627)^- = 18.1 \pm 5.4\text{(stat)} \pm 1.8\text{(sys) MeV}$. The relative production rate is reported to be $\mathcal{B}(\Xi_b(6627)^- \to \Lambda^b_0 K^-)/\mathcal{B}(\Xi_b(6627)^- \to \Xi^0_b \pi^-) \sim 1 \pm 0.5$. The new state can be either $\Xi_b(1P)^-$ or $\Xi_b(2S)^-$. More

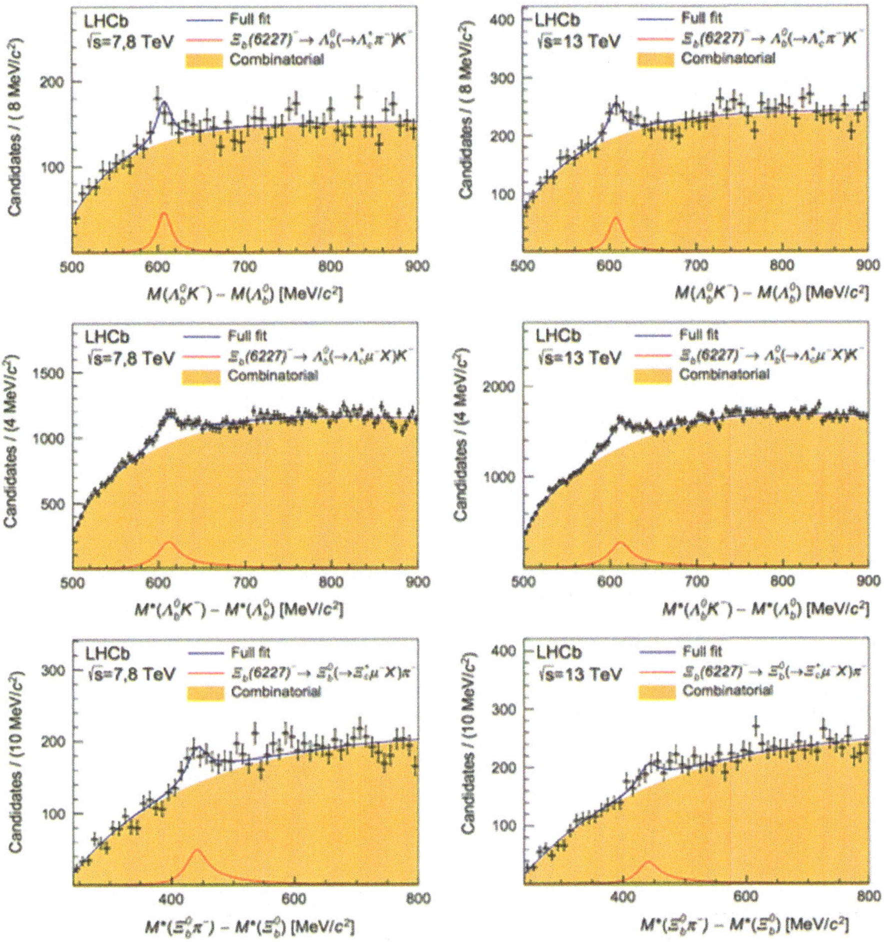

Fig. 22.8 Spectra of mass differences for $\Xi_b(6627)^-$ candidates, reconstructed in the different final states along with the results of the fits [13]. The blue line represents the full fit while the red line represents the signal component and the yellow shaded region represents the combinatorial background

precise measurements of the width and the relative branching fractions could help to determine the J^P quantum numbers of this state.

22.7 $\Lambda_b^0 \to \Lambda_c^+ p\bar{p}\pi^-$ Decay at LHCb

The decay $\Lambda_b^0 \to \Lambda_c^+ p\bar{p}\pi^-$ has been observed by LHCb with 3 fb^{-1} of pp data collected at $\sqrt{s} = 7$ and 8 TeV [14]. The ratio of branching fractions is measured to be:

$$\frac{\mathcal{B}(\Lambda_b^0 \to \Lambda_c^+ p\bar{p}\pi^-)}{\mathcal{B}(\Lambda_b^0 \to \Lambda_c^+\pi^-)} = 0.0540 \pm 0.0023 \pm 0.0032$$

Two resonances have been observed in the $\Lambda_c^+\pi^-$ invariant mass spectrum corresponding to $\Sigma_c(2455)^0$ and $\Sigma_c^*(2520)^0$ resonances (Fig. 22.9). The ratios of branching fractions of the resonant modes with respect to the total are:

$$\frac{\mathcal{B}(\Lambda_b^0 \to \Sigma_c^0 p\bar{p}) \times \mathcal{B}(\Sigma_c^0 \to \Lambda_c^+\pi^-)}{\mathcal{B}(\Lambda_b^0 \to \Lambda_c^+ p\bar{p}\pi^-)} = 0.089 \pm 0.015 \pm 0.006$$

$$\frac{\mathcal{B}(\Lambda_b^0 \to \Sigma_c^{*0} p\bar{p}) \times \mathcal{B}(\Sigma_c^{*0} \to \Lambda_c^+\pi^-)}{\mathcal{B}(\Lambda_b^0 \to \Lambda_c^+ p\bar{p}\pi^-)} = 0.119 \pm 0.020 \pm 0.014$$

However, no evidence of any peaking structure of a suspected dibaryon resonance [cd][ud][ud] has been found in the $\Lambda_c^+\pi^- p$ invariant mass spectrum (Fig. 22.10).

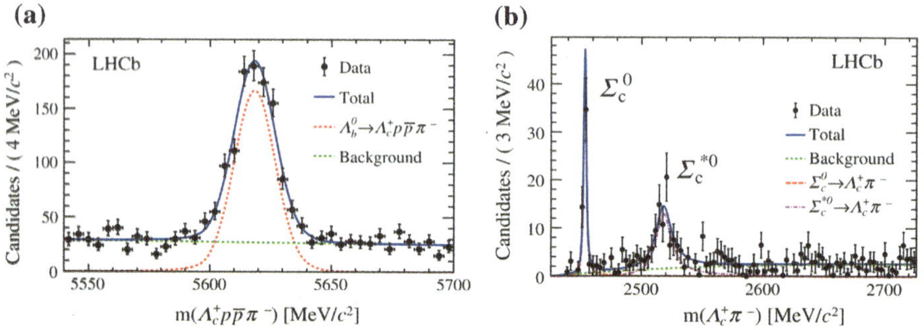

Fig. 22.9 Observation of the decay $\Lambda_b^0 \to \Lambda_c^+ p\bar{p}\pi^-$ by LHCb, **a** $\Lambda_c^+ p\bar{p}\pi^-$ invariant mass, **b** $\Lambda_c^+\pi^-$ invariant mass [14]

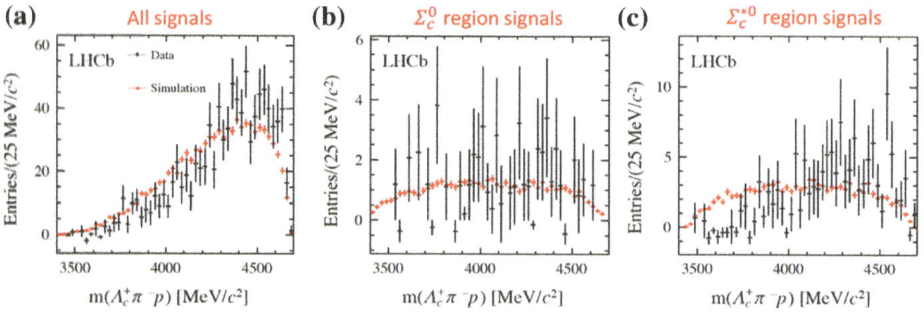

Fig. 22.10 Background-subtracted mass spectrum of the $\Lambda_c^+\pi^- p$ system [14]

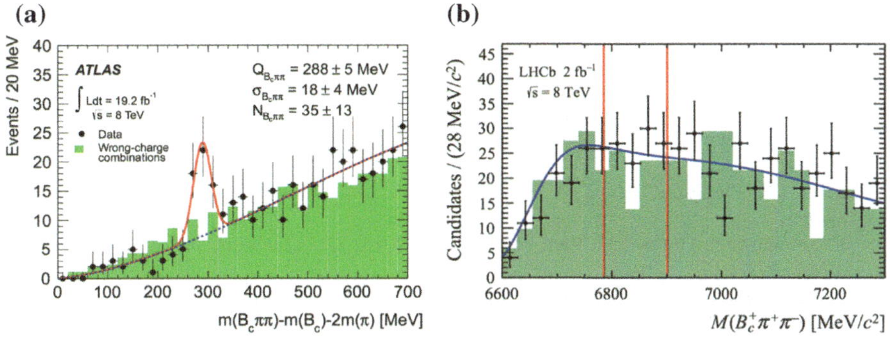

Fig. 22.11 Search for excited $B_c^{\pm}$ states at ATLAS and LHCb, **a** ATLAS result [15], **b** LHCb result [16]

22.8 Searches for Excited B_c States

A narrow resonance in the $B_c\pi^+\pi^-$ mass spectrum, interpreted as the $B_c(2S)$ state was first reported by ATLAS in 2014 [15]. The peak was seen in both 7 and 8 TeV data with M = 6842 ± 4 ± 5 MeV with a 5.2σ global significance (with Look Elsewhere Effect) (Fig. 22.11a). LHCb performed a new search for $B_c(2S)$ with 2 fb^{-1} of 8 TeV pp collision data but no evidence of a resonance was seen [16] (Fig. 22.11b). Upper limits on the ratio of $\sigma \times \mathcal{B}$ of the $B_c(2S)$ relative to the B_c cross section in the range [0.02, 0.14] @ 95% CL (as a function of mass) have been provided. The ATLAS result still awaits confirmation.

22.9 Summary

The LHC experiments are significantly contributing to exotic hadron spectroscopy and will continue to do so with Run-II data while facing new experimental challenges. Although designed for high p_T physics, ATLAS and CMS have produced valuable results in heavy flavour physics. Spectroscopy measurements keep filling in the missing tiles of the spectroscopy puzzle. Among the latest results are the observation of a new Ξ_b^- resonance and of a new Λ_b decay by LHCb, and of a new B_{S2}^* decay by CMS. The measurement of the $\chi_{b1}(3P) - \chi_{b2}(3P)$ mass splitting by CMS rules out most models and favours those predicting a direct mass hierarchy. The X(5568) state claimed to have been seen by DØ at the TeVatron has not been confirmed by ATLAS, CMS, and LHCb. An unresolved discrepancy remains between LHCb and ATLAS for the existence of the $B_c(2S)$ state. Many more fascinating results are anticipated from the LHC experiments in the near future.

References

1. ATLAS Collaboration, JINST **3**, S08003 (2008). https://doi.org/10.1088/1748-0221/3/08/S08003
2. CMS Collaboration, JINST **3**, S08004 (2008). https://doi.org/10.1088/1748-0221/3/08/S08004
3. LHCb Collaboration, JINST **3**, S08005 (2008). https://doi.org/10.1088/1748-0221/3/08/S08005
4. D0 Collaboration, Phys. Rev. Lett. **117**, 022003 (2016). https://doi.org/10.1103/PhysRevLett.117.022003
5. D0 Collaboration, Phys. Rev. D **97**, 092004 (2018). https://doi.org/10.1103/PhysRevD.97.092004
6. LHCb Collaboration, Phys. Rev. Lett. **117**, 152003 (2016). https://doi.org/10.1103/PhysRevLett.117.152003
7. ATLAS Collaboration, Phys. Rev. Lett. **120**, 202007 (2018). https://doi.org/10.1103/PhysRevLett.120.202007
8. CMS Collaboration, Phys. Rev. Lett. **120**, 202005 (2018). https://doi.org/10.1103/PhysRevLett.120.202005
9. CMS Collaboration, Phys. Rev. Lett. **121**, 092002 (2018). https://doi.org/10.1103/PhysRevLett.121.092002
10. M. Karliner et al., Phys. Rev. D **95**, 034011 (2017). https://doi.org/10.1103/PhysRevD.95.034011
11. LHCb Collaboration (2018). arXiv:1806.09707
12. CMS Collaboration (2018). http://cms-results.web.cern.ch/cms-results/public-results/publications/BPH-16-003/
13. LHCb Collaboration, Phys. Rev. Lett. **121**, 072002 (2018). https://doi.org/10.1103/PhysRevLett.121.072002
14. LHCb Collaboration (2018). arXiv:1804.09617
15. ATLAS Collaboration, Phys. Rev. Lett. **113**, 212004 (2014). https://doi.org/10.1103/PhysRevLett.113.212004
16. LHCb Collaboration, J. High Energ. Phys. **138** (2018). https://doi.org/10.1007/JHEP01(2018)138

Chapter 23
Heavy Neutral Lepton Search at NA62

Nora Patricia Estrada-Tristán

Abstract In this paper we present the results of the search for heavy neutral lepton (HNL) production in charged kaon decays using a data sample collected with a minimum bias trigger by the NA62 experiment at CERN in 2015. Upper limits at

---

For the NA62 Collaboration: R. Aliberti, F. Ambrosino, R. Ammendola, B. Angelucci, A. Antonelli, G. Anzivino, R. Arcidiacono, M. Barbanera, A. Biagioni, L. Bician, C. Biino, A. Bizzeti, T. Blazek, B. Bloch-Devaux, V. Bonaiuto, M. Boretto, M. Bragadireanu, D. Britton, F. Brizioli, M. B. Brunetti, D. Bryman, F. Bucci. T. Capussela, A. Ceccucci, P. Cenci, V. Cerny, C. Cerri, B. Checcucci, A. Conovaloff, P. Cooper, E. Cortina Gil, M. Corvino, F. Costantini, A. Cotta Ramusino, D. Coward, G. D'Agostini, J. Dainton, P. Dalpiaz, H. Danielsson, N. De Simone, D. Di Filippo, L. Di Lella, N. Doble, B. Dobrich, F. Duval, V. Duk, J. Engelfried, T. Enik, N. Estrada-Tristan, V. Falaleev, R. Fantechi, V. Fascianelli, L. Federici, S. Fedotov, A. Filippi, M. Fiorini, J. Fry, J. Fu, A. Fucci, L. Fulton, E. Gamberini, L. Gatignon, G. Georgiev, S. Ghinescu, A. Gianoli, M. Giorgi, S. Giudici, F. Gonnella, E. Goudzovski, C. Graham, R. Guida, E. Gushchin, F. Hahn, H. Heath, T. Husek, O. Hutanu, D. Hutchcroft, L. Iacobuzio, E. Iacopini, E. Imbergamo, B. Jenninger, K. Kampf, V. Kekelidze, S. Kholodenko, G. Khoriauli, A. Khotyantsev, A. Kleimenova, A. Korotkova, M. Koval, V. Kozhuharov, Z. Kucerova, Y. Kudenko, J. Kunze, V. Kurochka, V. Kurshetsov, G. Lanfranchi, G. Lamanna, G. Latino, P. Laycock, C. Lazzeroni, M. Lenti, G. Lehmann Miotto, E. Leonardi, P. Lichard, L. Litov, R. Lollini, D. Lomidze, A. Lonardo, P. Lubrano, M. Lupi, N. Lurkin, D. Madigozhin, I. Mannelli, G. Mannocchi, A. Mapelli, F. Marchetto, R. Marchevski, S. Martellotti, P. Massarotti, K. Massri, E. Maurice, M. Medvedeva, A. Mefodev, E. Menichetti, E. Migliore, E. Minucci, M. Mirra, M. Misheva, N. Molokanova, M. Moulson, S. Movchan, M. Napolitano, I. Neri, F. Newson, A. Norton, M. Noy, T. Numao, V. Obraztsov, A. Ostankov, S. Padolski, R. Page, V. Palladino, C. Parkinson, E. Pedreschi, M. Pepe, M. Perrin-Terrin, L. Peruzzo, P. Petrov, F. Petrucci, R. Piandani, M. Piccini, J. Pinzino, I. Polenkevich, L. Pontisso, Yu. Potrebenikov, D. Protopopescu, M. Raggi, A. Romano, P. Rubin, G. Ruggiero, V. Ryjov, A. Salamon, C. Santoni, G. Saracino, F. Sargeni, V. Semenov, A. Sergi, A. Shaikhiev, S. Shkarovskiy, D. Soldi, V. Sougonyaev, M. Sozzi, T. Spadaro, F. Spinella, A. Sturgess, J. Swallow, S. Trilov, P. Valente, B. Velghe, S. Venditti, P. Vicini, R. Volpe, M. Vormstein, H. Wahl, R. Wanke, B. Wrona, O. Yushchenko, M. Zamkovsky, A. Zinchenko.

N. P. Estrada-Tristán (✉)
Universidad de Guanajuato, Guanajuato, Mexico
e-mail: nora.patricia.estrada.tristan@cern.ch

Universidad Autónoma de San Luis Potosi, San Luis Potosi, Mexico

A. Giri and R. Mohanta (eds.), *16th Conference on Flavor Physics and CP Violation*, Springer Proceedings in Physics 234,
https://doi.org/10.1007/978-3-030-29622-3_23

the 10^{-7}–10^{-6} level are established on the elements $|U_{e4}|$ and $|U_{\mu 4}|$ of the extended neutrino mixing matrix for HNL masses in the range 170–448 MeV/c^2 and 250–376 MeV/c^2, respectively.

23.1 Introduction

The Standard Model (SM) of particle physics requires an extension to explain, among other things, neutral lepton masses, wich are very well established due to the observation of neutrino oscillations. One of this SM extensions, the Neutrino Minimal Standard Model (νMSM), propose three massive, right-handed "sterile" neutrinos, also called heavy neutral leptons (HNL), which mix with the ordinary light "active" neutrinos, explaining also dark matter and baryon asymmetry in the universe (BAU) [1]. The lightest HNL, with mass of $\mathcal{O}(10$ keV/c$^2)$, is a dark matter candidate. The other two HNL have masses of $\mathcal{O}(1$ GeV/c$^2)$, produce the SM neutrino masses through the see-saw mechanisms and introduce extra CP violating phases to account for BAU. A mixing matrix U describes the interactions between HNL and SM leptons. The mass range and the small mixing angles predicted in the νMSM make HNL long lived, with mean free paths of $\mathcal{O}(10$ km) or longer, and production branching fractions of $\mathcal{O}(10^{-10})$ or smaller.

The mixing between HNL (denoted N below) and active light neutrinos gives rise to HNL production in meson decays, including $K^+ \to \ell^+ N (\ell = e, \mu)$ [2, 3].

Here we report on a search for $K^+ \to \ell^+ N$ decays in the HNL mass range 170–448 MeV/c^2 using a data sample collected with a minimum bias trigger by the NA62 experiment at CERN during the first physics data-taking in 2015. The obtained upper limits on $|U_{\ell 4}|^2$ complement, and improve on those obtained in earlier HNL production searches [4–8].

23.2 NA62: Beam, Detector and Data

23.2.1 The NA62 Detector

NA62 is the lastest generation kaon experiment at the CERN SPS, aiming to measure the BR of the ultra-rare $K^+ \to \pi^+ \nu \bar{\nu}$ decay with 10% accuracy, which requires a high kinematic rejection power, effective photon and muon rejection and excellent particle identification. The layout of the NA62 detector is shown in Fig. 23.1. A primary beam of protons extracted from the CERN SPS is used to produce a secondary positive hadron beam with a central momentum of 75 GeV/c and 1% momentum spread (rms). Kaons from the secondary beam are tagged by a differential Cherenkov counter. Beam particle momenta are measured by a silicon pixel detector (GTK, under commissioning in 2015 and not used in this analysis). A 75 m long fiducial

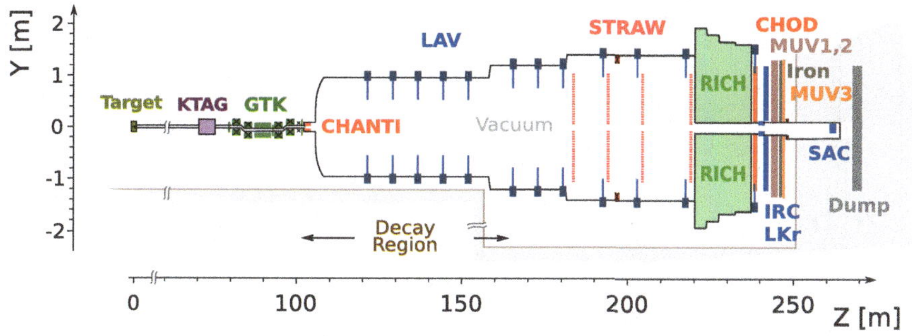

Fig. 23.1 NA62 schematic side view of beam line and detector

decay volume in vacuum (FV) follows the last GTK station. In the decay region, a spectrometer made of 4 chambers with straw tubes in vacuum and a dipole magnet measures track directions and momenta (STRAW). Pions and muons are separated by a ring imaging Cherenkov (RICH) detector and muon detectors (MUV1-3). A system of calorimeters (LKr, LAVs, IRC, SAC) detects photons at different acceptances. The nominal instantaneous beam particle rate is 750 MHz, mostly due to π^+ (70%), protons (23%) and K^+(6%). About 13% of the kaons decay in the FV, leading to about 5 MHz nominal K^+ rate in the FV. The beam intensity during the 2015 run was typically $\mathcal{O}(1\%)$ of the nominal value. Details of the NA62 beam and detector can be found in [9].

23.2.2 Selection Criteria

Assuming $|U_{\ell 4}|^2 < 10^{-4}$ and HNL decaying into SM particles [10], the smallest possible average decay length of a HNL produced in the $K^+ \to \ell^+ N$ decays in NA62 conditions is longer than 10 km. Under the above assumption, the probability of detecting a HNL decay in the FV can be neglected.

The $K^+ \to \ell^+ N$ decay is characterized by a single detected track in the final state, similarly to the SM $K^+ \to \ell^+ \nu$ decay. The principal selection criteria are listed below.

- We require a single positively charged track reconstructed in the spectrometer with momentum between 5 and 70 GeV/c, with no additional spectrometer tracks and LKr energy deposition clusters not geometrically compatible with the track within $\pm$100 ns of the track time measured by the CHOD and no activity in the large-angle and small-angle photon veto detectors and the CHANTI detector within $\pm$10 ns of the track time. Track impact points in the STRAW chambers, LKr calorimeter, CHOD and MUV1-3 detectors have to be within their geometrical acceptances.

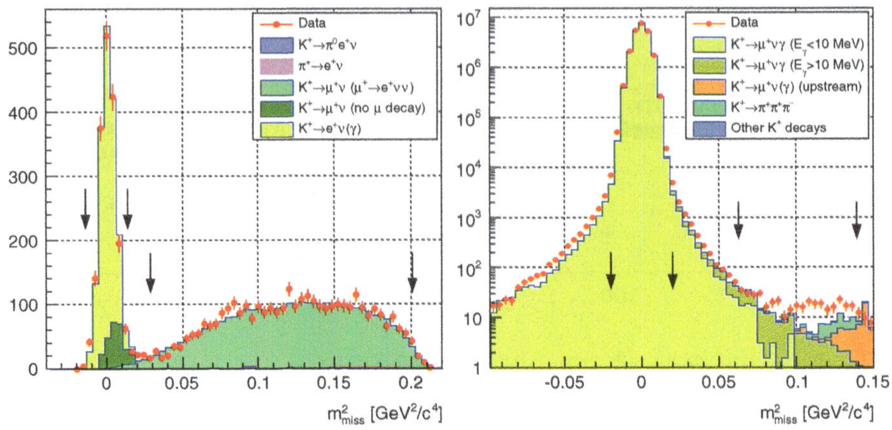

Fig. 23.2 Distributions of the m^2_{miss} variable for data and simulated events for the e^+ case (left) for the μ^+ case (right)

- The kaon decay vertex is reconstructed as the point of closest approach of the track and the beam axis (CDA), taking into account the measured stray magnetic field map in the vacuum tank, should be less than 25 mm and located in the FV.
- A signal in the KTAG is required within ± 10 ns of the track time measured by the CHOD
- The ratio E/p deposited in the calorimeter LKr must be in the range $0.9 < E/p < 1.15$ for positrons and $E/p < 0.2$ for muons.

The strategy of analysis was developed in 2007. The expected HNL signature is a peak above background in the distribution of the event squared missing mass computed as $m^2_{\mathrm{miss}} = (P_K - P_\ell)^2$, where P_K and P_ℓ are the kaon and lepton 4-momenta, respectively. P_K is obtained from the beam average 3-momentum in the K^+ mass hypothesis (verified with $K^+ \to \pi^+\pi^+\pi^-$ decays), while P_ℓ is evaluated from the reconstructed track 3-momentum in the corresponding ℓ^+ mass hypothesis.

Simulation of particle interactions with the detector and its response was done with Monte Carlo (MC). The m^2_{miss} spectra of the selected events from both data and simulation are displayed in Fig. 23.2. Signals from the SM leptonic decays $K^+ \to \ell^+\nu$ are observed as peaks at $m^2_{\mathrm{miss}} = 0$ with m^2_{miss} resolutions of 4.9×10^{-3} GeV^2/c^4 in the e^+ case and 4.7×10^{-3} GeV^2/c^4 in the μ^+ case.

23.2.3 Search for HNL Production

We did mass scans in the HNL signal regions with a step size of 1 MeV/c^2. The event selection used for each HNL mass hypothesis includes that the reconstructed missing mass should be within $\pm 1.5\sigma^\ell_m$ of the assumed HNL mass, where σ^ℓ_m is the

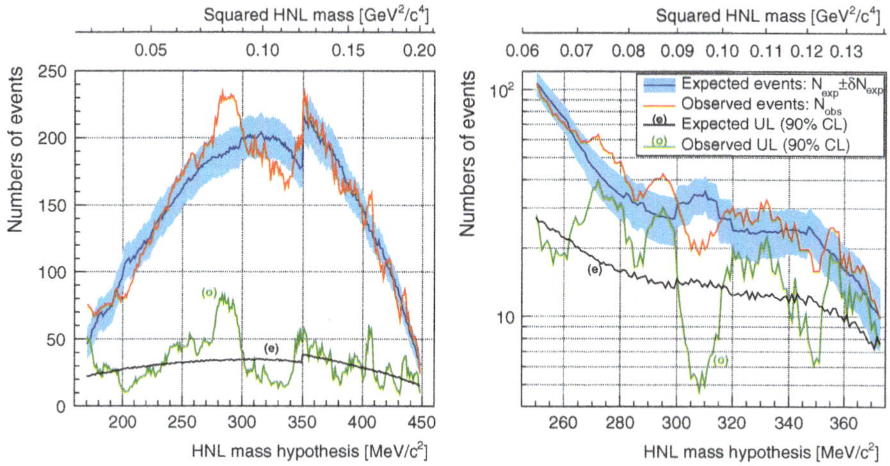

Fig. 23.3 Numbers of expected (N_{exp}) and observed (N_{obs}) events, together with the uncertainty δN_{exp}, shown by the blue band, for each NHL mass hypothesis. e^+ case left, μ^+ case right

mass resolution evaluated with MC. Also acceptances of the selections A_ℓ^N (including the $\pm 1.5\sigma_m^\ell$ mass cut) as functions of HNL mass were obtained with MC [11].

In each HNL mass hypothesis, the background was evaluated from sidebands of the data m_{miss} distribution using polynomial functions of third order. Statistical uncertainties δN_{exp} on the background estimates N_{exp} were computed by propagation of statistical errors on the fit function parameters. Systematic uncertainties on N_{exp} are negligible (typically 1%).

In each HNL mass hypothesis, the total number of observed events N_{obs} within the $\pm 1.5\sigma_m^\ell$ search window, the number of expected background events N_{exp} and its uncertainty δN_{exp} were used to compute confidence intervals for the number of observed $K^+ \to \ell^+ N$ decays N_S^ℓ using the Rolke-López method [12]. The values of N_{exp}, δN_{exp} and N_{obs} in each HNL mass hypothesis are shown in Fig. 23.3.

This method was exploited on a sample of $N_K \sim 4 \times 10^8$ kaon decays, searching for HNL in a mass signal region between 170 and 448 MeV/c^2 for the K_{e2} case and in a mass signal region between 250 and 373 MeV/c^2 for the $K_{\mu 2}$ case. No signal with significance above 3 sigma has been found. New upper limits of $\mathcal{O}(10^{-7})$–$\mathcal{O}(10^{-6})$ have been established on the mixing matrix parameters $|U_{e4}|^2$ and $|U_{\mu 4}|^2$, improving the previous ones Fig. 23.4.

23.3 Summary

A search for HNL production in $K^+ \to \ell^+ N$ decays was done with the NA62 data of 2015, at 1% of the nominal beam intensity with a minimum bias trigger. Upper limits have been established at the level between 10^{-7} and 10^{-6} on the HNL mixing

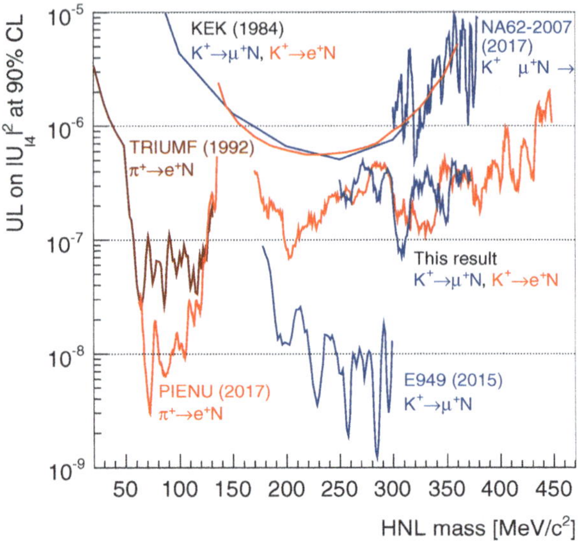

Fig. 23.4 Upper limits at 90% CL on $|U_{\ell 4}|^2$ obtained for each assumed HNL mass compared to the limits established by earlier HNL production searches in π^+ decays: TRIUMF (1992) [4], PIENU (2017) [5] and K^+ decays: KEK (1984) [6], E949 (2015) [7], NA62-2007 (2017) [8]

parameters $|U_{e4}|^2$ and $|U_{\mu 4}|^2$ in the ranges 170–448 MeV/c^2 and 250–373 MeV/c^2, respectively. This improves on the previous limits from HNL production searches over the whole mass range considered for $|U_{e4}|^2$, and above m_N = 300 MeV/c^2 for $|U_{\mu 4}|^2$.

References

1. T. Asaka, M. Shaposhnikov, Phys. Lett. B **620**, 17 (2005)
2. R. Shrock, Phys. Lett. B **96**, 159 (1980)
3. R. Shrock, Phys. Rev. D **24**, 1232 (1981)
4. D. Britton et al., Phys. Rev. D **46**, R885 (1992)
5. A. Aguilar-Arevalo et al. arXiv:1712.03275
6. T. Yamazaki et al., *Proceedings of Neutrino 1984 Conference*
7. A. Artamonov et al., Phys. Rev. D **91**, 052001 (2015)
8. C. Lazzeroni et al., Phys. Lett. B **772**, 712 (2017)
9. E. Cortina Gil et al., J. Instrum. **12**, P05025 (2017)
10. D. Gorbunov, M. Shaposhnikov, JHEP **0710**, 015 (2007)
11. E. Cortina Gil et al., Search for heavy neutral lepton production in K^+ decays. Phys. Lett. **B778**, 137–145 (2018)
12. W.A. Rolke, A.M. Lopez, Nucl. Instrum. Meth. **A458**, 745 (2001)

Chapter 24
Insight into Multiple Partonic Interactions and Production of Charmonia in $p + p$ Collisions at the LHC Energies

Raghunath Sahoo, Dhananjaya Thakur, Sudipan De and Soumya Dansana

Abstract At the LHC energies, the underlying observables are of major topics of interest in high multiplicity $p + p$ collisions. Multiple Parton Interactions (MPI) is one of them, in which several interactions occur in a single $p + p$ collision. It is believed that MPI is the main reason behind the high multiplicity in $p + p$ collisions at the LHC. MPIs not only affect the soft particle production, but also as per the recent ALICE results it can affect the production of hard-particles. In such cases, the self normalized yield of heavy particle like J/ψ shows an increasing trend with event multiplicity. In the present contribution, we discuss the energy and multiplicity dependence of charmonium production to understand the effects of MPI.

24.1 Introduction

The understanding of event structure in hadronic collisions is a very challenging task. It is said that the physics associated with it is sub divided into a number of components, like hard central interactions, fragmentation of beam remnant, multi-partonic interactions (MPI), and initial and final state radiation (ISR and FSR) etc. Technically, this is called as underlying event, which is the sum of all the processes that build up the final hadronic state in a collision. Among all, MPI is of great interest at the LHC energies. Earlier, it was thought that MPI can only affect soft-particle production. But, recent results of production of heavy particles like D-meson and J/ψ

R. Sahoo (✉) · D. Thakur · S. De
Discipline of Physics, School of Basic Sciences, Indian Institute of Technology Indore, Indore 453552, India
e-mail: Raghunath.Sahoo@cern.ch

S. Dansana
Department of Physical Sciences, Indian Institute of Science Education and Research, Kolkata 741246, India

A. Giri and R. Mohanta (eds.), *16th Conference on Flavor Physics and CP Violation*, Springer Proceedings in Physics 234,
https://doi.org/10.1007/978-3-030-29622-3_24

as a function of charged-particle multiplicity at $\sqrt{s} = 7$ and 13 TeV [1–3] reveal that it has also effect on hard-particle production. ALICE experiment has observed a linear increase of open-charm and J/ψ production as a function of multiplicity for $\sqrt{s} = 7$ TeV. Preliminary result of J/ψ via di-electron channel at $\sqrt{s} = 13$ TeV also shows a faster than liner increase of J/ψ production with charged-particle multiplicity. QCD inspired models like PYTHIA6 could not explain the behavior as MPI process therein can only affect soft-processes. An updated version of PYTHIA has been proposed, PYTHIA8 [4], where MPI plays an important role in the production of heavy quarks like charm and bottom. PYTHIA8 describes the increasing trend of heavy-flavor versus multiplicity at $\sqrt{s} = 7$ and 13 TeV. Along with MPI, color reconnection (CR) is an important ingredient in PYTHIA8, which describes the interactions between color field during hadronisation. CR is expected to occur in a significant rate at the LHC due to high number of color partons from MPI and parton shower. As we do not have J/ψ versus multiplicity results for all the LHC energies, we have made an attempt to study energy and multiplicity dependence of J/ψ production using PYTHIA8. In particular, we have studied the effect of MPI and CR on the production of J/ψ and its behavior with respect to charged-particle multiplicity and $\sqrt{s}$.

24.1.1 The Multiple Parton Interaction and J/ψ Production

At the LHC with very high $\sqrt{s}$, the number of interactions in proton+proton collisions depend on the impact parameter (b), where proton is thought to be an extended object. Therefore matter distribution inside hadron (proton) is introduced. So small impact parameter leads to large multiplicity and hence more MPIs. At the LHC, a new QCD regime can be reached, where MPIs occur with high rates, due to unprecedentedly high parton densities in the colliding hadrons. Hence MPI, in particular can be defined as soft or hard interaction, which can happen in parallel in a single $p + p$ collision. Particularly, the very first interaction is hard in nature and the subsequent collisions can be semi-hard or soft. A schematic picture of MPI structure is shown in the Fig. 24.1 (left) for a better visualization.

There is always a finite chance to produce heavy particles in first $2 \rightarrow 2$ collisions. But, The second hard can produce heavy quarks like charm and bottom, if it is hard enough and this leads to a correlation between heavy-flavor and charged-particle multiplicity. If former is true, then the self normalized yield of J/ψ with respect to minimumbias as a function of charged-particle multiplicity should show a decreasing trend, as observed in PYTHIA6 [1]. This is because, PYTHIA6 considers $2 \rightarrow 2$ interaction in hadronic collisions. If later is true, then only we can observe an increasing trend of self normalized J/ψ yield versus multiplicity.

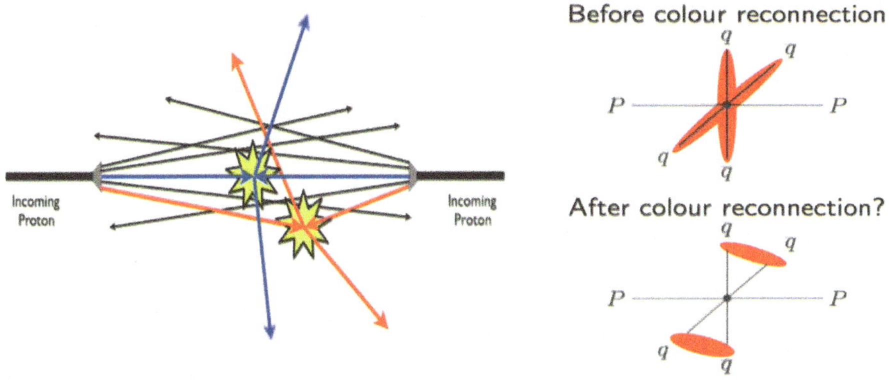

Fig. 24.1 Left: Event structure of a simple $p + p$ collision showing MPI. Right: An illustrative example of color reconnection mechanism with string minimisation

24.2 Analysis Procedure

Different theoretical models, which explain the physics of $p + p$ collisions, have MPI inbuilt in various ways. In our current study, we have used PYTHIA 8.2 tuned 4C [5]. As discussed in the introduction section, to achieve the objective of the study, we have included varying impact parameter (MultipartonInteractions:bProfile = 3) to allow all incoming partons to undergo hard and semi-hard interactions as well. We have used the MPI- based scheme of color reconnection (CR). In this scheme, the produced partons undergo a reconnection in which partons from lower-p_T MPI systems are added to the dipoles defined by the higher-p_T MPI system in such a way that minimizes total string length. An illustration to this is shown in a cartoon in Fig. 24.1 (right). After the color reconnection, all the produced partons, connected with strings, fragment into hadrons via the Lund string model.

We have performed this study by simulating the inelastic, non-diffractive component of the total cross section for all hard QCD processes (HardQCD: all = on), which includes the production of heavy quarks. A p_T cut of 0.5 GeV/c is used to avoid the divergences of QCD processes in the limit $p_T \rightarrow 0$. The charged-particle multiplicity measurement has been performed in the mid-rapidity ($|y| < 1.0$), where as J/ψ has been measured at forward rapidity ($2.5 < y < 4.0$), to cope with the ALICE measurement. Here to be noted that we are not directly studying yield of J/ψ versus charged-particle multiplicity, rather self-normalized yields, where the charged-particle and J/ψ yield in multiplicity bins are normalized to its minimumbias yield. The purpose of studying this kind of ratio is to see, how the physics of multiplicity classes are different from that of minimum bias. Hence one can comment on MPI and other UE observables. The relative charged-particle multiplicity yield is defined as $N_{\rm ch}/\langle N_{\rm ch}\rangle$. Where $N_{\rm ch}$ is the mean of the charged-particle multiplicity in a particular bin and $\langle N_{\rm ch}\rangle$ is the mean of the charged-particle multiplicity in minimum-bias events. The charged-particle multiplicity has been sliced taking care of significant number of J/ψ, and the self-normalized J/ψ yield is calculated as,

$$\frac{\frac{dN_{J/\psi}}{dy}}{\langle\frac{dN_{J/\psi}}{dy}\rangle} = \frac{N^{i}_{J/\psi}}{N^{\text{integrated}}_{J/\psi}} \times \frac{N^{\text{integrated}}_{\text{MB}}}{N^{i}_{\text{MB}}} \tag{24.1}$$

where ($N^{i}_{J/\psi}$, $N^{\text{integrated}}_{J/\psi}$) and ($N^{i}_{\text{MB}}$, $N^{\text{integrated}}_{\text{MB}}$) are the corrected number of J/ψ and number of minimum bias events in ith multiplicity bin and integrated multiplicity bins, respectively. The detailed about the uncertainty calculation of each quantity can be found in [6]. These uncertainties are propagated using the standard error propagation formula. We have tried to do multiplicity binning in such a way that, we can do direct comparison with experimental results of ALICE.

To check the compatibility of PYTHIA8 with experimental data, first we compare the basic distribution like integrated transverse momentum and rapidity spectra with experimental results using the same phase space cuts. We found a good agreement between Monte Carlo (MC) and experimental data. Rapidity spectra give around 1% maximum deviation. Where as for p_T-spectra the maximum deviation is ~50–60% for certain p_T bins, otherwise in most of the p_T bins, the deviation is around 10–20% for $\sqrt{s} = 0.9, 2.76, 5.02, 7$ and 13 TeV. After performing these compatibility studies, we compare the relative yield of J/ψ as a function of charged particle multiplicity for available experimental data at $\sqrt{s} = 7$ TeV. It can be seen from Fig. 24.2, that experimental data are well described by the MC.

These measurements provide us the confidence to extend the study of quarkonia production using PYTHIA8 in $p + p$ collisions at LHC energies and perform the multiplicity and energy dependence analysis.

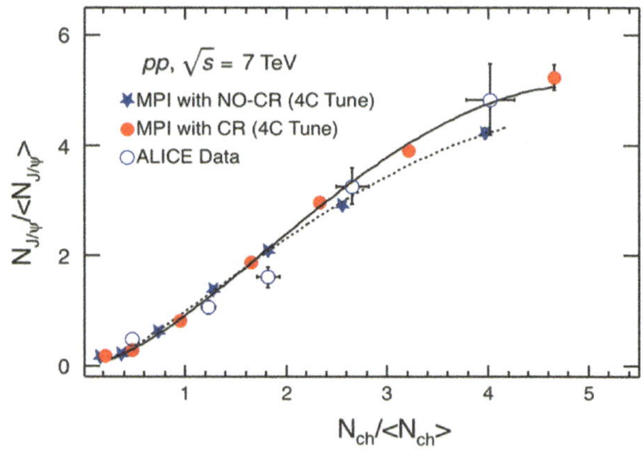

Fig. 24.2 Relative J/ψ yield as a function of relative charged-particle multiplicity at $\sqrt{s} = 7$ TeV at the forward rapidity ($2.5 < y < 4.0$). ALICE data [1] and PYTHIA8 (with CR, w/o CR) comparison is shown. The lines are the fitted curves using the percolation model inspired function [6]

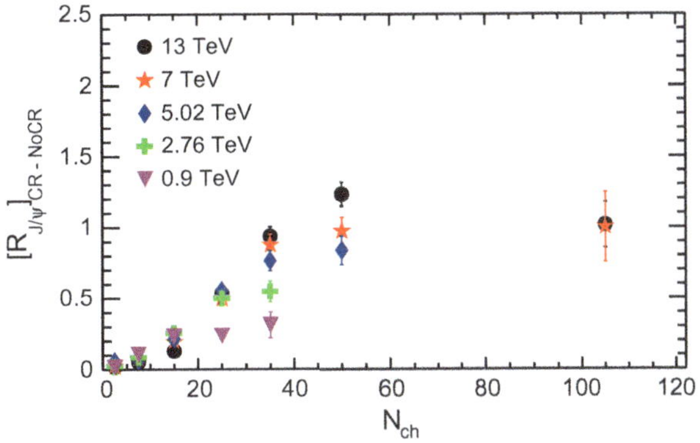

Fig. 24.3 The difference of relative J/ψ yield, $R_{J/\psi}$ with and without CR as a function of N_{ch} at the LHC energies

24.3 Results

After performing all the feasibility tests as discussed in the previous section, we have extended the event multiplicity dependence of J/ψ production to LHC energies: $\sqrt{s} = 0.9, 2.76, 5.02, 7$ and 13 TeV.

24.3.1 Multiplicity Dependence Study of J/ψ Production

We have simulated J/ψ at all the LHC energies using 4C tuned PYTHIA8 and studied the energy dependence behavior to understand effect of MPI on J/ψ production. This study is performed with two tunes of PYTHIA8: CR and No-CR, to see the final state effects on J/ψ production. To understand the CR effects on J/ψ production quantitatively, we have subtracted the yield of relative J/ψ production with no-CR from the yield with CR and plotted it as a function of charged-particle multiplicity for different center-of-mass energies, which is shown in Fig. 24.3.

It is found that the difference of the relative J/ψ yield between *with CR* and *without CR* increases with charged-particle multiplicity as well as with increasing energy. At LHC energies, due to high density of colored partons, there is a substantial degree of overlap of many colored strings in the position and momentum phase space. Hence, there is a higher probability of color reconnection. The partons from two different MPIs can reconnect via color strings with the minimization of the string length as discussed in the previous section. This study reveals that with the increase of MPIs the probability of color reconnection increases and hence the probability to combine charm and anticharm quark becomes higher, thereby producing a higher number of J/ψ particles.

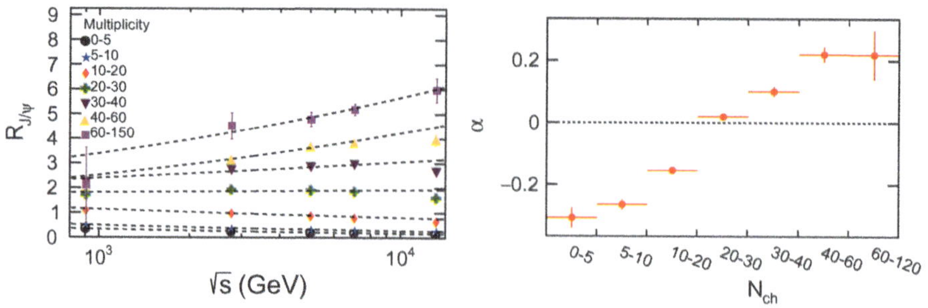

Fig. 24.4 Left panel: Relative J/ψ yield as a function of $\sqrt{s}$, using PYTHIA8 with CR. The dashed lines are the phenomenological $y = Ax^{\alpha}$ fitting. Right panel: The fitting parameter (α) as a function of charged-particle multiplicity

24.3.2 Energy Dependence Study of J/ψ Production

Let's now explore the effect of MPI and CR on charmonia production at various collision energies. Figure 24.4 shows the relative J/ψ yield as a function of the center-of-mass energy for different charged-particle multiplicity bins using PYTHIA8 with CR. It is found that the relative J/ψ yield increases with $\sqrt{s}$.

We have performed a quantitative study by fitting the results with a phenomenological function, $y = Ax^{\alpha}$, where A and α are the parameters. The parameter, α, represents the rate of increase of relative J/ψ as a function of center-of-mass energy for a particular multiplicity bin. From the left panel of Fig. 24.4, the α-parameter is found to increase with multiplicity. The values of α are negative up to (10–20) multiplicity bins and become positive towards higher-multiplicity bins. This indicates that MPI effects dominate for J/ψ production for $N_{\rm ch} > 20$.

24.4 Summary

In this contribution, energy and multiplicity dependence of J/ψ production has been presented using 4C tuned PYTHIA8. We have summarized below the important points drawn from the study.

- The difference between the relative J/ψ yield with CR and without CR increases with charged-particle multiplicity as well as with center-of-mass energy. As the difference is very less, one can infer that final state effect has very less contribution to J/ψ production.
- The relative J/ψ yield as function of $\sqrt{s}$ shows a monotonic increase, where as for $N_{\rm ch} \leq 20$, the behavior is opposite. This hints for dominance of MPI to the J/ψ production from $N_{\rm ch} \geq 20$.

The present studies are important in view of the interesting properties shown by high-multiplicity events in $p + p$ collisions at the LHC energies. It will be very interesting to get the experimental measurements, which can help to explore more into the multiplicity dependence of quarkonia production. This work has appeared as a regular publication in [6].

References

1. B. Abelev et al. (ALICE Collaboration), Phys. Lett. B **712**, 165 (2012)
2. S.G. Weber, (ALICE Collaboration), Nucl. Phys. A **967**, 333 (2017)
3. J. Adam et al. (ALICE Collaboration), JHEP **1509**, 148 (2015)
4. Pythia8 online manual, http://home.thep.lu.se/~torbjorn/pythia81html/Welcome.html
5. A study of different colour reconnection settings for Pythia8 generator using underlying event observables, (ATLAS Collaboration), ATL-PHYS-PUB-2017-008 (2017)
6. D. Thakur, S. De, R. Sahoo, S. Dansana, Phys. Rev. D **97**, 094002 (2018) (References there in)

Chapter 25
Prospects in Spectroscopy with Belle II

Vishal Bhardwaj

Abstract Belle played a leading role in shaping the spectroscopy sector for last decade. With 50 times more data than Belle, the Belle II experiment also expects to play crucial role in spectroscopy for the next decade. In this talk, a few chosen results one expects from Belle II will be discussed.

25.1 Introduction

The Belle II detector [1] is a general purpose detector built to test Standard Model mechanism by doing precision measurements. Belle II also provides a very clean environment and is an ideal place to carry quarkonium $q\bar{q}$ spectroscopy related studies. $q\bar{q}$ are produced through B decays, double charmonium production, two photon production, initial state radiation, and quarkonium decay/transitions.

For the last 15 years Belle [2] (predecessor of the Belle II detector with similar environment) had a very successful program on quarkonium ($q\bar{q}$). Many new $q\bar{q}$ (-like) states such as $\eta_c(2S)$, $X(3872)$, $X(3915)$, $Z(3930)$, $X(3940)$, $Z_1(4050)^+$, $Y(4260)$, $Z(4430)^+$, $Y(4660)$, $Z_b(10610)$, and $Z_b(10650)$ have been found. Many of these states cannot be accomodated by the conventional spectroscopy. Some states have non-zero charge which suggest that they are tetraquark/molecule-like states. Belle II (with the ability to accumulate 50 times more data in comparison to Belle) will be able to play important role in understanding the nature of these states. In this talk, I will try to give brief overview of the Belle II program for quarkonium. I should admit here that I have not done justice in this proceeding. Interested readers should refer to the Belle II Physics book [3].

Vishal Bhardwaj for the Belle II Collaboration.

V. Bhardwaj (✉)
Indian Institute of Science Education and Research Mohali, Punjab 140306, India
e-mail: vishstar@gmail.com

A. Giri and R. Mohanta (eds.), *16th Conference on Flavor Physics and CP Violation*, Springer Proceedings in Physics 234,
https://doi.org/10.1007/978-3-030-29622-3_25

25.2 Belle to Belle II

The Belle II experiment (situated in Tsukuba, Japan), is the upgraded successor of Belle. The detector's major upgrades in comparison to Belle are:

- A Vertex detector (VXD) consisting of two layers of DEPFET pixels (PXD) and four layers double-sided silicon strips (SVD), with improved resolution (to half) compared to Belle.
- A central drift chamber (CDC) with larger volume drift chamber, smaller drift cells and faster electronics.
- Completely new particle identification [time of propagation (barrel) and proximity-focusing Aerogel Ring-Imaging Cherenkov detector (end-cap)].
- Belle CsI (Tl) crystals are used for the electro-magentic calorimeter with modified waveform sampling electronics to reject pile-up events.
- Upgraded $K_L - \mu$ detection system (KLM) where resistive plate counter used in Barrel. Because of the projected inefficiency of RPCs at high ambient rate, Belle II end-caps are instrumented with scintillator strips.

25.3 Current Status of Belle II

Belle II successfully completed "Phase II" commissioning runs and accumulated 472 pb^{-1} of data. During Phase II, all the sub-detectors were in except the full vertex detector (partial vertex detector for a particular ϕ was in).

25.3.1 Re-discovery of "November Revolution"

Figure 25.1 shows the reconstructed $J/\psi \to \ell^+\ell^-$ demonstrating the capability of reconstructing lepton tracks. We see a clear peak of J/ψ to e^+e^- and $\mu^+\mu^-$ reconstruction.

25.3.2 Re-discovery of D and B Mesons

Figures 25.2 and 25.3 shows the reconstructed D and B mesons demonstrating the capability of reconstructing charged and neutral Kaon and pions.

As seen from the re-discovery plots of the J/ψ, D, and B, the Belle II detector is working as per expectation.

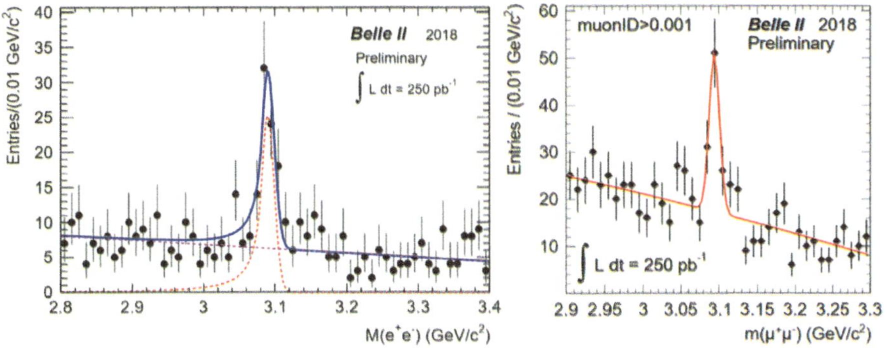

Fig. 25.1 Reconstructed invariant mass of $J/\psi \to e^+e^-$ (left) and $J/\psi \to \mu^+\mu^-$ (right) at Belle II using partial Phase II data. We also have the plots with full data at current date (however, the plots shown here are similar to what was shown at the conference). Plots with full data set can be found at [4]

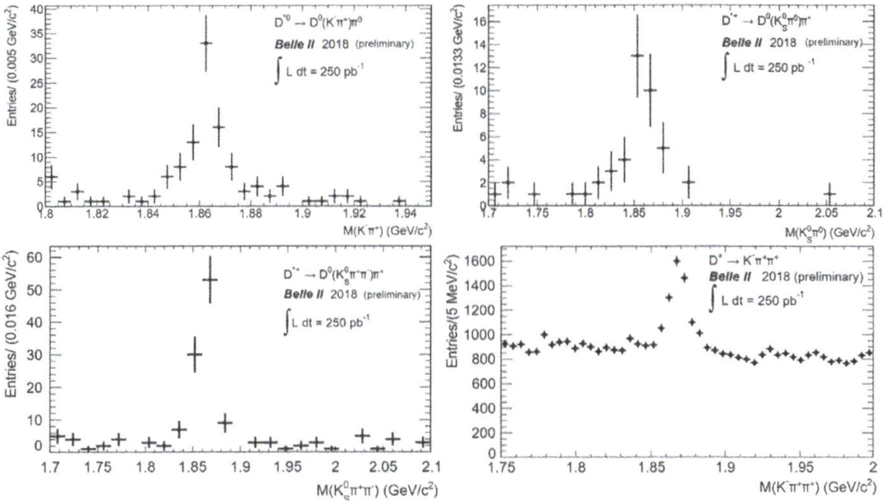

Fig. 25.2 Reconstructed invariant mass of D mesons from various decay modes

25.4 Prospects for $c\bar{c}$(-Like) States

$X(3872)$ was first observed in $B^+ \to (J/\psi\pi^+\pi^-)K^+$ process at Belle [5]. Soon after its discovery, $X(3872)$ was confirmed by CDF [6], D∅ [7], BaBar [8], LHCb [9] and CDF [10]. A lot of effort went into studying this particle, thanks to which now we know its precise mass, width, and J^{PC} to be (3871.69 ± 0.17) MeV/c^2 [11], <1.2 MeV [12], and 1^{++} [13], respectively. At Belle II, we expect 1500 signal events with 10 ab^{-1} of data (which is 1/5 of the total data Belle II aims to accumulate). Just to give an idea, the current yield of $B^+ \to \psi'(\to J/\psi\pi\pi)K^+$ is 3600 signal events

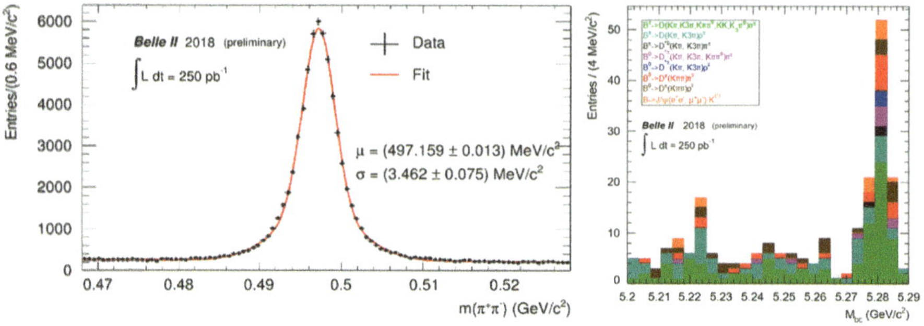

Fig. 25.3 Invariant mass of reconstructed $K_S^0 \to \pi^+\pi^-$ (left) and $M_{\rm bc}$ for reconstructed B meson from different modes

at Belle. This will help in measuring precisely $X(3872)$ mass and width. Within the first two years of data taking, one can expect that Belle II will accumulate 5–10 ab^{-1} of data.

The narrow width of $X(3872)$ and the proximity of its mass to the $D^0\bar{D}^*$ threshold makes it a good candidate for a $D^0\bar{D}^*$ molecule [14]. Currently the most probable explanation for the $X(3872)$ nature is a molecule with admixture of charmonium.

If $X(3872)$ is charmonium then one expects it to be χ'_{c1}. If so then it should decay to $\chi_{c1}\pi^+\pi^-$. Current search by the Belle has a negative result [15]. One can measure or expect a tighter constraint from Belle II.

Performing the study of $X(3872) \to \bar{D}^0 D^{*0}$ [16] with the full Belle II data will bring more information. Measuring the ratios of radiative decays [17] $\mathcal{B}(X(3872) \to \psi'\gamma)/\mathcal{B}(X(3872) \to J/\psi\gamma)$ with more data is what Belle II should do, as it is crucial for understanding the nature of $X(3872)$. If $X(3872)$ is a $D^0\bar{D}^{*0}$ molecule, then one expects that there may be other "X-like" particles with different quantum numbers that are bound states of $D^{(*)}$ mesons, such as a $(D^0\bar{D}^{*0} - \bar{D}^0 D^{*0})$ combination as the C-odd partner of $X(3872)$ with J^{PC} of $1^{\pm}$. C-odd search has been negative till now [18]. Searching for the the charged $X(3872) \to J/\psi\pi^+\pi^0$ [12] and C-odd partners such as $J\psi\eta$ at Belle II is interesting. If found, it will suggest a molecular/tetraquark nature of the $X(3872)$ [19]. On the other side, absence of charged partners suggest $X(3872)$ to be an iso-singlet state. This suggests $X(3872) \to J/\psi\pi^+\pi^-$ to be an iso-spin violating decay. BaBar has measured the ratio $\mathcal{B}(X(3872) \to J/\psi\omega(\to \pi^+\pi^-\pi^0))/\mathcal{B}(X(3872) \to J/\psi\pi^+\pi^-)) = 0.8 \pm 0.3$. Belle II can improve this ratio with much precision.

Absolute $\mathcal{B}(B \to X(3872)K^+)$ helps in measuring $\mathcal{B}(X(3872) \to$ final states). This measurement is only possible at the e^+e^- B factories. One has to reconstruct the missing mass recoiling against the K^+,

$$M_{\rm miss} = \sqrt{(p^*_{e^+e^-} - p^*_{\rm tag} - p^*_K)^2}/c \qquad (25.1)$$

where M_{miss} is the missing mass recoiling against the K^+, and $p^*_{e^+e^-}$, p^*_{tag}, and p^*_K are the four-momenta of the electron-positron initial state, B_{tag} (full reconstruction of one of the two charged B mesons via hadronic states) and kaon, respectively, in the center-of-mass frame. The M_{miss} peaks around the mass of the signal. Belle measured $\mathcal{B}(B^+ \to X(3872)K^+) < 2.6 \times 10^{-4}$ (@ 90% CL) [20]. With 50 times more data, Belle II can measure the branching fraction till 10^{-5} or less due to the improvement [21] in the full reconstruction algorithm (in comparison to Belle).

Not only decays, but also production of $X(3872)$ in the B decay provide information about the nature of $X(3872)$. Belle observed the $B^0 \to X(3872)K^+\pi^-$ decay mode having 7σ significance. In their study of the production dynamics of $B^0 \to X(3872)K^+\pi^-$, they found that $B^0 \to X(3872)K^*(892)^0$ does not dominate the $B^0 \to X(3872)K^+\pi^-$ decay, which is in contrast to the normal charmonium states (where $K^*(892)^0$ dominates) [22]. This suggest that $X(3872)$ doesn't behave like normal charmonium states. With 10 ab^{-1} of data collected with Belle II, we expect $B \to X(3872)K\pi$ to have the same number of events to what Belle has accumulated for $B \to \psi' K\pi$. Therefore, one can expect to have a more precise measurement.

In the two photon process, $\gamma\gamma \to J/\psi\phi$, Belle observed $X(4350)$ [23]. However, recently in the amplitude analysis of $B \to J/\psi\phi K$, LHCb found several structures [$Y(4140)$, $Y(4274)$, $X(4500)$, and $X(4700)$] but did not found $X(4350)$ [24]. Belle II should revisit with more data. Another area where Belle II can contribute is the $Y(4260)$ study. Belle II will compliment BESIII here. We expects improvement in mass resolution due to CDC improvements. Belle II with 50 ab^{-1} should be able to study the line-shape of $Y(4260)$. Another possible study one can think of is $e^+e^- \to Y(4260)(\to J/\psi\pi^0\pi^0)\gamma_{ISR}$ to search for a neutral partner. Also, measuring $\mathcal{B}(B \to Y(4260)K)$ at Belle II is an important in step to understand the nature of $Y(4260)$. The first charged $Z(4430)^+$ state was seen by Belle in the $B^0 \to (\psi'\pi^+)K^-$ decay mode [25]. Till recently this state was not well established due to non observation in other experiments. Recently, LHCb confirmed $Z(4430)^+$ and using an Argand diagram, they supported the resonance nature of this state [26]. Belle II can perform amplitude analyses with more statistics (similar to the one done at Belle [27, 28]) and help in understanding these states with precision. Other modes not feasible at Belle are also accessible at Belle II. For example only with 10 ab^{-1} of data at Belle II, one expects the yield of the $B^0 \to (\chi_{c2}\pi^-)K^+$ decay mode to become comparable to what Belle accumulated for $B^0 \to (\chi_{c1}\pi^-)K^+$ [29]. Not only that but Belle II can also search for the neutral partners using π^0 modes ($B^0 \to (c\bar{c})\pi^0K^+$).

25.5 Prospects for $b\bar{b}$(-Like) States

The bottomonium spectrum has found to be different from what we have understood in charmonium spectrum. Belle II is a unique place to carry out bottomonium related studies due to the energy accessible by SuperKEKB (expecting to reach $\Upsilon(5, 6S)$

energy). We know that Z_b states were found in the $\Upsilon(5S)$ decays by Belle and are clear signature of exotic states. Belle [30] found that

$$\frac{\Gamma(\Upsilon(5S) \to h_b(nP)\pi^+\pi^-)}{\Gamma(\Upsilon(5S) \to \Upsilon(2S)\pi^+\pi^-)} = \begin{cases} 0.45 \pm 0.08^{+0.07}_{-0.12}, \text{ for } h_b(1P) \\ 0.77 \pm 0.08^{+0.22}_{-0.17}, \text{ for } h_b(2P) \end{cases}. \tag{25.2}$$

While one expected the decay to h_b should be suppressed due to spin flip, its higher rate was something puzzling. The $\Upsilon(5S) \to h_b(nP)\pi^+\pi^-$ decay mechanism seems to be exotic. Belle found that $\Upsilon(5S) \to Z_b^+\pi^-$, then Z_b^+ decays to $h_b\pi^+$. $Z_b(10610)$ and $Z_b(10650)$ were found in $\Upsilon(1S)\pi^+\pi^-$, $\Upsilon(2S)\pi^+\pi^-$, $\Upsilon(3S)\pi^+\pi^-$, $h_b(1P)\pi^+\pi^-$, and $h_b(2P)\pi^+\pi^-$ decay with masses around the B^*B and B^*B^* thresholds [31]. With more data, Belle II expects to measure the mass and width more precisely. Further, Belle II can study neutral Z_b^0 in $\Upsilon(5S) \to \Upsilon(nS)\pi^0\pi^0$ [32] and confirm in other modes also.

Another study of interest to be done at Belle II is an energy scan. A previous energy scan of the $e + e^- \to h_b(nP)\pi^+\pi^-$ $(n = 1,\ 2)$ cross sections by Belle gave first evidence for $\Upsilon(6S) \to h_b(1P)\pi^+\pi^-$ and observation of $\Upsilon(6S) \to h_b(2P)\pi^+\pi^-$. While studying the resonant structure, they found evidence that it proceeds entirely via the intermediate iso-vector states $Z_b(10610)$ and $Z_b(10650)$ [33]. Currently only Belle II has the capability to do an $\Upsilon(nS)$ scan.

With a unique data set at $\Upsilon(6S)$, Belle II can study $\Upsilon(6S) \to h_b(nP)\pi^+\pi^-$, $\Upsilon(6S) \to \Upsilon(mS)\pi^+\pi^-$ $(n = 1,\ 2;\ m = 1,\ 2,\ 3)$. If Z_b is a molecular state, then Heavy Quark Spin symmetry suggests there should be 2 or 4 molecular partner bottomonium-like states (W_b): $\Upsilon(5S, 6S) \to W_{b0}\gamma$, and $\Upsilon(6S) \to W_{b0}\pi^+\pi^-$, where $W_{b0} \to \eta_b\pi, \to \chi_b\pi, \Upsilon\rho$. Figure 25.4 summarizes the possible decays via which one can access the molecular partners of bottomonium-like states [34].

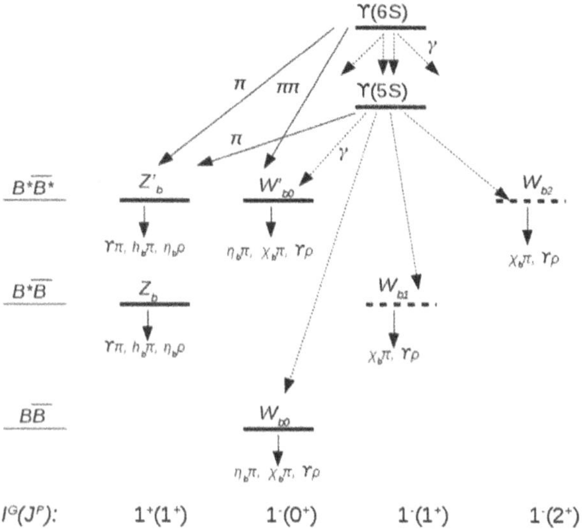

Fig. 25.4 Accessing molecular partner bottomonium-like states (W_b) via transitions from $\Upsilon(5, 6S)$

Acknowledgements I would like to thank colleagues from the Belle II Collaboration. This work is supported by the INSPIRE Faculty Award of the Department of Science and Technology (India).

References

1. T. Abe et al., *Belle II Technical Design Report* (2010). arXiv:1011.0352
2. A. Abashian et al. (Belle Collaboration), Nucl. Instrum. Methods Phys. Res., Sect. A **479**, 117 (2002); also see detector section in J. Brodzicka et al., Prog. Theor. Exp. Phys. 04D001 (2012)
3. E. Kou et al., *The Belle II Physics Book* (2018). arXiv:1808.10567 [hep-ex]
4. https://docs.belle2.org/collection/Belle%20II%20Notes%20%3A%20Plots?ln=en
5. S.K. Choi et al. (Belle Collaboration), Phys. Rev. Lett. **91**, 262001 (2003)
6. D. Acosta et al. (CDF Collaboration), Phys. Rev. Lett. **93**, 072001 (2004)
7. V.M. Abazov et al. (DO Collaboration), Phys. Rev. Lett. **93**, 162002 (2004)
8. B. Aubert et al. (BaBar Collaboration), Phys. Rev. D **71**, 071103 (2005)
9. R. Aaij et al. (LHCb Collaboration), Eur Phys. J. C **72**, 1972 (2012)
10. S. Chatrchyan et al. (CMS Collaboration), J. High Energy Phys. **04**, 154 (2013)
11. K.A. Olive et al. (Particle Data Group), Chin. Phys. C **38**, 090001 (2014)
12. S.-K. Choi et al. (Belle Collaboration), Phys. Rev. D **84**, 052004 (2011)
13. R. Aaij et al. (LHCb Collaboration), Phys. Rev. Lett. **110**, 222001 (2013)
14. E.S. Swanson, Phys. Lett. B **598**, 197 (2004); E.S. Swanson, Phys. Rep. **429**, 243 (2006)
15. V. Bhardwaj et al. (Belle Collaboration), Phys. Rev. Lett. **93**, 052016 (2016)
16. T. Aushev et al. (Belle Collaboration), Phys. Rev. D **81**, 031103 (2010)
17. B. Aubert et al. (The BABAR Collaboration), Phys. Rev. Lett. **102**, 132001 (2009); V. Bhardwaj et al. (Belle Collaboration), Phys. Rev. Lett. **107**, 091803 (2011); and R. Aaij et al. (LHCb Collaboration), Nucl. Phys. B **886**, 665 (2014)
18. A. Vinokurova et al. (Belle Collaboration), J. High Energy Phys. **1506**, 132 (2015); T. Iwashita et al. (Belle Collaboration), Prog. Theor. Exp. Phys. 043C01 (2014); and V. Bhardwaj et al. (Belle Collaboration), Phys. Rev. Lett. **111**, 032001 (2013)
19. L. Maiani et al., Phys. Rev. D **71**, 014028 (2005)
20. Y. Kato et al., Phys. Rev. D **97**, 012005 (2018)
21. T. Keck et al. arXiv:1807.08680
22. A. Bala et al. (Belle Collaboration), Phys. Rev. D **91**, 051101(R) (2015)
23. C.P. Shen et al. (Belle Collaboration), Phys. Rev. Lett. **104**, 112004 (2010)
24. R. Aaij et al. (LHCb Collaboration), Phys. Rev. D **95**, 012002 (2017)
25. S.-K. Choi et al. (Belle Collaboration), Phys. Rev. Lett. **100**, 142001 (2008)
26. R. Aaij et al. (LHCb Collaboration), Phys. Rev. Lett. **112**, 222002 (2014)
27. K. Chilikin et al. (Belle Collaboration), Phys. Rev. D **88**, 074026 (2013)
28. K. Chilikin et al. (Belle Collaboration), Phys. Rev. D **90**, 112009 (2014)
29. R. Mizuk et al. (Belle Collaboration), Phys. Rev. D **78**, 072004 (2008)
30. I. Adachi et al. (Belle Collaboration), Phys. Rev. Lett. **108**, 032001 (2012)
31. A. Bondar et al. (Belle Collaboration), Phys. Rev. Lett. **108**, 122001 (2012)
32. P. Krokovny et al. (Belle Collaboration), Phys. Rev. D **88**, 052016 (2013)
33. A. Garmash et al. (Belle Collaboration), Phys. Rev. Lett. **116**, 212001 (2016); and R. Mizuk et al. (Belle Collaboration), Phys. Rev. Lett. **117**, 142001 (2016)
34. M. Voloshin, Phys. Rev. D **84**, 031502(R) (2011)

Chapter 26
Light Hadron Spectroscopy and Decay at BESIII

Vindhyawasini Prasad

Abstract Light hadron spectroscopy plays an important role in understanding the decay dynamics of unconventional hadronic states, such as strangeonium and glueballs. BESIII provides an ideal avenue to search for these exotic states thanks to a huge amount of data recorded at various energy points in the tau-charm mass region including J/ψ resonance. This report summarizes recent results of the BESIII experiment related to the glueballs and strangeonium-like states.

26.1 Introduction

Quantum chromodynamics (QCD) describes hadrons as the bound states of quarks held together via the color force mediated by gluons. The non-Abelian nature of the QCD allows the self-interaction of gluons that can form the hadronic matter. These quarkless states are called 'glueballs'. The lattice QCD predicts that the glueballs having mass within the range of 1–2 (2–3) GeV/c^2 are scalar (tensor or pseudoscalar) [1]. The branching fractions for glueballs in the radiative decays of J/ψ are expected to be within the range of 10^{-2}–10^{-3} depending upon the exact nature of glueballs [2]. The large data samples collected at the center-of-mass (CM) energies between 2.0 and 4.6 GeV, more than 130 energy points, including the J/ψ resonance, by BESIII provide an ideal avenue to explore the possibilities of these glueballs. The QCD also allows other possible forms of multiquark and hybrid states, such as a tetraquark-like Z_c that was observed in the $\pi J/\psi$ mass spectrum in $\Upsilon(4260) \rightarrow \pi\pi J/\psi$ decays by the BESIII [3] and intermediately confirmed by Belle experiment [4]. Similar to Z_c [3, 4], a strangeonium-like state is also expected in the $\pi\phi$ spectrum via $\phi(2170) \rightarrow \pi\pi\phi$ decays, where $\phi(2170)$ [also denoted as Y(2175)] is analogous to Y(4260) [5]. The $\phi(2170)$ was observed by the BaBar collaboration via the initial-state-radiation pro-

V. Prasad (✉)
State Key Laboratory of Particle Detection and Electronics,
Department of Modern Physics, University of Science & Technology
of China, Hefei 230026, China
e-mail: vindy@ustc.edu.cn

A. Giri and R. Mohanta (eds.), *16th Conference on Flavor Physics and CP Violation*, Springer Proceedings in Physics 234,
https://doi.org/10.1007/978-3-030-29622-3_26

cess $e^+e^- \to \gamma\phi f_0(980)$ [6], and later confirmed by Belle [7], BESII [8] and BESIII collaborations [9]. Following sections summarize recent results of BESIII related to the gluonic and strangeonium-like states.

26.2 Search for Gluonic States

26.2.1 Anomalous Enhancement At the $p\bar{p}$ Mass-Threshold

An anomalous enhancement $X(p\bar{p})$ at $p\bar{p}$ mass-threshold was originally observed by BESII in $J/\psi \to \gamma p\bar{p}$ decay [10], and later confirmed by both BESIII [11] and CLEO experiments [12]. This enhancement might be due to an $X(1835)$ state, which was first observed by BESII in $J/\psi \to \gamma\eta'\pi^+\pi^-$ decays [13] and later confirmed by BESIII in the same process [14] and $J/\psi \to \gamma\eta K_S^0 K_S^0$ decays [15]. The observed spin-parity of $X(p\bar{p})$ and $X(3872)$ states is in favor of $J^P = 0^-$ [11, 15]. By analyzing the decay process of $J/\psi \to \gamma\eta'\pi^+\pi^-$ with 1.1 billion J/ψ events collected in 2012, BESIII has observed a significant abrupt change in the slope of the $\eta'\pi^+\pi^-$ mass spectrum at the $p\bar{p}$ mass threshold [16]. Two typical models are used to characterize the $\eta'\pi^+\pi^-$ line-shape around 1.85 GeV/c^2 (Fig. 26.1). The first incorporates the opening of the decay threshold in the mass spectrum (Flatté formula), while the second uses a coherent sum of two resonant amplitudes. Both the models describe the data, and well, suggesting the existence of either a $p\bar{p}$ molecule-like or bound state [16].

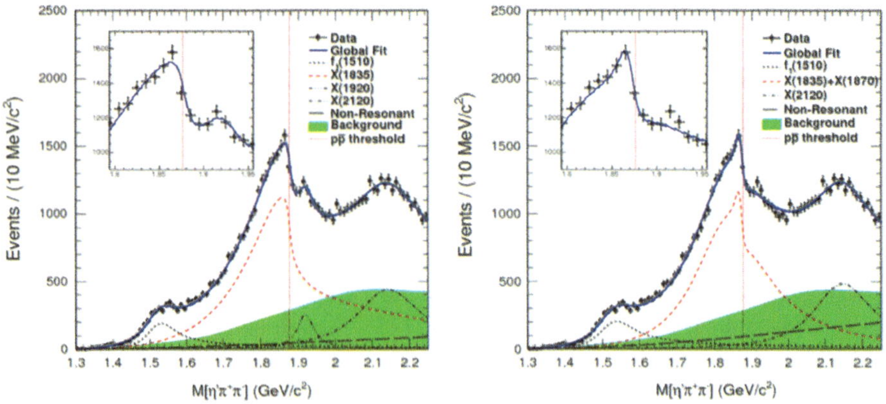

Fig. 26.1 Fit results based on the Flatté line-shape (left) and a coherent sum of two Breit-Wigner amplitudes. The position of $p\bar{p}$ mass-threshold is represented by a dashed vertical line, data by the points with error bars, signal and background contributions by various dashed curves, and the total fit by a solid blue curve

26.2.2 Observation of $\eta(1475)$ and $X(1835)$ in $J/\psi \to \gamma\gamma\phi$

The $\eta(1440)$ is a puzzling state first observed in $p\bar{p}$ annihilation at rest into $\eta(1440)\pi^+\pi^-$, $\eta(1440) \to K\bar{K}\pi$ [17], and later in the J/ψ radiative decays to $K\bar{K}\pi$ [18], $\gamma\rho$ [19] and $f_0(980)\pi^0$ [20]. Many experimental results reveal the existence of two different pseudoscalar states, the $\eta(1405)$ and $\eta(1475)$ [21]. The former is interpreted as an excellent candidate for $0^{\mp}$ glueball [21], and later as the first excitation of the η'. A triangle singularity is proposed to explain these anomalies by assuming both $\eta(1405)$ and $\eta(1475)$ to be a single state, the $\eta(1440)$, that appears as different line-shapes in different channels [22]. The $J/\psi \to \gamma\gamma\phi$ decay is studied using 1.3 billion J/ψ events collected by the BESIII detector [23]. Two resonant structures corresponding to $\eta(1475)$ and $X(1835)$ are observed in the $\gamma\phi$ invariant mass spectrum with a significance of 13.5σ and 6.3σ, respectively, by taking into account to interference between these two resonant structures during the fit (Fig. 26.2a, b). The angular distributions of these resonances are observed to be $J^{PC} = 0^{\mp}$ (Fig. 26.2c, d). These results reveal that both $\eta(1475)$ and $X(1835)$ contain a sizable $s\bar{s}$ component.

26.2.3 Observation of $X(2370)$ in $J/\psi \to \gamma KK\eta'$

The $X(2120)$ and $X(2370)$ states are observed in the $\pi^+\pi^-\eta'$ invariant mass spectrum through the decay of $J/\psi \to \gamma\pi^+\pi^-\eta'$ [13, 16]. The lattice QCD theory predicts the ratios of branching fractions of pseudoscalar glueball decays $\Gamma_{G\to KK\eta'}/\Gamma_G^{\text{tot}}$ and $\Gamma_{G\to\pi\pi\eta'}/\Gamma_G^{\text{tot}}$ to be 0.011 and 0.090, respectively, where the glueball mass is set at 2.37 GeV/c^2 [24]. The observation of these states in $K\bar{K}\eta'$ decays would support the hypothesis that they are glueballs. The search for $X(2120)$ and $X(2370)$ is performed via the decays of $J/\psi \to \gamma K^+K^-\eta'$ and $J/\psi \to \gamma K_S^0K_S^0\eta'$ using 1.3 billion J/ψ events. A structure around 2.34 GeV/c^2, the $X(2370)$, is observed in the $K\bar{K}\eta'$ spectra with a statistical significance of 7.6σ (Fig. 26.3). The product branching fractions for $J/\psi \to \gamma X(2370)$, $X(2370) \to K^+K^-\eta'$, and $J/\psi \to \gamma X(2370)$, $X(2370) \to K_S^0K_S^0\eta'$ are determined to be $[1.86 \pm 0.39$ (stat.) ± 0.29 (sys.)$] \times 10^{-5}$ and $[1.19 \pm 0.37$ (stat.) ± 0.18 (sys.)$] \times 10^{-5}$, respectively. No evidence for the $X(2120)$ production is found, and 90% confidence level (CL) upper limits on product branching fractions for $J/\psi \to \gamma X(2120) \to K^+K^-\eta'$ and $J/\psi \to \gamma X(2120) \to K_S^0K_S^0\eta'$ are set at 1.48×10^{-5} and 4.57×10^{-6}, respectively, for the first time. These results are preliminary.

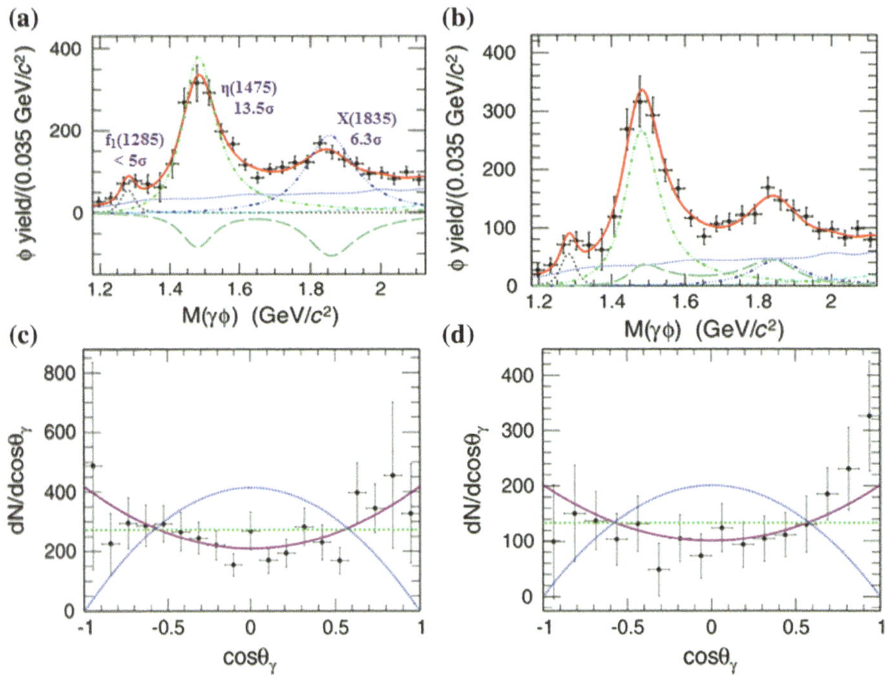

Fig. 26.2 (Top) fits to the $M_{\gamma\phi}$ distributions for the case of **a** constructive and **b** destructive interference, and (bottom) the efficiency corrected $\cos\theta_\gamma$ distributions for **c** $1.4 < M_{\gamma K^+K^-} < 1.6$ GeV/c^2 and **d** $1.75 < M_{\gamma K^+K^-} < 1.9$ GeV/c^2. The points with error bars are data, and solid and dashed curves represent the fit results. In $\cos\theta_\gamma$ distribution that is fitted by $1 + \alpha \cos\theta_\gamma^2$, the solid pink, dashed blue and dotted blue curves correspond to the hypotheses of angular distribution parameter $\alpha = 1, 0$ and -1, respectively

26.3 Search for Strangeonium-Like States

26.3.1 Search for Strangeonium-Like Structure Z_S at 2.125 GeV

A search for Z_S strangeonium-like structure is performed in the process $e^+e^- \to \phi\pi\pi$ using a data sample corresponding to an integrated luminosity of (108.49 ± 0.75) pb^{-1}, taken at the CM energy of 2.125 GeV by the BESIII detector [25]. A partial wave analysis of $e^+e^- \to \phi\pi\pi$ is performed to describe the di-pion invariant spectrum after applying all the selection criteria. The fit includes the amplitudes of $e^+e^- \to \phi\sigma$, $\phi f_0(980)$, $\phi f_0(1370)$ and $\phi f_2(1270)$ with a spin-parity of $J^P = 1^+$. No evidence for Z_S production is found in the invariant mass spectra of $\phi\pi^\pm$ and $\phi\pi^0$ around 1.4 GeV/c^2. The 90% CL upper limits on the cross-section of Z_S production are determined under different assumptions of mass, width and spin-parity

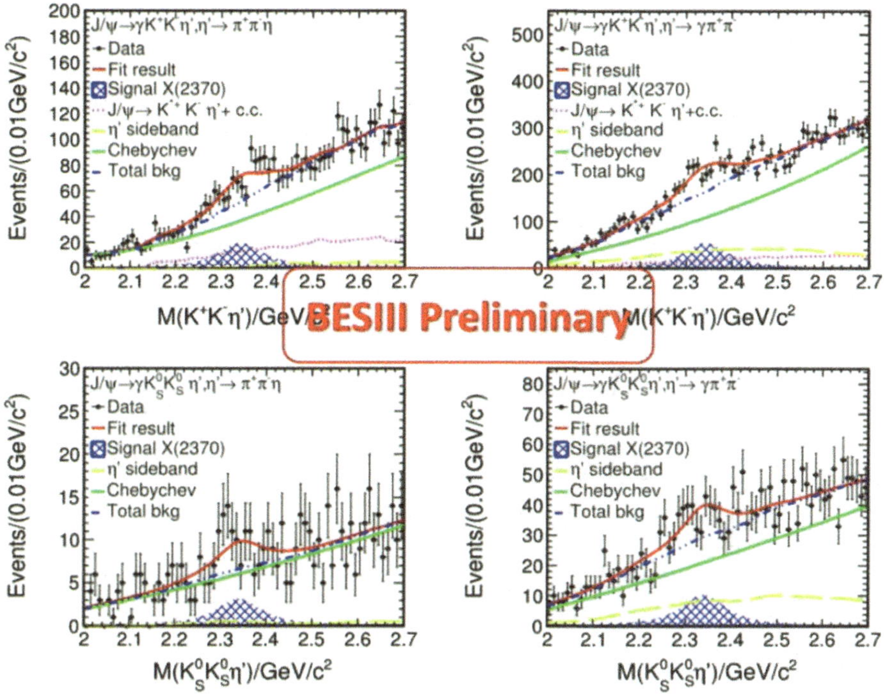

Fig. 26.3 Results of the fit for an $X(2370)$ signal on the invariant mass spectra of $K^+K^-\eta'$ (top) and $K^0_S K^0_S \eta'$ (bottom), where η' decays to $\pi^+\pi^-\eta(\to\gamma\gamma)$ (left) $\gamma\rho(\to\pi^+\pi^-)$ (right). The points with error bars are data, the blue grid area represents the signal, the blue dashed double dotted curves show the total background contributions of η' sideband (long dashed yellow curve), $J/\psi\to K^{*+}K^-\eta'$ (pink dashed curve) in $M(K^+K^-\eta')$ spectra only, and remaining backgrounds (solid green curve), and solid red curves show the total fit result

of Z_S (Fig. 26.4). In addition, the cross-section of $e^+e^-\to\phi\pi^+\pi^-$ and $e^+e^-\to\phi\pi^0\pi^0$ at 2.125 GeV/c^2 are measured as $(343.0\pm5.1\pm25.1)$ pb and $(208.3\pm7.6\pm13.6)$ pb, respectively. The first one slightly differs from BaBar [26] and Belle [7] measurements, being consistent within 3σ, whereas the second is consistent with the BaBar measurement [26].

26.3.2 Observation of $h_1(1380)$ in $J/\psi\to\eta' K\overline{K}\pi$

The $h_1(1380)$ is considered to be a strangeonium state and an $s\bar{s}$ partner of $J^{PC}=1^{\pm}$ axial-vector state $h_1(1170)$. Experimentally, this state was observed by the LASS [27] and Crystal Barrel [28] collaborations, and later confirmed by BESIII via $\psi(3686)\to\gamma\chi_{cJ}$, $\chi_{cJ}\to\phi h_1(1380)$, $h_1(1380)\to K^*(892)\overline{K}$ with $J=1,2$ [29]. BESIII has recently reported the first observation of $J/\psi\to\eta' h_1(1380)$, where

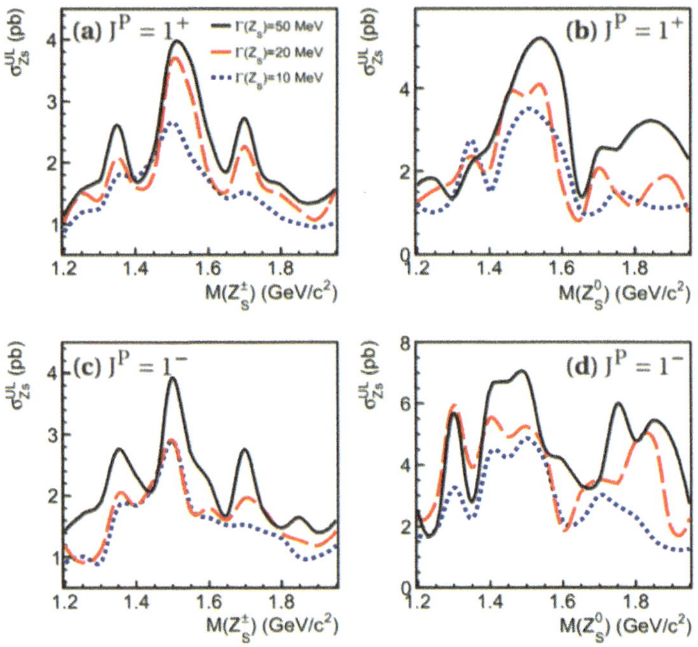

Fig. 26.4 90% CL upper limits on cross-section for different widths of Z_S as a function of assumed Z_S mass for the cases **a** $J^P = 1^+$ of $Z_S^{\pm}$, **b** $J^P = 1^+$ of Z_S^0, **c** $J^P = 1^-$ of $Z_S^{\pm}$ and **d** $J^P = 1^-$ of Z_S^0

$h_1(1380) \to K^*(892)\overline{K} + c.c \to K^+K^-\pi^0/K_S^0K^{\pm}\pi^{\mp}$, using a sample of 1.3 billion J/ψ events [30]. The $h_1(1380)$ resonance is observed in $K^+K^-\pi^0$ and $K_S^0K^{\pm}\pi^{\mp}$ mass spectra with a statistical significance larger than 10σ by performing a simultaneous fit (Fig. 26.5). An isospin symmetry violation is found in $h_1(1380)$ decays between $h_1(1380) \to K^*(892)^+K^-$ and $h_1(1380) \to K^*(892)^0K^0$. In addition, the mixing angle between the $h_1(1170)$ and $h_1(1380)$ is also determined to be $(35.9 \pm 2.6)^\circ$. This measured angle supports that the quark contents of the $h_1(1380)$ and $h_1(1170)$ are predominantly by $s\overline{s}$ and $u\overline{u} + d\overline{d}$, respectively.

26.4 Summary and Future Prospects

The BESIII has conducted a series of studies on gluonic and strangeonium-like states using the data samples collected at J/ψ, $\psi(3886)$ and $\phi(2170)$ resonances. Several such particles are reported to have been observed recently. The BESIII will accumulate additional 9 billion J/ψ events before the end of 2019, that will be utilized to improve the precision of measured states and possibly discover many new exotic states in the near future.

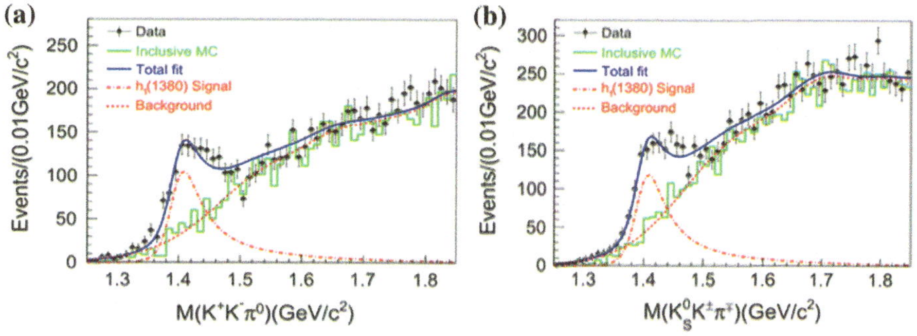

Fig. 26.5 Fits to $K^+K^-\pi^0$ (left) and $K_S^0K^\pm\pi^\mp$ (right) mass spectra with interference between signal and background. Points with error bars are data, red dash-dotted and dashed curves are signal and background contributions, and solid blue curves show the total fit

Acknowledgements This work has supported in part by the National Natural Science Foundation of China (NFSC) under contract No. 11705192.

References

1. Y. Chen et al., Phys. Rev. D **73**, 014516 (2006)
2. L.C. Gui et al., Phys. Rev. Lett. **110**, 021601 (2013); Y.B. Yang et al., Phys. Rev. Lett. **111**, 091601 (2013)
3. M. Ablikim et al. (BESIII Collaboration), Phys. Rev. Lett. **110**, 252001 (2013)
4. Z.Q. Liu et al. (Belle Collaboration), Phys. Rev. Lett. **110**, 252002 (2013)
5. B. Aubert et al. (BaBar Collaboration), Phys. Rev. Lett. **95**, 142001 (2005); M. Ablikim et al. (BESIII Collaboration), Phys. Rev. Lett. **118**, 092001 (2017)
6. B. Aubert et al. (BaBar Collaboration), Phys. Rev. D **74**, 091103(R) (2006)
7. C.P. Shen et al. (Belle Collaboration), Phys. Rev. D **80**, 031101(R) (2009)
8. M. Ablikim et al. (BES Collaboration), Phys. Rev. Lett. **100**, 102003 (2008)
9. M. Ablikim et al. (BESIII Collaboration), Phys. Rev. D **91**, 052017 (2015); Observation of $e^+e^- \to \eta Y(2175)$ at center-of-mass energies above 3.7 GeV (2017). arXiv:1709.04323
10. J.Z. Bai et al. (BES Collaboration), Phys. Rev. Lett. **91**, 022001 (2003)
11. M. Ablikim et al. (BESIII Collaboration), Chin. Phys. C **34**, 421 (2010); Phys. Rev. Lett. **108**, 112003 (2012)
12. J.P. Alexander et al. (CLEO Collaboration), Phys. Rev. D **82**, 092002 (2010)
13. M. Ablikim et al. (BES Collaboration), Phys. Rev. Lett. **95**, 262001 (2005)
14. M. Ablikim et al. (BESIII Collaboration), Phys. Rev. Lett. **106**, 072002 (2011)
15. M. Ablikim et al. (BESIII Collaboration), Phys. Rev. Lett. **115**, 091803 (2015)
16. M. Ablikim et al. (BESIII Collaboration), Phys. Rev. Lett. **117**, 042002 (2016)
17. P.H. Baillon et al., Nuovo Cimento A **50**, 393 (1967)
18. D.L. Scharre et al., Phys. Lett. B **97**, 329 (1980)
19. J.Z. Bai et al. (BES Collaboration), Phys. Lett. B **594**, 47 (2004)
20. M. Ablikim et al. (BESIII Collaboration), Phys. Rev. Lett. **108**, 182001 (2012)
21. L. Faddev, A.J. Niemi, U. Wiedner, Phys. Rev. D **70**, 114033 (2004)
22. J.J. Wu, X.H. Liu, Q. Zhao, B.S. Zou, Phys. Rev. Lett. **108**, 081803 (2012)
23. M. Ablikim et al. (BESIII Collaboration), Phys. Rev. D **97**, 051101(R) (2018)
24. W.I. Eshraim, S. Janowski, F. Giacosa, D.H. Rischke, Phys. Rev. D **87**, 054036 (2013)

25. M. Ablikim et al. (BESIII Collaboration), Phys. Rev. D **99**, 011101(R) (2018)
26. J.P. Lees et al. (BaBar Collaboration), Phys. Rev. D **86**, 012008 (2012)
27. D. Aston et al. (LASS Collaboration), Phys. Lett. B **201**, 573 (1988)
28. A. Abele et al. (Crystal Barrel Collaboration), Phys. Lett. B **415**, 280 (1997)
29. M. Ablikim et al. (BESIII Collaboration), Phys. Rev. D **91**, 112008 (2015)
30. M. Ablikim et al. (BESIII Collaboration), Phys. Rev. D **98**, 072005 (2018)

Chapter 27
Baryon Form Factors at BESIII

Lei Xia

Abstract Baryons, what they really are, is far from being understood. The Beijing Spectrometer (BESIII) is situated on the Beijing Electron Positron Collider (BEPCII), a e^+e^- collider running at center-of-mass energies ($\sqrt{s}$) between 2.0 and 4.6 GeV. This wide energy range allows the measurement of electromagnetic form factors (FFs) both from scan technique and initial-state radiation (ISR) technique. On the channel $e^+e^- \to p\overline{p}$, BESIII have published results based on 156.9 pb^{-1} of data. The preliminary results based on 688.5 pb^{-1} of data, a world-leading data sample for precision measurements have also been released. Preliminary results from the analysis of the ISR process $e^+e^- \to \gamma p\overline{p}$ based on 7.41 fb^{-1} of data have also been released. BESIII have also published the results on the channel $e^+e^- \to \Lambda\overline{\Lambda}$ as well as the channel $e^+e^- \to \Lambda_c^+\bar{\Lambda}_c^-$. Preliminary results of Λ Electromagnetic FFs relative phase are already available.

27.1 Introduction

Baryon mass is the main component of the mass of the universe which comes from the strong force, not from the Higgs mechanism. Many meson features come from QED to QCD, once α to α_s while there is no analogue in QED and unique QCD feature for baryon. For instance: a fermion with mass, magnetic moment and other parameters close to proton and neutron ones can be obtained as a soliton of a π point-like boson field, by means of a non linear Lagrangian with one free parameter only. The baryon spin is not due to the spins of the valence quarks. Therefore, it is meaningful to point out open questions, concerning baryon structure.

On behalf of BESIII Collaboration.

L. Xia (✉)
University of Science and Technology of China, Hefei 230026,
People's Republic of China

State Key Laboratory of Particle Detection and Electronics,
Beijing 100049, People's Republic of China
e-mail: jessemcc@mail.ustc.edu.cn

A. Giri and R. Mohanta (eds.), *16th Conference on Flavor Physics and CP Violation*, Springer Proceedings in Physics 234,
https://doi.org/10.1007/978-3-030-29622-3_27

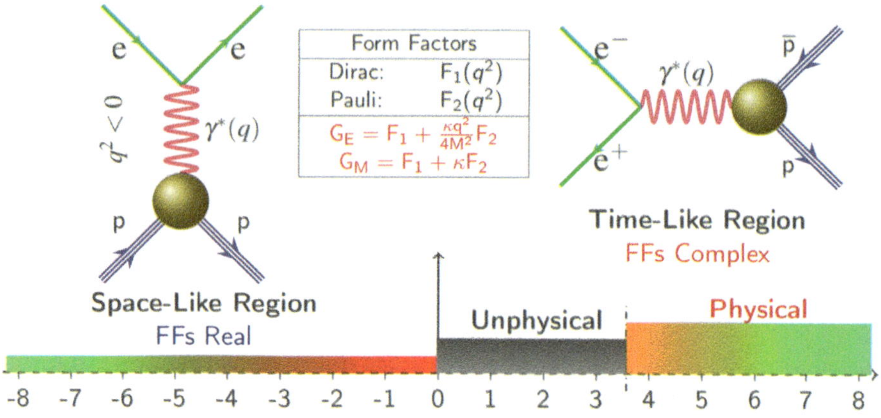

Fig. 27.1 Lowest-order Feynman diagrams for elastic electron-baryon scattering $e^-B(p) \rightarrow e^-B(p)$ (left), and for the annihilation process $e^-e^+ \rightarrow B\overline{B}$ ($p\overline{p}$) (right)

Figure 27.1 shows the lowest-order Feynman diagrams of the electron-hadron elastic scattering and the e^+e^- annihilation into a pair of baryons. In the case of spin 1/2 baryons, the hadronic vertex is described by a non-constant matrix:

$$\Gamma_\mu(p', p) = \gamma_\mu F_1(q^2) + \frac{i\sigma_{\mu\nu}q^\nu}{2m_B}F_2(q^2), \tag{27.1}$$

where F_1 and F_2 are the so called Dirac and Pauli FFs and m_B is the mass of the corresponding baryon. The Dirac FF, F_1, is related to the electric and the magnetic scattering from the baryon, while the Pauli FF, F_2, is related to the additional scattering contribution arising from the anomalous magnetic moment of the baryon. Instead of F_1 and F_2, the use of the so-called Sachs FFs has become conventional:

$$G_E(q^2) = F_1(q^2) + \tau\kappa_B F_2(q^2), G_M(q^2) = F_1(q^2) + \kappa_B F_2(q^2), \tag{27.2}$$

where $\tau = \frac{q^2}{4m_B^2}$, $\kappa = \frac{g-2}{2}$ and $g = \frac{\mu}{J}$ [1]. In the Breit frame, nucleon spin flip for G_M and non spin flip for G_E. The Born differential angular cross section in the e^+e^- center-of-mass (c.m.) reads [2]:

$$\frac{d\sigma_{B\overline{B}}(s)}{d\Omega} = \frac{\alpha^2\beta C}{4s}[|G_M(s)|^2(1+\cos^2\theta) + \frac{4m_B^2}{s}|G_E(s)|^2\sin^2\theta], \tag{27.3}$$

where s is the squared center-of-mass energy of $B\overline{B}$, θ is the polar angle of the baryon at c.m. and $\beta = \sqrt{1 - 4m_B^2/s}$. The Coulomb factor, $C = y/(1 - \exp(-y))$ with $y = \pi\alpha\sqrt{1-\beta^2}/\beta$, accounts for the electromagnetic $B\overline{B}$ interactions. The total cross section, integrated over the full solid angle is:

$$\sigma_{B\overline{B}}(s) = \frac{4\pi\alpha^2\beta C}{3s}[|G_M(s)|^2 + \frac{2m_B^2}{s}|G_E(s)|^2]. \tag{27.4}$$

An effective form factor (EFF) can be defined as

$$|G_{\text{eff}}(s)| = \sqrt{\frac{|G_M(s)|^2 + \frac{2m_B^2}{s}|G_E(s)|^2}{1 + \frac{2m_B^2}{s}}}, \tag{27.5}$$

which is equivalent to $|G_M(s)|$ under the working hypothesis $|G_E(s)| = |G_M(s)|$. However, the simultaneous extraction of $|G_E(s)|$ and $|G_M(s)|$ without any assumption is only possible by measuring the angular distributions of the outgoing particles (27.3).

An alternative approach to measure hadronic cross sections at high luminosity e^+e^- storage rings is the study of ISR processes. The differential cross section of the ISR process $e^+e^- \to B\overline{B}\gamma$ is related to the cross section of the non-radiative process $e^+e^- \to B\overline{B}$ through

$$\frac{d\sigma^{ISR}_{B\overline{B}\gamma}}{dq^2 d\theta_\gamma} = \frac{1}{s}W(s, x, \theta_\gamma)\sigma_{B\overline{B}}(q^2), \tag{27.6}$$

where $x = 2E_\gamma/\sqrt{s} = 1 - q^2/s$, and E_γ and θ_γ are the energy and the polar angle of the ISR photon in the e^+e^- c.m., respectively. The radiator function, $W(s, x, \theta_\gamma)$, describes the probability of the ISR photon emission [3].

27.2 Detector and Accelerator

The Beijing Electron-Positron Collider (BEPCII) [4] is a double-ring e^+e^- collider designed to provide a peak luminosity of 10^{33} cm^{-2}s^{-1} at $\sqrt{s} = 3770$ MeV. The BESIII [4] detector has a geometrical acceptance of 93% of the full solid angle and has four main components: (1) A small-cell, helium-based (60% He, 40% C_3H_8) main drift chamber (MDC). (2) A time-of-flight system (TOF). (3) An electromagnetic calorimeter (EMC). (4) The muon counter (MUC).

27.3 Measurement of Baryon Form Factors at BESIII

27.3.1 Measurement of Proton Form Factor

BESIII has published the measurement of the channel $e^+e^- \to p\overline{p}$ between 2.2324 and 3.6710 GeV [5], the results are shown in Fig. 27.2a together with previous experimental results and the EFF results are shown in Fig. 27.2b. These data were collected in 2011 and 2012 and correspond to a luminosity of 156.9 pb^{-1}. A world-leading

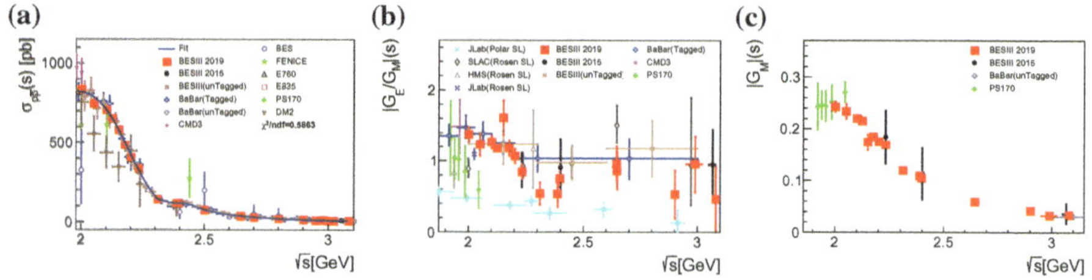

Fig. 27.2 Results from this analysis (black solid squares) including statistical and systematic uncertainties, together with other existing measurements for **a** the $e^+e^- \to p\overline{p}$ cross section; **b** the $|G_E/G_M|$ of the proton; **c** the magnetic FF of the proton $|G_M|$

data sample for precision measurements of baryon FFs, a luminosity of 688.5 pb^{-1} was collected in 2015, BESIII has preliminary result between 2.00 and 3.08 GeV, and the results are shown in Fig. 27.2a. The precision in the cross section measurements dominated by systematic, the best accuracy is between 3.0 and 4.2% at lower $\sqrt{s}$, less than 12% at higher $\sqrt{s}$. The $|G_E/G_M|$ and $|G_M|$ were extracted as Fig. 27.2b and c. The precision in the $|G_E/G_M|$ measurements dominated by statistics, the accuracy is around 10% for lower $\sqrt{s}$, especially the best accuracy is 3.4% at 2.125 GeV.

BESIII has preliminary result for untagged method while the results of tagged method for $e^+e^- \to \gamma p\overline{p}$ is under reviewing. Both of the method were collected at 7 c.m.energies correspond to a luminosity of 7.4 fb^{-1} above 3.773 GeV. The results of $\sigma(p\overline{p})$ are shown Fig. 27.2a and the results of $|G_E/G_M|$ are shown in Fig. 27.2b.

A step in the cross section is very likely due to Coulomb effect, since in the Coulomb factor there is a factor $1/\beta$ that cancels the factor beta in the cross section formula and produces a step Fig. 27.3a, which could be proved by tagged ISR technique. Some unexpected features are proved from our measurements. Plateau above threshold, corresponding to |G| close to 1, like a point-like fermion, similar features is also shown $e^+e^- \to \Lambda_c^+\bar{\Lambda}_c^-$ Fig. 27.3b. Andrea Bianconi and Egle Tomasi–Gustafsson discovered the oscillations in effective FF (|G|) from BABAR Fig. 27.4a and, which was confirmed by BESIII Fig. 27.4b [6].

27.3.2 Measurement of Λ Form Factor

BESIII has published the measurement of the channel $e^+e^- \to \Lambda\overline{\Lambda}$ around threshold [7], and the results are shown in Fig. 27.5a. These data were collected in 2012 and correspond to a luminosity of 2.63 pb^{-1}. Λ is neutral baryon, there is no Coulomb factor effect, but there is also a step at threshold. The observed threshold enhancement implies a more complicated underlying physics scenario. The Coulomb factor cancel the β for a charged $B\overline{B}$ pair, equals to 1 for a neutral $B\overline{B}$ pair. Help to understand the mechanism of baryon production and test the theory hypotheses based on the threshold enhancement effect.

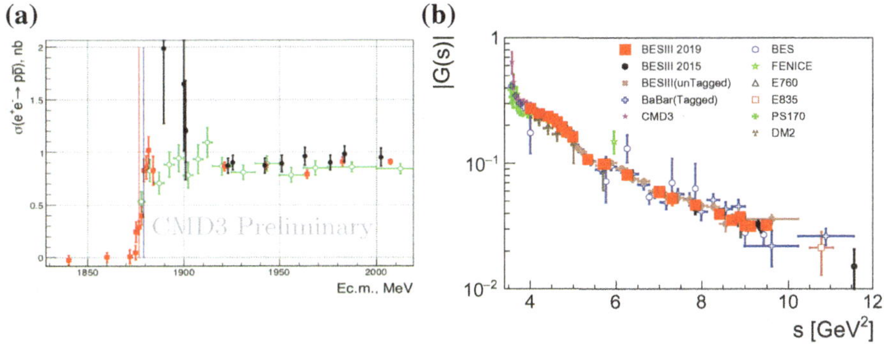

Fig. 27.3 **a** Cross sections of $e^+e^- \to p\overline{p}$ around threshold; **b** Effective FF of proton

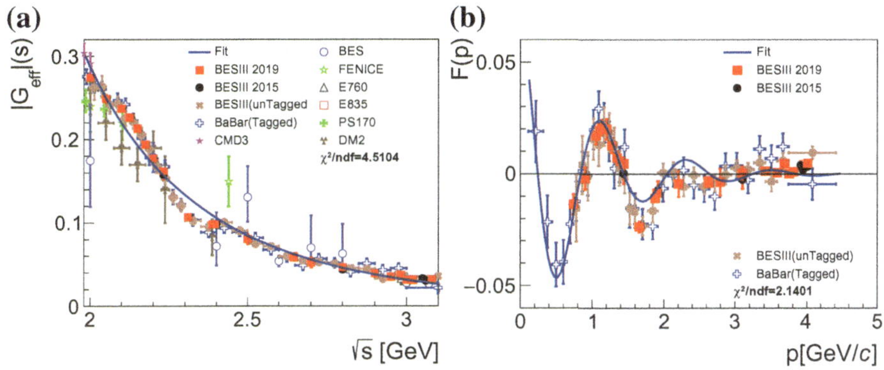

Fig. 27.4 The oscillations in effective FF (|G|) from **a** BABAR and **b** BESIII

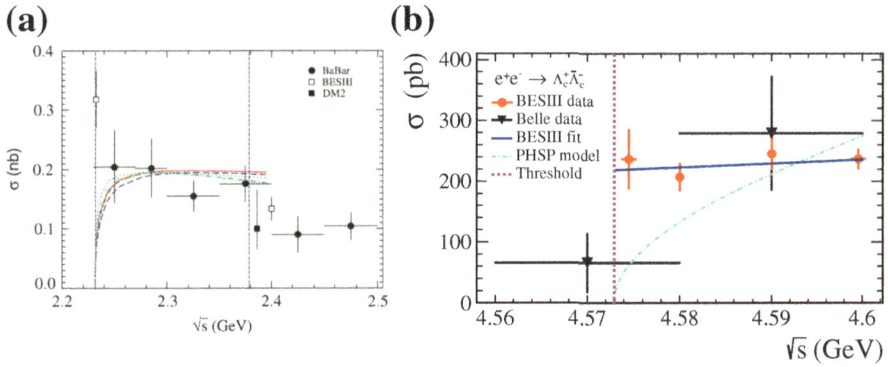

Fig. 27.5 The cross section of **a** $e^+e^- \to \Lambda\overline{\Lambda}$ and **b** $e^+e^- \to \Lambda_c^+\bar{\Lambda}_c^-$ around threshold

27.3.3 Measurement of Λ_c cross section

BESIII has published the measurement of the channel $e^+e^- \rightarrow \Lambda_c^+\bar{\Lambda}_c^-$ around threshold [8], and the results are shown in Fig. 27.5b. This analysis uses the data collected in 2014 and correspond to a luminosity of 2.63 pb^{-1}, ten modes of Λ_c^+ ($\bar{\Lambda}_c^-$) are reconstructed and the measurement of the born cross section at 4 energy points with unprecedented statistical accuracy. The σ_{Born} at near the threshold indicates the complexity of production behavior of the Λ_c. At threshold, there is again a step in $\sigma_{\Lambda_c^+\bar{\Lambda}_c^-}$ which is close to the point-like value, once the Coulomb factor is taken into account:

$$\sigma_{\Lambda_c^+\bar{\Lambda}_c^-(\mathrm{point-like})} \approx \frac{\pi^2\alpha^3}{2m_{\Lambda_c}} \approx 145\ \mathrm{pb}. \tag{27.7}$$

27.3.4 Measurement the G_E/G_M Phase of Λ

BESIII has preliminary result of the G_E/G_M phase of Λ, using data were collected in 2015 and correspond to a luminosity of 66.9 pb^{-1}.

$$|G_E/G_M| = 0.94 \pm 0.16 \pm 0.03 \pm 0.02(\alpha_\Lambda), \tag{27.8}$$
$$\Delta\phi = 42^\circ \pm 16^\circ \pm 8^\circ \pm 6^\circ(\alpha_\Lambda). \tag{27.9}$$

A non-zero phase has polarization effect on the baryons: $P_y \propto \sin\Delta\phi$. With hyperon weak decay to $B + P$, the polarization of hyperon can be measurement, so does the relative phase between G_E and G_M. Two projected distributions, i.e. the Λ scattering angle and the polarization as a function of the scattering angle, are shown in Fig. 27.6.

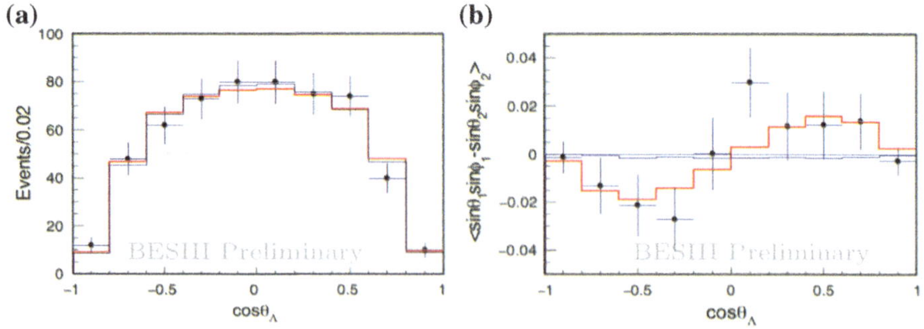

Fig. 27.6 **a** The acceptance corrected Λ scattering angle as obtained from the projection from the multidimensional log-likelihood fit. **b** The polarization as a function of the scattering angle

27.4 Summary

EFF provide a quantitative description of hadron structure and are basic observables of QCD. BESIII is unique in its capability to measure baryon FFs, from nucleons to Λ_c and use two complementary approaches: energy scan and ISR technique. Proton FFs have been measured using a test energy scan of 2012 and 2015, for 2012 data have published, for 2015 data have preliminary results, the precision greatly improved. In time-like region, results from BESIII are unprecedented precision, especially $|G_E/G_M|$ providing an uncertainty comparable to the space-like region for the first time. Very exciting results from tagged ISR on protons expected very soon, BESIII have preliminary results on untagged ISR techniques. BESIII have published results on cross section and FFs from Λ and Λ_c close to threshold. BESIII have preliminary results on the relative phase of Λ EFFs. More results, like neutron FFs, are coming soon.

Acknowledgements This work supported in part by the National Natural Science Foundation of China (NFSC) under contracts No. 11605196.

References

1. C. Morales, arXiv:1706.07674
2. A. Zichichi, S.M. Berman, N. Cabibbo, R. Gatto, Nuovo Cimento, **24**, 170 (1962)
3. G. Bonneau, F. Martin, Nucl. Phys. B **27**, 381 (1971)
4. M. Abilikim et al., BESIII Collaboration, Nucl. Instru. Meth. A **614**, 345 (2010)
5. M. Ablikim et al., BESIII Collaboration, Phys. Rev. D **91**, 112004 (2015)
6. A. Bianconi, E. Tomasi-Gustafsson, Phys. Rev. Lett. **114**, 232301 (2015)
7. M. Abilikim et al., BESIII Collaboration. Phys. Rev. D **97**, 032013 (2018)
8. M. Abilikim et al., BESIII Collaboration. Phys. Rev. Lett. **120**, 132001 (2018)

Chapter 28
Charm Mixing Study at Belle and Prospects at Belle II

Minakshi Nayak

Abstract We report D^0-mixing precision at Belle using a full data sample of 772×10^6 $B\bar{B}$ pairs collected at the $\Upsilon(4S)$ resonance with the Belle detector at the KEKB asymmetric energy e^+e^- collider and the expected precision at Belle II, with a total integrated luminosity of 50 ab^{-1} and using the golden modes: $D^0 \to K^+\pi^-, K_S\pi^+\pi^-, K^+K^-, \pi^+\pi^-, K^+\pi^-\pi^0$.

28.1 SuperKEKB and Belle II

SuperKEKB, the high luminosity frontier machine is the major upgrade of KEKB factory using the so-called Nano-Beam Scheme by reducing the vertical spot size to nearly 50 nm. It is designed to improve the beta function and beam current by a factor of twenty and two respectively, in order to achieve a peak luminosity of $\mathscr{L} = 8.0 \times\ 10^{35}\text{cm}^{-2}\text{s}^{-1}$, a factor of forty than KEKB.

Belle II is a major upgrade of Belle experiment [1] located at the collision point of SuperKEKB machine. It will start collecting data from early 2019, and will accumulate data of integrated luminosity of 50 ab^{-1} by 2025.

Given the clean environment of SuperKEKB collider and the hermiticity of the detector, Belle II is expected to have low background, high trigger efficiency, and excellent performances in the reconstruction of the neutral particles (e.g. γ, π^0, η). Belle II has successfully finished taking its first data in July 2018 without the vertex detector, and is ready for physics data taking with full detector in early 2019. By 2025, Belle II will collect 60 times higher charm sample than Belle which will allow for rich charm physics program.

On behalf of the Belle II Collaboration.

M. Nayak (✉)
Tel Aviv University, Tel Aviv-Yafo, Israel
e-mail: minakshi@tauex.tau.ac.il

A. Giri and R. Mohanta (eds.), *16th Conference on Flavor Physics and CP Violation*, Springer Proceedings in Physics 234,
https://doi.org/10.1007/978-3-030-29622-3_28

28.2 D^0 Decay Time Resolution

The Belle II vertex detector consists of four outer layers of double-sided silicon strip detector and two inner layers of pixel detectors. The inner most layer will sit close to the interaction point at a radius of 1.4 cm, and the outer most layer of the silicon detector will have a larger outer radius than that of Belle [1], and will cover the full Belle II angular acceptance. The vertex detector will contribute to better reconstruction efficiency with a large improvement of D^0 decay vertex and the interaction point resolutions compared with Belle and BaBar. This will precisely determine the D^0 decay time, which is essential in time-dependent measurements. According to Monte Carlo (MC) simulation for the processes $D^0 \to h^-h^+$ (where h = π, K), Belle II will have D^0 decay time resolution of 140 fs [2], which is a factor of two improvement over Belle and BaBar (270 fs).

28.3 Time Dependent $D^0 - \bar{D^0}$ Mixing Studies

The D^0 meson final state can be accessed either through D^0 via Doubly Cabibbo Suppressed (DCS) decay or $D^0 - \bar{D}^0$ mix and then $\bar{D}^0$ decays to the final state via Cabibbo Favored (CF) decay. By measuring the D^0 decay rate as a function of D^0 proper time, one can extract the sensitivity to mixing. With multibody final state, the Dalitz analysis allows to access to more than one channel at the same time. The sensitivity estimation at Belle II is extracted by scaling the Belle results. In detail, if σ_{stat} is the statistical error of the Belle measurements, σ_{syst} is the systematic error that scales with the luminosity and σ_{irred} is the systematic error that does not scale with luminosity (e.g. vertex resolution due to the detector misalignment), the expected uncertainty of Belle II with the full integrated luminosity is given by:

$$\sigma_{\text{Belle II}} = \sqrt{(\sigma^2_{\text{stat}} + \sigma^2_{\text{syst}}).\frac{\mathscr{L}_{\text{Belle}}}{50\ \text{ab}^{-1}} + \sigma^2_{\text{irred}}}. \tag{28.1}$$

28.4 D^0 Mixing Precision at Belle and Estimated Precision at Belle II

28.4.1 $D^0 \to K^+\pi^-$ Decay

To study the sensitivity of Belle II for the mixing parameters x, y, and CP violating parameters $\left|\frac{q}{p}\right|$, ϕ in $D^0 \to K^+\pi^-$ decays, we perform a toy MC study by generating separate samples of $D^{*+} \to D^0\pi_{\text{slow}}$, $D^0 \to K^+\pi^-$ and its charge conjugate decays corresponding to 5 ab^{-1}, 20 ab^{-1}, and 50 ab^{-1} of data. The probability density

Table 28.1 The sensitivity of the mixing parameters x', y' and CP violating parameters $|q/p|$, ϕ for $\overline{D}^0 \rightarrow K^-\pi^+$ decays at Belle II, under the condition of no CP Violation (CPV) and with CPV

Condition	Parameter	Belle measurement	Belle II estimation		
		976 fb^{-1}	5 ab^{-1}	20 ab^{-1}	50 ab^{-1}
No CPV	$\sigma x'^2$ (10^{-5})	22	7.5	3.7	2.3
	$\sigma y'$ (%)	0.34	0.11	0.056	0.035
CPV allowed	$\sigma x'$ (%)		0.37	0.23	0.15
	$\sigma y'$ (%)		0.26	0.17	0.10
	$\sigma \lvert q/p \rvert$		0.20	0.09	0.05
	$\sigma \phi$ (°)		16	9.2	5.7

functions (PDFs) used to generate such decays are contributed by two processes: a DCS process and a CF process following mixing. The PDFs are given by:

$$\frac{dN(D^0 \rightarrow f)}{\mathrm{dt}} = e^{-\overline{\Gamma}t}\left\{R_D + \left|\frac{q}{p}\right|\sqrt{R_D}(y'\cos\phi - x'\sin\phi)(\overline{\Gamma}t) + \left|\frac{q}{p}\right|^2 \frac{(x'^2+y'^2)}{4}(\overline{\Gamma}t)^2\right\} \tag{28.2}$$

$$\frac{dN(\overline{D}^0 \rightarrow \bar{f})}{\mathrm{dt}} = e^{-\overline{\Gamma}t}\left\{\overline{R}_D + \left|\frac{p}{q}\right|\sqrt{\overline{R}_D}(y'\cos\phi + x'\sin\phi)(\overline{\Gamma}t) + \left|\frac{p}{q}\right|^2 \frac{(x'^2+y'^2)}{4}(\overline{\Gamma}t)^2\right\}, \tag{28.3}$$

where $x' = x\cos\delta + y\sin\delta$, $y' = y\cos\delta - x\sin\delta$, and δ is the strong phase difference between $D^0 \rightarrow K^-\pi^+$ and $\overline{D}^0 \rightarrow K^-\pi^+$ amplitudes. The decay times are smeared by the expected decay time resolution of Belle II, and the resulting decays times are fitted with mixing parameters x, y in the case of no CP violation, or together with $\left|\frac{q}{p}\right|$, ϕ allowing CP violation. The fit results are then compared with the generated values. The RMS of the residual distributions are taken as the precision Belle II should achieve for these parameters. The sensitivity estimation from the toy study along with the previous Belle measurements [3] are listed in Table 28.1.

28.4.2 $D^0 \rightarrow K^+\pi^-\pi^0$ Decay

The wrong-sign (WS) decay $D^0 \rightarrow K^+\pi^-\pi^0$ proceeds directly via a DCS decay, and indirectly via mixing followed by a CF decay, where the later amplitude provides sensitivity to mixing. Assuming Belle II has similar efficiency as BaBar, the sensitivity estimation of the mixing parameters is performed with Toy MC study by generating an ensemble of 10 experiments, with each experiment consisting of 225000 $D^0 \rightarrow K^+\pi^-\pi^0$ decays corresponding to 50 ab^{-1} of Belle II data. The generated decay times are smeared by the expected Belle II resolution of 140 fs, and then

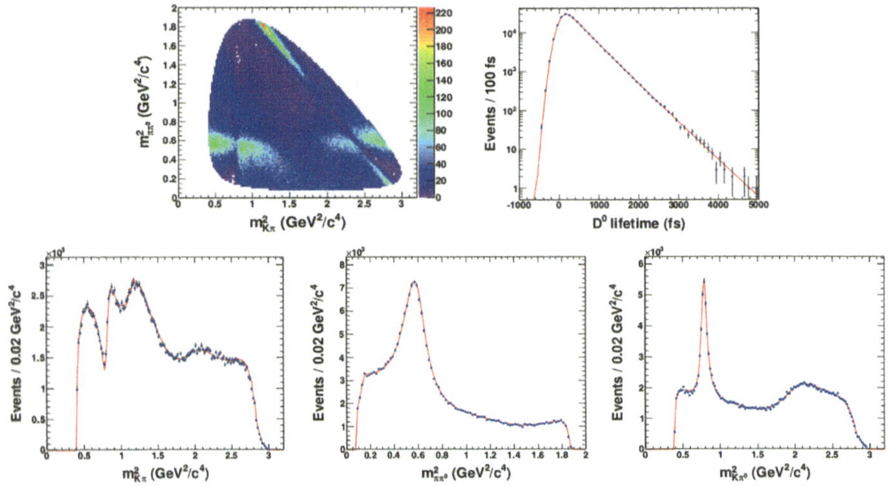

Fig. 28.1 The top row shows the time-dependent Dalitz plot fit and the decay time fit to the lifetime-smeared sample of wrong-sign $D^0 \to K^+\pi^-\pi^0$ decays. The bottom row shows projections of the fitted Dalitz variables [2]

fitted using a time-dependent fit to the $(m^2_{K^+\pi^-}, m^2_{K^+\pi^0})$ Dalitz plot as done in BaBar [4] to measure the effective mixing parameters $x'' = x.\cos\delta_{K\pi\pi^0} + y.\sin\delta_{K\pi\pi^0}$ and $y'' = y.\cos\delta_{K\pi\pi^0} - x.\sin\delta_{K\pi\pi^0}$. Here $\delta_{K\pi\pi^0}$ is the strong phase difference between the amplitudes for $D^0 \to K^+\rho^-$ and $\overline{D}^0 \to K^+\rho^-$. From the fit of ten experiments, the mixing sensitivity are obtained as $\sigma x = 0.057\%$ and $\sigma y = 0.049\%$ [5], which are almost an order of magnitude improvement than BaBar if the effect of background is not included. A typical time-dependent Dalitz plot fit is illustrated in Fig. 28.1.

28.4.3 $D^0 \to K_S\pi^+\pi^-$ Decay

Mixing parameters for $D^0 \to K_S\pi^+\pi^-$ decay is obtained by fitting the time-dependent Dalitz plot. Here by calculating the observables $m_+ = (P_{K^0_S} + P_{\pi^+})^2$ and $m_- = (P_{K^0_S} + P_{\pi^-})^2$ and by performing an unbinned maximum likelihood fit to m_+, m_- and the decay time t, one can extract the information of mixing parameters x and y. Belle II sensitivity to the above parameters for the above decay is estimated by scaling from the Belle measurement [6] are listed in Table 28.2.

Table 28.2 Precision obtained for x, y from Belle analysis of $D^0 \to K_S\pi^+\pi^-$ decays, and the expected precision for Belle II as obtained by scaling the Belle errors

Data	Stat.	Systematic		Total	Stat.	Systematic		Total
		red.	irred.			red.	irred.	
	$\sigma_x(\%)$				$\sigma_y(\%)$			
976 fb^{-1}	0.19	0.06	0.11	0.20	0.15	0.06	0.04	0.16
5 ab^{-1}	0.08	0.03	0.11	0.14	0.06	0.03	0.04	0.08
50 ab^{-1}	0.03	0.01	0.11	0.11	0.02	0.01	0.04	0.05

Table 28.3 The Expected Belle II sensitivity to the D mixing parameter, y_{cp} for $\overline{D}^0 \to K^+K^-, \pi^+\pi^-$ decays

Observable	Statistical	Systematic		Total
		red.	irred.	
$y_{\text{CP}}(\%)$				
976 fb^{-1}	0.22	0.07	0.07	0.24
5 ab^{-1}	0.10	0.03–0.04	0.07–0.04	0.11–0.12
50 ab^{-1}	0.03	0.01	0.07–0.04	0.05–0.08

28.4.4 $D^0 \to K^+K^-, \pi^+\pi^-$ Decay

Mixing in $D^0 \to CP$ eigenstates like K^+K^-, $\pi^+\pi^-$ gives an effective lifetime which differs from that decays in to flavor eigenstates such as $D^0 \to K^-\pi^+$. Hence the observable which represents the relative lifetime difference between decays to CP and flavor specific final state is obtained as:

$$y_{\text{cp}} = \eta_{\text{CP}} \left(\frac{\Gamma_{\text{cp}} - \Gamma_{K^\pm\pi^\mp}}{\Gamma_{K^\pm\pi^\mp}} \right) = \eta_{\text{CP}} \left(\frac{\tau(D \to K^\pm\pi^\mp)}{\tau(D \to CP)} - 1 \right), \tag{28.4}$$

where $\eta_{\text{CP}} = +1$ for CP-even final state. y_{cp} is related to mixing parameter x and y as:

$$y_{\text{cp}} = \frac{1}{2}\left(\left|\frac{q}{p}\right| + \left|\frac{p}{q}\right| \right) y\cos\phi - \frac{1}{2}\left(\left|\frac{q}{p}\right| - \left|\frac{p}{q}\right| \right) x\sin\phi \tag{28.5}$$

where $\phi = \arg(q/p)$. In the limit of CP conservation, $y_{\text{cp}} = y$. If CP is violated y_{CP} gets contribution from x. Belle with its full data sample has measured the $y_{\text{cp}} = (1.11 \pm 0.22 \pm 0.09)\%$ [7], and BaBar measurement of $y_{\text{cp}} = (0.72 \pm 0.18 \pm 0.12)\%$ [8]. Expected Belle II sensitivity by scaling Belle measurement has been summarized in Table 28.3.

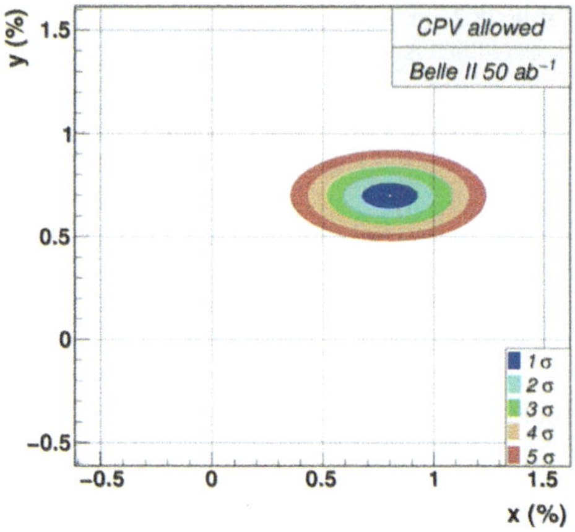

Fig. 28.2 Expected Belle II precision at 50 ab^{-1} data

28.5 Expected Belle II Precision Versus Current World Average

The experimental data consistently indicate that the D^0 and $\overline{D}^0$ do mix and current measurement provides constraints on many new physics models like fourth generation, extra gauge bosons, left right symmetric models [9]. Current world average values of the mixing parameters x and y are $x = (0.32 \pm 0.14)\%$, and $y = (0.69^{+0.06}_{-0.07})\%$. The expected Belle II precision at 50 ab^{-1} is extrapolated to be $x = (0.8 \pm 0.09)\%$, and $y = (0.7 \pm 0.04)\%$ as shown in Fig. 28.2.

References

1. T. Abe et al., Belle II technical design report, KEK-REPORT-2010-1 (2010), arXiv: 1011.0352
2. E. Kou et al., The Belle II Physics book, 2018, arXiv:1808.10567 [hep-ex]
3. B.R. Ko et al., (Belle Collaboration), Observation of $D^0 - \overline{D}^0$ Mixing in e^+e^- Collisions. Phys. Rev. Lett. **112**, 111801 (2014)
4. B. Aubert et al., (BABAR Collaboration), Measurement of $D^0 - \overline{D}^0$ mixing from a time-dependent amplitude analysis of $D^0 \to K^+\pi^-\pi^0$ decays. Phys. Rev. Lett. **103**, 211801 (2009)
5. L.-K. Li et al., $D^0 - \overline{D}^0$ mixing sensitivity estimation at Belle II in wrong-sign decays $D^0 \to K^+\pi^-\pi^0$ via time-dependent amplitude analysis. Chin. Phys. C **41**, 023001 (2017)
6. T. Peng et al., (Belle Collaboration), Measurement of $D^0 - \overline{D}^0$ mixing and search for indirect *CP* violation using $D^0 \to K^0_S\pi^+\pi^-$ decays. Phys. Rev. D **89**, 091103 (2014)
7. M. Staric et al., (Belle Collaboration), Measurement of D^0-mixing and search for *CP* violation in $D^0 \to K^+K^-, \pi^+\pi^-$ decays with the full Belle data set. Phys. Lett. B **753**, 412 (2016)
8. J.P. Lees et al., (BABAR Collaboration), Measurement of $D^0 - \overline{D}^0$ mixing and *CP* violation in two-body D^0 decays. Phys. Rev. D **87**, 012004 (2013)
9. E. Golowich, J.A. Hewett, S. Pakvasa, A.A. Petrov, Implications of $D^0 - \overline{D}^0$ mixing for new physics. Phys. Rev. D **76**, 095009 (2007)

Chapter 29
Direct *CP* Violation in $K \to \mu^+\mu^-$

T. Kitahara

Abstract The upgrade of the LHCb experiment will probe a rare decay $K_S \to \mu^+\mu^-$ precisely. Although the process is dominated by *CP*-conserving decay, one can extract *CP*-violating pieces via the precise measurement. In this contribution, I introduce an interference contribution between K_L and K_S that is sensitive to a direct *CP* violation. It is found that the interference contribution can affect $\mathrm{Br}(K_S \to \mu^+\mu^-)$ Standard Model prediction up to 60 %. It is also found that an unknown sign of an amplitude of $K_L \to \gamma\gamma$ can be determined by measuring the interference effect, which can much reduce a theoretical uncertainty of $\mathrm{Br}(K_L \to \mu^+\mu^-)$. I also show that the interference contributions in new physics models and their correlations with other kaon rare decays, especially ε'/ε. This contribution is based on D'Ambrosio and Kitahara in Phys Rev Lett 119:201802, 2017, Chobanova et al. in JHEP 1805:024, 2018, Endo et al. in JHEP 1804:019, 2018.

29.1 Introduction

The upgrade of the LHCb experiment is expected to be able to probe short-distance (SD) physics using the $K^0 \to \mu^+\mu^-$ decay. In the Standard Model (SM), $K_S \to \mu^+\mu^-$ has not been observed yet, and is significantly dominated by a *P*-wave *CP*-conserving long-distance (LD) contribution, while a *S*-wave *CP*-violating SD contributions from the *Z*-penguin and *W*-box are small [1–3]:

$$\mathrm{Br}(K_S \to \mu^+\mu^-)_{\mathrm{SM}} = \left[(4.99 \pm 1.50)_{\mathrm{LD}} + (0.19 \pm 0.02)_{\mathrm{SD}}\right] \times 10^{-12}. \quad (29.1)$$

T. Kitahara (✉)
Institute for Advanced Research, Nagoya University, Nagoya 464-8602, Japan
e-mail: teppeik@kmi.nagoya-u.ac.jp

Kobayashi-Maskawa Institute for the Origin of Particles and the Universe, Nagoya University, Nagoya 464-8602, Japan

Physics Department, Technion–Israel Institute of Technology, 3200003 Haifa, Israel

A. Giri and R. Mohanta (eds.), *16th Conference on Flavor Physics and CP Violation*, Springer Proceedings in Physics 234,
https://doi.org/10.1007/978-3-030-29622-3_29

The large uncertainty comes from the LD contribution which has been computed by Chiral Perturbation Theory (ChPT) [2]. This uncertainty is expected to be reduced by a dispersive treatment of $K_S \to \gamma^*\gamma^*$ [4], where $K_S \to \gamma\gamma$, $K_S \to \mu^+\mu^-\gamma$, $K_S \to \mu^+\mu^-e^+e^-$ and $K_S \to \mu^+\mu^-\mu^+\mu^-$ are measurable in LHCb experiment and KLOE-2 experiment at DAΦNE [5].

In contrast to the $K_L \to \mu^+\mu^-$ decay which has been measured, the following points are important. The $K_S \to \mu^+\mu^-$ provides another sensitive probe of imaginary parts of SD couplings and consequently is very sensitive to new sources of *CP* violation, while the $K_L \to \mu^+\mu^-$ is sensitive to the real couplings. Besides, as seen in (29.1) the LD and SD contributions to the total rate are added incoherently [2, 3], which represents a big theoretical advantage over $K_L \to \mu^+\mu^-$ where LD and SD amplitudes interfere. This means that when the SD contribution is significantly enhanced in Beyond the Standard Models (BSMs), the LD uncertainty in $K_S \to \mu^+\mu^-$ ceases to be important, and theoretically clean tests of new-physics (NP) scenarios are possible. In particular, being the SD contribution dominant, correlations of $K_S \to \mu^+\mu^-$ with ε'/ε and also $K_L \to \pi^0\nu\bar{\nu}$ are investigated within many new-physics models.

Indeed, within concrete BSMs, Br($K_S \to \mu^+\mu^-$) can be modified substantially, for instance Br $\sim \mathcal{O}(10^{-10})$ in the leptoquark models [6] and Br $\sim \mathcal{O}(10^{-11})$, or even saturate the current experimental bound in certain Minimal Supersymmetric Standard Model (MSSM) parameter space [7], albeit somewhat fine-tuned. It turns out then already the present upper bound from LHCb can have some impact on the allowed parameter range of certain models. This shows that future improvement of this bound can have important impact on various NP scenarios.

The LHCb capabilities for strange decays were first demonstrated in [8], which achieved world's best result in $K_S \to \mu^+\mu^-$ even though the trigger efficiency on well reconstructed decays was only 1 % (to be compared to 90 % for $B_s \to \mu^+\mu^-$). The LHCb full Run1 analysis has set the upper bound on $K_S \to \mu^+\mu^-$ [9],

$$\mathrm{Br}(K_S \to \mu^+\mu^-)_{\mathrm{LHCb\ Run1}} < 0.8\,(1.0) \times 10^{-9} \text{ at } 90\,\%\,(95\,\%)\,\mathrm{C.L.}, \tag{29.2}$$

which is 2 orders of magnitude larger than the SM sensitivity.

In Run 2 of LHCb, dedicated trigger lines have been implemented, selecting muons down to 80 MeV in transverse momentum and increasing the trigger efficiency by one order of magnitude for strangeness decays to dimuons. The main limitation is the hardware trigger. In Run 3, the LHCb trigger will be fully software-based, which can in principle allow for efficiencies as high as those achieved for B's. The LHCb Upgrade II will reach sensitivities for Br($K_S \to \mu^+\mu^-$) below the 10^{-11} level if it keeps the performance of the current detector, taking into account that the full software trigger will allow for very high trigger efficiencies [10].

29.2 Interference Between K_L and K_S

A crucial aspect of the $K^0 \to \mu^+\mu^-$ decay is a flavor-tagged measurement which can probe CP-violating SD contributions directly. Its numerical effect is $\mathcal{O}(1)$ compared to the prediction in (29.1) even in the SM [1]. While K_L decays typically outside the LHCb fiducial volume, for K_S the interference between K_L and K_S affects the number of signal events as $\Gamma_{\text{int.}} \propto \mathcal{A}(K_L \to \mu^+\mu^-)\mathcal{A}(K_S \to \mu^+\mu^-)^*$ when the flavor tagging, K^0 or $\overline{K}^0$ at $t = 0$, is performed. An effective branching ratio into $\mu^+\mu^-$, which includes the interference correction and would correspond to experimental event numbers after the removal of $K_L \to \mu^+\mu^-$ background, is given by [1],

$$\begin{aligned} \text{Br}(K_S \to \mu^+\mu^-)_{\text{eff}} = \tau_S \left[\int_{t_{\min}}^{t_{\max}} \text{d}t e^{-\frac{t}{\tau_S}} \varepsilon(t)\right]^{-1} \int_{t_{\min}}^{t_{\max}} \text{d}t \Big\{ \Gamma(K_S \to \mu^+\mu^-) e^{-\frac{t}{\tau_S}} \\ + \frac{D f_K^2 m_K^3 \beta_\mu}{8\pi} \text{Re}\left[i\left(A_S A_L - \beta_\mu^2 B_S^* B_L\right) e^{-i\Delta m_K t}\right] e^{-\frac{t}{2\tau_S}\left(1+\frac{\tau_S}{\tau_L}\right)} \Big\} \varepsilon(t), \end{aligned} \tag{29.3}$$

with

$$\Gamma(K_{\text{S,L}} \to \mu^+\mu^-) = \frac{1}{16\pi} f_K^2 m_K^3 \beta_\mu \left(|A_{\text{S,L}}|^2 + \beta_\mu^2 |B_{\text{S,L}}|^2\right), \tag{29.4}$$

where final-state muon polarizations are summed over, $t_{\min}$ to $t_{\max}$ corresponds to a range of detector for K_S tagging, $\varepsilon(t)$ is the decay-time acceptance of the detector, $\beta_\mu = \sqrt{-4m_\mu^2/m_K^2}$, and $f_K = (155.9 \pm 0.4)$ MeV [11]. The $K_L \to \mu^+\mu^-$ background can be subtracted by a combination of the simultaneous measurement of $K_S \to \pi^+\pi^-$ and the knowledge of the observed value of $\text{Br}(K_L \to \mu^+\mu^-)$ [1]. The dilution factor D is a measure of the initial K^0–$\overline{K}^0$ asymmetry,

$$D = \frac{K^0 - \overline{K}^0}{K^0 + \overline{K}^0}. \tag{29.5}$$

The $A_{\text{S,L}}$ and $B_{\text{S,L}}$ are the S-wave and P-wave contributions in $K_{\text{S,L}} \to \mu^+\mu^-$ transitions, respectively. The expressions for them are given in [7] using the general $\Delta S = 1$ effective Hamiltonian. Note that A_S and B_L are real, while B_S and A_L are complex in general BSMs.

The interference contribution is proportional to the dilution factor D, which requires flavor tagging. This can be done by detecting the accompanying K^- in the process $pp \to K^0K^-X$, Λ^0 in the process $pp \to K^0\Lambda^0X$, or π^+ in the process $pp \to K^{*+}X \to K^0\pi^+X$ [1].

In the SM, the effective branching ratio in (29.3) can be reduced to [1, 7]

$$A_S A_L - \beta_\mu^2 B_S^* B_L = \frac{4G_F^2 M_W^2 m_\mu^2}{m_K^2 \pi^2} \underbrace{\text{Im}C_{\text{A,SM}}}_{\text{SD (CPV)}} \Bigg(\underbrace{A_{L\gamma\gamma}^\mu}_{\text{LD (CPC)}} - \frac{\pi^2}{G_F^2 M_W^2} \underbrace{\text{Re}C_{\text{A,SM}}}_{\text{SD (CPC)}} \Bigg). \tag{29.6}$$

The Wilson coefficient C_A is defined by

$$\mathscr{H}_{\text{eff}} = -C_A(\bar{s}\gamma^\mu P_L d)(\bar{\ell}\gamma_\mu\gamma_5\ell) + \text{h.c.}, \tag{29.7}$$

and

$$C_{A,\text{SM}} = -\frac{[\alpha_2(M_Z)]^2}{2M_W^2}\left(V_{\text{ts}}^* V_{\text{td}} Y_t + V_{\text{cs}}^* V_{\text{cd}} Y_c\right), \tag{29.8}$$

where $\alpha_2 = g^2/(4\pi)$, $Y_t = 0.950 \pm 0.049$ and $Y_c = (2.95 \pm 0.46) \times 10^{-4}$ [16]. The large CP-conserving LD two-photon contribution to K_L is [3, 17]

$$A^\mu_{L\gamma\gamma} = \pm 2.01(1) \times 10^{-4} \times [0.71(101) - i\,5.21], \tag{29.9}$$

where the sign has not been determined theoretically and experimentally. The large imaginary term in $A^\mu_{L\gamma\gamma}$, which corresponds to the absorptive component, can amplify the small CP-violating SD contribution in the effective branching ratio of $K_S \to \mu^+\mu^-$ in (29.3). Figure 29.1 shows the effective branching ratio and time distributions in the SM, where the interference of the CP-violating contribution can affect $K_S \to \mu^+\mu^-$ decay up to 60 %.

This quantity is sensitive to NP contributions to the electroweak penguin. This is of current interest to the possible discrepancy between the SM prediction and the data of ε'/ε (direct CP violation in $K_L^0 \to \pi\pi$) [13, 14, 18–20]. Also, a comparison with the future measurement of $K_L \to \pi^0\nu\bar{\nu}$ at the KOTO experiment will be interesting [21].

Using the effective branching ratio in (29.3), one can define the flavor-tagging asymmetry in $K_S \to \mu^+\mu^-$ by [7]

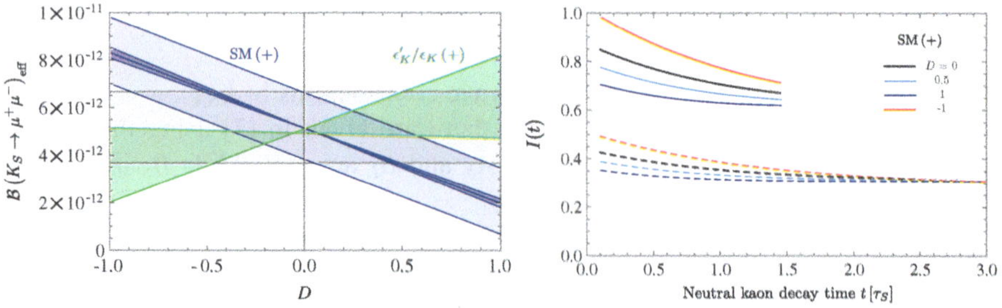

Fig. 29.1 *Left panel*: The effective branching ratios in (29.3) as function of D in (29.5), where the blue (gray) band denoting the SM prediction with (without) taking into account the interference contribution. The green band is predicted in a NP scenario [12] that can explain the ε'/ε discrepancy [13, 14] (based on a lattice result [15]). *Right panel*: The $K \to \mu^+\mu^-$ time distributions are shown in the SM for several choices of D. The $D = 0$ decay intensity is normalized, either over the interval 0.1–1.45 τ_S (solid lines) or 0.1–3 τ_S (dashed lines). In both panels, the positive sign of $A^\mu_{L\gamma\gamma}$ is assumed in (29.9). A detailed explanation and results for the negative sign of $A^\mu_{L\gamma\gamma}$ are given in [1]

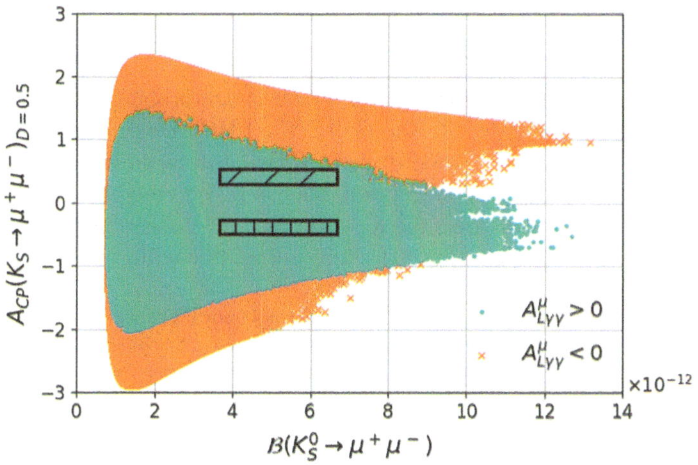

Fig. 29.2 Allowed regions of $A_{CP}(K_S \to \mu^+\mu^-)_{D,D'}$ and $\mathrm{Br}(K_S \to \mu^+\mu^-)$ in the MSSM where $(\delta_d^{LL})_{12} \neq 0$ and $(M_3 \cdot \mu) < 0$. The plot is done for $D = -D' = 0.5$. The cyan dots correspond to $A^{\mu}_{L\gamma\gamma} > 0$ and the orange crosses to $A^{\mu}_{L\gamma\gamma} < 0$ in (29.9). The vertically hatched areas correspond to the SM prediction for $A^{\mu}_{L\gamma\gamma} > 0$ and the inclined hatched areas to the SM prediction for $A^{\mu}_{L\gamma\gamma} < 0$. A detailed explanation and results for the different scenarios are given in [7]

$$A_{\mathrm{CP}}(K_S \to \mu^+\mu^-)_{D,D'} = \frac{\mathrm{Br}(K_S \to \mu^+\mu^-)_{\mathrm{eff}}(D) - \mathrm{Br}(K_S \to \mu^+\mu^-)_{\mathrm{eff}}(D')}{\mathrm{Br}(K_S \to \mu^+\mu^-)_{\mathrm{eff}}(D) + \mathrm{Br}(K_S \to \mu^+\mu^-)_{\mathrm{eff}}(D')}, \tag{29.10}$$

where D' is obtained by requiring an opposite flavor tagging. This asymmetry is a theoretically clean quantity that emerges from a genuine direct CP violation in general new physics models. In the SM, $A_{\mathrm{CP}}(K_S \to \mu^+\mu^-)^{\mathrm{SM}}_{D,-D} = \mathcal{O}(0.7) \times D$ is predicted in the case of $D' = -D$ and it is significantly sensitive to new sources of CP violation in BSMs [7].

The detailed studies about investigating the correlation among $K \to \mu^+\mu^-$, ε'/ε, and $K_L \to \pi^0\nu\bar{\nu}$ on the MSSM parameter space have been shown in [7, 22]. Figure 29.2 exhibits one of the results: Allowed regions of $A_{\mathrm{CP}}(K_S \to \mu^+\mu^-)_{D,D'}$ and $\mathrm{Br}(K_S \to \mu^+\mu^-)$ are shown in a scenario of the MSSM [7]. It is found that the CP asymmetry of $K_S \to \mu^+\mu^-$ can be significantly modified by the MSSM contributions, being up to eight times bigger than the SM asymmetry.

In a similar way, the CP asymmetry of $B_{d,s} \to \mu^+\mu^-$ has been studied [23, 24]. However, for the B_s system the situation differs substantially from K_S, since the LD contributions are negligible and the life-time difference between the two mass eigenstates is small compared to $K_{S,L}$. The CP asymmetry in $B_{d,s} \to \mu^+\mu^-$ vanishes in the SM, but is also sensitive to a new CP-violating phase.

29.3 Summary

The precise measurements of the kaon rare decays can probe CP-violating flavour changing neutral currents from various ways. The LHCb Upgrade can probe $K_S \to \mu^+\mu^-$ around SM sensitivity and could open a SD window by the interference effect in $K \to \mu^+\mu^-$. The interference contribution in $K \to \mu^+\mu^-$ emerges from the genuine direct CP violation and can affect the $K_S \to \mu^+\mu^-$ CP-conserving prediction up to 60 %. These measurements give the interesting correlation with ε'/ε and $K_L \to \pi^0\nu\bar{\nu}$.

Acknowledgements I would like to thank Veronika Chobanova, Giancarlo D'Ambrosio Motoi Endo, Toru Goto, Miriam Lucio Martinez, Diego Martinez Santos, Satoshi Mishima, Isabel Suarez Fernandez, Daiki Ueda, and Kei Yamamoto for fruitful collaborations on the presented works. I also want to warmly thank the organizers of 16th Conference on Flavor Physics and CP Violation (FPCP 2018) for inviting and giving me the opportunity to present this result in a great conference.

References

1. G. D'Ambrosio, T. Kitahara, Phys. Rev. Lett. **119**(20), 201802 (2017)
2. G. Ecker, A. Pich, Nucl. Phys. B **366**, 189 (1991)
3. G. Isidori, R. Unterdorfer, JHEP **0401**, 009 (2004)
4. G. Colangelo, R. Stucki, L.C. Tunstall, Eur. Phys. J. C **76**(11), 604 (2016)
5. G. Amelino-Camelia et al., Eur. Phys. J. C **68**, 619 (2010)
6. C. Bobeth, A.J. Buras, JHEP **1802**, 101 (2018)
7. V. Chobanova, G. D'Ambrosio, T. Kitahara, M. Lucio Martinez, D. Martinez Santos, I.S. Fernandez, K. Yamamoto, JHEP **1805**, 024 (2018)
8. R. Aaij et al., LHCb collaboration. JHEP **1301**, 090 (2013)
9. R. Aaij et al., LHCb collaboration, Eur. Phys. J. C **77**(10), 678 (2017)
10. A.A. Alves Jr. et al., arXiv:1808.03477 [hep-ex]
11. M. Tanabashi et al., Particle data group, Phys. Rev. D **98**(3), 030001 (2018)
12. M. Endo, T. Kitahara, S. Mishima, K. Yamamoto, Phys. Lett. B **771**, 37 (2017)
13. A.J. Buras, M. Gorbahn, S. Jäger, M. Jamin, JHEP **1511**, 202 (2015)
14. T. Kitahara, U. Nierste, P. Tremper, JHEP **1612**, 078 (2016)
15. Z. Bai et al., RBC and UKQCD collaborations, Phys. Rev. Lett. **115**(21), 212001 (2015)
16. M. Gorbahn, U. Haisch, Phys. Rev. Lett. **97**, 122002 (2006)
17. F. Mescia, C. Smith, S. Trine, JHEP **0608**, 088 (2006)
18. E. Pallante, A. Pich, Phys. Rev. Lett. **84**, 2568 (2000)
19. A.J. Buras, J.M. Gérard, JHEP **1512**, 008 (2015)
20. H. Gisbert, A. Pich, Rept. Prog. Phys. **81**(7), 076201 (2018)
21. J.K. Ahn et al., KOTO collaboration, arXiv:1810.09655 [hep-ex]
22. M. Endo, T. Goto, T. Kitahara, S. Mishima, D. Ueda, K. Yamamoto, JHEP **1804**, 019 (2018)
23. K. De Bruyn, R. Fleischer, R. Knegjens, P. Koppenburg, M. Merk, A. Pellegrino, N. Tuning, Phys. Rev. Lett. **109**, 041801 (2012)
24. A.J. Buras, R. Fleischer, J. Girrbach, R. Knegjens, JHEP **1307**, 77 (2013)

Chapter 30
J/ψ Production as a Function of Charged-Particle Multiplicity with ALICE at the LHC

Dhananjaya Thakur

Abstract At LHC energies, the charged-particle multiplicity dependence of particle production is a topic of considerable interest in pp collisions. It has been argued that multiple partonic interactions play an important role in particle production mechanisms, not only affecting the soft processes but also the hard processes. Recently, ALICE has measured J/ψ production as a function of charged-particle multiplicity to study the correlation between soft and hard processes. In this contribution, we present the J/ψ production versus multiplicity for pp and p–Pb collisions measured by ALICE. We compare the results with different theoretical models.

30.1 Introduction

Understanding the mechanism of charmonium production is one of the major challenges in pp collisions. There are many theoretical models that try to explain heavy-flavor production in hard scattering processes. Examples are the Color Singlet Model [1], non-relativistic QCD (NRQCD) [2] and the Color Evaporation Model [3]. The production of charm and anti-charm quark pairs are described by perturbative Quantum Chromodynamics (pQCD) and their binding into charmonium states by non-perturbative QCD. Charmonium suppression is a universally accepted probe for the de-confined medium in heavy-ion collisions. To understand the suppression, it is necessary to understand J/ψ production in pp collisions and also potential cold nuclear matter effects in p–Pb collisions. Recently, ALICE has measured J/ψ as a function of charged-particle multiplicity in $J/\psi \rightarrow \mu^+\mu^-$ and $J/\psi \rightarrow e^+e^-$ at $\sqrt{s}$ = 7 TeV, and observed an increasing trend with respect to charged-particle multiplicity [4]. A similar result has been found for D-meson production. This reveals that the multiple partonic interaction (MPI) which was thought to affect only soft processes

Dhananjaya Thakur for the ALICE Collaboration.

D. Thakur (✉)
Discipline of Physics, School of Basic Sciences, Indian Institute of Technology Indore, Indore 453552, India
e-mail: Dhananjaya.Thakur@cern.ch

A. Giri and R. Mohanta (eds.), *16th Conference on Flavor Physics and CP Violation*, Springer Proceedings in Physics 234,
https://doi.org/10.1007/978-3-030-29622-3_30

can also affect the hard processes and hence J/ψ production. With the help of new data at $\sqrt{s} = 13$ TeV, we can measure the trend with multiplicity more precisely than what was previously possible at $\sqrt{s} = 7$ TeV. Therefore, to have a clear view of the observed picture, ALICE has extended this analysis to pp collisions at $\sqrt{s} = 13$ TeV and p–Pb collisions at $\sqrt{s_{NN}} = 5.02$ TeV. We present the results of this analysis and compare them to available theoretical models.

30.2 Experimental Setup and Analysis Procedure

ALICE is one of the four major experiments at the LHC. Details about ALICE can be found in [5]. For the present work, two different spectrometers have been used for J/ψ reconstruction. The central barrel detector covers the rapidity range $|y| < 0.9$ and includes the Inner Tracking System (ITS) and Time Projection Chamber (TPC). It is used for the reconstruction of J/ψ via the di-electron decay channel. The forward muon spectrometer is used for the reconstruction of J/ψ via the di-muon decay channel in the rapidity range $-4.0 < y < -2.5$. Two V0 scintillator arrays used for triggering and are located at $-3.7 < \eta < -1.7$ and at $2.8 < \eta < 5.1$. The V0 detectors are also used as a high-multiplicity trigger.

The charged-particle pseudo-rapidity density ($\mathrm{d}N_{\mathrm{ch}}/\mathrm{d}\eta$) is measured at mid-rapidity ($|\eta| < 1.0$) from the information of track segments (tracklets) in the Silicon Pixel Detector (SPD). Several cuts are applied to determine the accurate position of the z-coordinate of the vertex (z_{vtx}). Tracklets are measured within $|\eta| < 1.0$ and $|z_{\mathrm{vtx}}| < 10.0$ cm. This account for the SPD acceptance. A z_{vtx}−dependent correction is applied using data driven method [6]. This also take into account the SPD inefficiency and acceptance. The correction factor is randomized on an event-by-event basis using a Poisson distribution, for the matching of true charged-particle multiplicity and the tracklet multiplicities.

The self-normalized J/ψ yield in bins of charged-particle multiplicity is calculated as:

$$\frac{\frac{\mathrm{d}N_{\mathrm{J}/\psi}}{\mathrm{d}y}}{\langle\frac{\mathrm{d}N_{\mathrm{J}/\psi}}{\mathrm{d}y}\rangle} = \frac{N_{\mathrm{J}/\psi}^{\mathrm{corr,\ i}}}{N_{\mathrm{J}/\psi}^{\mathrm{corr,\ integrated}}} \times \frac{N_{\mathrm{MB}}^{\mathrm{integrated}}}{N_{\mathrm{MB}}^{\mathrm{i}}},$$

where ($N_{\mathrm{J}/\psi}^{\mathrm{corr,i}}$, $N_{\mathrm{J}/\psi}^{\mathrm{corr,integrated}}$) and ($N_{\mathrm{MB}}^{\mathrm{i}}$, $N_{\mathrm{MB}}^{\mathrm{integrated}}$) are the corrected number of J/ψ and number of minimum bias events in ith multiplicity bin and integrated over all multiplicity bins, respectively. In p–Pb collisions, the mean p_{T} of J/ψ, $\langle p_{\mathrm{T}}^{J/\psi}\rangle$, is found by fitting the mean p_{T} of unlike-sign pairs of muons as a function of the dimuon invariant mass.

30.3 Results

Figure 30.1 shows the relative yield of inclusive J/ψ at mid-rapidity as a function of charged-particle multiplicity for integrated p_{T} and in p_{T} slices. It can be seen from the figure that a stronger than linear increase of yield is observed as compared to the charged-particle multiplicity. These data are more precise and extend the multiplicity reach with respect to the results at $\sqrt{s} = 7$ TeV. The result is compared to four models: Ferreiro et al. [7], EPOS3 [8], PYTHIA 8 [9] and Kopeliovich et al. [10].

- The model of **Ferreiro et al.**, which is able to well-explain *pp* result at $\sqrt{s} = 7$ TeV [7], overestimates the J/ψ yield at high multiplicities in *pp* at $\sqrt{s} = 13$ TeV. This model assumes that in high energy hadronic collisions, all the interacting partons have finite spatial extension and thus collide at finite impact parameter by means of parton-parton collisions. It considers *color strings* as the fundamental degrees of freedom. According to this model, J/ψ multiplicity is proportional to number of strings produced (N_{S}), whereas charged-particle multiplicity behaves roughly as $\sqrt{N_{\mathrm{S}}}$, due to the interaction among the strings.
- **EPOS3** includes MPI and hydrodynamical expansion of the system, describes well the azimuthal correlation of D-meson with charged-particle [11], is also describes the multiplicity dependence of J/ψ production. The good agreement of the EPOS3 model with data shows that the energy density reached in *pp* collisions at the LHC might be high enough to be described by a hydrodynamical evolution.
- **PYTHIA 8** has MPI and color reconnection in the final state underestimates the data towards the higher multiplicity bins.
- The model of **Kopeliovich et al.** assumes higher Fock states in protons, which contain increased number of gluons. Inelastic collisions of the Fock components lead to high hadron multiplicity and the relative production of J/ψ is enhanced in such gluon-rich collisions.

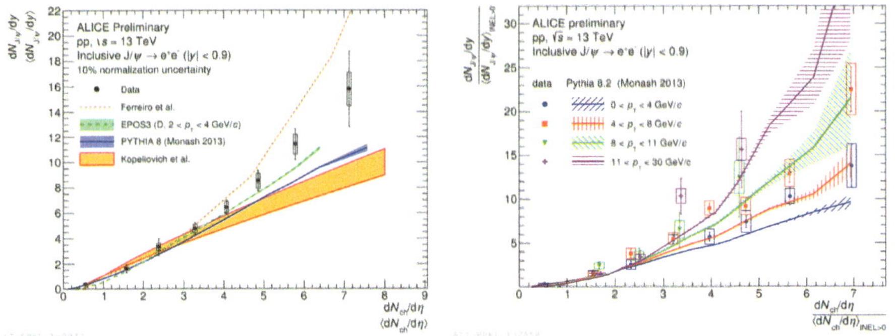

Fig. 30.1 Self-normalized yield of inclusive J/ψ production as a function of multiplicity at mid-rapidity for *pp* collisions at $\sqrt{s} = 13$ TeV. The result for integrated transverse momentum compared with predictions from different models (left panel) and result in transverse momentum slices compared with PYTHIA 8 (right panel)

All the models containing MPI, qualitatively reproduce the multiplicity dependence of J/ψ production, which reveals the importance of MPI in pp collisions and in particular for heavy-flavor production. Among all, EPOS is describing the data best. The self normalized J/ψ yield as a function of charged-particle multiplicity studied in four p_{T} intervals is shown in the right panel of Fig. 30.1. The results are compared with PYTHIA8. It reproduces the multiplicity and p_{T} dependence well, further highlighting the importance of MPI. The enhancement is strongest at high p_{T}, indicating that the effect of MPI is more important at higher p_{T} in the production of J/ψ.

The multiplicity dependence of J/ψ production as a function of charged-particle multiplicity has also been studied in p–Pb collisions at $\sqrt{s_{NN}}$ = 5.02 TeV [12]. ALICE has performed the study in different rapidity ranges by inverting the directions of the lead and proton beams. The measurement has been performed in three rapidity regions, forward ($2.03 < y_{\mathrm{cms}} < 3.53$), backward ($-4.46 < y_{\mathrm{cms}} < -2.96$) and mid-rapidity ($-1.37 < y_{\mathrm{cms}} < 0.43$), as shown in the left panel of Fig. 30.2. The mid-rapidity and backward-rapidity data show a linear increase of self-normalized J/ψ yield with multiplicity. At forward rapidity a saturation towards higher multiplicity is observed. In this kinematic region, the proton probes the small Bjorken-x region of the Pb-nucleus, where cold nuclear effect, gluon shadowing and saturation effects are expected.

ALICE has also measured $\langle p_T^{J/\psi} \rangle$ as a function of multiplicity at forward and backward rapidities, which is shown in the right panel of the Fig. 30.2. It can be seen from the figure that both rapidity regions show a similar trend. The $\langle p_T \rangle$ is the same within uncertainty. The $\langle p_T \rangle$ of charged hadrons is represented by a dashed band. The $\langle p_T^{J/\psi} \rangle$ shows a similar trend as that observed in Pb–Pb collisions for charged particles [13], possibly hinting at collective effects in p–Pb collisions.

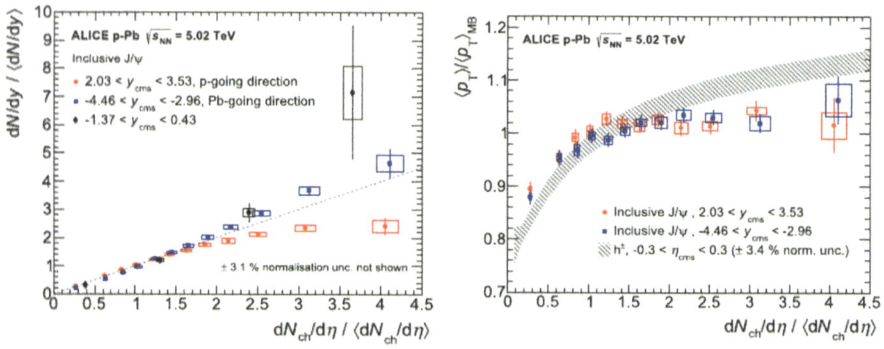

Fig. 30.2 Self-normalized yield and mean transverse momentum of J/ψ as a function of self-normalized charged-particle multiplicity for p–Pb collisions at $\sqrt{s_{NN}}$ = 5.02 TeV. The result for self-normalized yield of inclusive J/ψ production at forward ($2.03 < y_{\mathrm{cms}} < 3.53$), backward ($-4.46 < y_{\mathrm{cms}} < -2.96$) and mid-rapidity ($-1.37 < y_{\mathrm{cms}} < 0.43$) are shown in left pannel. The result for self-normalized mean transverse momentum of J/ψ at forward ($2.03 < y_{\mathrm{cms}} < 3.53$) and backward rapidity ($-4.46 < y_{\mathrm{cms}} < -2.96$) are shown in right panel

30.4 Summary

In this contribution, the multiplicity dependence of J/ψ production has been presented for pp collisions at $\sqrt{s}$ = 13 TeV and p–Pb collisions at $\sqrt{s_{NN}}$ = 5.02 TeV. Also, $\langle p_T^{J/\psi} \rangle$ as a function of multiplicity for $p - Pb$ collisions at $\sqrt{s_{NN}}$ = 5.02 TeV has been presented at forward and backward rapidities. Preliminary results of J/ψ production as a function of charged-particle multiplicity for $J/\psi \rightarrow e^+e^-$ are presented. Similar work at forward rapidity looking at the decay channel $J/\psi \rightarrow \mu^+\mu^-$ is ongoing. The J/ψ yield versus charged-multiplicity study for p–Pb results will help to understand differences in the production mechanisms between pp and p–Pb collisions.

References

1. M.B. Einhorn, S.D. Ellis, Phys. Rev. D **12**, 2007 (1975)
2. G.T. Bodwin, E. Braaten, G.P. Lepage, Phys. Rev. D **51**, 1125 (1995) Erratum: [Phys. Rev. D **55**, 5853 (1997)]
3. F. Halzen, Phys. Lett. **69B**, 105 (1977)
4. B. Abelev et al., ALICE collaboration. Phys. Lett. B **712**, 165 (2012)
5. K. Aamodt et al., ALICE collaboration. JINST **3**, S08002 (2008)
6. J.M. Blanco, Study of J/ψ production dependence with the charged particle multiplicity in p-Pb collisions at $\sqrt{s_{\rm NN}} = 5.02\ TeV$ and pp collisions at $\sqrt{s} = 8\ TeV$ with the ALICE experiment at the LHC. Ph.D. thesis, http://inspirehep.net/record/1503635/files/CERN-THESIS-2016-070.pdf
7. E.G. Ferreiro, C. Pajares, Phys. Rev. C **86**, 034903 (2012)
8. H.J. Drescher, M. Hladik, S. Ostapchenko, T. Pierog, K. Werner, Phys. Rept. **350**, 93 (2001)
9. T. Sjostrand, S. Mrenna, P.Z. Skands, Comput. Phys. Commun. **178**, 852 (2008)
10. B.Z. Kopeliovich, H.J. Pirner, I.K. Potashnikova, K. Reygers, I. Schmidt, Phys. Rev. D **88**, 116002 (2013)
11. J. Adam et al., ALICE collaboration. Eur. Phys. J. C **77**, 245 (2017)
12. D. Adamová et al., ALICE collaboration. Phys. Lett. B **776**, 91 (2018)
13. B.B. Abelev et al., ALICE collaboration. Phys. Lett. B **727**, 371 (2013)
14. S. Acharya et al., ALICE collaboration. Eur. Phys. J. C **77**, 392 (2017)
15. J. Adam et al., ALICE collaboration. JHEP **1509**, 148 (2015)

Part III
Flavour Mixing, CP and CPT

Chapter 31
Time Dependent CPV in the Beauty Sector

Sean Benson

Abstract Time-dependent *CP* violating phenomena probe the time evolution and the decay of beauty hadrons and allow access to the underlying dynamics. We present various time-dependent *CP* violation measurements in the B^0 and B^0_s meson systems that are performed using pp collisions by LHCb and using e^+e^- collisions by the B factory experiments BaBar and Belle. For LHCb, we report measurements of the quantum-loop mediated $B^0_s \to (K^-\pi^+)(K^+\pi^-)$ and $B^0_s \to \phi\phi$ decays, and measurements of $B^0 \to D^{\mp}\pi^{\pm}$ and $B^0_s \to h^+h^-$ decays. For Belle, we report new results for the charm-less $B^0 \to \pi^0\pi^0 K^0_S$ and $B^0 \to K^0_S\eta\gamma$ decays. In addition, we present a measurement of $\cos 2\beta$ obtained by a time-dependent Dalitz plot analysis of $B^0 \to D^*h^0$ with $D \to K^0_S\pi^+\pi^-$ that combines the final BaBar and Belle data sets.

31.1 Introduction

Measurements of *CP* asymmetries in neutral *B* mesons probe the interference of the mixing and decay Feynman diagrams. The *CP* asymmetry as a function of time can be described as

$$A_{CP} = \frac{\Gamma_{\overline{B}^0_{(d,s)}\to f} - \Gamma_{B^0_{(d,s)}\to f}}{\Gamma_{\overline{B}^0_{(d,s)}\to f} + \Gamma_{B^0_{(d,s)}\to f}} = \frac{-C_f\cos(\Delta m_{d,s}t) + S_f\sin(\Delta m_{d,s}t)}{\cos h(\Delta\Gamma_{d,s}t/2) + A_f^{\Delta\Gamma}\sin h(\Delta\Gamma_{d,s}t/2)}, \quad (31.1)$$

where

$$C_f \equiv \frac{1-|\lambda_f|}{1+|\lambda_f|},\ C_f \equiv \frac{2Im\lambda_f}{1+|\lambda_f|},\ A_f^{\Delta\Gamma} \equiv -\frac{2Re\lambda_f}{1+|\lambda_f|},\ \lambda_f \equiv \frac{q}{p}\frac{\overline{A}_f}{A_f}.$$

On behalf of the LHCb, BaBar, and Belle collaborations.

S. Benson (✉)
Nikhef National Institute for Subatomic Physics, Amsterdam, The Netherlands
e-mail: sean.benson@cern.ch

A. Giri and R. Mohanta (eds.), *16th Conference on Flavor Physics and CP Violation*, Springer Proceedings in Physics 234,
https://doi.org/10.1007/978-3-030-29622-3_31

In the above equation, f represents the final state, A_f the $B_{s,d}$ decay amplitude, and $\overline{A}_f$ the $\overline{B}_{s,d}$ decay amplitude.

In the Standard Model (SM), the weak phase from B^0_s–$\overline{B}^0_s$ mixing is given by $\phi_s^{c\bar{c}s} \approx -2\beta_s = 2arg(-V_{ts}V^*_{tb}/V_{cs}V^*_{cb})$. The SM prediction for $\phi_s^{c\bar{c}s}$ has been obtained from global fits to experimental data yielding a value of -0.0360 ± 0.0006 rad [1]. There are however beyond-the-SM theories that could provide additional contributions to B^0_s mixing diagrams which alter this value [2, 3].

In addition to decays with a tree-level contribution, measurements of decays in which the leading order contribution is a flavour changing neutral current (FCNC) interaction are important in the search for physics beyond the SM. The CP-violating phase measured in the $B^0_s \to \phi\phi$ decay results from $b \to s\bar{s}s$ transitions and is therefore expected to be close to zero in the SM due to the effective cancellation of the CP-violating weak phase between the B^0_s mixing diagrams and the penguin decay diagrams [4, 5]. The $B^0_s \to K^{*0}\overline{K}^{*0}(892)$ decay is an example of a $b \to d\bar{d}s$ transition and the associated CP-violating phase is expected to be zero for similar reasons as the $B^0_s \to \phi\phi$ decay.

Time-dependent CP violation measurements in the B^0 system are mainly performed using b-factory data. The high flavour tagging efficiency of the b-factories makes many analyses of decay channels with event yields of as little as a hundred events possible. For time-dependent analyses in the B^0 system at the b-factories, the decay time difference with respect to the other $\overline{B}^0$ decay produced in the event is measured and denoted as Δt. The time-dependent decay rate is then of the form

$$\mathcal{P}(\Delta t) = \frac{e^{-|\Delta t|/\tau_{B^0}}}{4\tau_{B^0}}\{1 + q[\mathcal{S}\sin(\Delta m_d \Delta t) + \mathcal{A}\cos(\Delta m_d \Delta t)]\}, \tag{31.2}$$

where q is the B^0 meson flavour, $\mathcal{S}$ and $\mathcal{A}$ are the S_f and C_f parameters in (31.1), respectively. Time-dependent analyses in the B^0 system are sensitive to the CKM angle $\beta \equiv \phi_1$, the current central value of $\sin(2\beta)$ is -0.699 ± 0.017 [6]. Measurements of $\sin(2\beta)$ result in an ambiguity in the sign of $\cos(2\beta)$.

The following sections summarise updated and most accurate measurements of the CP-violating weak phases in $b \to c\bar{c}s$ and $b \to s\bar{s}s$ transitions [7–9], in addition to a first measurement of the weak phase in $b \to d\bar{d}s$ transitions from LHCb data [9]. Decay time dependent measurements of $B^0 \to \pi^0\pi^0K^0_S$ and $B^0 \to K^0_S\eta\gamma$ decays are presented using Belle data [10, 11], along with a new measurement of $\cos(2\beta)$ using a combination of BaBar and Belle data that provides first evidence that $\cos(2\beta) > 0$ [12].

31.2 Time-Dependent *CP* Violation in B_s^0 Decays

31.2.1 The $B_s^0 \to J/\psi K^+K^-$ Decay

The $B_s^0 \to J/\psi K^+K^-$ decay requires the use of decay angles in the helicity basis to disentangle the four polarisation amplitudes contributing to the decay (3 *P*-wave and an *S*-wave). The four observables are shown in Fig. 31.1.

A good understanding of the efficiencies as a function of each observable is important in keeping measurements dominated by statistical uncertainties. The measurement uses 95690 candidates obtained from 3 fb^{-1} of LHCb data collected in 2011 and 2012 [7]. The decay gives access to the B_s^0–$\overline{B}_s^0$ oscillation frequency, Δm_s, the decay width and decay width difference, Γ_s and $\Delta\Gamma_s$, in addition to the decay amplitudes, $|A_0|^2$ and $|A_\perp|^2$, and the *CP*-violating parameters, $\phi_s^{c\bar{c}s}$ and $|\lambda|$. The results of the measurement are provided in Table 31.1. Results are consistent with SM predictions and result in stringent constraints on physics beyond the SM [13].

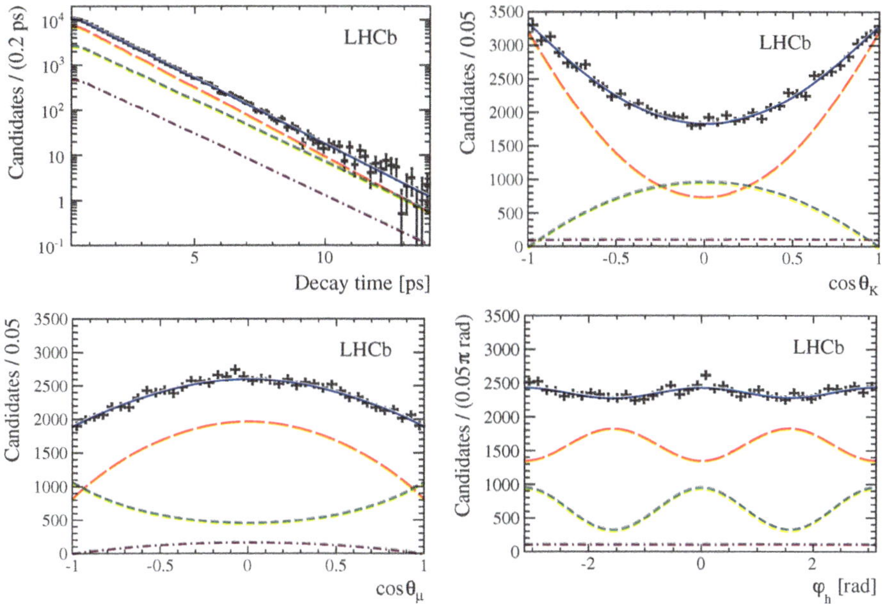

Fig. 31.1 Decay-time and helicity-angle distributions for $B_s^0 \to J/\psi K^+K^-$ decays (data points) with the one-dimensional fit projections overlaid. The solid blue line shows the total signal contribution, which is composed of *CP*-even (long-dashed red), *CP*-odd (short-dashed green) and S-wave (dotted-dashed purple) contributions [7]

Table 31.1 Values of the principal physics parameters determined from the polarisation-independent fit. The first uncertainty is statistical and the second systematic

Parameter	Value
Γ_s [ps^{-1}]	$0.6603 \pm 0.0027 \pm 0.0015$
$\Delta\Gamma_s$ [ps^{-1}]	$0.0805 \pm 0.0091 \pm 0.0032$
$\lvert A_\perp \rvert^2$	$0.2504 \pm 0.0049 \pm 0.0036$
$\lvert A_0 \rvert^2$	$0.5241 \pm 0.0034 \pm 0.0067$
$\phi_s^{c\bar{c}s}$ [rad]	$-0.058 \pm 0.049 \pm 0.006$
$\lvert\lambda\rvert$	$0.964 \pm 0.019 \pm 0.007$
Δm_s [ps^{-1}]	$17.711\ ^{+0.055}_{-0.057} \pm 0.011$

31.2.2 The $B_s^0 \to \phi\phi$ Decay

The $B_s^0 \to \phi\phi$ decay requires an angular analysis to disentangle the polarisation states, as for the $B_s^0 \to J/\psi K^+K^-$ decay. In total, approximately 9000 decay candidates are used to make the measurement, from 5.0 fb^{-1} of data collected between 2011 and 2016 [8]. The number of candidates is much smaller than that used for the measurement of $B_s^0 \to J/\psi K^+K^-$ decays, therefore external constraints are taken from the best known values of Δm_s, Γ_s, and $\Delta\Gamma_s$. The four observables describing the decay are shown in Fig. 31.2 along with the corresponding fit projections. Numerical results for the polarisation amplitudes, $|A_0|^2$ and $|A_\perp|^2$, are given in Table 31.2, along with the *CP*-violating parameters, $\phi_s^{s\bar{s}s}$ and $|\lambda|$. Results are consistent with *CP* conservation and consequently SM predictions.

31.2.3 The $B_s^0 \to K^+\pi^-K^-\pi^+$ Decay

A further example of a decay channel that requires an angular analysis is given by the $B_s^0 \to K^+\pi^-K^-\pi^+$ decay. The decay mode is further complicated by the many contributions to the $K\pi$ mass spectrum. These contributions consist of non-resonant *S*-wave, K^{*0} *P*-wave, in addition to that of a tensor $\overline{K}_2^*(1430)^0$ along with associated interferences. In order to properly disentangle so many contributions, the decay-time dependent fit incorporates the $K\pi$ invariant mass as an observable. The $K\pi$ invariant mass along with the associated fit is shown in Fig. 31.3. The dataset used for the analysis consists of 3 fb^{-1} of LHCb data collected in 2011 and 2012 [9].

The results of the *CP*-violation parameters, $\phi_s^{d\bar{d}s}$ and $|\lambda|$, are given in Table 31.3. *CP* violation is consistent with the SM expectation. The analysis is dominated by systematic uncertainties, the largest contribution of which originates from the size of the simulation sample.

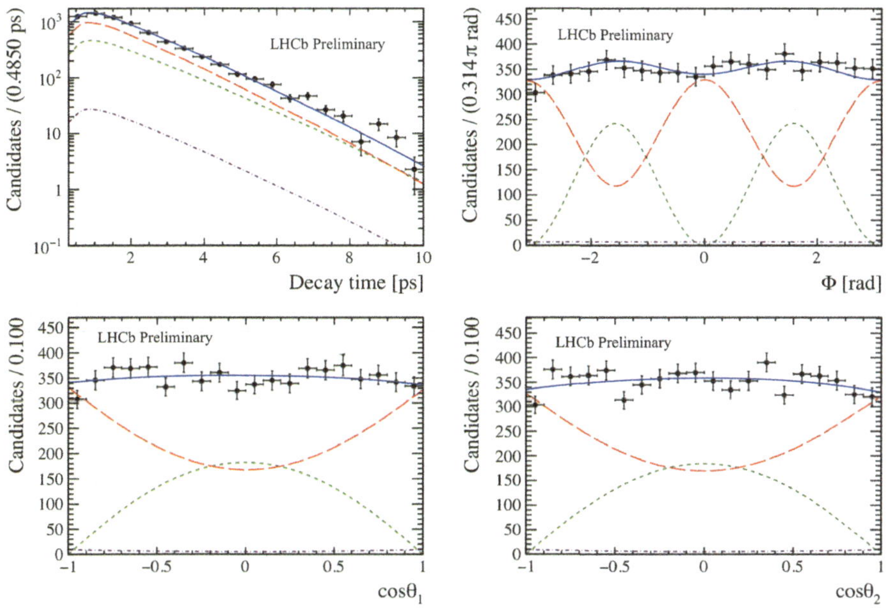

Fig. 31.2 Decay-time and helicity-angle distributions for $B_s^0 \to \phi\phi$ decays (data points) with the one-dimensional fit projections overlaid. The solid blue line shows the total signal contribution, which is composed of *CP*-even (long-dashed red), *CP*-odd (short-dashed green) and S-wave (dotted-dashed purple) contributions [8]

Table 31.2 Values of the principal physics parameters determined from the polarisation-independent fit. The first uncertainty is statistical and the second systematic

Parameter	Value
$\phi_s^{s\bar{s}s}$ [rad]	$-0.07 \pm 0.13 \pm 0.03$
$\lvert\lambda\rvert$	$1.02 \pm 0.05 \pm 0.03$
$\lvert A_0\rvert^2$	$0.382 \pm 0.008 \pm 0.011$
$\lvert A_\perp\rvert^2$	$0.287 \pm 0.008 \pm 0.005$

31.2.4 *CP* Violation in $B \to hh$ Decays

The measurement of *CP* violation in $B \to hh$ decays is a measurement of time-dependent *CP* violation in both B_s^0 and B^0 decays, as different mesons are the main contributors to different final states. The $\pi\pi$ final state is studied with B^0 decays and the KK final state with B_s^0 decays. The $B^0 \to K^-\pi^+$ decay is used to understand the decay time acceptance and calibrate the incorrect flavour assignment probability. The fit to all final states simultaneously allows for mis-identified pions and kaons to be accounted for. The analysis used $3.0\,\text{fb}^{-1}$ of data collected in 2011 and 2012 [14].

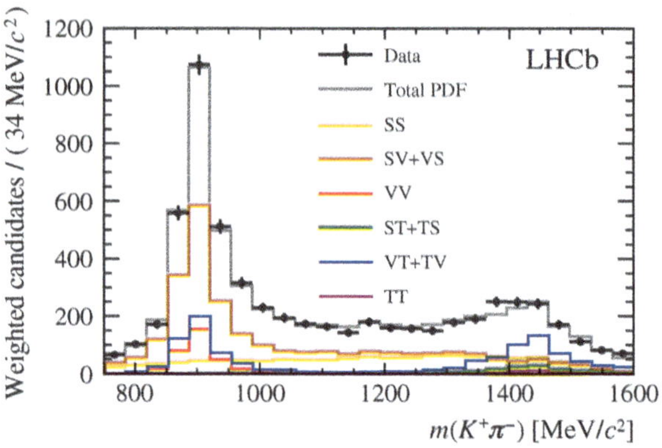

Fig. 31.3 One-dimensional projections of the decay-time-dependent, flavour-tagged fit to (black points) the sPlot weighted data for the $K^+ \pi^-$ invariant mass. The solid gray line represents the total fit model along with the i*CP*-averaged components for each contributing decay [9]

Table 31.3 Values of the *CP* violation parameters determined from the polarisation-independent fit. The first uncertainty is statistical and the second systematic

Parameter	Value
$\phi_s^{d\bar{d}s}$ [rad]	$-0.10 \pm 0.13 \pm 0.14$
$\lvert\lambda\rvert$	$1.035 \pm 0.034 \pm 0.089$

Table 31.4 Values of the principal physics parameters determined from the polarisation-independent fit. The first uncertainty is statistical and the second systematic

Parameter	Value
$C_{\pi\pi}$	$-0.34 \pm 0.06 \pm 0.01$
$S_{\pi\pi}$	$-0.63 \pm 0.05 \pm 0.01$
C_{KK}	$0.20 \pm 0.06 \pm 0.02$
S_{KK}	$0.18 \pm 0.06 \pm 0.02$
$A_{KK}^{\Delta\Gamma}$	$-0.79 \pm 0.07 \pm 0.10$
$A_{CP}^{B^0}$	$-0.084 \pm 0.004 \pm 0.003$
$A_{CP}^{B_s^0}$	$0.213 \pm 0.015 \pm 0.007$

The results of the *CP* violation parameters defined in (31.1) are shown in Table 31.4. The significance for $(C_{KK}, S_{KK}, A_{CP}^{B_s^0})$ do deviate from $(0, 0, -1)$ has been determined from a χ^2 test statistic to be 4.0 standard deviations. This result constitutes the strongest evidence for *CP* violation in the B_s^0 system to date.

31.3 Time-Dependent *CP* Violation in B^0 Decays

31.3.1 The $B^0 \to K^0_S\pi^0\pi^0$ Decay

The known energy of electron-positron collisions results in a fully kinematically constrained decay. This allows the use of the beam-constrained mass, $M_{\rm bc} \equiv \sqrt{(E_{\rm beam}/c^2)^2 - |\mathbf{p}_B^{CM}/c|^2}$, and the energy difference $\Delta E \equiv E_{\rm beam} - E_B^{CM}$, to isolate the signal candidates from the background candidates, where $\mathbf{p}_B^{CM}$ and E_B^{CM} are the B momentum and energy respectively in the centre-of-mass (CM) frame. A common variable used in addition to isolate signal is a likelihood ratio calculated from modified Fox-Wolfram moments, $\mathcal{R}_{s/b}$ [11].

The signal and background distributions of $M_{\rm bc}$, ΔE, and $\mathcal{R}_{s/b}$ are shown in Fig. 31.4, where 335 $B^0 \to K^0_S\pi^0\pi^0$ signal candidates are seen from a three dimensional fit. The total dataset consists of 742×10^6 $B\overline{B}$ pairs collected with the Belle detector at the $\Upsilon(4S)$ resonance. The high effective flavour tagging power of $29.8 \pm 0.4\,\%$ makes a decay-time dependent fit possible with the number of candidates found. The resulting decay-time dependent fit measures a values of $\mathcal{S} = -0.92^{+0.31}_{-0.27}(\text{stat}) \pm 0.11(\text{syst})$ and $\mathcal{A} = -0.28 \pm 0.21(\text{stat}) \pm 0.04(\text{syst})$.

31.3.2 The $B^0 \to K^0_S\eta\gamma$ Decay

A similar decay time dependent analysis is performed using $B^0 \to K^0_S\eta\gamma$ decays as that described in the previous section. The total dataset consists of 772×10^6 $B\overline{B}$ pairs collected with the Belle detector at the $\Upsilon(4S)$ resonance. A total of 92 candidates are used to make the measurement. The *CP* violation parameters are determined to be $\mathcal{S} = -1.32 \pm 0.77(\text{stat}) \pm 0.36(\text{syst})$ $\mathcal{A} = -0.48 \pm 0.41(\text{stat}) \pm 0.07(\text{syst})$.

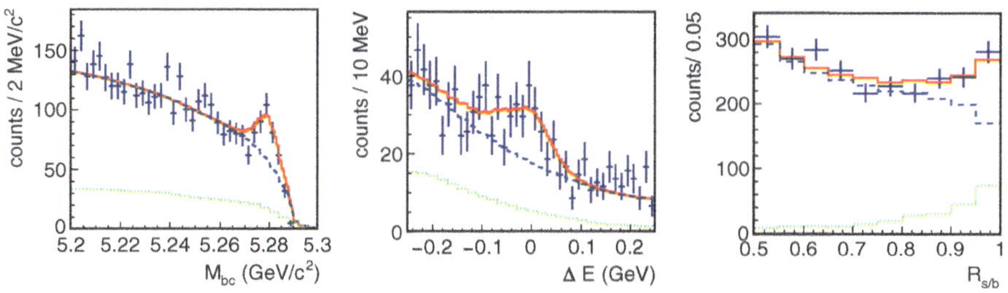

Fig. 31.4 $M_{\rm bc}$, ΔE and $\mathcal{R}_{s/b}$ distributions (points with uncertainties) using signal-enhanced selections $M_{\rm bc} > 5.27\,\text{GeV}/c^2$, âĹŠ0.15 GeV $< \Delta E < 0.10$ GeV, and $\mathcal{R}_{s/b} > 0.9$ except for the variable displayed. The fit result is illustrated by the solid curve, while the total and $B\overline{B}$ backgrounds are shown by broken and dotted curves, respectively [11]

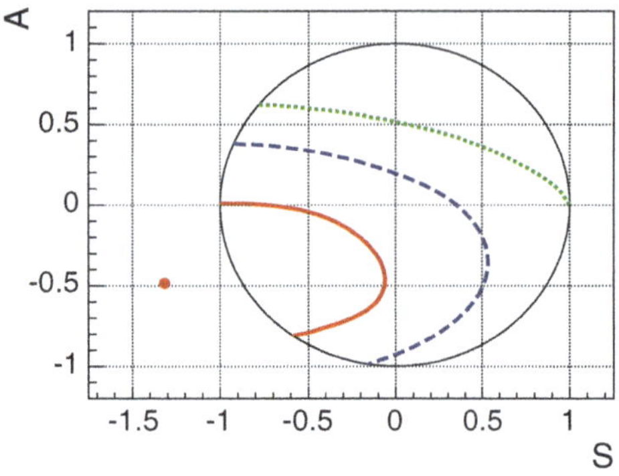

Fig. 31.5 The solid red, dashed blue, and dotted green curves show the 1σ, 2σ, 3σ confidence contours, respectively. The red dot shows the fit result. The physical boundary $S^2 + A^2 = 1$ is drawn with a thin solid black curve. Our result is consistent with a null asymmetry within 2σ [10]

The 1, 2, and 3σ confidence contours of the $\mathcal{S}$ and $\mathcal{A}$ parameters are shown in Fig. 31.5. The result is consistent with null asymmetry within 2σ.

31.3.3 Measurement of $\cos(2\beta)$ Using $B^0 \to D^*h^0$ decays

The first evidence for the resolution of the sign of $\cos(2\beta)$ has required the combination of 471×10^6 $B\overline{B}$ pairs collected with the BaBar detector and 772×10^6 $B\overline{B}$ pairs collected with the Belle detector [12]. The measurement of $\cos(2\beta)$ is achieved through the interference of the D^* and $\overline{D}^*$ decay amplitudes, $\mathcal{A}_{D^*}$ and $\mathcal{A}_{\overline{D}^*}$, which can be written as

$$\mathcal{I}m(e^{-i2\beta}\mathcal{A}_{D^*}\mathcal{A}_{\overline{D}^*}) = \mathcal{I}m(\mathcal{A}_{D^*}\mathcal{A}_{\overline{D}^*})\cos(2\beta) - \mathcal{R}e(\mathcal{A}_{D^*}\mathcal{A}_{\overline{D}^*})\sin(2\beta). \quad (31.3)$$

This then gives access to both $\sin(2\beta)$ and $\cos(2\beta)$, which are measured to be $0.80 \pm 0.14 \pm 0.06 \pm 0.03$ and $0.91 \pm 0.22 \pm 0.09 \pm 0.07$, respectively, where the first uncertainty is statistical, the second is systematic and the third is related to the choice of fit model. The significance of which $\cos(2\beta)$ is greater than zero is determined to be 3.7σ, representing the first evidence.

31.4 Summary

The most accurate single measurement of CP violation in B^0_s mixing has been presented along with an updated measurement of the weak phase in the $B^0_s \to \phi\phi$ decay and the first measurement of the weak phase in $B^0_s \to K^+\pi^-K^-\pi^+$ decays using LHCb data. An LHCb analysis has also revealed the strongest evidence yet of CP violation in the B^0_s sector, measured from $B^0_s \to K^+K^-$ decays to be at the 4.0σ level. Two new decay time dependent measurements have been presented using data collected with the Belle detector, $B^0 \to K_S^0\pi^0\pi^0$ and $B^0 \to K_S^0\eta\gamma$. In addition, first evidence has been presented that $\cos(2\beta) > 0$, with a significance of 3.7σ, potentially resolving the long standing ambiguity in the value of the CKM angle β .

References

1. J. Charles et al., Phys. Rev. D **91**, 073007 (2015). https://doi.org/10.1103/PhysRevD.91.073007
2. P. Ball, R. Fleischer, Eur. Phys. J. C **48**, 413 (2006). https://doi.org/10.1140/epjc/s10052-006-0034-4
3. A. Lenz, Phys. Rev. D **76**, 065006 (2007). https://doi.org/10.1103/PhysRevD.76.065006
4. M. Raidal, Phys. Rev. Lett. **89**, 231803 (2002). https://doi.org/10.1103/PhysRevLett.89.231803
5. B. Bhattacharya, A. Datta, M. Duraisamy, D. London, Phys. Rev. D **88**, 016007 (2013). https://doi.org/10.1103/PhysRevD.88.016007
6. M. Tanabashi et al., Phys. Rev. D **98**(3), 030001 (2018). https://doi.org/10.1103/PhysRevD.98.030001
7. R. Aaij et al., Phys. Rev. Lett. **114**(4), 041801 (2015). https://doi.org/10.1103/PhysRevLett.114.041801
8. R. Aaij et al., Measurement of CP violation in $B_s^0 \to \phi\phi$ decays (2018)
9. R. Aaij et al., JHEP **03**, 140 (2018). https://doi.org/10.1007/JHEP03(2018)140
10. H. Nakano et al., Phys. Rev. D **97**(9), 092003 (2018). https://doi.org/10.1103/PhysRevD.97.092003
11. Y. Yusa et al., Submitted to: Phys. Rev. D (2018)
12. I. Adachi et al. (2018)
13. L. Di Luzio, M. Kirk, A. Lenz, Phys. Rev. D **97**(9), 095035 (2018). https://doi.org/10.1103/PhysRevD.97.095035
14. R. Aaij et al., Phys. Rev. D **98**(3), 032004 (2018). https://doi.org/10.1103/PhysRevD.98.032004
15. M. Bartsch, G. Buchalla, C. Kraus (2008)
16. H.Y. Cheng, C.K. Chua, Phys. Rev. D **80**, 114026 (2009). https://doi.org/10.1103/PhysRevD.80.114026

Chapter 32
Lattice QCD Impact on Determination of the CKM Matrix

Steven Gottlieb

Abstract We review many lattice QCD calculations that impact the precise determination of the CKM matrix. We focus on decay constants and semileptonic form factors of both light (π and K) and heavy-light ($D_{(s)}$ and $B_{(s)}$) mesons. Implication of Λ_b form factors will be shown. When combined with experimental results for branching fractions and differential decay rates, the above calculations strongly constrain the first two rows of the CKM matrix. We discuss a long standing difference between $|V_{ub}|$ and $|V_{cb}|$ as determined from exclusive or inclusive decays.

32.1 Introduction

Lattice QCD contributes strongly to understanding the CKM matrix and the search for beyond the Standard Model (SM) physics. To begin, I would like to relate a little about my background. I am a member of the Flavour Lattice Averaging Group (FLAG) and participate in the D and B semileptonic decays working group. However, this is not a FLAG approved talk. The most recent FLAG review [1] dates from 2016, and I will concentrate on more recent plots. The closing date for papers to appear in the next review is a couple of months after FPCP 2018, so the next FLAG plots and averages are not yet available.

For more than a decade, my own research has been in the context of the MILC collaboration and the Fermilab Lattice/MILC Collaborations (FNAL/MILC). Much of our work is directed toward more precisely determining the CKM matrix, and looking for discrepancies that would indicate physics beyond the SM, I will liberally use plots from my own collaborations when they contain results not yet reviewed by FLAG.

I also happen to be a member of the local organizing committee for Lattice 2018, which takes place the week after FPCP 2018. I can assure you that topics discussed

S. Gottlieb (✉)
Indiana University, Bloomington 47405, IN, USA
e-mail: sg@indiana.edu
URL: http://physics.indiana.edu/~sg/

A. Giri and R. Mohanta (eds.), *16th Conference on Flavor Physics and CP Violation*, Springer Proceedings in Physics 234,
https://doi.org/10.1007/978-3-030-29622-3_32

here are very active areas within the lattice QCD community. For Lattice 2018, there were 37 abstracts in the weak matrix element category. Of them, five dealt with decay constants, 12 with *K*, *D*, or *B* meson semileptonic decay, and seven with nucleon or nuclear matrix elements.

32.1.1 CKM Matrix

In expression (32.1), CKM matrix elements are in bold and below each of them are one or two processes that can be used to determine that element. Below the last row, the matrix elements represent the $B_{(s)}\bar{B}_{(s)}$ mixing phenomena that depend on V_{td} and V_{ts} through loop diagrams.

$$\begin{pmatrix} \mathbf{V_{ud}} & \mathbf{V_{us}} & \mathbf{V_{ub}} \\ \pi \to l\nu & K \to l\nu & B \to l\nu \\ & K \to \pi l\nu & B \to \pi l\nu \\ \mathbf{V_{cd}} & \mathbf{V_{cs}} & \mathbf{V_{cb}} \\ D \to \pi l\nu & D \to Kl\nu & B \to D^{(*)}l\nu \\ D \to l\nu & D_s \to l\nu & \Lambda_b \to \Lambda_c l\nu \\ \mathbf{V_{td}} & \mathbf{V_{ts}} & \mathbf{V_{tb}} \\ \langle B_d|\overline{B}_d\rangle & \langle B_s|\overline{B}_s\rangle & \end{pmatrix} \tag{32.1}$$

The CKM matrix is unitary, so each row and column is a complex unit vector and each row (or column) is orthogonal to the other two. Violation of unitarity would be evidence for non-SM physics. Since different decays can depend on the same CKM matrix element, if the value of the matrix element inferred from the different decays do not agree, that would be evidence for new physics.

Consider the branching fraction $\mathcal{B}$ for leptonic decay of a D or D_s meson.

$$\mathcal{B}(D_{(s)} \to \ell\nu_\ell) = \frac{G_F^2|V_{cq}|^2\tau_{D_{(s)}}}{8\pi} f_{D_{(s)}}^2 m_\ell^2 m_{D_{(s)}} \left(1 - \frac{m_\ell^2}{m_{D_{(s)}}^2}\right)^2 \tag{32.2}$$

where V_{cq} is the (unknown) CKM matrix element with $q = d$, or s, $f_{D_{(s)}}$ is the decay constant of the meson that we calculate using lattice QCD, and the other factors on the RHS are well known. For semileptonic decays, the LHS would be a differential decay rate, and the RHS would involve a CKM matrix element and form factors describing the transition matrix element between the initial and final state hadrons induced by the weak current responsible for the decay. In both cases, the experimental measurement and hadronic input from lattice QCD allow determination of the CKM matrix element.

32.2 First Row: Light Quarks

We start with decays of π and K mesons. As implied by (32.1) and subsequent discussion generalized to other mesons, we need f_π and f_K to describe the leptonic decays and a form factor to describe the kaon semileptonic decay. The two decay constants can each be calculated; however, f_π is often used to set the lattice scale, so the ratio $f_{K^\pm}/f_{\pi^\pm}$ which has the advantage of smaller systematic errors is a key quantity. From experiment [2] it is known that

$$\left|\frac{V_{\text{us}}}{V_{\text{ud}}}\right| \frac{f_{K^\pm}}{f_{\pi^\pm}} = 0.2760(4). \tag{32.3}$$

Thus, knowledge of the decay constant ratio, allows us to determine the ratio of the first two elements of the CKM matrix. The decay constant ratio has recently been updated by FNAL/MILC [3]. The result is $f_{K^\pm}/f_{\pi^\pm} = 1.1950(^{+15}_{-22})$ which may be compared with the FLAG 2016 [1] $N_{\text{f}} = 2+1+1$ average 1.193(3). N_{f} is the number of dynamical sea quarks in the calculation. Figure 32.1 summarizes the most relevant calculations including those with $N_{\text{f}} = 2+1$.

For kaon semileptonic decay, $p_K = p_\pi + q_\ell + q_\nu$ by energy-momentum conservation. The relevant variable for the form factors is q^2 with $q = q_\ell + q_\nu$ the 4-momentum of the leptons. One could, in principle, determine the shape of the vector form factor $f_+(q^2)$ to predict the shape of the differential cross section. However, it is a bit easier to just calculate $f_+(q^2 = 0)$ using lattice QCD and take the experimental measurement [4] that determines $|V_{\text{us}}|f_+(0) = 0.21654(41)$. The latest result for the form factor is [5]

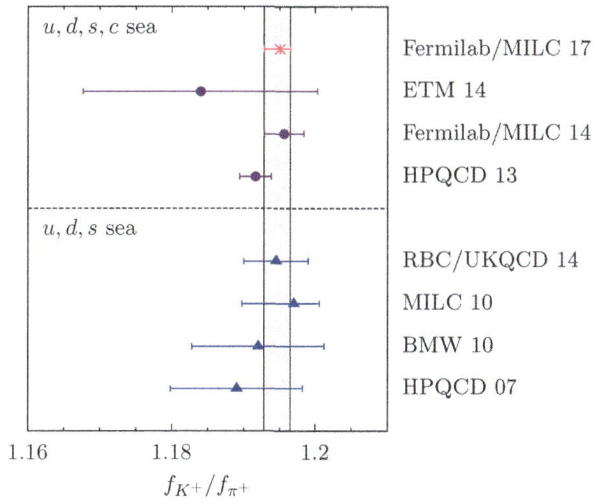

Fig. 32.1 Comparison of calculations of decay constant ratio f_K/f_π with $N_{\text{f}} = 2+1+1$ and $2+1$ sea quark flavors. From [3]

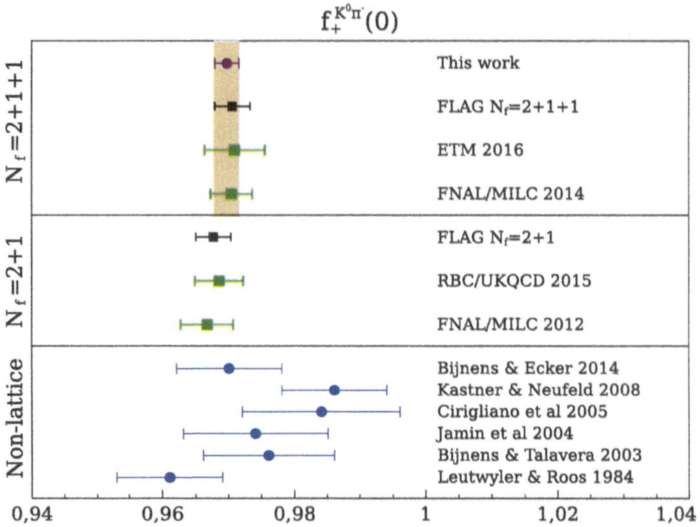

Fig. 32.2 Comparison of calculations of the kaon form factor $f_+^K(q^2 = 0)$ with $N_f = 2+1+1$, or $2+1$ sea quark flavors, and non-lattice QCD calculations [5]

$$f_+(0) = 0.9696(15)_{\rm stat}(11)_{\rm sys} = 0.9696(19), \tag{32.4}$$

separating statistical and systematic errors before combining in quadrature. The total theoretical error is 0.19%, the same size as the experimental error. Figure 32.2 taken from [5] shows the result just quoted (denoted "This work") along with FLAG 2016 [1] averages in black and the results included in the averages as green squares. Blue circles come from non-lattice QCD calculations. See [5] for details. Using the previously quoted experimental value [4], we find

$$|V_{\rm us}| = 0.22333(43)_{f_+(0)}(42)_{\rm exp} = 0.22333(60). \tag{32.5}$$

We consider the implications of these results on first row unitarity in Fig. 32.3. Since $|V_{\rm ub}|$ is so small, we can neglect it. The vertical band labeled $0^+ \to 0^+$ comes from analysis of superallowed nuclear β-decays [6] and is independent of lattice QCD input. The diagonal band labeled K_{l2} comes from the ratio of decay constants f_K/f_π, and the horizontal band labeled K_{l3} comes from the kaon semileptonic form factor. The diagonal band, the vertical band, and the unitarity curve nicely intersect. However, there is tension with unitary when we look at the kaon semileptonic decay. The small blue ellipse uses the result in (32.5) and the value of $|V_{\rm ud}|$ from [6], and we see that it does not intersect the unitarity curve. We find that

$$|V_{\rm ud}|^2 + |V_{\rm us}|^2 + |V_{\rm ub}|^2 - 1 = -0.00104(27)_{V_{\rm us}}(41)_{V_{\rm ud}} \tag{32.6}$$

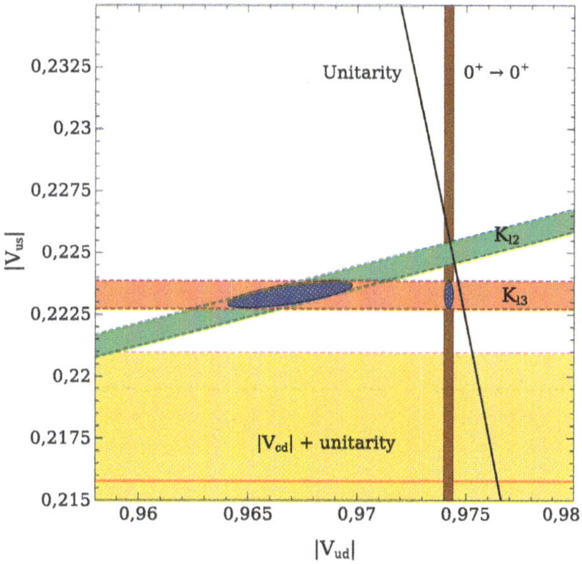

Fig. 32.3 Constraints in the $|V_{ud}|$–$|V_{us}|$ plane from leptonic (K_{l2}) and semileptonic (K_{l3}) decays. In addition, we have the result of nuclear β-decay on $|V_{ud}|$, unitarity, and a wide horizontal band from $|V_{cd}|$ combined with unitarity. (see [5])

which is 2.1σ from zero. The large blue ellipse does not rely on V_{ud} from β-decay and only uses results from pion leptonic decay and kaon leptonic and semileptonic decay. There is a clear tension with unitarity. In this case, we have

$$|V_{ud}|^2 + |V_{us}|^2 + |V_{ub}|^2 - 1 = -0.0151(38)_{f_+(0)}(35)_{f_K^\pm/f_\pi^\pm}(36)_{\text{exp}}(27)_{\text{EM}} \quad (32.7)$$

which is 2.2σ different from zero. The tension between leptonic and semileptonic determination of $|V_{us}|$ and $|V_{ud}|$ can also be seen in the FLAG 2016 [1] Fig. 9 summary plot for $|V_{us}|$ and $|V_{ud}|$ in which semileptonic results are triangles and leptonic results are squares. The tension is most noticeable for the $N_f = 2 + 1 + 1$ calculations where the precision is higher.

32.3 Second Row of CKM Matrix

We now turn to the decay of charm mesons to determine $|V_{cd}|$ and $|V_{cs}|$. In 2005, the initial $N_f = 2 + 1$ calculations of decay constants were done with an accuracy of roughly 10%. The latest results for f_{D^+} and f_{D_s} now have errors $< 0.3\%$. We have [3]

$$f_{D^+} = 212.7(0.6)\text{MeV}, \qquad f_{D_s} = 249.9(0.4)\text{MeV}. \quad (32.8)$$

Prior to that calculation, the FLAG [1] and Particle Data Group (PDG) [2] average values had errors of roughly 1–3 MeV. Figure 32.4 summarizes the best recent calculations [3].

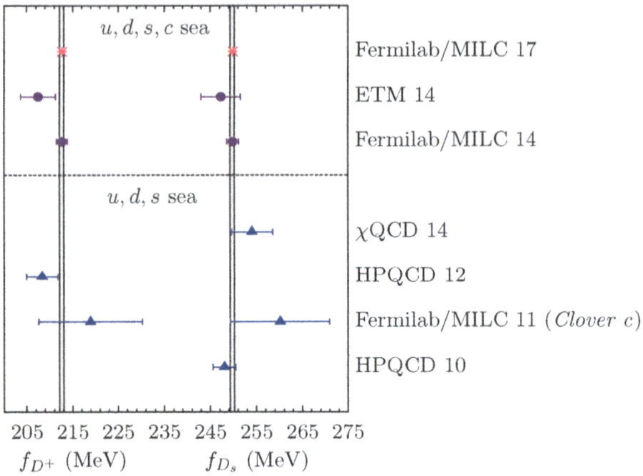

Fig. 32.4 Comparison of recent calculations of f_{D^+} and f_{D_s} with $N + f = 2 + 1 + 1$ and $2 + 1$. From [3]

We use experimental results from the PDG [2] to determine the CKM matrix elements from the lattice QCD decay constants. They have $f_D|V_{\rm cd}| = 45.91(1.05)$MeV and $f_{D_s}|V_{\rm cs}| = 250.9(4.0)$MeV. The experimental error is 1.6–2.3%. For the CKM matrix element, we have

$$|V_{\rm cd}|_{{\rm SM},\ f_D} = 0.2152(5)_{f_D}(49)_{\rm expt}(6)_{\rm EM}, \tag{32.9}$$

$$|V_{\rm cs}|_{{\rm SM},\ f_{D_s}} = 1.001(2)_{f_{D_s}}(16)_{\rm expt}(3)_{\rm EM}, \tag{32.10}$$

where the errors are from lattice decay constant, experiment, and a structure dependent electromagnetic correction. The experimental errors are dominant.

Turning to the charm meson semileptonic decays, the FLAG 2016 form factor average is based on HPQCD results from 2010 [7] and 2011 [8]. There are new results from ETM Collaboration [9] and JLQCD [10]. In addition, FNAL/MILC is completing an analysis that should soon have errors smaller than those of HPQCD. I have taken a rough average of the three results above, even though they mix results with $N_{\rm f} = 2 + 1$ and $2 + 1 + 1$. I find $f_+^{D\to\pi}(0) = 0.637(20)$ and $f_+^{D\to K}(0) = 0.745(15)$. These values may be a little aggressive. The FLAG 2016 values are 0.666(29) and 0.747(19), respectively. Using the HFLAV 2016 values [11] $f_+^{D\pi}|V_{\rm cd}| = 0.1426(26)$ and $f_+^{DK}|V_{\rm cs}| = 0.7226(34)$, we obtain $|V_{\rm cd}| = 0.2239(76)$ and $|V_{\rm cs}| = 0.970(20)$ corresponding to errors of 3.4 and 2.1%. In each case, the error is dominated by the error in the lattice form factor input. I updated the experimental input after FPCP 2018, so the values here are different from those in my slides.

We can test second row unitarity using a variety of determinations of $|V_{\rm cd}|$ and $|V_{\rm cs}|$. In this case, $|V_{\rm cb}| \approx 0.0414(8)$ contributes about 0.0017(6) to the unitarity sum. We consider in Table 32.1 the latest result using leptonic decay constants from [3], the FLAG 2016 $N_{\rm f} = 2 + 1 + 1$ result, the latest ETMC semileptonic result [12], and my rough average of semileptonic results. We find a slight (1.5σ) tension from

Table 32.1 Tests of second row unitarity from various determinations of $|V_{cd}|$ and $|V_{cs}|$

$\lvert V_{cd}\rvert^2 + \lvert V_{cs}\rvert^2 + \lvert V_{cb}\rvert^2 - 1$	Input
$0.049(2)_{\lvert V_{cd}\rvert}(32)_{\lvert V_{cs}\rvert}(0)_{\lvert V_{cb}\rvert}$	FNAL/MILC leptonic [3]
0.06(3)	FLAG 2016 leptonic
−0.004(64)	ETMC semileptonic [12]
0.005(53)	my semileptonic average

the leptonic decay determination and none from the semileptonic. The semileptonic error is dominated by experimental error and the semileptonic by theory. However, both will improve.

32.4 Decays of Hadrons with *b* Quarks

Decays of hadrons containing b quarks have been studied in order to determine $|V_{ub}|$ and $|V_{cb}|$. Mesonic decays have been extensively studied by a number of groups. Recently, Meinel and his collaborators have been looking at several decays of baryons with b or c quarks [14]. Rare decays involving flavor changing neutral currents (FCNC) are a good place to look for new physics as FCNC processes vanish at the tree level. These processes also may involve third row CKM matrix elements and provide an alternative to $B_{(s)}$ meson mixing for determining $|V_{td}|$ and $|V_{ts}|$. Meson mixing is covered by FLAG.

Reference [3] provides the best values for $B_{(s)}$ meson decay constants. (See Fig. 32.5). Errors are < 1.3 MeV or 0.7%. There is good agreement with earlier calculations that have errors as small as 5–7 MeV. To exploit these results to get $|V_{ub}|$, we await precise results from Belle II for $B \to \tau\nu$, as the difference between BaBar and Belle is large [2].

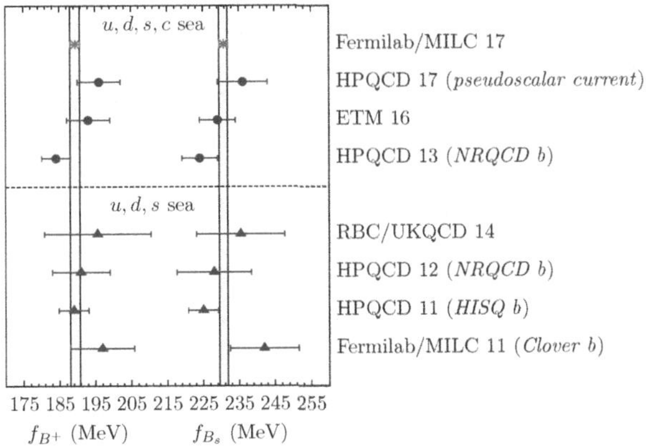

Fig. 32.5 Comparison of recent calculations of f_{B^+} and f_{B_s} with $N_f = 2 + 1 + 1$ and $2 + 1$. From [3]

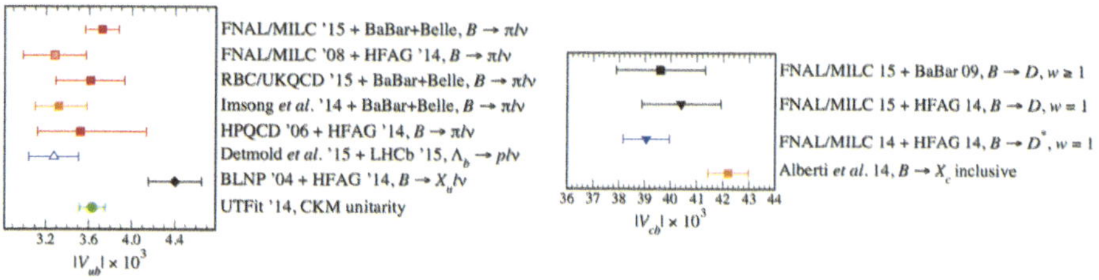

Fig. 32.6 **L** Comparison of several determinations of $|V_{\rm ub}|$ [13]. **R** Comparison of several determinations of $|V_{\rm cb}|$ [15]

Turning to bottom hadron semileptonic and rare decays, there are many possible channels. For $|V_{\rm ub}|$ these include $B \to \pi\ell\nu$, $B_s \to K\ell\nu$, $B_s \to K^*\ell\nu$, and $\Lambda_b \to p\ell\nu$. For $|V_{\rm cb}|$, we might study $B \to D\ell\nu$, $B \to D^*\ell\nu$, $B_s \to D_s^{(*)}\ell\nu$, and $\Lambda_b \to \Lambda_c\ell\nu$. We can test lepton universality as ℓ can be e, μ, or τ. There are also interesting rare decays such as $B^0 \to \mu^+\mu^-$, $B_s \to \mu^+\mu^-$, $B_s \to \phi\ell\nu$, and $B \to K\ell^+\ell^-$. Unfortunately, there is not enough time to cover all of these decays. Let's consider the long standing difference between CKM matrix elements as determined in exclusive and inclusive decay measurements. In 2015, the form factors needed for $B \to \pi\ell\nu$ were updated [13] and the resulting value of $|V_{\rm ub}| = 3.72(16) \times 10^{-3}$, was in somewhat better agreement with the inclusive value as seen in Fig. 32.6 (L). The figure shows several different determinations of $|V_{\rm ub}|$ including one based on Λ_b decay (triangle) [14]. The inclusive result is plotted as a diamond, unitarity as a circle. The exclusive value of $|V_{\rm ub}|$ is in good agreement with unitarity, but the inclusive one is not. As mentioned above, the leptonic decay $B \to \tau\nu$ could shed light on $|V_{\rm ub}|$, but we'll need to wait for Belle II results for that.

As of 2015, the situation for $|V_{\rm cb}|$ is depicted in Fig. 32.6 (R). At that time, exclusive decay processes $B \to D^*\ell\nu$ and $B \to D\ell\nu$ were both being studied. The experimental error was larger in the D channel (3.9%) whereas it was just 1.4% for the D^* channel. In the figure, w is an alternate kinematic variable equivalent to q^2. The form factor at $w = 1$ (zero recoil) can be calculated using lattice QCD; however, it is difficult to get the corresponding experimental value as the differential decay rate vanishes there, so it is necessary to fit the experimental results as a function of w. For the D channel, the theoretical form factors were available for a range of w. The notation HFAG 2014 in the figure indicates that the experimental input for $w = 1$ came from the fit of the Heavy Flavor Averaging Group (now HFLAV). In 2016, Bigi and Gambino [16] used updated Belle data for $B \to D\ell\nu$ and the BGL parameterization [17] to obtain $|V_{\rm cb}| = (40.49 \pm 0.97) \times 10^{-3}$. In 2017, Bigi, Gambino, Schacht [18]; and Grinstein and Kobach [19] examined new Belle data [20] for $B \to D^*\ell\nu$ and found a 10% difference when changing between CLN [21] and BGL parameterizations of the experimental data. Using CLN, they found $(38.2 \pm 1.5) \times 10^{-3}$, and for BGL, they found $(41.7 \pm 2.0) \times 10^{-3}$. The PDG inclusive value for $|V_{\rm cb}| = (42.2 \pm 0.8) \times 10^{-3}$. We see that for $B \to D^*$, exclusive and inclusive $|V_{cb}|$ values are totally compatible, and for $B \to D$, the difference between inclusive and exclusive determinations is only 1.36σ. Thus, for $|V_{cb}|$, the issue appeared to be resolved at the time of FPCP 2018. However, after the conference there were new fits to experimental results from BaBar and Belle. Please see Sect. 8.8 in the latest FLAG review [22], where relevant experimental and theoretical references can be found.

32.5 Conclusions and Outlook

There has been very significant progress using lattice QCD to calculate hadronic matrix elements needed for precise evaluation of SM contributions to numerious decay processes. This theoretical input is essential to determine the CKM matrix. A number of quantities can now be calculated to sub-percent accuracy. The interplay between theory and experiment will continue to yield increasingly stringent tests of the SM. In semileptonic decays, we see some tension with unitarity in the first and second rows. In the first row, we see slightly $>2\sigma$ tension with unitarity from semileptonic K decay. There is some tension between leptonic and semileptonic determinations of $|V_{ud}|$ and $|V_{us}|$. The tests of unitarity in the second row are not as stringent. The difference between exclusive and inclusive determination of $|V_{cb}|$ may be due to how the experimental data had been fit; however, for $|V_{ub}|$ a difference remains. Although I had hoped to cover some of the other recent observations that are in tension with the SM predictions, there was not enough time. Future results from Belle II, BES III, and LHCb, combined with increasingly precise calculations from lattice QCD, will provide more critical tests of the SM and opportunities to find evidence of new physics.

Acknowledgements I thank the FPCP organizers for their wonderful hospitality and a stimulating conference. I gratefully acknowledge my colleagues in the Fermilab Lattice and MILC Collaborations for wonderful working relationships and friendships. I also thank FLAG members who contribute countless hours to making lattice QCD results more easily available to the wider community. This work was supported by the US DOE grant DE-SC0010120.

References

1. S. Aoki et al., Eur. Phys. J. C **77**(2), 112 (2017). https://doi.org/10.1140/epjc/s10052-016-4509-7, arXiv:1607.00299 [hep-lat]
2. C. Patrignani et al., [Particle Data Group], Chin. Phys. C **40**(10), 100001 (2016). https://doi.org/10.1088/1674-1137/40/10/100001
3. A. Bazavov et al., Phys. Rev. D **98**(7), 074512 (2018). https://doi.org/10.1103/PhysRevD.98.074512, arXiv:1712.09262
4. M. Moulson, PoS CKM **2016**, 033 (2017). https://doi.org/10.22323/1.291.0033, arXiv:1704.04104 [hep-ex]

5. A. Bazavov et al., Fermilab Lattice and MILC Collaborations. Phys. Rev. D **99**(11), 114509 (2019). https://doi.org/10.1103/PhysRevD.99.114509, arXiv:1809.02827
6. J.C. Hardy, I.S. Towner, arXiv:1807.01146 [nucl-ex]
7. H. Na, C.T.H. Davies, E. Follana, G.P. Lepage, J. Shigemitsu, Phys. Rev. D **82**, 114506 (2010). https://doi.org/10.1103/PhysRevD.82.114506, arXiv:1008.4562 [hep-lat]
8. H. Na, C.T.H. Davies, E. Follana, J. Koponen, G.P. Lepage, J. Shigemitsu, Phys. Rev. D **84**, 114505 (2011). https://doi.org/10.1103/PhysRevD.84.114505, arXiv:1109.1501 [hep-lat]
9. V. Lubicz et al., [ETM Collaboration], Phys. Rev. D **96**(5), 054514 (2017). https://doi.org/10.1103/PhysRevD.96.054514, arXiv:1706.03017 [hep-lat]
10. T. Kaneko et al., [JLQCD Collaboration], EPJ Web Conf. **175**, 13007 (2018). https://doi.org/10.1051/epjconf/201817513007, arXiv:1711.11235 [hep-lat]
11. Y. Amhis et al., [HFLAV Collaboration], Eur. Phys. J. C **77**(12), 895 (2017). https://doi.org/10.1140/epjc/s10052-017-5058-4, arXiv:1612.07233 [hep-ex]
12. L. Riggio, G. Salerno, S. Simula, Eur. Phys. J. C **78**(6), 501 (2018). https://doi.org/10.1140/epjc/s10052-018-5943-5, arXiv:1706.03657 [hep-lat]
13. J.A. Bailey et al., [Fermilab Lattice and MILC Collaborations], Phys. Rev. D **92**(1), 014024 (2015). https://doi.org/10.1103/PhysRevD.92.014024, arXiv:1503.07839 [hep-lat]
14. W. Detmold, C. Lehner, S. Meinel, Phys. Rev. D **92**(3), 034503 (2015). https://doi.org/10.1103/PhysRevD.92.034503, arXiv:1503.01421 [hep-lat]
15. J.A. Bailey et al., [MILC Collaboration], Phys. Rev. D **92**(3), 034506 (2015). https://doi.org/10.1103/PhysRevD.92.034506, arXiv:1503.07237 [hep-lat]
16. D. Bigi, P. Gambino, Phys. Rev. D **94**(9), 094008 (2016). https://doi.org/10.1103/PhysRevD.94.094008, arXiv:1606.08030 [hep-ph]
17. C.G. Boyd, B. Grinstein, R.F. Lebed, Phys. Rev. D **56**, 6895 (1997). https://doi.org/10.1103/PhysRevD.56.6895, arXiv:9705252 [hep-ph]
18. D. Bigi, P. Gambino, S. Schacht, Phys. Lett. B **769**, 441 (2017). https://doi.org/10.1016/j.physletb.2017.04.022, arXiv:1703.06124 [hep-ph]
19. B. Grinstein, A. Kobach, Phys. Lett. B **771**, 359 (2017). https://doi.org/10.1016/j.physletb.2017.05.078, arXiv:1703.08170 [hep-ph]
20. A. Abdesselam et al., [Belle Collaboration], arXiv:1702.01521 [hep-ex]
21. I. Caprini, L. Lellouch, M. Neubert, Nucl. Phys. B **530**, 153 (1998). https://doi.org/10.1016/S0550-3213(98)00350-2, arXiv:9712417
22. S. Aoki et al., Flavour lattice averaging group (20 Feb, 2019), p. 537. arXiv:1902.08191 [hep-lat]

Chapter 33
Measurement of CP Violation in $B_s^0 \to J/\psi\,\phi$ Decays

Konstantin Gizdov

Abstract Precise measurements of CP violation provide stringent tests of the Standard Model towards a search for signs of New Physics. The decay-time dependent CP asymmetry in $B_s^0 \to J/\psi\,\phi$ decays has been measured using proton-proton, pp, collision data, collected with the LHCb detector. The CP-violating phase ϕ_s is measured, along with the difference in decay widths of the light and heavy mass eigenstates of the $B_s^0 - \overline{B}_s^0$ system, and the average B_s^0 decay width. Discussed are also data-driven corrections to simulated event samples, and the control of systematic effects using data control sample. Prospects for the sensitivity that can be achieved with future LHCb data taking periods will be also given.

33.1 Introduction

The CP-violating phase ϕ_s originates from the interference between the mixing and direct decay of the B_s^0 mesons to CP eigenstates. Ignoring high-order contributions, it is predicted within the Standard Model (SM) to be $-2\beta_s$, where $\beta_s = \arg(-V_{ts}V_{tb}^*/V_{cs}V_{cb}^*)$ [1]. An indirect determination of $-2\beta_s = 0.0376^{+0.0008}_{-0.0007}$ rad is obtained using a global fit to experimental data [3]. Deviation from this prediction would be a clear sign of so-called New Physics effects, strongly motivating the need for precise experimental measurements of this quantity [4]. Discussed are the measurements of CP-violating phase ϕ_s independently performed using $B_s^0 \to J/\psi\,\phi$, $B_s^0 \to J/\psi\,K^+K^-$ as well as measurements of special higher order contributions to ϕ_s from $B_s^0 \to J/\psi\,\overline{K}^{*0}$ and $B_s^0 \to J/\psi\,\rho^0$ decay channels. All measurements shown here use 3 fb^{-1} of data collected by the LHCb experiment [2] in pp collisions during Run 1.

On behalf of LHCb Collaboration.

K. Gizdov (✉)
University of Edinburgh, Edinburgh, UK
e-mail: k.gizdov@ed.ac.uk

A. Giri and R. Mohanta (eds.), *16th Conference on Flavor Physics and CP Violation*, Springer Proceedings in Physics 234,
https://doi.org/10.1007/978-3-030-29622-3_33

33.2 Status of ϕ_s Measurements

33.2.1 Measurement of ϕ_s with $B_s^0 \to J/\psi K^+K^-$ in ϕ (1020) Region

For neutral B mesons the mass and flavour eigenstates do not coincide. Thus, B_s^0 mesons exist as a time-evolving oscillating mixture $|B_{L,H}\rangle = p|B_s^0\rangle \pm q|\overline{B}_s^0\rangle$ of light, B_L, and heavy, B_H, mass eigenstates, where p and q are complex parameters, such that $p^2 + q^2 = 1$. The $B_s^0 \to J/\psi K^+K^-$ decay final state is decomposed into four amplitudes: three P-waves, A_0, $A_\parallel$, $A_\perp$ and one S-wave, A_S accounting for the non-resonant K^+K^- configuration. Moreover, the final state is a mixture of CP-even states, $\eta_i = +1$ for $i \in \{0, \parallel\}$ and CP-odd states, $\eta_i = -1$ for $i \in \{\perp, S\}$. Thus, in order to extract the CP-violating phase ϕ_s, a tagged time-dependent angular fit to $B_s^0 \to J/\psi K^+K^-$ candidates is applied [7]. The phase ϕ_s is determined by $\phi_s = -\arg(\lambda)$ where $\lambda = \lambda_i/\eta_i$ and $\lambda_i = \frac{q}{p}\frac{\overline{A}_i}{A_i}$. The angular decomposition of the different CP states uses the decay helicity angles as defined in Fig. 33.1. The $B_s^0 \to J/\psi K^+K^-$ candidates are reconstructed using the decay $J/\psi \to \mu^+\mu^-$ and $\phi \to K^+K^-$. After applying a full online, trigger and event reconstruction, as well as offline selection, 95 690 ± 350 signal candidates (Fig. 33.2) of the $B_s^0 \to J/\psi K^+K^-$ are obtained [7]. The fitting procedure takes into account the flavour tagging efficiency, decay time resolution as well as decay time and angular acceptances. The decay time resolution is estimated using a sample of prompt $J/\psi K^+K^-$ combinations produced directly in the pp interactions and it is found to be 46 fs. A sample obtained through a prescaled unbiased trigger is used in conjunction with the tag-and-probe method to determine the decay time acceptance from data. The angular acceptance is determined using simulated events. The combined effective tagging power is $(3.73 \pm 0.15)\%$ [7]. A weighted unbinned maximum likelihood fit is performed using a signal-only (background subtracted) Probability Density Function (PDF), as described in [8]. Figure 33.3 shows the projections of the decay time and angular distributions. The final results are $\phi_s = -0.058 \pm 0.049 \pm 0.006$ rad, $\Gamma_s = 0.6603 \pm 0.0027 \pm 0.0015\,\text{ps}^{-1}$ and

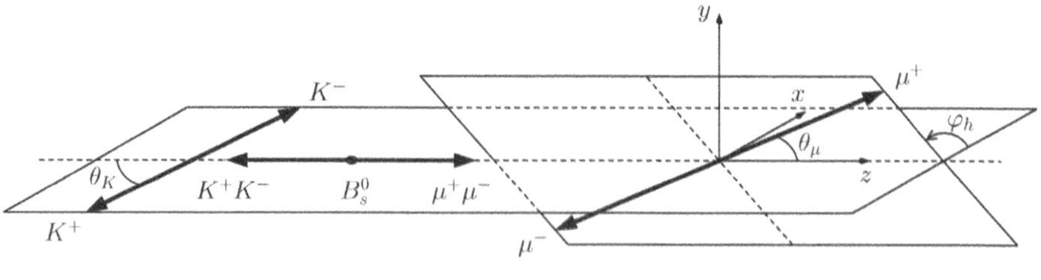

Fig. 33.1 Decay angles of $B_s^0 \to J/\psi\,(\phi \to K^+K^-)$: θ_μ is the angle between direction of μ in J/ψ rest frame and J/ψ direction in B_s^0 rest frame; θ_K is the angle between direction of K in ϕ rest frame and ϕ direction in B_s^0 rest frame φ_h is the angle between decay plane of $J/\psi \to \mu\mu$ and $\phi \to KK$ decay plane [17]

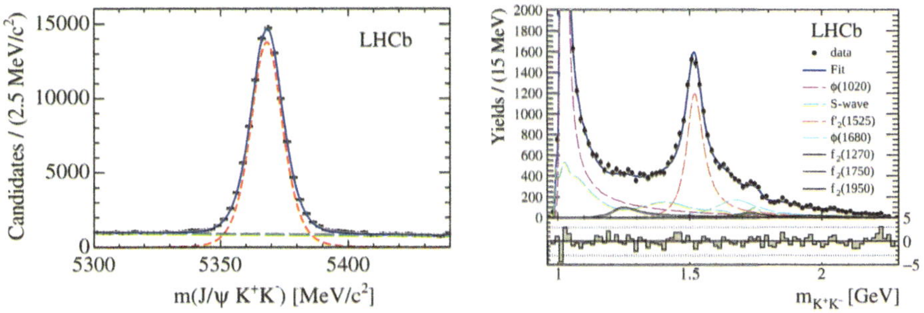

Fig. 33.2 Left: Distribution of the $m(J/\psi\,K^+K^-)$ invariant mass with $m(K^+K^-)$ in $\phi(1020)$ region [7]—black markers are data, blue line is total fit, red dashed line is signal and green dashed is combinatorial background. Right: Projection of time-dependent amplitude analysis fit in $m(K^+K^-)$ invariant mass above $\phi(1020)$ region with contributing components [9]

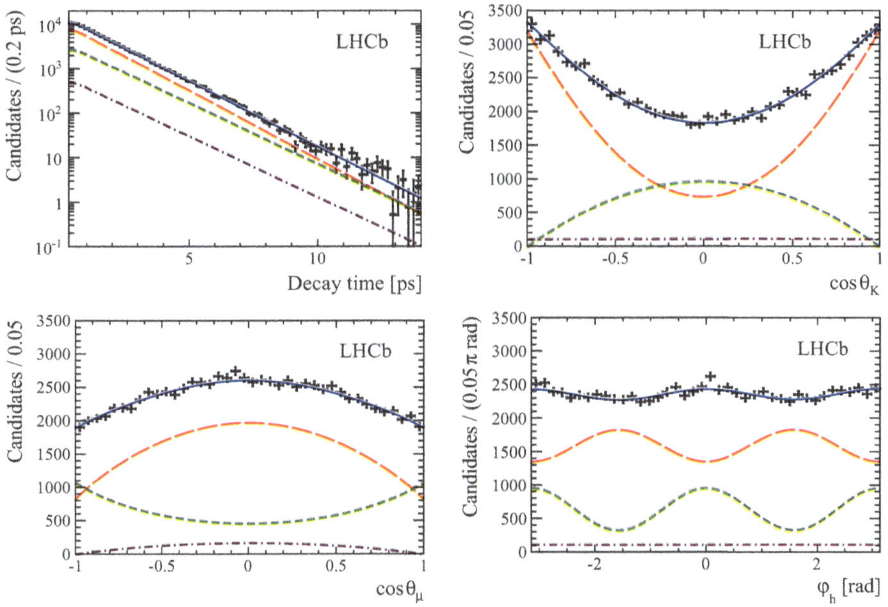

Fig. 33.3 Decay time and angle distributions for $B_s^0 \to J/\psi\,K^+K^-$ decays (black markers) with the one-dimensional projections of the PDF. The solid blue line shows the total signal contribution, which is composed of CP-even (long-dashed red), CP-odd (short-dashed green) and S-wave (dotted-dashed purple) contributions [7]

$\Delta\Gamma_s = 0.0805 \pm 0.0091 \pm 0.0032\,\mathrm{ps}^{-1}$ [7]. The dominant contribution to the systematic uncertainty of ϕ_s is given by the decay time and angular efficiency and background subtraction.

33.2.2 Measurement of ϕ_s with $B_s^0 \to J/\psi K^+K^-$ with $m(K^+K^-)$ Above ϕ (1020) Region

The measurement of the phase ϕ_s has been performed in the $B_s^0 \to J/\psi K^+K^-$ decay with K^+K^- invariant mass larger than $1050\,\text{MeV}/c^2$ [9], which is above the $\phi(1020)$ resonance region. The important difference between both decay analyses is the modelling of the $m(K^+K^-)$ distribution, which is included to distinguish different resonant and non-resonant contributions. The decay time acceptance is determined with the same method as described in [10] by using the control channel $B^0 \to J/\psi K^{*0}$. The K^+K^- mass spectrum is fitted by considering the different contributions found in the time-dependent amplitude analysis as shown in Fig. 33.2. The final fit has been performed allowing eight independent sets of CP-violating parameters: three corresponding to $\phi(1020)$ transversity states, K^+K^- S-wave, $f_2(1270)$, $f_2^{'}(1525)$, $\phi(1680)$ and the combination of the two high-mass $f_2(1750)$ and $f_2(1950)$ states. The CP-violating parameters measurement of $B_s^0 \to J/\psi K^+K^-$ in high $m(K^+K^-)$ region is $\phi_s = 0.119 \pm 0.107 \pm 0.034\,\text{rad}$, $\Gamma_s = 0.650 \pm 0.006 \pm 0.004\,\text{ps}^{-1}$ and $\Delta\Gamma_s = 0.066 \pm 0.018 \pm 0.006\,\text{ps}^{-1}$. The largest contribution to systematic uncertainty results from the resonance fit model. The combination with the B_s^0 decay fit results in the $\phi(1020)$ region gives $\phi_s = -0.025 \pm 0.045 \pm 0.008\,\text{rad}$ that improves a precision of the ϕ_s measurement by more than 9%.

33.2.3 Measurement of Penguin Diagram Contributions to ϕ_s

Additional contributions to the leading $b \to c\bar{c}s$ tree Feynman diagram, as shown in Fig. 33.4, are thought to be negligible. However, the shift in ϕ_s due to these contributions, called hereafter "penguin pollution", is difficult to compute due to the non-perturbative nature of the quantum chromodynamics (QCD) processes involved. Penguin pollution must be measured or constrained in order to exclude its effect on ϕ_s from any BSM searches. As described in [13], U-spin and SU(3) related modes can be used for this purpose by choosing decay modes in which the tree decay diagram is suppressed and the "penguin pollution" enhanced. The decay modes of $B_s^0 \to J/\psi \overline{K}^{*0}$ and $B^0 \to J/\psi \rho^0$ have been used to measure and put constraints on the

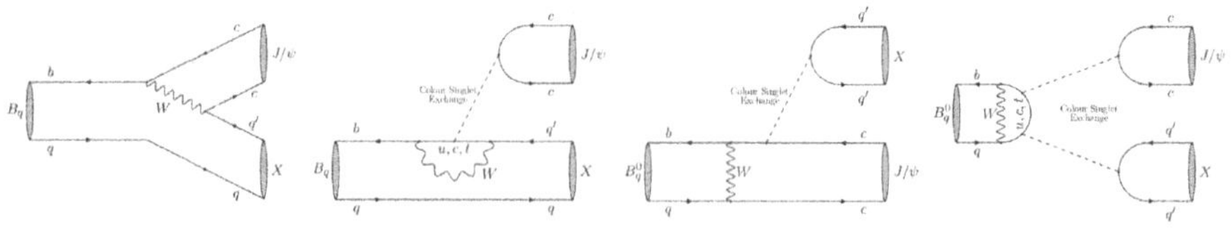

Fig. 33.4 Feynman diagrams of possible decays of $B_q^0 \to J/\psi X$, from left to right: tree-level, penguin, exchange, penguin-annihilation diagrams

amount of penguin diagram contributions [15, 16]. The decay mode $B_s^0 \to J/\psi\,\overline{K}^{*0}$ has been normalized to the $B^0 \to J/\psi\,K^{*0}$ mode to determine the branching fraction and a time-integrated angular fit is performed to measure the ϕ_s "penguin pollution" - $\Delta\phi_s$. The $B^0 \to J/\psi\,\rho^0$ decay uses $B_s^0 \to J/\psi\,K_s^0$ as normalization channel and a combination of time-dependent and time-integrated angular fits to measure and disentagle total and partial $\Delta\phi_s$. Specifically in the combination of the two modes, the $B_s^0 \to J/\psi\,\overline{K}^{*0}$ decay includes no penguin annihilation nor exchange diagrams (Fig. 33.4), but the $B^0 \to J/\psi\,\rho^0$ does, thus the channels can isolate the individual penguin components. Combining both, the contributions affecting ϕ_s have been constrained to be: $\Delta{\phi_s}^0 = 0.000^{+0.009}_{-0.011}\,{}^{+0.004}_{-0.009}$ rad, $\Delta{\phi_s}^{\parallel} = 0.001^{+0.010}_{-0.014} \pm 0.008$ rad, $\Delta{\phi_s}^{\perp} = 0.003^{+0.010}_{-0.014} \pm 0.008$ rad. It can be seen that with the current uncertainty they are still consistent with zero.

33.2.4 Global Combination

The CP-violating phase and lifetime parameters have been measured by several experiments over the years, including $B_s^0 \to J/\psi\,\phi$ analyses from CDF, D0, ATLAS and CMS collaborations [18] as well as analyses using several different final states performed by the LHCb collaboration, two of which discussed here. The world average result of ϕ_s and $\Delta\Gamma_s$ measurements from the Heavy Flavour Averaging Group [18] is shown in Fig. 33.5. They find $\phi_s = -0.021 \pm 0.031$ rad and

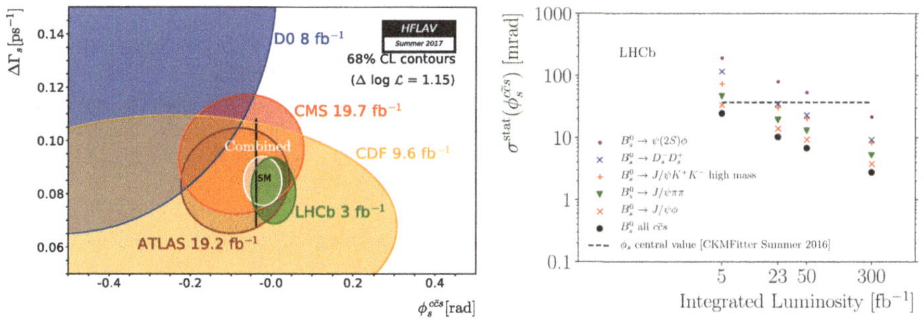

Fig. 33.5 Left: 68% confidence level regions in $\Delta\Gamma_s$ and ϕ_s plane obtained from individual contours of CDF, D0, CMS, ATLAS and LHCb measurements and the combined contour [18]. The expectation within the SM [3] is shown as a black thin rectangle. Right: Comparison of ϕ_s sensitivity from different decay modes [19]

$\Delta\Gamma_s = 0.085 \pm 0.006\,\text{ps}^{-1}$ that is dominated by the measurements of ϕ_s using $B_s^0 \to J/\psi K^+K^-$ and $B_s^0 \to J/\psi \pi^+\pi^-$[11] from LHCb collaboration and it is consistent with the SM predictions.

33.3 Summary

Using data collected during Run 1 by the LHCb experiment, the most precise measurements to date of CP-violating phase ϕ_s and the lifetime parameters of the $B_s^0 - \bar{B}_s^0$ system are presented. Several measurements are discussed including the precision dominating $B_s^0 \to J/\psi\,\phi$, multiple-resonance $B_s^0 \to J/\psi K^+K^-$ and rare decays constraining the so-called "penguin pollutions". Currently, the results are compatible with the SM predictions. Ultimately, in order to reduce the experimental uncertainties and obtain a measurement comparable or better then the theoretical predictions, new modes are not sufficient. More data has to be included and combined with current measurements. Thus, with the coming LHCb Upgrade II, detector performance and luminosity will increase delivering an estimated ϕ_s uncertainty ~ 0.004 rad from $B_s^0 \to J/\psi K^+K^-$ only and ~ 0.003 rad from all modes combined at 300 fb^{-1} [19]. At the same time, higher order effects such as "penguin pollution" will be constrained to ~ 0.0015 rad.

References

1. M. Kobayashi, T. Maskawa, CP violation in the renormalizable theory of weak interaction. Prog. Theor. Phys. **49**, 652 (1973). https://doi.org/10.1143/PTP.49.652
2. A.D. Sakharov, Violation of CP invariance, C asymmetry, and baryon asymmetry of the universe. Pisma Zh. Eksp. Teor. Fiz. **5**, 32 (1967). https://doi.org/10.1070/PU1991v034n05ABEH002497. [JETP Lett. **5**, 24 (1967)] [Sov. Phys. Usp. **34**(5), 392 (1991)] [Usp. Fiz. Nauk **161**(5), 61 (1991)]
3. J. Charles et al., Current status of the standard model CKM fit and constraints on $\Delta F = 2$ new physics. Phys. Rev. D **91**(7), 073007 (2015). https://doi.org/10.1103/PhysRevD.91.073007, arXiv:1501.05013
4. Z. Ligeti, M. Papucci, G. Perez, Implications of the measurement of the $B_s^0 - \bar{B}_s^0$ mass difference. Phys. Rev. Lett. **97**, 101801 (2006). https://doi.org/10.1103/PhysRevLett.97.101801, arXiv:0604112
5. R. Aaij, et al., [LHCb Collaboration]: precision measurement of the B_s^0-$\bar{B}_s^0$ oscillation frequency with the decay $B_s^0 \to D_s^-\pi^+$. New J. Phys. **15**, 053021 (2013). https://doi.org/10.1088/1367-2630/15/5/053021, arXiv:1304.4741
6. A. Simon, Recent Improvements and Prospects with Flavour Tagging at LHCb [LHCb-TALK-2018-304] (2018). http://cds.cern.ch/record/2630477
7. R. Aaij, et al., [LHCb Collaboration]: precision measurement of CP violation in $B_s^0 \to J/\psi K^+K^-$ decays. Phys. Rev. Lett. **114**(4), 041801 (2015). https://doi.org/10.1103/PhysRevLett.114.041801, arXiv:1411.3104
8. Y. Xie, sFit: a method for background subtraction in maximum likelihood fit, arXiv:0905.0724

9. R. Aaij et al., [LHCb Collaboration]: resonances and CP violation in B_s^0 and $\overline{B}_s^0 \to J/\psi K^+K^-$ decays in the mass region above the $\phi(1020)$. JHEP **1708**, 037 (2017). https://doi.org/10.1007/JHEP08(2017)037, arXiv:1704.08217
10. R. Aaij et al., [LHCb Collaboration]: first study of the CP -violating phase and decay-width difference in $B_s^0 \to \psi(2S)\phi$ decays. Phys. Lett. B **762**, 253 (2016). https://doi.org/10.1016/j.physletb.2016.09.028. arXiv:1608.04855
11. R. Aaij et al., [LHCb Collaboration]: measurement of the CP-violating phase ϕ_s in $\overline{B}_s^0 \to J/\psi\pi^+\pi^-$ decays. Phys. Lett. B **713**, 378 (2012). https://doi.org/10.1016/j.physletb.2012.06.032, arXiv:1204.5675
12. R. Aaij et al., [LHCb Collaboration]: LHCb detector performance. Int. J. Mod. Phys. A **30**(07), 1530022 (2015). https://doi.org/10.1142/S0217751X15300227, arXiv:1412.6352
13. S. Faller, M. Jung, R. Fleischer, T. Mannel, The golden modes B0 —> J/psi K(S,L) in the era of precision flavour physics. Phys. Rev. D **79**, 014030 (2009). https://doi.org/10.1103/PhysRevD.79.014030, arXiv:0809.0842
14. K. De Bruyn, R. Fleischer, A roadmap to control penguin effects in $B_d^0 \to J/\psi K_S^0$ and $B_s^0 \to J/\psi\phi$. JHEP **1503**, 145 (2015). https://doi.org/10.1007/JHEP03(2015)145, arXiv:1412.6834
15. R. Aaij, et al., [LHCb Collaboration]: measurement of CP violation parameters and polarisation fractions in $B_s{}^0 \to J/\psi\overline{K}^{*0}$ decays. JHEP **1511**, 082 (2015). https://doi.org/10.1007/JHEP11(2015)082, arXiv:1509.00400
16. R. Aaij et al., [LHCb Collaboration]: measurement of the CP-violating phase β in $B^0 \to J/\psi\pi^+\pi^-$ decays and limits on penguin effects. Phys. Lett. B **742**, 38 (2015). https://doi.org/10.1016/j.physletb.2015.01.008, arXiv:1411.1634
17. R. Aaij et al., [LHCb Collaboration]: measurement of CP violation and the B_s^0 meson decay width difference with $B_s^0 \to J/\psi K^+K^-$ and $B_s^0 \to J/\psi\pi^+\pi^-$ decays. Phys. Rev. D **87**(11), 112010 (2013). https://doi.org/10.1103/PhysRevD.87.112010, arXiv:1304.2600
18. Y. Amhis et al., [HFLAV Collaboration]: averages of b-hadron, c-hadron, and τ-lepton properties as of summer 2016. Eur. Phys. J. C **77**(12), 895 (2017). https://doi.org/10.1140/epjc/s10052-017-5058-4, arXiv:1612.07233
19. I. Bediaga et al., [LHCb Collaboration]: physics case for an LHCb upgrade II -opportunities in flavour physics, and beyond, in the HL-LHC era, arXiv:1808.08865

Chapter 34
Measurement of *CPV* in Charmless Two-Body *B*-Hadron Decays to Charged Hadrons

Emilie Bertholet

Abstract We present the latest results from the LHCb experiment on the measurement of the time-dependent *CP* asymmetries in $B^0 \to \pi^+\pi^-$ and $B^0_s \to K^+K^-$ in addition to the time-integrated *CP* asymmetries in $B^0 \to K^+\pi^-$ and $B^0_s \to \pi^+K^-$. The analysis uses a data sample of *pp* collisions corresponding to an integrated luminosity of 3.0 fb^{-1}, collected with the LHCb detector at center-of-mass energies of 7 and 8 TeV. Evidence for *CP* violation in $B^0_s \to K^+K^-$ is found for the first time with a significance of 4.0 standard deviations.

34.1 Introduction

The study of *CP* violation in charmless *B*-mesons decays offers many opportunities for testing the standard model and probing for New Physics (NP). In particular, measurements of *CP* violation in two-body $B^0_{(s)}$-meson decays to charged pions and kaons give valuable input for constraining the angles γ and α of the Unitarity Triangle and the mixing phases ϕ_d and ϕ_s. The non-negligible contribution of loop (penguin) processes in these modes introduces hadronic uncertainties but makes the observables potentially sensitive to NP. Comparison with results dominated by tree-level processes provides tests of the SM and constraints on NP. The results presented here are measurements of the time-dependent *CP* asymmetries in $B^0 \to \pi^+\pi^-$ and $B^0_s \to K^+K^-$ decays and of the time-integrated *CP* asymmetries in $B^0 \to K^+\pi^-$ and $B^0_s \to \pi^+K^-$. The analysis is based on a data sample of *pp* collisions corresponding to an integrated luminosity of 3.0 fb^{-1}, collected with the LHCb detector at center-of-mass energies of 7 and 8 TeV and was published in [1].

On behalf of the LHCb collaboration.

E. Bertholet (✉)
LPNHE, Sorbonne Universite, Paris Diderot Sorbonne Paris Cite, CNRS/IN2P3, Paris, France
e-mail: eberthol@cern.ch

A. Giri and R. Mohanta (eds.), *16th Conference on Flavor Physics and CP Violation*, Springer Proceedings in Physics 234,
https://doi.org/10.1007/978-3-030-29622-3_34

34.2 Observables

The eigenstates of the effective Hamiltonian describing neutral meson mixing can be expressed as an admixture of flavour eigenstates

$$|B_{\mathrm{H/L}}\rangle = p|B^0_{(\mathrm{s})}\rangle \pm q|\overline{B}^0_{(\mathrm{s})},\rangle \tag{34.1}$$

where p and q are complex parameters which satisfy $|p|^2 + |q|^2 = 1$. It is convenient to define

$$\lambda_f \equiv \frac{q}{p}\frac{\overline{A}_{\mathrm{f}}}{A_{\mathrm{f}}}, \tag{34.2}$$

where A_{f} ($\overline{A}_{\mathrm{f}}$) is the amplitude for the process $B^0_{(\mathrm{s})} \to f$ ($\overline{B^0}_{(\mathrm{s})} \to f$).

The time-dependent *CP* asymmetry can be expressed, assuming *CPT* invariance, as a function of $\Delta m_{\mathrm{d,s}}$ and $\Delta\Gamma_{\mathrm{d,s}}$ of the mass and width differences of the mass eigenstates in neutral *B*-meson system

$$A_{\mathrm{CP}}(t) = \frac{\Gamma_{\overline{B}^0_{(\mathrm{s})}\to f}(t) - \Gamma_{B^0_{(\mathrm{s})}\to f}(t)}{\Gamma_{\overline{B}^0_{(\mathrm{s})}\to f}(t) + \Gamma_{B^0_{(\mathrm{s})}\to f}(t)} = \frac{-C_{\mathrm{f}}\cos(\Delta m_{\mathrm{d,s}}t) + S_f\sin(\Delta m_{\mathrm{d,s}}t)}{\cosh\left(\frac{\Delta\Gamma_{\mathrm{d,s}}}{2}t\right) + A_f^{\Delta\Gamma}\sinh\left(\frac{\Delta\Gamma_{\mathrm{d,s}}}{2}t\right)}, \tag{34.3}$$

where the observables C_{f}, S_{f} and $A_{\mathrm{f}}^{\Delta\Gamma}$ are defined as

$$C_{\mathrm{f}} \equiv \frac{1-|\lambda_f|^2}{1+|\lambda_f|^2}, \qquad S_{\mathrm{f}} \equiv \frac{2\Im\lambda_f}{1+|\lambda_f|^2}, \qquad A_{\mathrm{f}}^{\Delta\Gamma} \equiv -\frac{2\Re\lambda_f}{1+|\lambda_f|^2}, \tag{34.4}$$

and satisfy the constraint

$$(C_{\mathrm{f}})^2 + (S_{\mathrm{f}})^2 + (A_{\mathrm{f}}^{\Delta\Gamma})^2 = 1. \tag{34.5}$$

Assuming no *CP* violation in the mixing, $|\frac{q}{p}| = 1$, C_{f} corresponds to *CP* violation in the decay and S_{f} is *CP* violation in the interference between mixing and decay. In the case of the B^0 meson, the value of $\Delta\Gamma_{\mathrm{d}}$ is neglected according to current experimental results [2]. As a consequence, for B^0 mesons, (34.3) simplifies and the denominator is equal to one. The time-integrated *CP* asymmetry for $B^0_{(\mathrm{s})}$ mesons decaying to a flavour specific final state f, such as $B^0 \to K^+\pi^-$ and $B^0_{\mathrm{s}} \to \pi^+K^-$, is given by

$$A_{\mathrm{CP}} = \frac{|\overline{A}_{\overline{\mathrm{f}}}|^2 - |A_{\mathrm{f}}|^2}{|\overline{A}_{\overline{\mathrm{f}}}|^2 + |A_{\mathrm{f}}|^2}. \tag{34.6}$$

LHCb preformed previous measurements of $C_{\pi^+\pi^-}$, $S_{\pi^+\pi^-}$, $C_{K^+K^-}$, $S_{K^+K^-}$ [3], $A_{\mathrm{CP}}^{B^0}$ and $A_{\mathrm{CP}}^{B^0_{\mathrm{s}}}$ [4] with an integrated luminosity of 1.0 fb^{-1} at center-of-mass energy of 7 TeV. The observables $C_{\pi^+\pi^-}$, $S_{\pi^+\pi^-}$ have also been measured by BABAR [5] and Belle [6] and $A_{\rm CP}^{B^0}$ and $A_{\rm CP}^{B_s^0}$ have been measured by the B-factories [5, 7] and CDF [8]. All the results are in good agreement with each other.

34.3 Analysis Strategy

34.3.1 Event Selection and Backgrounds

The events are selected online by the trigger. After applying some offline filtering to reduce the amount of combinatorial background, the candidates are classified into mutually exclusive samples: $\pi^+\pi^-$, K^+K^- and $K^\pm\pi^\mp$. The three main sources of background are illustrated in Fig. 34.1. The one marked as three-body background comes from mis-reconstructed three-body decays into two-body decays. This happens, for example, if one of the daughters goes out of the acceptance of the detector. This background peaks at a lower mass compared to the signal peak. Another source of background originates from the cross-feeds, which consist in two-body final states with misidentified particles; for example, in the case of $B^0 \to \pi^+\pi^-$, when one of the π mesons is mis-reconstructed as a K meson. The particle identification requirement is optimised to reduce the amount of cross-feed to approximately 10% of the signal yields. The combinatorial background, which comes from random pairing of tracks of opposite charge, is further reduced using a Boosted Decision Tree algorithm.

34.3.2 Fit Model

The *CP* asymmetries are obtained by performing a simultaneous unbinned maximum likelihood fit to the invariant mass, m, the decay time, t, and the predicted decay time error, δt, the flavour tagging assignment, ξ, and the predicted mistag probability, η, for the distributions of candidates reconstructed in the $\pi^+\pi^-$, K^+K^- and $K^\pm\pi^\mp$ samples. The parameters $\Delta m_{\rm d,s}$, $\Gamma_{\rm d,s}$ and $\Delta\Gamma_{\rm s}$ are fixed in the fits to HFLAV values and $\Delta\Gamma_{\rm d}$ is fixed to 0, refer to Table 4 in [1]. The probability density functions (PDFs) are functions of the observables described in Sect. 34.2 and take into account detector effects.

In *pp* collisions, the production rates of b and $\overline{b}$ quarks are not supposed to be identical due to the fact that the initial *pp* state is not a *CP* eigenstate. This effect is known as production asymmetry and it is crucial to determine it with precision when measuring *CP* violation. In this analysis, the production asymmetry is extracted from the fit with no need of any external input and is $(0.19 \pm 0.60)\%$ for B^0 mesons and $(2.4 \pm 2.1)\%$ for $B_{\rm s}^0$ mesons.

Another effect to be assessed is the detection asymmetry, which describes the fact that the efficiencies in reconstructing the final state or its charge conjugate differ. This asymmetry does not affect the modes $B^0 \to \pi^+\pi^-$ and $B_{\rm s}^0 \to K^+K^-$ since in both cases, the final state is the same for B^0 and $\overline{B}^0$ but it has to be taken into account for $B^0 \to K^+\pi^-$ and $B_s^0 \to \pi^+K^-$ decays.

Since we need to differentiate between B^0 and $\overline{B}^0$, the flavour tagging is another important ingredient in the fit. In LHCb, the flavour of the B-meson is tagged at production. Considering that in pp collisions the b-quarks are produced in pairs of $b\overline{b}$, the tagging can either be done by exploiting the hadronisation tracks of the signal b-hadron (same side taggers, SS) or the decay products of the other b-hadron (opposite side taggers, OS). Each class of taggers has its proper algorithm and performance. For example, the OS taggers act in the same way on B^0 and B_s^0, which is not the case for the SS taggers. The mistag probability has an influence on the amplitude of the asymmetry as a function of the decay time, the higher the probability of mistag the smaller the amplitude of the asymmetry. It is estimated by an artificial neural network and then calibrated via control channels. Combining the different taggers, the tagging-power in this analysis is of the order of 4% for $B^0 \to \pi^+\pi^-$ and 3.7% for $B_s^0 \to K^+K^-$.

The decay time resolution, σ_t, has an impact on the sensitivity to the CP observables. It induces a dilution factor on C_f and S_f given by

$$D_{\sigma_t} = e^{-\frac{1}{2}\Delta m_{d,s}^2 \sigma_t^2}. \tag{34.7}$$

For B^0 decays, Δm_d is small and the dilution factor can be neglected. On the contrary, for B_s^0, Δm_s is large and the dilution can no longer be neglected. The resolution function is modelled as a sum of two gaussians and the different parameters are taken from data and from simulation.

34.3.3 Systematic Uncertainties

Several sources of systematic uncertainties are considered in this analysis. The impact of the knowledge of external inputs on the values of the CP asymmetries is evaluated by repeating the fit to data each time modifying the values of these parameters. The systematic uncertainties on the fit are estimated by generating and fitting several pseudo-experiments according to the baseline model. The detailed list of systematic uncertainties can be found in Table 5 of the paper [1].

34.4 Results and Conclusions

The distributions of the different fit variables for the $\pi^+\pi^-$ sample are given in Fig. 34.1. The time-dependent CP asymmetries are obtained by considering candidates in an interval around the mass of the $B_{(s)}^0$ meson. The results are shown for the $\pi^+\pi^-$ sample in Fig. 34.2, split between OS and SS taggers. For the other samples refer to the paper [1].

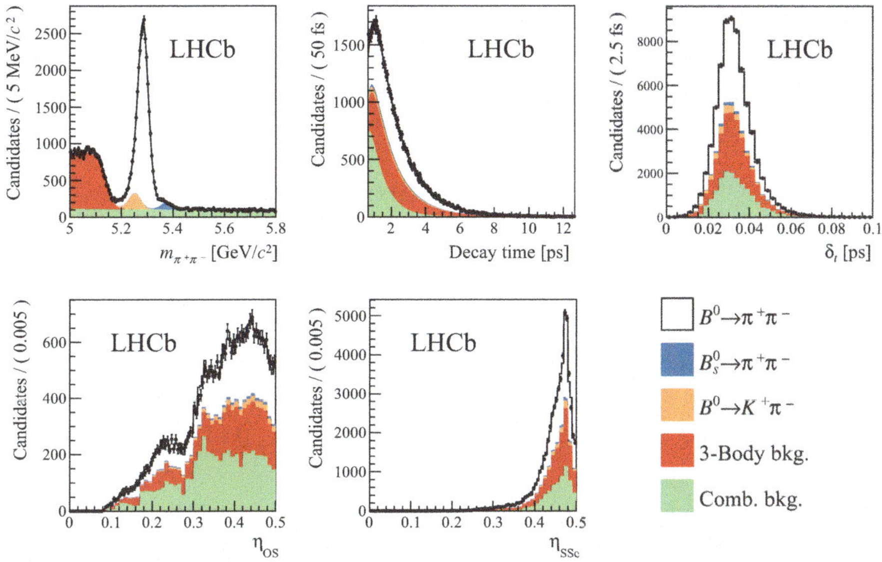

Fig. 34.1 Distributions of (top left) invariant mass, (top right) decay time, (middle left) decay-time uncertainty, (middle right) η_{OS}, and (bottom) η_{SSc} for candidates in the $\pi^+\pi^-$ sample. The result of the simultaneous fit is overlaid. The individual components are also shown

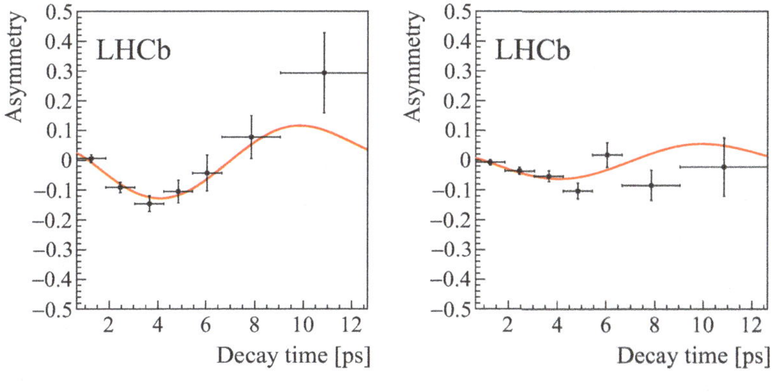

Fig. 34.2 Time-dependent asymmetries for (top) $\pi^+\pi^-$ candidates with mass values in the intervals $5.20 < \mathrm{m} < 5.35\,\mathrm{GeV}/\mathrm{c}^2$ (left) using the OS-tagging decision and (right) the SS-tagging decision. The result of the simultaneous fit is overlaid

34.5 Final Results and Conclusion

The results for the time-dependent *CP* violation in $B^0 \to \pi^+\pi^-$ and $B^0_s \to K^+K^-$ decays and for *CP* asymmetries in $B^0 \to K^+\pi^-$ and $B^0_s \to \pi^+K^-$ are

$$
\begin{aligned}
A_{\mathrm{CP}}^{B^0} &= -0.084 \pm 0.004 \pm 0.003,\\
A_{\mathrm{CP}}^{B^0_s} &= 0.213 \pm 0.015 \pm 0.007,\\
C_{\pi^+\pi^-} &= -0.34 \pm 0.06 \pm 0.01,\\
S_{\pi^+\pi^-} &= -0.63 \pm 0.05 \pm 0.01,\\
C_{K^+K^-} &= 0.20 \pm 0.06 \pm 0.02,\\
S_{K^+K^-} &= 0.18 \pm 0.06 \pm 0.02,\\
A^{\Delta\Gamma}_{K^+K^-} &= -0.79 \pm 0.07 \pm 0.10,
\end{aligned}
$$

where the first uncertainties are statistical and the second systematic. The parameter $A^{\Delta\Gamma}_{K^+K^-}$ was measured for the first time. These results are consistent with those of the previous LHCb analysis [3, 4] and are of greater precision. LHCb now dominates the world average for $C_{\pi^+\pi^-}$, $S_{\pi^+\pi^-}$, $A_{\mathrm{CP}}^{B^0}$ and $A_{\mathrm{CP}}^{B^0_s}$.

The significance for $(C_{K^+K^-}, S_{K^+K^-}, A^{\Delta\Gamma}_{K^+K^-})$ to differ from $(0, 0, -1)$ is estimated to be 4.0 standard deviations, which consists in the strongest evidence for time-dependent *CP* violation in the B^0_{rms} sector to date.

This analysis was performed with 3 fb^{-1} of run 1 data. Improving the sensitivity, by using a larger dataset, may lead to a discovery of *CP* violation in B^0_s sector.

References

1. Roel Aaij et al., Measurement of *CP* asymmetries in two-body $B^0_{(s)}$-meson decays to charged pions and kaons. Phys. Rev. D **98**(3), 032004 (2018)
2. Y. Amhis et al., Averages of *b*-hadron, *c*-hadron, and τ-lepton properties as of summer 2016. Eur. Phys. J. C **77**(12), 895 (2017)
3. R. Aaij et al., First measurement of time-dependent *CP* violation in $B^0_s \to K^+K^-$ decays. JHEP **10**, 183 (2013)
4. R. Aaij et al., First observation of *CP* violation in the decays of B^0_s mesons. Phys. Rev. Lett. **110**(22), 221601 (2013)
5. J.P. Lees et al., Measurement of CP asymmetries and branching fractions in charmless two-body *B*-meson decays to pions and kaons. Phys. Rev. D **87**(5), 052009 (2013)
6. I. Adachi et al., Measurement of the CP violation parameters in $B^0 \to \pi^+\pi^-$ decays. Phys. Rev. D **88**(9), 092003 (2013)
7. Y.T. Duh et al., Measurements of branching fractions and direct CP asymmetries for BK, B and BKK decays. Phys. Rev. D **87**(3), 031103 (2013)
8. T.A. Aaltonen et al., Measurements of direct CP-violating asymmetries in charmless decays of bottom baryons. Phys. Rev. Lett. **113**(24), 242001 (2014)

Chapter 35
Precision Measurements of the CKM Parameters (Mainly γ/ϕ_3 Measurements)

Prasanth Krishnan

Abstract The CKM angle γ/ϕ_3 is the only one that is accessible with tree level decays in a theoretically clean way such that it provides a precision test of CP violation in the standard model. The Belle II experiment is a substantial upgrade of the Belle detector and will operate at the SuperKEKB asymmetric-energy e^+e^- collider. The accelerator has already successfully completed the first phase of commissioning in 2016 and first e^+e^- collisions in Belle II happened during April 2018. The design luminosity of SuperKEKB is 8×10^{35} cm^{-2}s^{-1} and the Belle II experiment aims to record 50 ab^{-1} of data, a factor of 50 more than its predecessor (Belle). The key method to measure ϕ_3 is through interference between the $B^- \to D^0 K^-$ and $B^- \to \bar{D}^0 K^-$ decays which occurs if the final state of the charm-meson decay is accessible to both the D^0 and $\bar{D}^0$ mesons. To achieve the best sensitivity, a large variety of D and B decay modes are required, which is possible at the Belle II experiment as almost any final state can be reconstructed including those with photons. With the ultimate Belle II data sample of 50 ab^{-1}, a determination of ϕ_3 with a precision of $1°$ or better is foreseen. We explain herein the details of the planned measurement at Belle II.

35.1 Introduction

At present, the current uncertainty on γ/ϕ_3 is approximately 5°, still worse by a factor 10 with respect to $\phi_1 = (21.9 \pm 0.7)°$ [1]. From CKMfitter [1], we find that the uncertainties on the CKM parameters measured from tree-level processes are larger than those from measurements of loop level diagrams. Thus, it is important to reduce the error on ϕ_3 to test the validity of the standard model (SM). One of the

On behalf of the Belle II Collaboration.

P. Krishnan (✉)
Tata Institute of Fundamental Research, Mumbai, India
e-mail: prasanth.k@tifr.res.in

A. Giri and R. Mohanta (eds.), *16th Conference on Flavor Physics and CP Violation*, Springer Proceedings in Physics 234,
https://doi.org/10.1007/978-3-030-29622-3_35

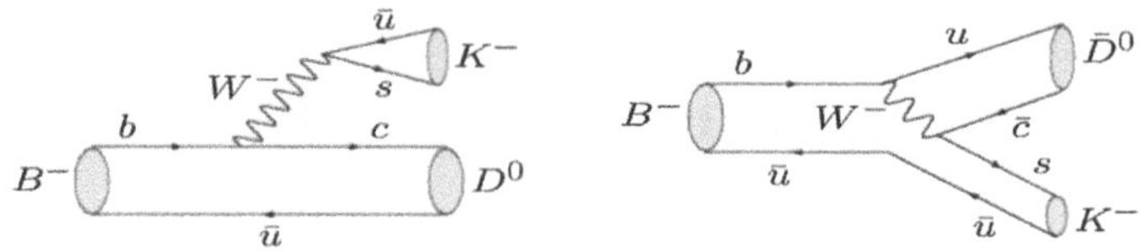

Fig. 35.1 Color-favored (left) and -suppressed (right) $B^- \to DK^-$ processes

reasons for this is the relative small branching fraction of the decays involved in the measurement owing to non-diagonal CKM matrix elements, since

$$\phi_3 \equiv \arg\left(-\frac{V_{ud}V_{ub}^*}{V_{cd}V_{cb}^*}\right), \tag{35.1}$$

where V_{ij} is the weak vertex factor for a quark transition $i \to j$. Thus, with more data, we can improve the dominant statistical uncertainty of the measurement.

The angle ϕ_3 can be extracted via interference between the color-favored $B^- \to D^0K^-$ and color-suppressed $B^- \to \bar{D}^0K^-$ decays as shown in Fig. 35.1. These are pure tree-level processes, hence theoretically clean. The correction to these processes is only $\mathcal{O}(10^{-7})$ [2]. If the amplitude for the color-favored decay is A_{fav}, then for the color-suppressed one (A_{sup}), it is $A_{fav}r_Be^{i(\delta_B-\phi_3)}$, where δ_B is the strong phase difference between the decay processes, ϕ_3 is the weak phase, and

$$r_B = \frac{|\, A_{sup} \,|}{|\, A_{fav} \,|}. \tag{35.2}$$

For $B \to DK$ decays, $r_B \sim 0.1$, whereas for $B \to D\pi$, $r_B \sim 0.05$. Though $B \to D\pi$ decays are not very sensitive to r_B and ϕ_3, we can use them as the control sample for $B \to DK$ to eliminate most of the systematic uncertainties.

The remainder of this document is structured as follows: Sect. 35.2 describes the methods and constraints to extract ϕ_3, Sect. 35.3 describes the potential ϕ_3 sensitivity from the Belle II experiment, as well as preliminary results from the Belle II phase 2 data. Section 35.4 gives the summary.

35.2 Methods for ϕ_3 Extraction

We classify the methods used to extract ϕ_3 according to the D meson final state:

(i) **GLW** [3] method: CP eigenstates such as K^+K^-, $\pi^+\pi^-$, $K_S^0\pi^0$,
(ii) **ADS** [4] method: doubly-Cabibbo-suppressed states such as K^+X^-, where X^- can be π^-, $\pi^-\pi^0$, $\pi^-\pi^-\pi^+$, and
(iii) **GGSZ** [5] method: self-conjugate multibody states such as $K_S^0\pi^+\pi^-$, $K_S^+K^+K^-$, $K_S^0\pi^+\pi^-\pi^0$.

In both GLW and ADS methods, ϕ_3-sensitive parameters can be extracted by taking a ratio between the suppressed and favored decay rates and a measurement of asymmetries between them. We obtain four GLW parameters $R^{\pm}_{CP} = 1 + r_B^2 \pm 2r_B\delta_B \cos\phi_3$ and $A^{\pm}_{\rm CP} = \pm 2r_B \sin\delta_B \sin\phi_3/R^{\pm}_{CP}$, and two ADS parameters $R_{\rm ADS} = r_B^2 + r_D^2 + 2r_B r_D \cos(\delta_B + \delta_D)\cos\phi_3$ and $A_{\rm ADS} = 2r_B r_D \sin(\delta_B\delta_D)\sin\phi_3 / R_{\rm ADS}$ for the ϕ_3 extraction. Here, r_D and δ_D are the ratio of the amplitudes of the suppressed and favored D decays, and the D strong phase, respectively. These are external inputs from charm measurements.

An inclusive approach leads to almost zero sensitivity for the GGSZ modes. Thus, we bin the Dalitz space into region with differing strong phases, which allows ϕ_3 to be determined from a single channel in a model-independent manner. This eliminates the model-dependent systematic uncertainty in the measurement. Fraction of D^0 and $\bar{D}^0$ events in bin i, called K_i and $\bar{K}_i$ can be obtained from $D^{*\pm} \to D\pi^{\pm}$ decays at B-factories, which reduces the statistical uncertainty thanks to their large data samples. But, we need the information of these strong phases, c_i and s_i, as external inputs from the charm factory experiments CLEO-c or BESIII, where the quantum-entangled $D^0\bar{D}^0$ pairs are produced via $e^+e^- \to \psi(3770) \to D^0\bar{D}^0$. Here, c_i and s_i correspond to the amplitude weighted average cosine and sine of the strong phase difference between D^0 and $\bar{D}^0$ decay in the ith bin. An optimal binning scheme is needed to obtain the maximal sensitivity. Figure 35.2 shows the Dalitz plot and the c_i and s_i values for the golden mode $B^{\pm} \to D(K^0_S\pi^+\pi^-)K^{\pm}$ [6].

The Belle combined measurement is $\phi_3 = (73^{+13}_{-15})^{\circ}$ [1], which is dominated by the GGSZ method. Similarly, the BaBar Collaboration combined all their measurements to give a value $\phi_3 = (69^{+17}_{-16})^{\circ}$ [7]. The LHCb Collaboration finds $\phi_3 = (74.0^{+5.0}_{-5.8})^{\circ}$ [8] by combining all their measurements. While B-factories used their full data sets, LHCb results are based on the Run I data. Combining these three results, we obtain $\phi_3 = (73.5^{+4.2}_{-5.1})^{\circ}$ [1], which is currently dominated by the results from LHCb.

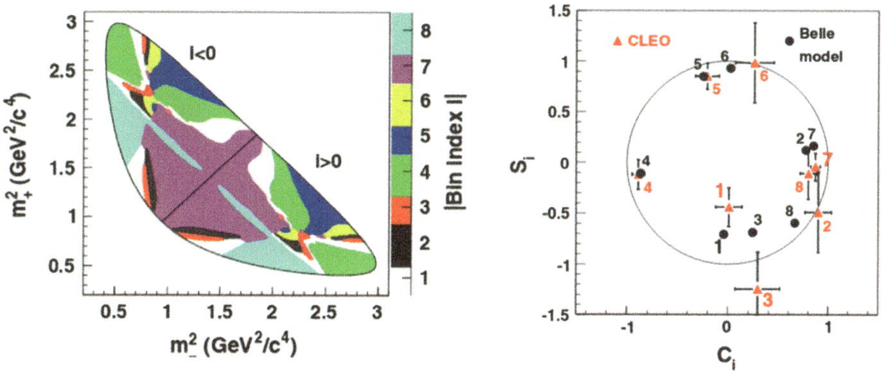

Fig. 35.2 Binned Dalitz plot and corresponding c_i, s_i values for $D \to K^0_S\pi^+\pi^-$ decays

35.3 Sensitivity from the Belle II Experiment

The Belle II experiment will start collecting data from early 2019 with all its sub-detectors. It will accumulate 50 ab^{-1} data, about 50 times that of its predecessor, with an instantaneous luminosity of 8×10^{35} $cm^{-2}s^{-1}$. Belle II will also have better K/π separation capability with the central drift chamber, imaging time-of-propagation and ring imaging Cherenkov counters, which work in different K/π momentum ranges. An improved K_S^0 reconstruction efficiency is expected, mainly due to the larger acceptance of the silicon vertex detector. All this will result in a substantially improved precision measurements. More details can be found in [9].

Currently, the ϕ_3 sensitivity is dominated by the statistical uncertainty from the number of reconstructed B decays. Thus, by going from a Belle integrated luminosity of 711 fb^{-1} to 50 ab^{-1}, the sensitivity is expected to get a significant boost. The major background is continuum events, coming from $e^+e^- \to q\bar{q}$ $(q = u, d, s, c)$ due to their large cross-section. As signal hides behind this large background, we need to eliminate the latter as much as we can to get a better sensitivity. The aim is to reach a precision of 1° with the full data sample, combining improvements obtained from K/π separation, K_S^0 reconstruction, and continuum suppression. This would allow us to probe for possible new physics contributions, that can potentially cause a shift in the value of ϕ_3 by $\pm 4°$ [10]. We perform a toy study with the golden mode $B^\pm \to D(K_S^0\pi^+\pi^-)K^\pm$, which results in the expected sensitivity as a function of integrated luminosity in Fig. 35.3. By adding more modes, such as $B \to D^*K$, $B^\pm \to D(K_S^0K^+K^-)K^\pm$, we can further improve the sensitivity and approach the 1° precision.

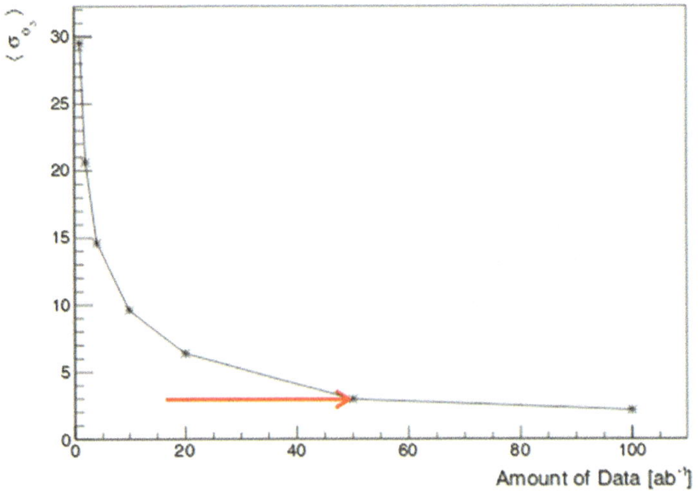

Fig. 35.3 ϕ_3 sensitivity with the amount of data collected at Belle II. The red arrow indicates the expected sensitivity from the 50 ab^{-1} sample

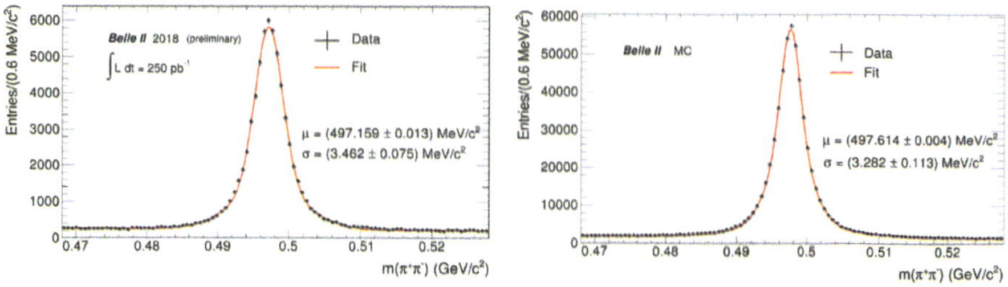

Fig. 35.4 Comparison of K^0_S reconstruction with the Belle II data and MC events

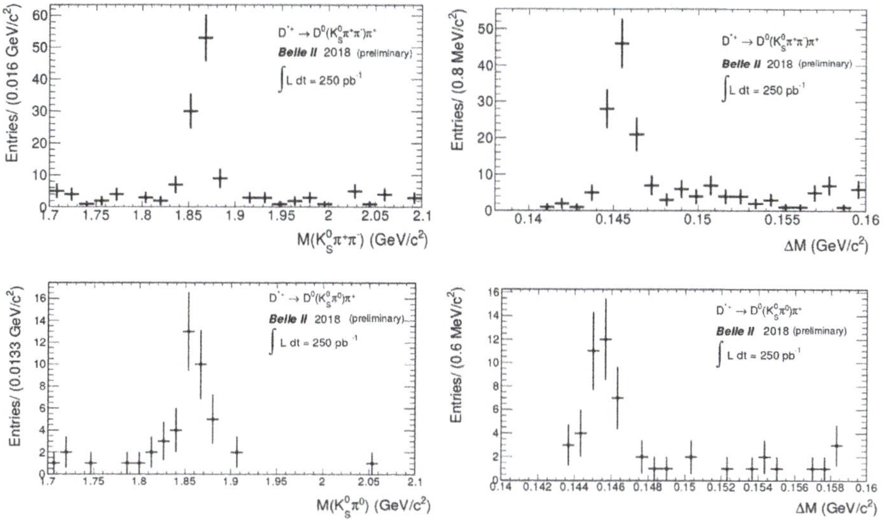

Fig. 35.5 M_i (left) and ΔM (right) distributions for the D^* tagged modes $D \to K^0_S\pi^+\pi^-$ (top) and $D \to K^0_S\pi^0$ (bottom)

Recently, Belle II successfully recorded data with all but for the vertex detector during April-July 2018. It has accumulated a data sample corresponding to an integrated luminosity of 472 pb^{-1}. Figure 35.4 shows a comparison between data and Monte Carlo (MC) simulations for the reconstruction of K^0_S candidates in Belle II. The invariant mass resolution is already showing a good agreement between data and MC events.

We perform the reconstruction of $D^{*\pm} \to D\pi^\pm$, $D \to K^0_S\pi^+\pi^-$ and the CP mode $D^0 \to K^0_S\pi^0$, which are shown in Fig. 35.5. The plotted variables are M_i ($i = K^0_S\pi^+\pi^-$or $K^0_S\pi^0$), where M_i is the invariant mass of the final state i, and ΔM, the reconstructed mass difference between $M_{(i)\pi^\pm}$ and M_i. The signal-to-background ratio is already good in both cases, and the reconstruction of $D \to K^0_S\pi^0$ shows the capability of Belle II for neutral particle reconstruction (Fig. 35.5).

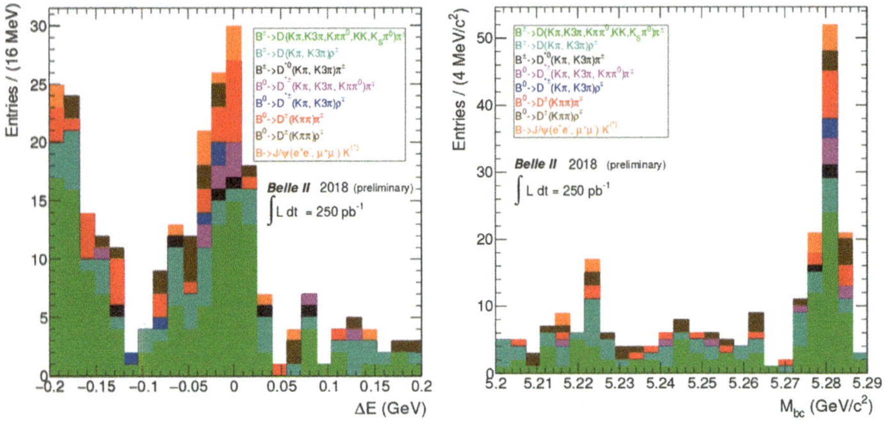

Fig. 35.6 ΔE (left) and M_{bc} (right) distributions for various B modes

We perform the "rediscovery" of the B meson from these data. We have accumulated about 100 B candidates in which the majority are from the $B^{\pm} \to D\pi^{\pm}$ mode, which is the control channel for the ϕ_3 extraction; corresponding distributions of fit variables are shown in Fig. 35.6. The variables are the energy difference $\Delta E = E_B^* - E_{\text{beam}}^*$ and the beam-energy constrained mass $M_{\text{bc}} = \sqrt{(E_{\text{beam}}^*/c^2)^2 - (p_B^*/c)^2}$, where E_B^* and p_B^* are the energy and momentum of the B meson candidate, and E_{beam}^* is the beam energy, all calculated in the center-of-mass frame.

35.4 Summary

At Belle and Belle II, the GGSZ method will have the largest impact on the ϕ_3 sensitivity. Simulation studies show that a precision of 3° is achievable in Belle II with the $D \to K_S^0\pi^+\pi^-$ mode alone, even without the full benefit particle identification, K_S^0 finding and continuum suppression. By combining it with all other modes and improved reconstruction algorithm, we can go closer to the goal of 1° precision. Preliminary results from the recently concluded run of Belle II without the vertex detector show promising results.

References

1. J. Charles et al., (CKMfitter Group). Eur. Phys. J. C **41**, 1–131 (2005), arXiv:0406184. updated results and plots http://ckmfitter.in2p3.fr
2. J. Brod, J. Zupan, The ultimate theoretical error on gamma from $B \to DK$ decays, arxiv:1308.5663

3. M. Gronau, D. London, How to determine all the angles of the unitarity triangle from $B_d^0 \rightarrow DK_S$ and $B_s^0 \rightarrow D\phi$. Phys. Lett. B **253**, 483 (1991); M. Gronau, D. Wyler, On determining a weak phase from charged B decay asymmetries. Phys. Lett. B **265**, 172 (1991)
4. D. Atwood, I. Dunietz, A. Soni, Enhanced CP violation with $B \rightarrow KD^0(\bar{D^0})$ modes and extraction of the Cabibbo-Kobayashi-Maskawa angle γ. Phys. Rev. Lett. **78**, 3257 (1997)
5. A. Giri, Y. Grossman, A. Soffer, J. Zupan, Determining γ using $B^{\pm} \rightarrow DK^{\pm}$ with multibody D decays. Phys. Rev. D **63**, 054018 (2003)
6. H. Aihara et al., (Belle Collaboration): First measurement of ϕ_3 with a model-independent Dalitz plot analysis of $B^{\pm} \rightarrow DK^{\pm}$, $D \rightarrow K_S^0\pi^+\pi^-$ decay. Phys. Rev. D **85**, 112014 (2012)
7. J.P. Lees et al, Observation of direct CP violation in the measurement of the Cabibbo-Kobayashi-Maskawa angle γ with $B^{\pm} \rightarrow D^{(*)}K^{(*)\pm}$ decays. Phys. Rev. D **87**, 052015 (2013)
8. The LHCb Collaboration, LHCb-CONF-2018-002
9. T. Abe et al., Belle II Technical Design Report, arXiv:1011.0352
10. J. Brod, A. Lenz, G. Tetlalmatzi-Xolocotzi, M. Wiebusch, New physics effects in tree-level decays, arXiv:1412.1446; E. Kou et al., The Belle II Physics Book, arXiv:1808.10567

Chapter 36
Input from the Charm Threshold for the Measurement of the CKM Angle γ

P. K. Resmi

Abstract A brief overview of the inputs from charm threshold that are essential to the determination of one of the Unitarity Triangle angles, γ is presented. The focus is on the measurements of four-body final states that have not previously been considered: $D^0 \to K_S^0\pi^+\pi^-\pi^0$ and $D^0 \to \pi^+\pi^-\pi^+\pi^-$.

36.1 Introduction

Among the three CKM [1, 2] angles, the uncertainty on γ is much worse than that on β. This is due to the small branching fraction of decays sensitive to γ. An improved measurement of γ is essential for testing the standard model description of *CP* violation. The decays $B^\pm \to DK^\pm$, where D indicates a neutral charm meson reconstructed in a final state common to both D^0 and $\bar{D}^0$, provide *CP*-violating observables and they can be used for measuring γ by analysing data collected at detectors such as BaBar, Belle, LHCb or the future Belle II experiment. These are tree-level decays and hence the theoretical uncertainty is $\mathcal{O}(10^{-7})$ [3].

There are different methods of measuring γ depending on the D final state. If both D^0 and $\bar{D}^0$ decay to a *CP* eigenstate such as $K_S^0\pi^0$ or K^+K^-, then the GLW formalism [4, 5] is used for the measurement. When the D meson decays into Cabibbo favoured and doubly Cabibbo suppressed final states like $K^\mp\pi^\pm$, the ADS method [6] is used for the extraction of γ. In these methods, asymmetry parameters and charge-averaged rates are measured from which γ is extracted. Multibody D decays like $K^\mp\pi^\pm\pi^0$ or $K^\mp\pi^\pm\pi^\pm\pi^\mp$ can be analysed using this method if coherence factor κ is known [7]. The GGSZ framework [8] is used when D decays to multibody self-conjugate final states like $K_S^0\pi^+\pi^-$, $K_S^0K^+K^-$, $K_S^0\pi^+\pi^-\pi^0$ or $\pi^+\pi^-\pi^+\pi^-$. The framework is implemented in a model-independent method via a binned Dalitz plot analysis of the D final state.

P. K. Resmi (✉)
Indian Institute of Technology Madras, Chennai, India
e-mail: resmipk@physics.iitm.ac.in

A. Giri and R. Mohanta (eds.), *16th Conference on Flavor Physics and CP Violation*, Springer Proceedings in Physics 234,
https://doi.org/10.1007/978-3-030-29622-3_36

36.2 Charm Inputs

The global average of γ is driven mostly by inputs from ADS and GGSZ measurements. The ADS method needs inputs from D decays: r_D and δ_D, the ratio of suppressed and favoured D decay amplitudes and D strong-phase respectively. For the ADS analysis of multibody D decays, the coherence factor κ is needed as input. It modulates the interference terms in the asymmetry parameters and varies between zero and one depending on whether there are many overlapping resonances contributing or a few isolated resonances. D decay inputs c_i and s_i, the amplitude weighted average of the cosine and sine of the strong phase difference between D^0 and $\bar{D}^0$ in different regions of phase space, are needed for a GGSZ style γ extraction. New multibody D modes can be explored in GLW framework, if CP content F_+ is known. The interference terms in the asymmetry parameters will be modulated by $(2F_+ - 1)$.

As the results are statistically limited, measuring these charm inputs from B data leads to further loss in precision. So it is essential to measure them at a charm factory like CLEO-c where quantum-correlated $D\bar{D}$ mesons are produced in e^+e^- collisions at an energy corresponding to $\psi(3770)$ resonance. The current CLEO-c inputs contributes 2° to the γ uncertainty [9]. In future, when B statistics is expected to increase, the inputs from BES III experiment will be imperative.

36.2.1 Quantum Correlated D Mesons at CLEO-c

The D meson pairs are produced coherently in a $C = -1$ state from $\psi(3770)$ decay. Thus the wave function becomes antisymmetric. The decay rate depends on the CP eigenvalue of each D final state. If they have opposite CP , then there will be two-fold enhancement in the yield and the yield will be zero when both are of same CP. It changes with them being quasi-CP states or self-conjugate states.

CLEO-c detector has a good 4π solid angle coverage and hence full reconstruction of a $D\bar{D}$ event is possible. High efficiencies for track and photon reconstruction are also a feature of CLEO-c.

36.3 Results

A number of D final states have been studied using a data sample corresponding to an integrated luminosity of 0.8 fb^{-1} collected by CLEO-c. The CP content, c_i and s_i values and coherence factor κ have been measured for various multibody D decay modes.

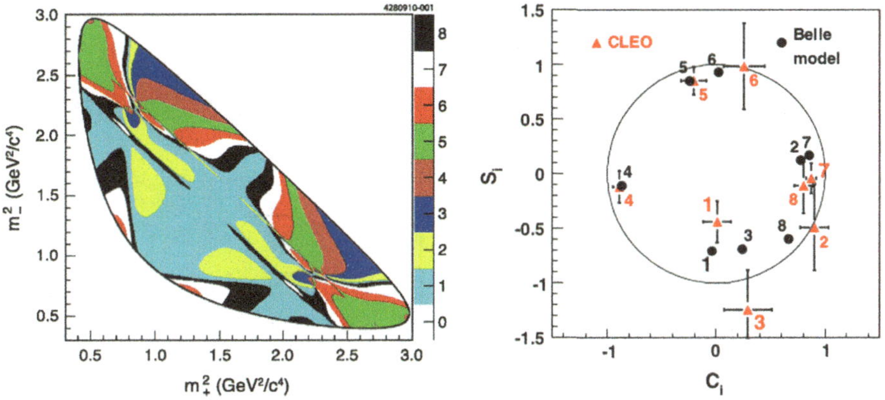

Fig. 36.1 Binned Dalitz plot for $D \to K_S^0\pi^+\pi^-$ (left) and measured c_i and s_i values (right). The circle encloses the region given by $c_i^2 + s_i^2 = 1$ and the bin number is specified for each point

36.3.1 $D \to K_S^0\pi^+\pi^-$

This is the golden mode to measure γ via the GGSZ formalism, especially at the B factories. The D Dalitz plot is binned and c_i and s_i values have been extracted in each of them as shown in Fig. 36.1 [10]. Optimal binning, in terms of sensitivity to γ, is done with guidance from an amplitude model [11]. Eight symmetric bins have been constructed with respect to $m_+^2 = m_-^2$ line, where $m_\pm$ corresponds to invariant mass of $K_S^0\pi^\pm$.

These inputs have been used by both Belle and LHCb for γ measurements [12, 13]. In the model independent approach, the uncertainty due to modelling is replaced by the statistical uncertainty from CLEO-c sample.

36.3.2 $D \to K_S^0\pi^+\pi^-\pi^0$

The decay $D \to K_S^0\pi^+\pi^-\pi^0$ has a relatively large branching fraction of 5.2% [14]. This has been analysed against several tag (other D) modes that are *CP* eigenstates, self-conjugate states etc. The yields with *CP*-odd and *CP*-even tags have been measured as N^+ and N^- as given in Fig. 36.2.

The *CP* content is defined as

$$F_+ = \frac{N^+}{N^+ + N^-} \qquad (36.1)$$

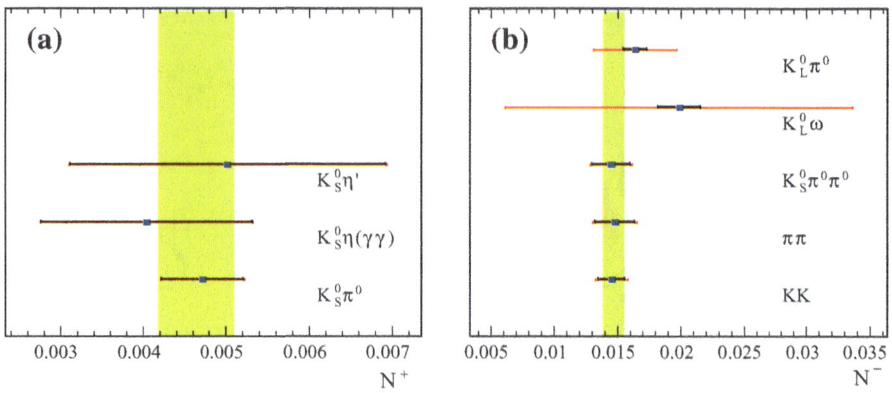

Fig. 36.2 N^+ (left) and N^- (right) for $D \to K^0_S\pi^+\pi^-\pi^0$ with CP-odd and even tags, respectively. Yellow region shows the average value

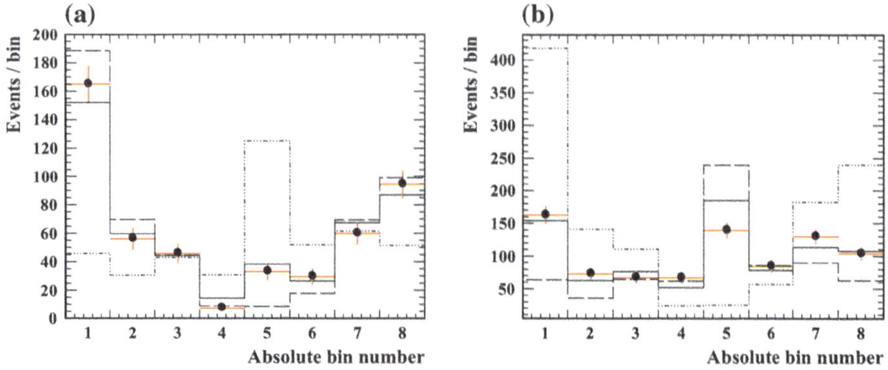

Fig. 36.3 Measured and expected yields for $D \to K^0_S\pi^+\pi^-\pi^0$ in $K^0_S\pi^+\pi^-$ (left) and $K^0_L\pi^+\pi^-$ (right) tag bins. The solid histogram shows the expected yield with best fit F_+ and the data points show the measured yields. The dashed and dotted lines represent the expected yields with $F_+ = 0$ and $F_+ = 1$, respectively

for CP eigenstate tags. $K^0_{S,L}\pi^+\pi^-$ modes are used as self-conjugate tags. The yield measured in bins of tag Dalitz space is proportional to $1 - (F_+^{\rm sig} - 1)(F_+^{\rm tag} - 1)$. The measured and expected yields in $K_{S,L}\pi^+\pi^-$ bins are given in Fig. 36.3.

From both the methods, the average $F_+ = 0.238 \pm 0.018$ [15], where the uncertainty includes both statistical and systematic contributions. This suggests that $K^0_S\pi^+\pi^-\pi^0$ is significantly CP-odd.

The multibody decay proceeds via some interesting resonance substructures. They are CP eigenstates like $K^0_S\omega$ (GLW like), Cabibbo favoured states like $K^{*-}\pi^+\pi^0$ (ADS like) etc. The strong-phase information can be extracted from the five dimensional phase space. In the absence of an amplitude model, the phase space is binned around these resonances. The results are given in Table 36.1 and Fig. 36.4 [15].

Table 36.1 c_i and s_i results in various bins for $K_S^0\pi^+\pi^-\pi^0$. Bins 1, 6 and 9 are *CP* self-conjugate, which implies $s_i = 0$

Bin	Resonance	c_i	s_i
1	ω	$-1.11 \pm 0.09^{+0.02}_{-0.01}$	0.00
2	$K^{*-}\rho^+$	$-0.30 \pm 0.05 \pm 0.01$	$-0.03 \pm 0.09^{+0.01}_{-0.02}$
3	$K^{*+}\rho^-$	$-0.41 \pm 0.07^{+0.02}_{-0.01}$	$0.04 \pm 0.12^{+0.01}_{-0.02}$
4	K^{*-}	$-0.79 \pm 0.09 \pm 0.05$	$-0.44 \pm 0.18 \pm 0.06$
5	K^{*+}	$-0.62 \pm 0.12^{+0.03}_{-0.02}$	$0.42 \pm 0.20 \pm 0.06$
6	K^{*0}	$-0.19 \pm 0.11 \pm 0.02$	0.00
7	ρ^+	$-0.82 \pm 0.11 \pm 0.03$	$-0.11 \pm 0.19^{+0.04}_{-0.03}$
8	ρ^-	$-0.63 \pm 0.18 \pm 0.03$	$0.23 \pm 0.41^{+0.04}_{-0.03}$
9	Remainder	$-0.69 \pm 0.15^{+0.15}_{-0.12}$	0.00

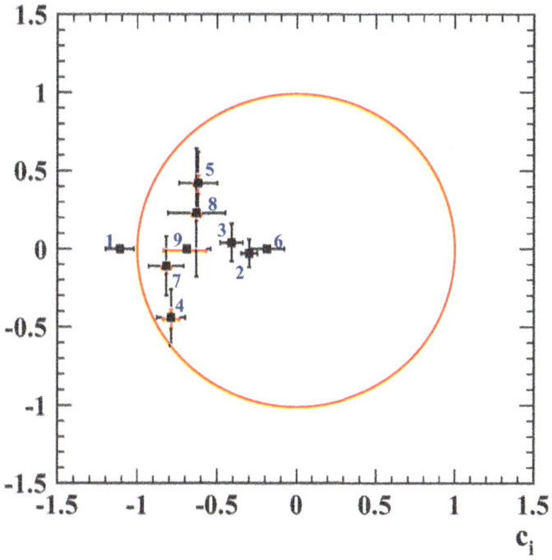

Fig. 36.4 c_i and s_i results in various bins for $K_S^0\pi^+\pi^-\pi^0$

The γ sensitivity with these results have been estimated for the expected full data set of 50 ab^{-1} at Belle II. With $B^\pm \to D(K_S^0\pi^+\pi^-\pi^0)K^\pm$ decays, it is possible to reach $\sigma_\gamma = 4.4°$ (see Fig. 36.5) [15]. Here, it is assumed that branching fraction × efficiency is similar to that of $K_S^0\pi^+\pi^-$. Improvements are possible with optimal binning using the knowledge of an amplitude model or finer binning from a large statistics sample at BES III.

36.3.3 $D \to \pi^+\pi^-\pi^+\pi^-$

The all charged final state of $D \to \pi^+\pi^-\pi^+\pi^-$ makes the detection easier at LHCb. The phase space is analysed to extract c_i and s_i results. Here, binning is done based on amplitude model [16]. The prominent contributions are $a_1(1260)^+$ and $\rho(770)^0$. The invariant mass projections are shown in Fig. 36.6.

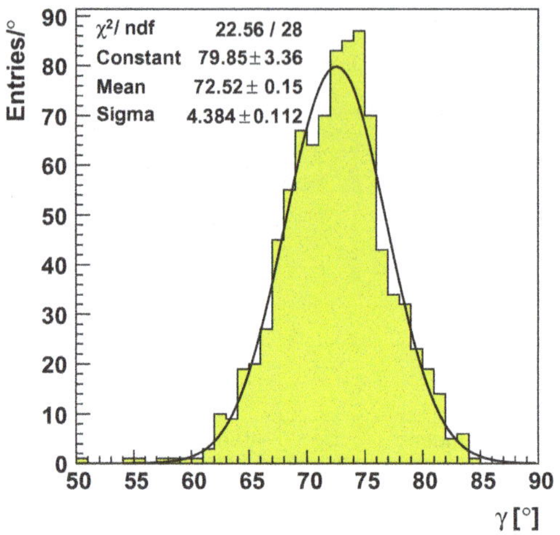

Fig. 36.5 γ sensitivity with 50 ab^{-1} dataset at Belle II

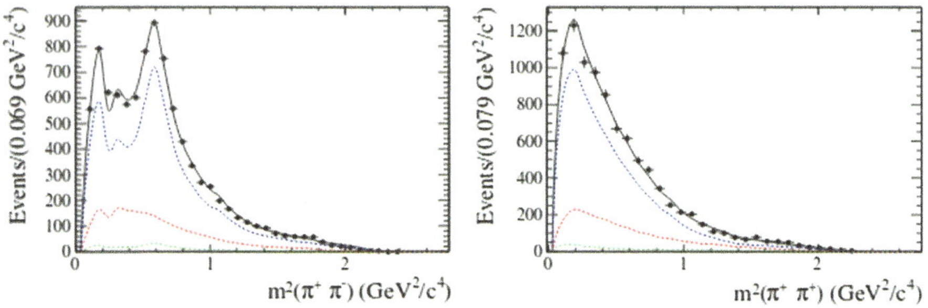

Fig. 36.6 Invariant mass projections for $\pi\pi$ combinations. Black points with error bars show data and black solid line shows the fit projection. The signal component is shown in blue, the background component in red and the wrongly tagged contribution in green dashed lines

The five dimensional phase space is binned with the variables m_+, m_-, $\cos\theta_+$, $\cos\theta_-$ and ϕ, where $m_+(m_-)$ is the invariant mass of $\pi^+\pi^+(\pi^-\pi^-)$ pair, $\theta_+(\theta_-)$ is the helicity angle of $\pi^+\pi^+(\pi^-\pi^-)$ pair and ϕ is the angle between $\pi^+\pi^+$ and $\pi^-\pi^-$ decay planes. The results with optimal binning are given in Fig. 36.7 [17]. These results will contribute to 5° uncertainty on γ with 50 fb^{-1} data after LHCb phase I upgrade along with 2° from B sample statistics. The sensitivity for different scenarios are presented in Fig. 36.8 [17].

The *CP* content F_+ has also been measured for this mode with *CP* eigenstates and $K^0_{\text{S,L}}\pi^+\pi^-$ modes as tags. The N^+ and N^- results are shown in Fig. 36.9. The average F_+ result obtained is 0.737 ± 0.028 [18]. Consistent results have been obtained using amplitude model as well as c_i and s_i results [16, 17].

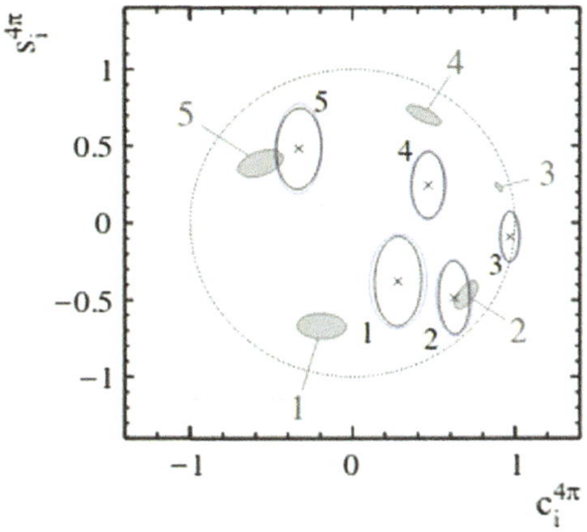

Fig. 36.7 c_i and s_i results for $D \to \pi^+\pi^-\pi^+\pi^-$. The grey regions shows the model predictions and black ellipses are measured values with statistical uncertainties

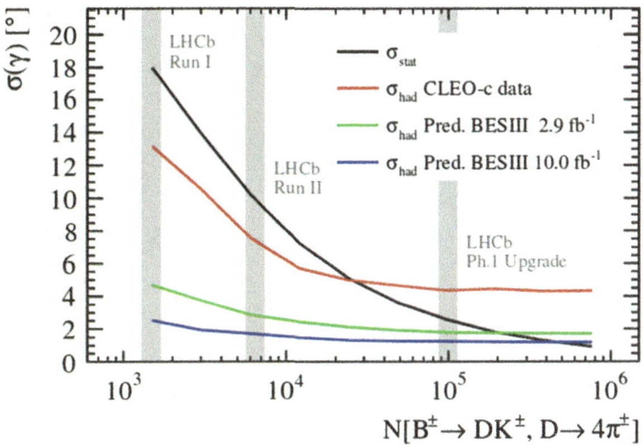

Fig. 36.8 Sensitivity predictions with c_i and s_i results for $D \to \pi^+\pi^-\pi^+\pi^-$

36.3.4 $D \to K^-\pi^+\pi^-\pi^+$

The decay $D \to K^-\pi^+\pi^-\pi^+$ can be analysed in ADS formalism to extract γ if coherence factor κ or $R_{K3\pi}$ is measured. It will then be treated like a two-body with single effective strong phase δ_D. The results obtained with CLEO-c data are shown in Fig. 36.10 [7].

It has been also shown that $D - \bar{D}$ mixing results can be used as input for γ measurements [19]. So charm mixing results for this mode at LHCb [20] are combined with these to obtain more precise values. There exists an amplitude model for $D \to K^-\pi^+\pi^-\pi^+$ decay [21] and this will allow for a more precise measurement of γ by binning the five dimensional phase space. There is also scope for improvement to have a much more precise measurement of the coherence factors in those bins with more data at BES III.

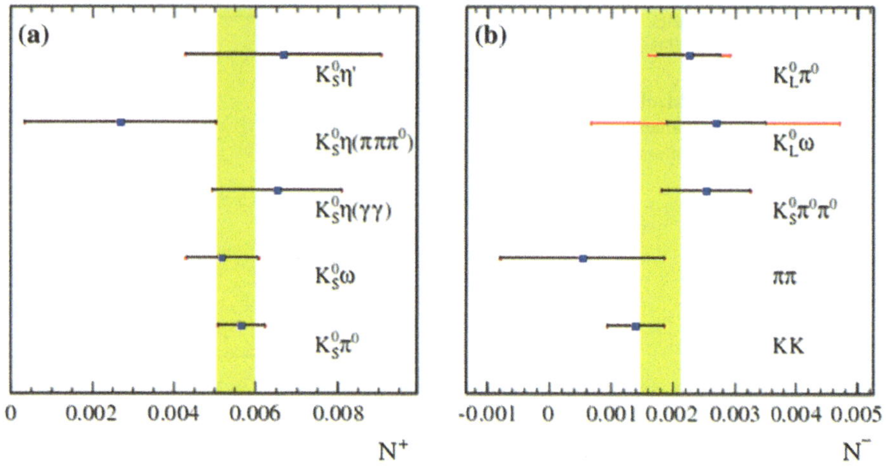

Fig. 36.9 N^+ (left) and N^- (right) for $D \to \pi^+\pi^-\pi^+\pi^-$ with *CP*-odd and even tags, respectively. Yellow region shows the average value

Fig. 36.10 Scans of $\Delta\chi^2$ for the fit to CLEO-c observables in $R_{K3\pi}$, $\delta_D^{K3\pi}$ parameter space

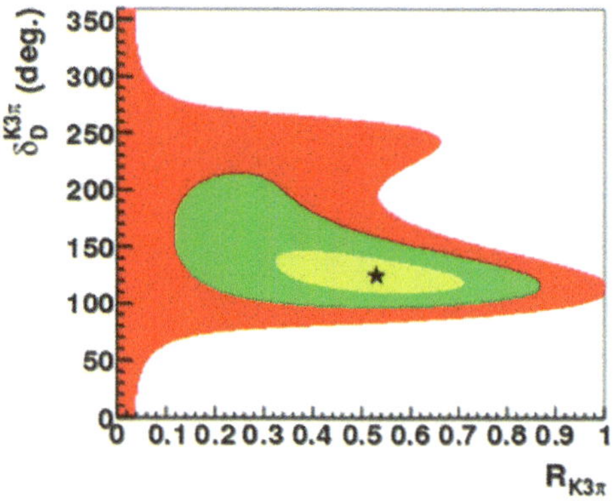

36.3.5 $D \to h^+h^-\pi^0$ ($h = \pi, K$)

The $D \to \pi^+\pi^-\pi^0$ Dalitz plot is symmetric and it suggests that it is isospin I = 0 state. Further, *G*-parity suggests that it is an almost pure *CP*-even eigenstate. So this could potentially be used for γ extraction in GLW method. It is analysed in CLEO-c data with *CP* eigenstates and $K^0_{S,L}\pi^+\pi^-$ modes as tags. The measured value of F_+ is 0.973 ± 0.017 [18, 22], which confirms that it is an almost pure *CP*-even eigenstate. Similar measurement has been done for $D \to K^+K^-\pi^0$ resulting in $F_+ = 0.732 \pm 0.055$ [18, 22].

The coherence factor has been measured for $D \to K^-\pi^+\pi^0$ decay mode to be $R_{K\pi\pi^0} = 0.82 \pm 0.06$ [7]. Since the value is close to 1, the dilution in the ADS observables due to strong phase from D decay multi-particle phase space is quite small.

36.4 Conclusions

Quantum correlated D decays at CLEO-c have been exploited to measure charm inputs that are used to determine γ from B meson decays. Inputs for GGSZ framework, c_i and s_i, have been measured for $K_S\pi^+\pi^-$, $K_S^0\pi^+\pi^-\pi^0$ and $\pi^+\pi^-\pi^+\pi^-$ final states. Also, the coherence factor is measured for $K^-\pi^+\pi^+\pi^-$, which can be used in an ADS formalism to extract γ. The *CP* content of decays like $\pi^+\pi^-\pi^0$, $K^+K^-\pi^0$ and $\pi^+\pi^-\pi^+\pi^-$ allow for new additions in GLW method. The precision on γ will reach $\mathcal{O}(1°)$ with LHCb upgrade and Belle II. Therefore, inputs from BES III are required to prevent these measurements being systematically limited by uncertainties in the strong-phase measurements of D decays.

Acknowledgements Acknowledge the erstwhile CLEO collaboration members for the privilege of using the data for the results presented.

References

1. N. Cabibbo, Unitary symmetry and leptonic decays. Phys. Rev. Lett. **10**, 531 (1963)
2. M. Kobayashi, T. Maskawa, CP violation in the renormalizable theory of weak interaction. Prog. Theor. Phys. **49**, 652 (1973)
3. J. Brod, J. Zupan, The ultimate theoretical error on gamma from $B \to DK$ decays. J. High Energ. Phys. **01**, 051 (2014)
4. M. Gronau, D. London, How to determine all the angles of the unitarity triangle from $B_d^0 \to DK_S$ and $B_s^0 \to D\phi$. Phys. Lett. B **253**, 483 (1991)
5. M. Gronau, D. Wyler, On determining a weak phase from charged B decay asymmetries. Phys. Lett. B **265**, 172 (1991)
6. D. Atwood, I. Dunietz, A. Soni, Enhanced *CP* violation with $B \to KD^0(\bar{D^0})$ modes and extraction of the Cabibbo-Kobayashi-Maskawa angle γ. Phys. Rev. Lett. **78**, 3257 (1997)
7. T. Evans et al., Improved determination of the $D \to K^-\pi^+\pi^+\pi^-$ coherence factor and associated hadronic parameters from a combination of $e^+e^- \to \psi(3770) \to c\bar{c}$ and $pp \to c\bar{c}X$ data. Phys. Lett. B **765**, 402 (2017)
8. A. Giri, Y. Grossman, A. Soffer, J. Zupan, Determining γ using $B^\pm \to DK^\pm$ with multibody D decays. Phys. Rev. D **63**, 054018 (2003)
9. LHCb-PUB-2016-025
10. J. Libby et al. (CLEO Collaboration), Model-independent determination of the strong-phase difference between D^0 and $\bar{D^0} \to K^0_{S,L}h^+h^-(h = \pi, K)$ and its impact on the measurement of the CKM angle γ/ϕ_3. Phys. Rev. D **82**, 112006 (2010)
11. B. Aubert et al. (BaBar Collaboration), Improved measurement of the CKM angle γ in $B^\mp \to D^{(*)}K^{(*\mp)}$ decays with a Dalitz plot analysis of D decays to $K_S^0\pi^+\pi^-$ and $K_SK^+K^-$. Phys. Rev. D **78**, 034023 (2008)
12. H. Aihara et al. (Belle Collaboration), First measurement of ϕ_3 with a model-independent Dalitz plot analysis of $B^+ \to DK^+$, $D \to K_S^0\pi^+\pi^-$ decay. Phys. Rev. D **85**, 112014 (2012)
13. R. Aaij et al. (LHCb Collaboration), Measurement of the CKM angle γ using $B^\pm \to DK^\pm$ with $D \to K_S^0\pi^+\pi^-$, $K_S^0K^+K^-$ decays. J. High Energy Phys. **10**, 097 (2014)
14. M. Tanabashi et al. (Particle Data Group Collaboration), Review of particle physics. Phys. Rev. D **98**, 030001 (2018)
15. P.K. Resmi et al., Quantum-correlated measurements of $D \to K_S^0\pi^+\pi^-\pi^0$ decays and consequences for the determination of the CKM angle γ. J. High Energy Phys. **01**, 82 (2018)

16. P. d'Argent et al., Amplitude analyses of $D^0 \to \pi^+\pi^-\pi^+\pi^-$ and $D^0 \to K^+K^-\pi^+\pi^-$ decays. J. High Energy Phys. **05**, 143 (2017)
17. S. Harnew et al., Model-independent determination of the strong phase difference between D^0 and $\bar{D^0} \to \pi^+\pi^-\pi^+\pi^-$ amplitudes. J. High Energy Phys. **01**, 144 (2018)
18. S. Malde et al., First determination of the CP content of $D \to \pi^+\pi^-\pi^+\pi^-$ and updated determination of the CP contents of $D \to \pi^+\pi^-\pi^0$ and $D \to K^+K^-\pi^0$. Phys. Lett. B **747**, 9 (2015)
19. S. Harnew, J. Rademacker, Charm mixing as input for model-independent determinations of the CKM phase γ. Phys. Lett. B **728**, 296 (2014)
20. R. Aaij et al. (LHCb Collaboration), First observation of $D^0 - \bar{D^0}$ oscillations in $D^0 \to K^+\pi^-\pi^+\pi^-$ decays and measurement of the associated coherence parameters. Phys. Rev. Lett. **116**, 24 (2016)
21. R. Aaij et al. (LHCb Collaboration), Studies of the resonance structure in $D^0 \to K^\mp\pi^\pm\pi^\pm\pi^\mp$ decays. Eur. Phys. J. C **78**, 443 (2018)
22. M. Nayak et al., First determination of the CP content of $D \to \pi^+\pi^-\pi^0$ and $D \to K^+K^-\pi^0$. Phys. Lett. B **740**, 1 (2015)

Chapter 37
Estimation of T and CPT Violation in Neutral B Meson Mixing from Indirect CP Asymmetry

Anirban Karan and Abinash Kumar Nayak

Abstract CPT invariance is believed to be a sacred symmetry of nature as well as quantum field theory (QFT). Therefore, it is very important to test the validity of this assumption experimentally. We have found that the time-dependent indirect CP asymmetry (A^f_{CP}) involving B decays to a CP eigenstate contains enough information to measure T and CPT violation in $B^0 - \bar{B}^0$ mixing, in addition to the standard CP -violating weak phases. The advantages of this method are following: (1) Entangled $B^0 - \bar{B}^0$ states are not required and hence this analysis can be carried out at LHCb, as well as at the B factories, (2) Penguin pollutions need not be neglected, (3) This approach can be used in both B^0_d and B^0_s systems.

37.1 Introduction

CPT invariance is assumed to be one of the sacred principles of nature. Our quantum field theory is CPT invariant by construction. According to CPT theorem, any interaction described by Lorentz invariant local gauge theory must be CPT invariant and hence CPT violation would have a profound impact on physics in general [1, 2]. So, much attention has been devoted to test the validity of CPT invariance experimentally.

One strong reason to believe in CPT invariance is the striking equality between masses or life times of any particle and its antiparticle [3]. But, as these quantities are mostly dominated by the strong or electromagnetic interactions, one can argue that there could be very small weak effects which are hard to measure in the differences of masses or lifetimes of any particle and its antiparticle. Nevertheless, one must agree that if some CPT violating effects are there in nature at all, they must be very small otherwise they would have been detected else where. In this regard, neutral

A. Karan (✉) · A. K. Nayak
The Institute of Mathematical Sciences, HBNI, Taramani, Chennai 600113, India
e-mail: kanirban@imsc.res.in

A. K. Nayak
e-mail: abinashkn@imsc.res.in

A. Giri and R. Mohanta (eds.), *16th Conference on Flavor Physics and CP Violation*, Springer Proceedings in Physics 234,
https://doi.org/10.1007/978-3-030-29622-3_37

pseudoscalar meson mixing is a promising area for testing CPT violation [4] as the mixing is a second-order electroweak process which are dominated by box diagrams. In addition, as the most general $B^0 - \bar{B}^0$ mixing matrix involves T and CP violation too, we must consider the effects of CP, T and CPT violation together.

The usual method of measuring CPT violation was developed in [5–12] where entangled $B^0\bar{B}^0$ states are produced from the decay of $\Upsilon(4S)$, with one meson decaying to a CP eigenstate ($J/\psi K_S$ or $J/\psi K_L$) and the other one used to tag the flavour. True T- and CPT-violating asymmetries are measured using this method. The BaBar Collaboration had implemented this strategy [13, 14], culminating in a measurement of T violation [15]. Though all the experimental results for CPT violating parameters, till date, are consistent with zero, an huge improvement in statistics is expected at LHCb and Belle II. However a different approach is needed for LHCb as entangled $B^0\bar{B}^0$ states cannot be produced there.

In this work, we show that the time-dependent, indirect CP asymmetry (A^f_{CP}) involving B decays to a CP eigenstate contains enough information to measure T and CPT violation in $B^0 - \bar{B}^0$ mixing as well as the standard CP-violating weak phases. However, this is an indirect determination of the T- and CPT-violating parameters as no true T- and CPT-violating asymmetries are measured.

One important point to mention is that our analysis is restricted to T and CPT violation arising from the $B^0 - \bar{B}^0$ mixing matrix alone and not from the decay. If there are new-physics contributions to B decays, we assume they are CPT-conserving. Similar approach with slight modification can be applied to B^0_s system also.

37.2 Theoretical Formalism

We write the flavour eigenstates as $|B^0\rangle = \begin{pmatrix}1\\0\end{pmatrix}$ and $|\bar{B}^0\rangle = \begin{pmatrix}0\\1\end{pmatrix}$. Then a general state can be written as $|\psi(t)\rangle = a(t)|B^0\rangle + b(t)|\bar{B}^0\rangle = \begin{pmatrix}a(t)\\b(t)\end{pmatrix}$. The time evolution of this state is given by Schrödinger equation: $i\hbar\,\frac{\partial}{\partial t}|\psi(t)\rangle = \mathscr{H}|\psi(t)\rangle$. After incorporating CP, T and CPT violation, the mixing matrix $\mathscr{H}$ becomes non-hermitian but can be expressed in terms of two 2×2 hermitian matrices $\boldsymbol{M}$ and $\boldsymbol{\Gamma}$, respectively the mass and decay matrices defined in the $(B^0, \bar{B}^0)$ basis, as $\mathscr{H} = \boldsymbol{M} - (i/2)\boldsymbol{\Gamma}$. The physical states $|B_L\rangle$ and $|B_H\rangle$ are the eigenstates of this mixing matrix $\mathscr{H}$. Now, any 2×2 matrix can be expressed in terms of the three Pauli matrices σ_i and the unit matrix with complex coefficients:

$$\boldsymbol{M} - \frac{i}{2}\boldsymbol{\Gamma} = E_1\sigma_1 + E_2\sigma_2 + E_3\sigma_3 - iD\mathbf{1}\,. \tag{37.1}$$

Comparing both sides of this equation, we obtain

$$E_1 = \text{Re}\, M_{12} - \frac{i}{2}\text{Re}\, \Gamma_{12}\,, \qquad E_3 = \frac{1}{2}(M_{11} - M_{22}) - \frac{i}{4}(\Gamma_{11} - \Gamma_{22})\,,$$
$$E_2 = -\text{Im}\, M_{12} + \frac{i}{2}\text{Im}\, \Gamma_{12}\,, \qquad D = \frac{i}{2}(M_{11} + M_{22}) + \frac{1}{4}(\Gamma_{11} + \Gamma_{22})\,. \quad (37.2)$$

Using spherical polar coordinate system, we can define complex numbers E, θ and ϕ as follows:

$$E = \sqrt{E_1^2 + E_2^2 + E_3^2},\ \ E_1 = E \sin\theta \cos\phi,\ \ E_2 = E \sin\theta \sin\phi,\ \ E_3 = E \cos\theta. \quad (37.3)$$

Now, the physical states can be written in terms of the flavour states as

$$|B_L\rangle = p_1|B^0\rangle + q_1|\bar{B}^0\rangle, \qquad |B_H\rangle = p_2|B^0\rangle - q_2|\bar{B}^0\rangle. \quad (37.4)$$

where $p_1 = \cos\frac{\theta}{2}$, $q_1 = e^{i\phi}\sin\frac{\theta}{2}$, $p_2 = \sin\frac{\theta}{2}$ and $q_2 = e^{i\phi}\cos\frac{\theta}{2}$. Defining $g_\pm = \frac{1}{2}(e^{-iH_L t} \pm e^{-iH_H t})$, the time evolution of the flavour eigenstates is given by

$$|B^0(t)\rangle = (g_+ + g_- \cos\theta)|B^0\rangle + e^{i\phi} g_- \sin\theta|\bar{B}^0\rangle\,, \quad (37.5)$$
$$|\bar{B}^0(t)\rangle = e^{-i\phi} g_- \sin\theta|B^0\rangle + (g_+ - g_- \cos\theta)|\bar{B}^0\rangle\,,$$

Now, we consider a final state f to which both B^0 and $\bar{B}^0$ can decay and express the time-dependent decay amplitudes as $\mathscr{A}(B^0(t) \to f) = \langle f|\mathscr{H}_{\Delta F=1}|B^0(t)\rangle$ and $\mathscr{A}(\bar{B}^0(t) \to f) = \langle f|\mathscr{H}_{\Delta F=1}|\bar{B}^0(t)\rangle$. We define $\mathscr{A}_f \equiv \langle f|\mathscr{H}_{\Delta F=1}|B^0\rangle$, $\bar{\mathscr{A}}_f \equiv \langle f|\mathscr{H}_{\Delta F=1}|\bar{B}^0\rangle$, $M \equiv (M_H + M_L)/2$, $\Delta M \equiv M_H - M_L$, $\Gamma \equiv (\Gamma_H + \Gamma_L)/2$ and $\Delta\Gamma \equiv \Gamma_H - \Gamma_L$. Then, we calculate the differential decay rates $d\Gamma_f/dt(B^0(t) \to f)$ and $d\Gamma_f/dt(\bar{B}^0(t) \to f)$, given in the Appendix section.

37.3 T and CPT Violation

The properties of $\boldsymbol{M}$ and $\boldsymbol{\Gamma}$ have been discussed in [16] (pp. 349–358) in light of CPT and T violation.

Theorem 1 *Independent of T symmetry, if CPT invariance holds, then*

$$M_{11} = M_{22}\ \text{and}\ \ \Gamma_{11} = \Gamma_{22} \implies E_3 = 0 \implies \theta = \frac{\pi}{2}\,. \quad (37.6)$$

Theorem 2 *If T invariance holds, then, independently of CPT symmetry,*

$$\frac{\Gamma_{12}^*}{\Gamma_{12}} = \frac{M_{12}^*}{M_{12}} \implies \text{Im}(E_1 E_2^*) = 0 \implies \text{Im}\,\phi = 0\,. \quad (37.7)$$

Theorem 3

$$\zeta \text{ is } \begin{cases} \textit{real} & \textit{if CPT holds,} \\ \textit{imaginary} & \textit{if T holds.} \end{cases} \tag{37.8}$$

$$\textit{where,} \quad \zeta = \langle B_L | B_H \rangle = \left[\cos\frac{\theta}{2} \sin\frac{\theta^*}{2} - e^{i(\phi-\phi^*)} \sin\frac{\theta}{2} \cos\frac{\theta^*}{2} \right] \tag{37.9}$$

The above theorems signify that $[\mathrm{Re}(\theta) - \pi/2]$ and $\mathrm{Im}(\theta)$ are CPT violating parameters whereas $\mathrm{Im}(\phi)$ is T violating parameter. Note that it is usually said that the absence of CP violation implies $|e^{i\phi}| = 1$ (i.e. $\mathrm{Im}(\phi) = 0$). However, strictly speaking, this is due to the absence of T violation. The two arguments are equivalent only if CPT is conserved. In the absence of both T and CPT violation in $B^0-\bar{B}^0$ mixing, the parameters θ and ϕ take the values $\theta = \frac{\pi}{2}$ and $\phi = -2\beta^{mix}$ where β^{mix} is the weak phase describing $B^0 - \bar{B}^0$ mixing. In the standard model, $\beta^{mix} = \beta$ for the B_d^0 meson. But in the presence of T and CPT violation, the parameters θ and ϕ will deviate from these values. Now, we introduce CPT violating parameters $\varepsilon_{1,2}$ as well as T violating parameter ε_3 and express the mixing parameters θ and ϕ in terms of them as

$$\theta = \frac{\pi}{2} + \varepsilon_1 + i\,\varepsilon_2\,, \quad \phi = -2\beta^{mix} + i\,\varepsilon_3\,. \tag{37.10}$$

The values for ε_1, ε_2 and ε_3 have been reported by the BaBar and Belle Collaborations [17, 18]. Their notation is related to ours as follows:

$$\cos\theta \leftrightarrow -z\,, \quad \sin\theta \leftrightarrow \sqrt{1-z^2}\,, \quad e^{i\phi} \leftrightarrow \frac{q}{p}\,, \tag{37.11}$$

$$\Longrightarrow \quad \varepsilon_1 = \mathrm{Re}(z)\,, \quad \varepsilon_2 = \mathrm{Im}(z)\,, \quad \varepsilon_3 = 1 - \left|\frac{q}{p}\right|. \tag{37.12}$$

ε_1 and ε_2 are expected to be very small, as they are CPT-violating parameters. As for ε_3, note that $|q/p|$ has been measured at the $\Upsilon(4S)$ using the same-sign dilepton asymmetry, assuming CPT conservation [19]:

$$\left|\frac{q}{p}\right| = 1.0010 \pm 0.0008 \quad \Longrightarrow \quad \varepsilon_3 = -(1.0 \pm 0.8) \times 10^{-3}\,. \tag{37.13}$$

Thus, ε_3 is also very small. The value of $y_d = \Delta\Gamma_d/2\Gamma_d$ has been measured to be small: $y_d = -0.003 \pm 0.015$ with the B_d^0 lifetime of 1.520 ± 0.004 ps [20]. This means that we can approximate $\sinh(\Delta\Gamma t/2) \simeq \Delta\Gamma t/2 = y_d\Gamma_d t$ and $\cosh(\Delta\Gamma t/2) \simeq 1$. In principle, for large enough times, this approximation will break down. However, even at time scales of $\mathcal{O}(10)$ ps, the approximation holds to $\sim 10^{-4}$ and by this time most of the B_d^0s will have decayed.

37.4 Time-Dependent Indirect *CP* Asymmetry

Time-dependent indirect CP asymmetry $A^f_{CP}(t)$ involving B-meson decays to CP eigenstate is defined as

$$A^f_{CP}(t) = \frac{d\Gamma/dt(\bar{B}^0_d(t) \to f_{CP}) - d\Gamma/dt(B^0_d(t) \to f_{CP})}{d\Gamma/dt(\bar{B}^0_d(t) \to f_{CP}) + d\Gamma/dt(B^0_d(t) \to f_{CP})} . \tag{37.14}$$

In the limit of CPT and T conservation in the mixing, and $\Delta\Gamma = 0$, one has the familiar expression

$$A^f_{CP}(t) = S\sin(\Delta M_d t) - C\cos(\Delta M_d t), \tag{37.15}$$

$$\text{where } \varphi \equiv -2\beta^{mix} - \arg[\mathscr{A}_f] + \arg[\overline{\mathscr{A}_f}],\ C \equiv \frac{|\mathscr{A}_f|^2 - |\overline{\mathscr{A}_f}|^2}{|\mathscr{A}_f|^2 + |\overline{\mathscr{A}_f}|^2},\ S \equiv \sqrt{1 - C^2}\ \sin\varphi \tag{37.16}$$

Here, C is the direct CP asymmetry and φ is the measured weak phase. If there is no CPT conserving New Physics or penguin pollution, then φ cleanly measures a weak phase and $C = 0$.

However, in the presence of T and CPT violation in the mixing (37.15), does not hold. After expanding $A^f_{CP}(t)$, keeping only the linear terms in the small quantities $\varepsilon_{1,2,3}$ and y_d, we obtain,

$$\begin{aligned} A^f_{CP/CPT}(t) \simeq\ & c_0 + c_1\cos(\Delta M_d t) + c_2\cos(2\Delta M_d t) + s_1\sin(\Delta M_d t) + s_2\sin(2\Delta M_d t) \\ & + c'_1\,\Gamma_d\, t\cos(\Delta M_d t) + s'_1\,\Gamma_d\, t\sin(\Delta M_d t)\,, \end{aligned} \tag{37.17}$$

where the coefficients are given by

$$\begin{aligned} c_0 &= \varepsilon_1\cos\varphi + \varepsilon_3 - \frac{1}{2}\varepsilon_3\sin^2\varphi\,, \quad s_1 = \sqrt{1 - C^2}\sin\varphi - \varepsilon_2\cos^2\varphi - \varepsilon_3 C\sin\varphi\,, \\ c_1 &= -C - \varepsilon_3 - \varepsilon_1\cos\varphi - \varepsilon_2 C\sin\varphi\,, \quad s_2 = -\frac{1}{2}\varepsilon_2\sin^2\varphi + \varepsilon_3 C\sin\varphi\,, \\ c_2 &= \frac{1}{2}\varepsilon_3\sin^2\varphi + \varepsilon_2 C\sin\varphi\,, \quad c'_1 = C\, y_d\, \cos\varphi\,, \quad s'_1 = -\frac{1}{2}\, y_d\, \sin 2\varphi\,. \end{aligned} \tag{37.18}$$

The seven pieces have different time dependences so that, by fitting $A^f_{CP/CPT}(t)$ to the seven time-dependent functions, all coefficients can be extracted.

The five observables c_0, c_1, c_2, s_1 and s_2 can be used to solve for the five unknown parameters C, φ and $\varepsilon_{1,2,3}$. The parameter C is simply given by

$$C = -(c_0 + c_1 + c_2)\,. \tag{37.19}$$

The solution for $\sin\varphi$ is obtained by solving the following quartic equation:

$$\sin^4\varphi - 2\left[\frac{s_1 + 2s_2}{2 - C^2}\right]\sin^3\varphi + 4C\left[C + \frac{c_2}{2 - C^2}\right]\sin^2\varphi \\ - 4\left[\frac{2C^2(s_1 + s_2) - s_2}{2 - C^2}\right]\sin\varphi - \left[\frac{8C\,c_2}{2 - C^2}\right] = 0\,. \quad (37.20)$$

Of course, there are four solutions, but, since the ε_i are small, the correct solution is the one that is roughly $s_1/\sqrt{1 - C^2}$. Finally, $\varepsilon_1, \varepsilon_2, \varepsilon_3$ are given by

$$\varepsilon_1 = c_0\sec\varphi - \frac{(2 - \sin^2\varphi)(c_2\sin\varphi + 2C\,s_2)}{(4C^2 + \sin^2\varphi)\,\sin\varphi\,\cos\varphi}\,,$$
$$\varepsilon_2 = \frac{2\,(2C\,c_2 - s_2\,\sin\varphi)}{(4C^2 + \sin^2\varphi)\,\sin\varphi}\,, \quad \varepsilon_3 = \frac{2\,(c_2\,\sin\varphi + 2C\,s_2)}{(4C^2 + \sin^2\varphi)\,\sin\varphi}\,. \quad (37.21)$$

Hence, it is possible to measure the parameters describing T and CPT violation in B_d^0–$\bar{B}_d^0$ mixing using the time-dependent indirect CP asymmetry, and this can be carried out at LHCb as well as at B factories. Knowing φ, the value of y_d can be found from measurements of c_1' and s_1'. Note that, even if the width difference $\Delta\Gamma_d$ between the two B-meson eigenstates vanishes, the T-violating parameter ε_3 can still be extracted, which is contrary to the claim of [7, 10].

In the case of B_s^0 mesons, $\Delta\Gamma_s$ is not that small, so the functions $\sinh(\Delta\Gamma_s t/2)$ and $\cosh(\Delta\Gamma_s t/2)$ must be kept throughout or one should truncate the series of $\sinh(\Delta\Gamma_s t/2)$ and $\cosh(\Delta\Gamma_s t/2)$ accordingly. This modifies the forms of 37.17, but the idea does not change.

37.5 *CPT* Conserving Scenario

Currently, we know that $\varepsilon_3 = -(1.0 \pm 0.8) \times 10^{-3}$ (37.13). Now, assuming no CPT violation (i.e., $\varepsilon_1 = \varepsilon_2 = 0$), the coefficients c_0, c_2 and s_2 can be expressed in terms of the measured quantities c_1, s_1 and ε_3 as follows:

$$c_0 = \varepsilon_3\left[1 - \frac{2s_1^2}{(2 - c_1^2 + \varepsilon_3^2)^2}\right], \quad c_2 = \frac{2s_1^2\,\varepsilon_3}{(2 - c_1^2 + \varepsilon_3^2)^2}\,, \quad s_2 = -\frac{2s_1\,(c_1 + \varepsilon_3)\,\varepsilon_3}{(2 - c_1^2 + \varepsilon_3^2)}\,. \quad (37.22)$$

The values of c_1 and s_1 have been measured for several B_d^0 decays to CP eigenstates [19], and the value of ε_3 is independent of the decay mode. Using these values, we can estimate c_0, c_2 and s_2 from (37.22), which assumes that CPT is conserved. As an example, for the final state $J/\psi K_S$, we find

$$c_0 = (-15.18 \pm 15.50) \times 10^{-4}\,,\; c_2 = (-4.31 \pm 4.41) \times 10^{-4}\,,\\ s_2 = (0.29 \pm 0.43) \times 10^{-4}\,. \tag{37.23}$$

Should the measurements of c_0, c_2 and s_2 deviate significantly from the above values, this would indicate the presence of CPT violation in B_d^0–$\bar{B}_d^0$ mixing.

37.6 Conclusion

In conclusion, we have shown that the time-dependent indirect CP asymmetries involving B^0, $\bar{B}^0 \to f_{CP}$ contain enough information to extract not only the CP-violating weak phases, but also the parameters describing T and CPT violation in B^0-$\bar{B}^0$ mixing. These measurements can be made at the $\Upsilon(4S)$ (e.g., BaBar, Belle) or at high energies (e.g., LHCb). Penguin pollutions in the decay need not be neglected and the method can be applied to B_d^0- as well as B_s^0-meson decays.

Acknowledgements The work reported was done in collaboration with Rahul Sinha and David London. We thank them for the collaboration.

Appendix

The full expressions for differential decay rates are given by

$$\begin{aligned}\frac{d\Gamma}{dt}(B^0(t) \to f) = \frac{1}{2}e^{-\Gamma t}\Big[&\sinh(\Delta\Gamma t/2)\Big\{2\mathrm{Re}\Big(\cos\theta|\mathscr{A}_\mathrm{f}|^2 + \mathrm{e}^{\mathrm{i}\phi}\sin\theta\mathscr{A}_\mathrm{f}^*\bar{\mathscr{A}_\mathrm{f}}\Big)\Big\}\\ &+ \cosh(\Delta\Gamma t/2)\Big\{|\mathscr{A}_f|^2 + |\cos\theta|^2|\mathscr{A}_f|^2 + |e^{i\phi}\sin\theta|^2|\bar{\mathscr{A}_f}|^2\\ &+ 2\mathrm{Re}\Big(\mathrm{e}^{\mathrm{i}\phi}\cos\theta^*\sin\theta\mathscr{A}_\mathrm{f}^*\bar{\mathscr{A}_\mathrm{f}}\Big)\Big\} + \cos(\mathrm{Mt})\Big\{|\mathscr{A}_\mathrm{f}|^2 - |\cos\theta|^2|\mathscr{A}_\mathrm{f}|^2\\ &- |e^{i\phi}\sin\theta|^2|\bar{\mathscr{A}_f}|^2 - 2\mathrm{Re}\Big(\mathrm{e}^{\mathrm{i}\phi}\cos\theta^*\sin\theta\mathscr{A}_\mathrm{f}^*\bar{\mathscr{A}_\mathrm{f}}\Big)\Big\}\\ &- \sin(\Delta M t)\Big\{2\mathrm{Im}\Big(\cos\theta|\mathscr{A}_\mathrm{f}|^2 + \mathrm{e}^{\mathrm{i}\phi}\sin\theta\mathscr{A}_\mathrm{f}^*\bar{\mathscr{A}_\mathrm{f}}\Big)\Big\}\Big]\end{aligned} \tag{37.24}$$

$$\begin{aligned}\frac{d\Gamma}{dt}(\bar{B}^0(t) \to f) = \frac{1}{2}e^{-\Gamma t}\Big[&\sinh(\Delta\Gamma t/2)\Big\{2\mathrm{Re}\Big(-\cos\theta^*|\bar{\mathscr{A}_\mathrm{f}}|^2 + \mathrm{e}^{\mathrm{i}\phi^*}\sin\theta^*\mathscr{A}_\mathrm{f}^*\bar{\mathscr{A}_\mathrm{f}}\Big)\Big\}\\ &+ \cosh(\Delta\Gamma t/2)\Big\{|\overline{\mathscr{A}_f}|^2 + |\cos\theta|^2|\overline{\mathscr{A}_f}|^2 + |e^{-i\phi}\sin\theta|^2|\mathscr{A}_f|^2\\ &- 2\mathrm{Re}\Big(\mathrm{e}^{\mathrm{i}\phi^*}\cos\theta\sin\theta^*\mathscr{A}_\mathrm{f}^*\bar{\mathscr{A}_\mathrm{f}}\Big)\Big\} + \cos(\mathrm{Mt})\Big\{|\overline{\mathscr{A}_\mathrm{f}}|^2 - |\cos\theta|^2|\overline{\mathscr{A}_\mathrm{f}}|^2\\ &- |e^{-i\phi}\sin\theta|^2|\mathscr{A}_f|^2 + 2\mathrm{Re}\Big(\mathrm{e}^{\mathrm{i}\phi^*}\cos\theta\sin\theta^*\mathscr{A}_\mathrm{f}^*\mathscr{A}_\mathrm{f}\Big)\Big\}\\ &+ \sin(\Delta M t)\Big\{2\mathrm{Im}\Big(-\cos\theta^*|\bar{\mathscr{A}_\mathrm{f}}|^2 + \mathrm{e}^{\mathrm{i}\phi^*}\sin\theta^*\mathscr{A}_\mathrm{f}^*\bar{\mathscr{A}_\mathrm{f}}\Big)\Big\}\Big]\,.\end{aligned} \tag{37.25}$$

References

1. O.W. Greenberg, *CPT* violation implies violation of Lorentz invariance. Phys. Rev. Lett. **89**, 231602 (2002). https://doi.org/10.1103/PhysRevLett.89.231602 [hep-ph/0201258]
2. V.A. Kostelecky, Gravity, Lorentz violation, and the standard model. Phys. Rev. D **69**, 105009 (2004). https://doi.org/10.1103/PhysRevD.69.105009 [hep-th/0312310]
3. A measurement of *CPT* violation using mass or lifetime differences can be found in R. Aaij *et al.* [LHCb Collaboration], Search for violations of Lorentz invariance and *CPT* symmetry in $B^0_{(s)}$ mixing. Phys. Rev. Lett. **116**(24), 241601 (2016). https://doi.org/10.1103/PhysRevLett.116.241601, arXiv:1603.04804 [hep-ex]
4. L. Lavoura, J.P. Silva, Disentangling violations of *CPT* from other new physics effects. Phys. Rev. D **60**, 056003 (1999). https://doi.org/10.1103/PhysRevD.60.056003 [hep-ph/9902348]
5. M.C. Bañuls, J. Bernabéu, *CP*, *T* and *CPT* versus temporal asymmetries for entangled states of the B_d system. Phys. Lett. B **464**, 117 (1999). https://doi.org/10.1016/S0370-2693(99)01043-6 [hep-ph/9908353]
6. M. C. Bañuls and J. Bernabéu, "Studying indirect violation of *CP*, *T* and *CPT* in a *B* factory," Nucl. Phys. B **590**, 19 (2000) https://doi.org/10.1016/S0550-3213(00)00548-4 [hep-ph/0005323]
7. E. Alvarez, J. Bernabéu, Correlated neutral *B* meson decays into *CP* eigenstates. Phys. Lett. B **579**, 79 (2004). https://doi.org/10.1016/j.physletb.2003.10.114 [hep-ph/0307093]
8. E. Alvarez, J. Bernabéu, N. E. Mavromatos, M. Nebot, J. Papavassiliou, *CPT* violation in entangled B^0-$\bar{B}^0$ states and the demise of flavor tagging. Phys. Lett. B **607**, 197 (2005). https://doi.org/10.1016/j.physletb.2004.12.032 [hep-ph/0410409]
9. E. Alvarez, A. Szynkman, Direct test of time reversal invariance violation in *B* mesons. Mod. Phys. Lett. A **23**, 2085 (2008). https://doi.org/10.1142/S021773230802728X [hep-ph/0611370]
10. J. Bernabéu, F. Martinez-Vidal, P. Villanueva-Perez, Time reversal violation from the entangled B^0-$\bar{B}^0$ system. JHEP **1208**, 064 (2012). https://doi.org/10.1007/JHEP08(2012)064, arXiv:1203.0171 [hep-ph]
11. E. Applebaum, A. Efrati, Y. Grossman, Y. Nir, Y. Soreq, Subtleties in the *BABAR* measurement of time-reversal violation. Phys. Rev. D **89**(7), 076011 (2014). https://doi.org/10.1103/PhysRevD.89.076011, arXiv:1312.4164 [hep-ph]
12. J. Bernabéu, F.J. Botella, M. Nebot, Genuine *T*, *CP*, *CPT* asymmetry parameters for the entangled B_d system. JHEP **1606**, 100 (2016). https://doi.org/10.1007/JHEP06(2016)100, arXiv:1605.03925 [hep-ph]
13. B. Aubert et al., [BaBar Collaboration], Limits on the decay-rate difference of neutral *B* mesons and on *CP*, *T*, and *CPT* violation in $B^0\bar{B}^0$ oscillations. Phys. Rev. Lett. **92**, 181801 (2004). https://doi.org/10.1103/PhysRevLett.92.181801 [hep-ex/0311037]
14. B. Aubert et al., [BaBar Collaboration], Limits on the decay rate difference of neutral−*B* mesons and on *CP*, *T*, and *CPT* violation in $B^0\bar{B}^0$ oscillations. Phys. Rev. D **70**, 012007 (2004). https://doi.org/10.1103/PhysRevD.70.012007 [hep-ex/0403002]
15. J.P. Lees et al. [BaBar Collaboration], Observation of time reversal violation in the B^0 meson system. Phys. Rev. Lett. **109**, 211801 (2012). https://doi.org/10.1103/PhysRevLett.109.211801, arXiv:1207.5832 [hep-ex]
16. T.D. Lee, Particle physics and introduction to field theory. Contemp. Concepts Phys. **1**, 1 (1981)
17. T. Higuchi et al., Search for time-dependent *CPT* violation in hadronic and semileptonic *B* decays. Phys. Rev. D **85**, 071105 (2012). https://doi.org/10.1103/PhysRevD.85.071105, arXiv:1203.0930 [hep-ex]

18. J.P. Lees et al. [BaBar Collaboration], Study of CP asymmetry in B^0-$\bar{B}^0$ mixing with inclusive dilepton events. Phys. Rev. Lett. **114**(8), 081801 (2015). https://doi.org/10.1103/PhysRevLett.114.081801, arXiv:1411.1842 [hep-ex]
19. Y. Amhis et al. [Heavy Flavor Averaging Group (HFAG) Collaboration], Averages of b-hadron, c-hadron, and τ-lepton properties as of summer 2016. arXiv:1612.07233, http://www.slac.stanford.edu/xorg/hfag [hep-ex]
20. C. Patrignani et al. [Particle Data Group], Review of particle physics. Chin. Phys. C **40**(10), 100001 (2016). https://doi.org/10.1088/1674-1137/40/10/100001

Chapter 38
Measurement of the CKM Angle ϕ_3 Using $B \to DK$ at Belle II

M. Kumar, K. Lalwani, K. Trabelsi and K. Prasanth

Abstract We present the preliminary Monte Carlo (MC) study of the $B^{\pm} \to D^0 K^{\pm}$, $D^0 \to K_S^0 \pi^+ \pi^-$ decays to extract ϕ_3 at the Belle II along. We discuss here the improvement in ϕ_3 measurement one may expect at Belle II with 50 ab^{-1}. We also present preliminary reconstruction of the K_S^0 and $D^0 \to K_S^0 \pi^+ \pi^-$ using the Phase II data and compared with MC.

38.1 Introduction

The CKM angle ϕ_3 is one of the least well constrained parameters of the Unitarity Triangle [1, 2]. The precise measurement of ϕ_3 is highly desirable to scrutinise the consistency of the Standard Model and to detect presence of new physics. The measurement that currently dominates sensitivity to ϕ_3 uses $B^{\pm} \to D^0 K^{\pm}$ decays with the neutral D mesons decaying to different final states such as $K^+ K^-$, $K^+ \pi^-$, $K_S^0 \pi^+ \pi^-$ etc.

As the sensitivity of ϕ_3 comes from the interference of $b \to c\bar{u}s$ and $b \to u\bar{c}s$, therefore, measurement of ϕ_3 is performed by exploiting the difference between $K_S^0 \pi^+ \pi^-$ Dalitz plots for D mesons from B^+ and B^- decay. The measurement of ϕ_3 from $B^{\pm} \to D^0 K^{\pm}$ and $B^{\pm} \to \overline{D}^0 K^{\pm}$ decays is theoretically clean as they occur at the tree level as shown in Fig. 38.1. Various methods [4, 5] for extracting ϕ_3 have been proposed, from which the Dalitz plot analysis method [6] is one of the novel methods. There have been many efforts by BaBar, Belle and LHCb collaborations

On behalf of the Belle II Collaboration.

M. Kumar (✉) · K. Lalwani
Malaviya National Institute of Technology, Jaipur, India
e-mail: 2016rpy9052@mnit.ac.in

K. Trabelsi
High Energy Accelerator Research Organization (KEK), Tsukuba, Japan

K. Prasanth
Tata Institute of Fundamental Research, Mumbai, India

A. Giri and R. Mohanta (eds.), *16th Conference on Flavor Physics and CP Violation*, Springer Proceedings in Physics 234,
https://doi.org/10.1007/978-3-030-29622-3_38

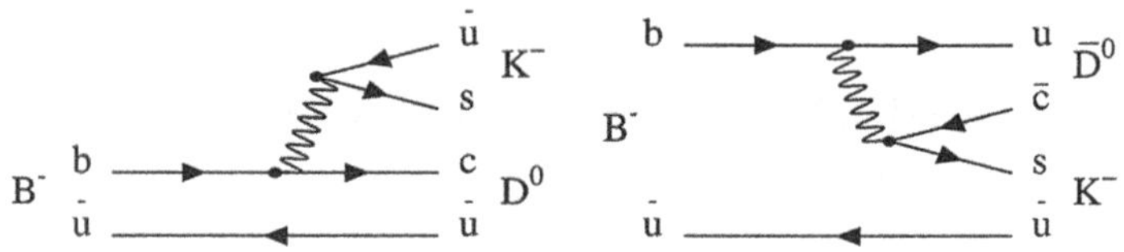

Fig. 38.1 Feynman diagram for $B^- \to D^0 K^-$ (left) and $B^- \to \overline{D}^0 K^-$ (right) [3]

Table 38.1 Previous results for the measurement of ϕ_3

Sr. No.	Experiment	Measurement of ϕ_3
1	Belle	$\left(73^{+13}_{-15}\right)^\circ$ [7]
2	BaBar	$\left(69^{+17}_{-16}\right)^\circ$ [8]
3	LHCb	$\left(74^{+5.0}_{-5.8}\right)^\circ$ [9]

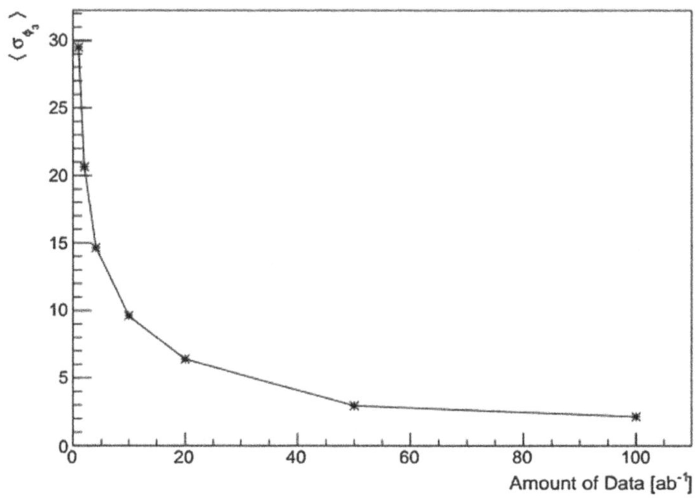

Fig. 38.2 The distribution shows the expected uncertainty($<\sigma_{\phi_3}>$) versus luminosity on ϕ_3 [12], which explains the expected uncertainty on the measurement of ϕ_3 at integrated luminosity of 50 ab^{-1} is 3°. It may also improve by adding more modes, such as $B^\pm \to D^0(K_S^0 K^+ K^-)K^\pm$

to measure this angle but due to the small data samples so far produced, ϕ_3 is poorly determined (Table 38.1). Therefore, an independent measurement with high statistics is required to measure ϕ_3, as the measurement is dominated by the statistical uncertainty. In this work, we present the preliminary MC study of $B^\pm \to D^0 K^\pm$ to extract ϕ_3 using the 50 ab^{-1} data to be accumulated by the Belle II detector. The Belle II [10] experiment at the SuperKEKB asymmetric e$^+$e$^-$ collider [11], will accumulate the collision data at an unprecedented instantaneous luminosity of 8×10^{35} cm^{-2}s^{-1}, which is 40 times larger than Belle experiment. Figure 38.2 shows how the expected uncertainty on ϕ_3 scale with luminosity based on toy Monte Carlo studies. It shows that the expected uncertainty with integrated luminosity of 50 ab^{-1} is approximately 3°. In addition, Belle II will also have a better particle identification with the Central Drift Chamber (CDC), Time of Propagation (TOP) and the Aerogel Ring Imaging Cherenkov Counter (ARICH), which is functioning in different momentum ranges [10]. Due to larger acceptance of the detector, an improved reconstruction efficiency of K_S^0 is anticipated. Recently, Belle II successfully completed its Phase II and collected 478 pb^{-1}. Here, Phase II data is incorporating single ladder per layer of the vertex detector and all other subdetectors. However, when the poster was shown, available data was 250 pb^{-1}. Therefore, the plots here are with 250 pb^{-1}.

38.2 Preliminary Results from Phase II Data

The analysis begins with the reconstruction of K_S^0 from the two charged tracks of π^+ and π^-. The invariant mass of K_S^0 is shown in Fig. 38.3 (left) with MC samples collected at integrated luminosity 1 fb^{-1} and with Phase II data shown in Fig. 38.3 (right). Here, black points are data and red line is fitting. As can be seen from the figure, the invariant mass resolution shows the good aggrement between data and MC. Further, the D^0 is reconstructed from one K_S^0 and two charged tracks of π^+ and π^- followed by the inclusive decay of $D^{*\pm} \to D^0(K_S^0\pi^+\pi^-)\pi^\pm$. Invariant mass of D^0 is shown in Fig. 38.4 (left) and ΔM is shown in Fig. 38.4 (right). Here, ΔM is the mass difference between the $D^{*\pm}$ and D candidates. The reconstruction of B meson is in progress by using the two important variables, energy difference, $\Delta E = \sum E_i - E_{beam}$ and the beam constrained mass, $M_{bc} = \sqrt{(E_{beam})^2 - \sum(\overrightarrow{p_i})^2}$, where E_{beam} is the center-of-mass (CM) beam energy, E_i and p_i are the CM energies and momenta of B candidates decay product.

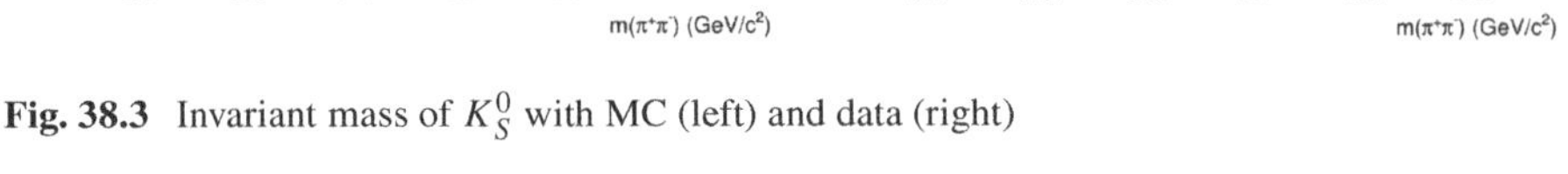

Fig. 38.3 Invariant mass of K_S^0 with MC (left) and data (right)

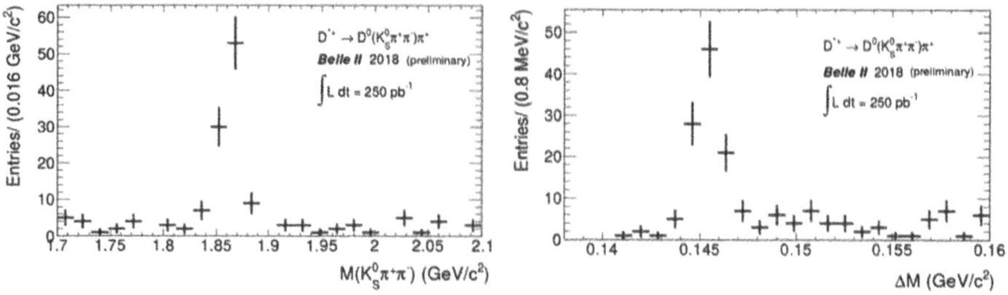

Fig. 38.4 Invariant mass (left) and ΔM (right) for D^* tagged mode $D^0 \to K_S^0\pi^+\pi^-$ shown with Phase II data

38.3 Summary

The large statistics with Belle II at SuperKEKB will provide a substantial improvement in the precision of ϕ_3 with the full 50 ab^{-1} data sample. Extrapolated results are expected to provide a more precise measurement of the CKM angle ϕ_3 at integrated luminosity 50 ab^{-1}. Invariant masses of K_S^0 and D^0 are reconstructed and compared with MC and Phase II data. The reconstruction of B meson is in progress.

References

1. M. Kobayashi, T. Maskawa, Prog. Theor. Phys. **49**, 652 (1973)
2. N. Cabibbo, Phys. Rev. Lett. **10**, 531 (1963)
3. J. Brodzicka et al., Physics achievements from the Belle experiment, PTEP 2012, 04D001
4. Gronau, M., London, D, How to determine all the angles of the unitarity triangle from $B_d^0 \to DK_S$ and $B_s^0 \to D\phi$. Phys. Lett. B**253**, 483(1991); Gronau, M., Wyler, D., On determining a weak phase from charged B decay asymmetries. Phys. Lett. B **265**, 172 (1991)
5. D. Atwood, I. Dunietz, A. Soni, Enhanced CP violation with $B \to KD^0(\overline{D}^0$ modes and extraction of the Cabibbo-Kobayashi-Maskawa angle γ. Phys. Rev. Lett. **78**, 3257 (1997)
6. H. Aihara, K. Arinstein et al., First measurement of ϕ_3 with a model-independent dalitz plot analysis of $B^\pm \to DK^\pm$, $D \to K_S^0\pi^+\pi^-$ decay. Phys. Rev. D **85**, 112014 (2012)
7. K. Trabelsi, World average and experimental overview of ϕ_3, CKM-CONF-2014
8. J.P. Lees et al., The BABAR Collaboration. Phys. Rev. D **87**, 052015 (2013)
9. The LHCb Collaboration, CERN-LHCb-CONF-2018-002
10. T. Abe et al., Belle II technical design report (2010). arXiv:1011.0352
11. B. Golob, Super KEKB/Belle II Project, (The Belle II Collaboration). Nuovo Cim. **C33**, 319–326 (2010)
12. E. Kou et al., *The Belle II Collaboration* (The Belle II Physics Book, PTEP, 2018)

Part IV
Neutrino Physics, Leptogenesis and Baryogenesis

Chapter 39
NOvA Oscillation Results

V. Bhatnagar

Abstract The NOvA experiment is a long baseline oscillation experiment designed to measure the rates of electron neutrino appearance and muon neutrino disappearance. It uses the NuMI beam that has recently been upgraded to 700 kW as the neutrino beam source. The NOvA near detector is placed onsite at Fermilab, 800 m from the NuMI target while the far detector is placed 810 km away at a site near Ash River, Minnesota, both the detectors being functionally identical and 14.6 mrad off-axis w.r.t the NuMI beam. The primary physics goal of the experiment include precise measurement of θ_{23}, CP violating phase δ_{CP} and the neutrino mass hierarchy. This work is a proceeding to a talk describing the experiment and the latest $\nu_\mu \rightarrow \nu_\mu$ and $\nu_\mu \rightarrow \nu_e$ oscillation results based on the exposure equivalent to 8.85×10^{20} for the neutrino and 6.9×10^{20} for the antineutrino beam.

39.1 Introduction

The discovery of neutrino oscillations has shown that neutrinos are massive and paved the way for the measurements of the neutrino masses and their mixing expressed in terms of $\theta_{12}, \theta_{23}, \theta_{13}$ and CP violating phase δ_{CP}. Neutrino experiments have previously measured three mixing angles and the squared mass differences between the mass eigenstates: $m_1^2 - m_2^2$, $m_2^2 - m_3^2$. The unresolved problems are those of the CP violation parameter and the neutrino mass hierarchy. The NOvA experiment is looking for ν_e in a predominantly ν_μ beam, with a baseline of 810 km together with the upgrades in the NuMI beam intensity, it is capable of measuring the CP violating phase δ_{CP}, neutrino mass hierarchy along with providing strong constraints on the θ_{23} octant and Δm^2_{32}.

V. Bhatnagar (✉)
Department of Physics, Panjab University, Chandigarh, India
e-mail: vipin@pu.ac.in

A. Giri and R. Mohanta (eds.), *16th Conference on Flavor Physics and CP Violation*, Springer Proceedings in Physics 234,
https://doi.org/10.1007/978-3-030-29622-3_39

39.2 The NOvA Experiment

The NOvA experiment has its 0.3 kt near detector placed close to the NuMI neutrino source observing an intense beam of neutrinos and has a massive 14 kt far detector placed 810 km away ensuring enough statistics and neutrino interactions for making the oscillation parameter measurements.

39.2.1 NOvA Physics Goals

The NOvA experiment measures the $\nu_\mu(\overline{\nu_\mu}) \rightarrow \nu_\mu(\overline{\nu_\mu})$ disappearnce probability enabling the precise measurement of the atmospheric oscillation parameters Δm^2_{32} and θ_{23}. The appearance probability $\nu_\mu(\overline{\nu_\mu}) \rightarrow \nu_e(\overline{\nu_e})$ helps inverstigating (a) CP violating phase δ_{CP}, (b) the neutrino mass hierarchy, if normal (NH) or inverted (IH) (c) the mixing angle θ_{13} and the θ_{23} octant. The appearnce probability of the electron neutrino in a muon neutrino beam is given by

$$P(\nu_\mu \rightarrow \nu_e) \approx \left| \sqrt{P_{atm}} e^{-i\left(\frac{\Delta m^2_{32} L}{4E} + \delta_{CP}\right)} + \sqrt{P_{sol}} \right|^2 \tag{39.1}$$

where $P_{atm} = \sin\theta_{23} \sin^2 2\theta_{13} \sin^2 \frac{\Delta m^2_{32} L}{4E}$.
The above (39.1) shows that the ν_e appearance probability and hence the precision of the measurement of δ_{cp} and mass hierarchy depends on the mixing angle θ_{23}. The electron neutrino (ν_e) can undergo a W exchange with the electrons in the matter they pass through giving rise to an extra interaction potential V_W proportional to G_F, the Fermi coupling constant and N_e, the number of electrons per unit volume.

$$V_W = +\sqrt{2} G_F N_e \tag{39.2}$$

This interaction potential shown in (39.2) changes sign if the beam has $\overline{\nu_e}$ instead of ν_e. This extra interaction potential modifying the oscillation probabilty in matter compared to that in vacuum depends on the neutrino mass hierarchy and provides an extra handle on the same. Because of both the detectors being extremely huge, it can be also used to study the neutrino cross-sections, sterile neutrinos, supernovae and exotic searches.

39.2.2 The NuMI Beam

The neutrino beam characteristics play an important role in deciding the physics reach of an experiment. The NuMI beam (neutrinos at the Main Injector) provides

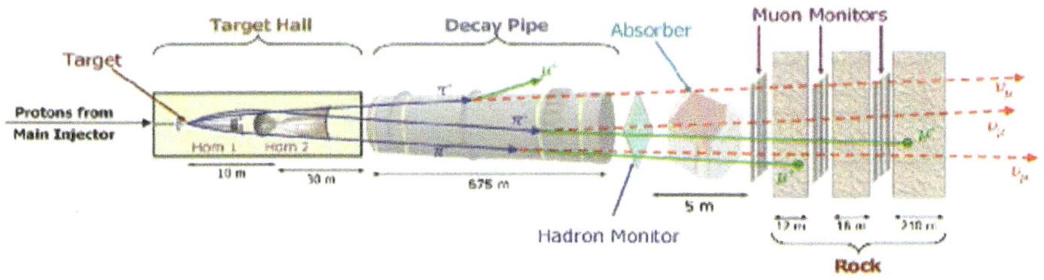

Fig. 39.1 Schematic of the NuMI Beam. Protons from the main injector impinge on the graphite target. The secondary mesons ($\pi^{\pm}$ and $K^{\pm}$) are focused by two magnetic horns induced with a current of 200 kA. Reversing the polarity of the horns focusses the hadron with opposite charge-sign, that enables running in both neutrino and anti-neutrino mode [1]

the neutrino beam to both the NOvA detectors. With the recent upgrades from to 700 kW in the beam power, it is currently the most powerful source of neutrino beam. The 120 GeV proton beam from the Main Injector collides with a graphite target producing mesons mainly pions and kaons, subsequently decaying to muon and muon neutrinos (Fig. 39.1).

The NOvA detectors placed 14 mrad off-axis w.r.t the NuMI beam will observe a very narrow band beam peaked at 2 GeV which is close to $\nu_\mu \rightarrow \nu_e$ oscillation maximum (1.6 GeV at 810 km).

39.2.3 NOvA Detectors

The NOvA detectors are highly segmented and 65% active tracking calorimeters. The extruded PVC cells filled with liquid scintillator make the basic unit of both the near and far detectors. These PVC cells with cross-sectional size (6 cm) × (4 cm) extending to the full width or height of the detectors (4.1 m in ND and 15.6 m in FD) provide segmentation and overall structure to the detectors. The cells are filled with liquid scintillator which is mostly mineral oil with 5% pseudocumene (the main scintillant). The particles traversing the cell excite the scintillator and the emitted light from the scintillator is transported from the cell by a wavelength shifting fiber to a pixel of 32-channel avalanche photo diode (APD) for the readout. The far detector has 344,000 PVC cells divided in 896 planes where as the near detector has 20,192 cells arranged in 214 planes. The planes are orthogonal to each other allowing 3D reconstruction of the particle trajectories. The detectors' granularity is well suited for electro-magnetic shower identification as the PVC cell dimensions are much smaller than the radiation length (38 cm) in the detector (Fig. 39.2).

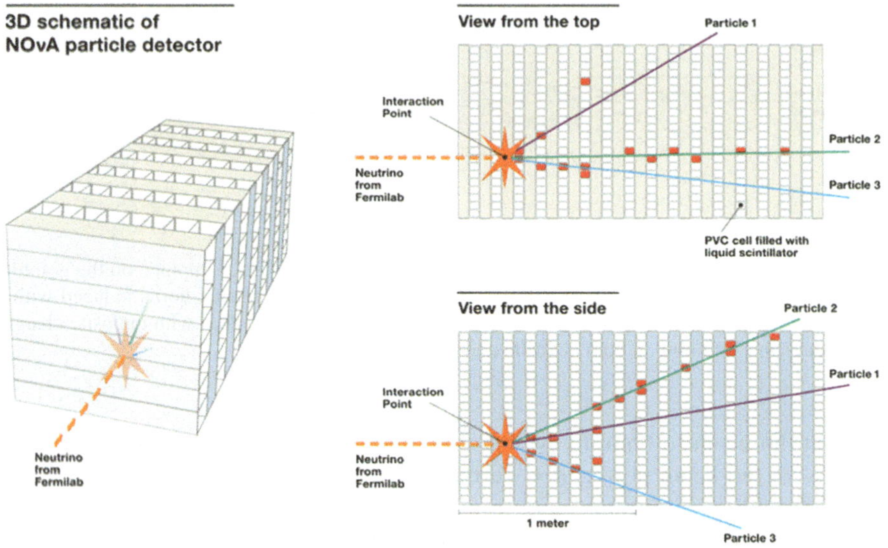

Fig. 39.2 The 3D schematic of the NOvA particle detector showing the top and side view. The readout is done in both the top and side views allowing the 3D reconstruction of the particle trajectories [2]

39.2.4 The Neutrino Interaction Classifier-CVN

Convolutional Neural network (CNNs) is widely used in the field of computer vision for image recognition. NOvA has pioneered the use of CNNs for particle classification in neutrino physics. NOvA has developed its own CNN, called CVN (convolutional visual network) for identifying the neutrino interaction [2]. Each interaction topology is treated as an image with cells as pixels and charge as color value. The convolutional layers optimally extract features from the images providing good seperation between the different interaction modes. The CVN characterizes the neutrino events into ν_μ CC, ν_e CC, NC and cosmic interaction types (Fig. 39.3).

39.2.5 The Neutrino Interaction Tuning

NOvA uses GENIE for neutrino event generation. The NOvA's energy range 1–4 GeV lies in a region which allows interplay between different inetraction modes e.g. Resonance, quasi-elastic, Deep Inelastic Scattering (DIS), and 2p-2h Meson Exchange Current (MEC). Each of these channels have different selection efficiency and reconstructed energy biases. The relative contribution of each of the interactions in the simulations should be correct for the oscillation analyses. The hadronic energy distribution for $\nu_\mu + \overline{\nu_\mu}$ CC events show disagreement with the default genie prediction. The event excess observed in the data was explained by the MEC events that

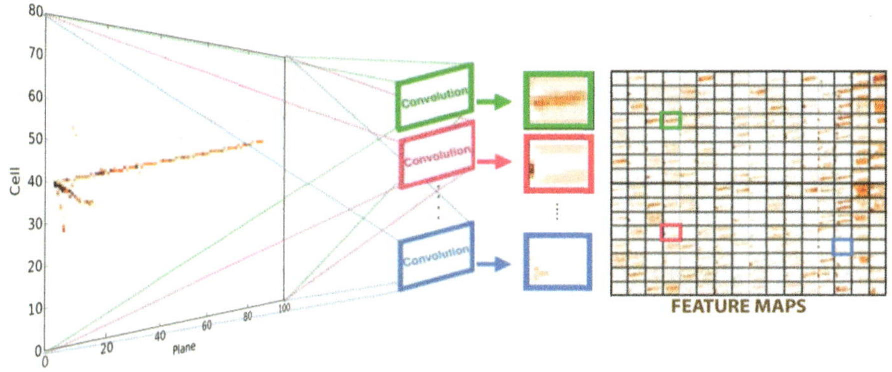

Fig. 39.3 Simulated ν_μ CC event. Some track activity is visible in one of the feature maps [3]

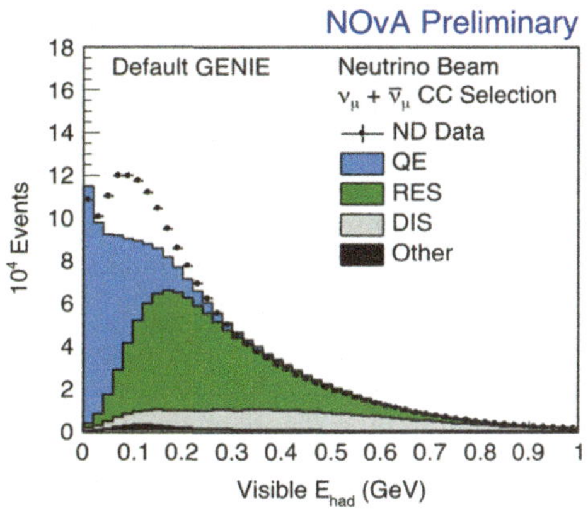

Fig. 39.4 Visible hadronic energy distribution for the near detector $\nu_\mu + \overline{\nu_\mu}$ CC events showing discrepancy between the data and the GENIE prediction [4]

improved the discrepancy significantly. It was used for the first time in the second oscillation analyses in 2016, the treatment of 2p-2h has grown more sophisticated since then and new 2p-2h models have been added to GENIE (Fig. 39.4).

The understanding of the nuclear effects and neutrino-nuclues interactions needs to be enhanced for limiting the systematic errors on the oscillation parameters. NOvA uses external information from other experiments as well its own near detector's data to tune the neutrino interaction models.

39.3 Experimental Results

The NOvA experiment recently published its oscillation results for $\nu_\mu \rightarrow \nu_\mu$ and $\nu_\mu \rightarrow \nu_e$ corresponding to an exposure of 8.85×10^{20} for the neutrino beam and 6.9×10^{20} for the anti-neutrino beam.

Table 39.1 Best Fit values of the oscillation parameters in ν_μ and ν_e joint fit

Oscillation Parameter	Values
Δm^2_{32}	$2.51 + 0.02(-0.08) \times 10^{-3} eV^2$
$\sin^2\theta_{23}$	0.58 ± 0.03 (Upper Octant)
$\delta_{CP}(\pi)$	0.17π

39.3.1 ν_μ Disappearance and ν_e Appearance

In the ν_μ disappearance analysis, 113 events are observed in the neutrino mode (expected 730 + 38(−49)(systematics) without oscillations) and 65 events are observed in the antineutrino mode (expected 266 + 12(−14)(systematics) without oscillations).The combined data of neutrino and antineutrino beams are fitted assuming CPT invariance. We observe 113 events and expect 126 at this combined best fit for the neutrino beam mode and observe 65 events and expect 52 at the best fit in antineutrino beam mode. For the $\nu_e + \overline{\nu_e}$ appearance analysis, NOvA observes 58 (expects 30 ($\pi/2$, IH)-75 ($3\pi/2$, NH) events) for neutrino mode and 18 events (expects 10 ($3\pi/2$, NH) to 22 ($\pi/2$, IH) events) for antineutrino mode. Out of 58 events observed for the neutrino beam, 15(11 beam, 3 cosmic and <1 wrong sign) are expected to be background interactions. The antineutro results give greater than 4σ evidence of $\overline{\nu_e}$ appearance in a $\overline{\nu_\mu}$ beam. For the antineutrino beam, out of 18 observed, 5.3 (3.5 beam, 1 wrong sign and <1 cosmic) are expected background interactions. The ν_μ disappeance and ν_e appearance joint fit in $\nu + \overline{\nu}$ beam prefers normal hierarchy and non maximal mixing at 1.8σ, excluding lower octant at similar level. Table 39.1 shows the value of allowed value of the oscillation parameters. The $\delta_{CP} = (\pi/2)$ is excluded at $>3\sigma$ in the inverted hierarchy (Figs. 39.5, 39.6 and 39.7).

39.4 Summary and Future Prospects

NOvA is currently running in the antineutrino mode, planning to run with an equal exposure in both the neutrino and antineutrino modes and expect to extend the runing upto 2024 with proposed improvemnets in the accelerator. NOvA is planning a test beam program starting from 2019 that will enhance the ultimate physics reach addressing the detector response and calibration uncertainties. It expects to reach 3σ sensitivity for mass hierarchy (if NH and $\delta_{CP} = 3\pi/2$) for allowed range of θ_{23} by 2020 and 3σ sensitivity for 30 − 50% (depending on octant) of δ_{CP} range by 2024 (Fig. 39.8).

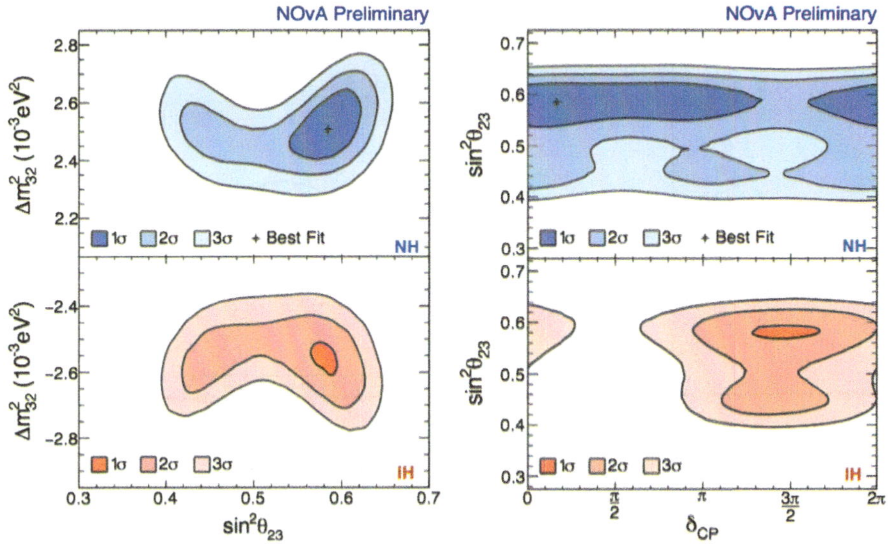

Fig. 39.5 The top and bottom panel show the allowed oscillation parameters corresponding to normal and inverted mass hierarchy respectively. The plots show parameter space of Δm^2_{32}. Versus $\sin^2\theta_{23}$ (left) and $\sin^2\theta_{23}$ versus δ_{CP} (right) at various levels of significance [4]

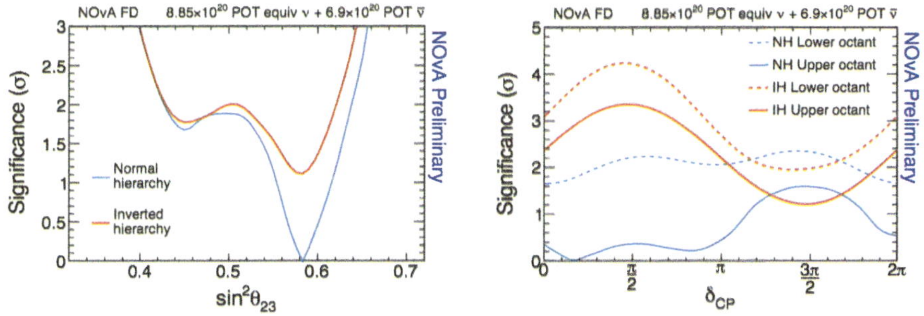

Fig. 39.6 (left) Significance at which each value of $\sin^2\theta_{23}$ is disfavored in the normal (blue) or inverted (red) mass hierarchy. (right) Significance at which each value of δ_{CP} is disfavored in the normal (blue) or inverted (red) mass hierarchy. The normal and inverted mass hierarchy is divided into upper (solid) and lower (dashed) θ_{23} octants corresponding to the near degeneracy in $\sin^2\theta_{23}$ [4]

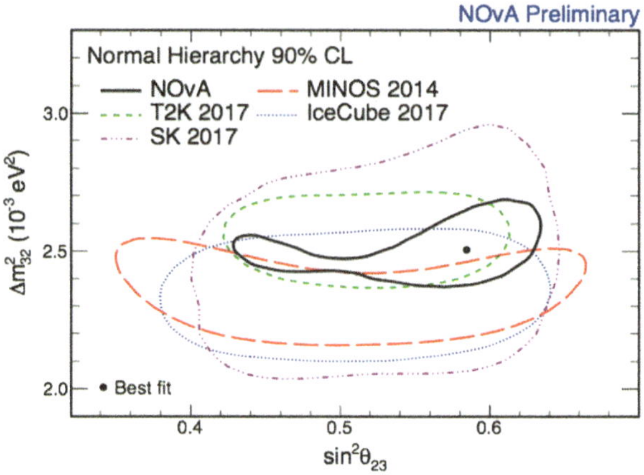

Fig. 39.7 Comparison of the measured 90% confidence level contours for the Δm^2_{32}. Versus $\sin^2\theta_{23}$ for NOvA experiment with T2K (green dashed), IceCube (blue dash-dotted), Super-Kamiokande (purple-dashed), MINOS (red dashed) [4]

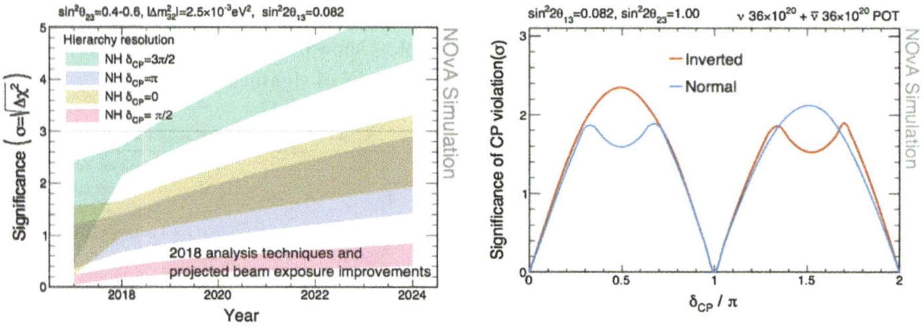

Fig. 39.8 (left) Significance at which each value of $\sin^2\theta_{23}$ is disfavored in the normal (blue) or inverted (red) mass hierarchy. (right) Significance at which each value of δ_{CP} is disfavored in the normal (blue) or inverted (red) mass hierarchy. The normal and inverted mass hierarchy is divided into upper (solid) and lower (dashed) θ_{23} octants corresponding to the near degeneracy in $\sin^2\theta_{23}$ [4]

References

1. P. Adamson et al., The NuMI Neutrino Beam (2015). arXiv:1507.06690v2 [physics.acc-ph]
2. A. Aurisano et al., A convolutional neural network neutrino event classifier (2016). arXiv:1604.01444 [hep-ex]
3. Fernanda Psihas and NOvA Collaboration, J. Phys.: Conf. Ser. **888**, 012063 (2017)
4. M. Sanchez, NOvA Results and Prospects, NOvA Document, DocDB-30249-v1 (2018)

Chapter 40
Baryogenesis and Leptogenesis

P. S. Bhupal Dev

Abstract This is a short review on the mechanism of *leptogenesis*, which provides an attractive link between two important pieces of evidence for beyond Standard Model physics, namely, observed neutrino masses and matter-antimatter asymmetry in the Universe. We give special emphasis on TeV-scale leptogenesis models and provide a concrete example, which is testable in foreseeable laboratory experiments.

40.1 Introduction

Our present observable Universe seems to be populated exclusively with matter and no ambient antimatter. Although this asymmetry is maximal today, at high temperatures ($T \gtrsim 1$ GeV) when quark-antiquark pairs were abundant in the thermal plasma, the baryon asymmetry observed today corresponds to a tiny asymmetry at recombination [1]:

$$\eta_B \equiv \frac{n_B - n_{\overline{B}}}{n_\gamma} = \left(6.105^{+0.086}_{-0.081}\right) \times 10^{-10} , \tag{40.1}$$

where $n_{B(\overline{B})}$ is the number density of baryons (antibaryons) and $n_\gamma = 2T^3\zeta(3)/\pi^2$ is the number density of photons, $\zeta(x)$ being the Riemann zeta function, with $\zeta(3) \approx 1.20206$. *Baryogenesis* is the mechanism by which the observed baryon asymmetry of the Universe (BAU) given by (40.1) can arise dynamically from an initially baryon-symmetric phase of the Universe, or even irrespective of any initial asymmetry. This necessarily requires the fulfilment of three basic Sakharov conditions: (i) baryon number (B) violation, which is essential for the Universe to evolve from a state with net baryon number $B = 0$ to a state with $B \neq 0$; (ii) C and CP violation, which allow particles and anti-particles to evolve differently so that we can have an

P. S. Bhupal Dev (✉)
Department of Physics and McDonnell Center for the Space Sciences,
Washington University, St. Louis, MO 63130, USA
e-mail: bdev@wustl.edu

A. Giri and R. Mohanta (eds.), *16th Conference on Flavor Physics and CP Violation*, Springer Proceedings in Physics 234,
https://doi.org/10.1007/978-3-030-29622-3_40

asymmetry between them; (iii) departure from thermal equilibrium, which ensures that the asymmetry does not get erased completely. The Standard Model (SM) has, in principle, all these basic ingredients, namely, (i) the triangle anomaly violates B through a non-perturbative instanton effect, which leads to effective $(B+L)$-violating sphaleron transitions for $T \gtrsim 100$ GeV; (ii) there is maximal C violation in weak interactions and CP violation due to the Kobayashi–Maswaka phase in the quark sector; (iii) the departure from thermal equilibrium can be realized at the electroweak phase transition (EWPT) if it is sufficiently first order. However, the SM CP violation turns out to be too small to account for the observed BAU. In addition, for the observed value of 125 GeV for the Higgs mass, the EWPT is not first order, but only a smooth crossover. Therefore, *the observed BAU provides a strong evidence for the existence of new physics beyond the SM.*

Many interesting scenarios for baryogenesis have been proposed in beyond SM theories; see e.g. [2]. Here we will focus on the mechanism of *leptogenesis* [3], which is an elegant framework to explain the BAU, while connecting it to another seemingly disparate evidence for new physics beyond the SM, namely, non-zero neutrino masses; for a recent review on various aspects of leptogenesis, see e.g. [4]. The minimal version of leptogenesis is based on the *type I seesaw* mechanism, which requires heavy SM gauge-singlet Majorana neutrinos N_α (with $\alpha = 1, 2, 3$) to explain the observed smallness of the three active neutrino masses at tree-level. The out-of-equilibrium decays of these heavy Majorana neutrinos in an expanding Universe create a lepton asymmetry, which is reprocessed into a baryon asymmetry through the equilibrated $(B+L)$-violating electroweak sphaleron interactions.

In the original scenario of thermal leptogenesis [3], the heavy Majorana neutrino masses are typically close to the Grand Unified Theory (GUT) scale, as suggested by natural GUT embedding of the seesaw mechanism. In fact, for a hierarchical heavy neutrino spectrum, i.e. $m_{N_1} \ll m_{N_2} < m_{N_3}$, the light neutrino oscillation data impose a *lower* limit on $m_{N_1} \gtrsim 10^9$ GeV [5]. As a consequence, such 'vanilla' leptogenesis scenarios [6] are very difficult to test in any foreseeable experiment. Moreover, these high-scale thermal leptogenesis scenarios run into difficulties, when embedded within supergravity models of inflation. In particular, it leads to a potential conflict with an *upper* bound on the reheat temperature of the Universe, $T_R \lesssim 10^6$–10^9 GeV, as required to avoid overproduction of gravitinos whose late decays may otherwise ruin the success of Big Bang Nucleosynthesis [7].

An attractive scenario that avoids the aforementioned problems is *resonant leptogenesis* (RL) [8], where the ε-type CP asymmetries due to the self-energy effects in the heavy Majorana neutrino decays get resonantly enhanced. This happens when the masses of at least two of the heavy neutrinos become quasi-degenerate, with a mass difference comparable to their decay widths [8]. The resonant enhancement of the CP asymmetry allows one to avoid the lower bound on $m_{N_1} \gtrsim 10^9$ GeV [5] and have successful leptogenesis at an experimentally accessible energy scale [8], while retaining perfect agreement with the light-neutrino oscillation data. The level of testability is further extended in the scenario of Resonant l-Genesis (RL$_l$), where the final lepton asymmetry is dominantly generated and stored in a *single* lepton flavor l [9, 10]. In such models, the heavy neutrinos could be as light as the electroweak scale, while still having sizable couplings to other charged-lepton flavors $l' \neq l$, thus giving rise to potentially large lepton flavor violating effects. Here we will mainly focus on low-scale type-I seesaw leptogenesis scenarios, which may be directly tested at the Energy [11] and Intensity [12] frontiers. For other leptogenesis scenarios, see e.g. [13].

40.2 Basic Picture

Our starting point is the minimal type I seesaw extension of the SM Lagrangian:

$$\mathscr{L} = \mathscr{L}_{\rm SM} + i\overline{N}_{{\rm R},\alpha}\gamma_\mu\partial^\mu N_{{\rm R},\alpha} - h_{l\alpha}\overline{L}_l\widetilde{\Phi}N_{{\rm R},\alpha} - \frac{1}{2}\overline{N}^C_{{\rm R},\alpha}(M_N)_{\alpha\beta}N_{{\rm R},\beta} + {\rm H.c.}\,, \tag{40.2}$$

where $N_{{\rm R},\alpha} \equiv \frac{1}{2}(1+\gamma_5)N_\alpha$ are the right-handed (RH) heavy neutrino fields, $L_l \equiv (\nu_l \;\; l)^{\rm T}_L$ are the $SU(2)_L$ lepton doublets (with $l = e, \mu, \tau$) and $\widetilde{\Phi} \equiv i\sigma_2\Phi^*$, Φ being the SM Higgs doublet and σ_2 being the second Pauli matrix. The complex Yukawa couplings $h_{l\alpha}$ induce CP-violating decays of the heavy Majorana neutrinos, if kinematically allowed: $N_\alpha \to L_l\Phi$ with a decay rate $\Gamma_{l\alpha}$ and the CP-conjugate process $N_\alpha \to L^c_l\Phi^c$ with a decay rate $\Gamma^c_{l\alpha}$ (the shorthand notation c denotes CP). The flavor-dependent CP asymmetry can be defined as

$$\varepsilon_{l\alpha} = \frac{\Gamma_{l\alpha} - \Gamma^c_{l\alpha}}{\sum_k\left(\Gamma_{k\alpha} + \Gamma^c_{k\alpha}\right)} \equiv \frac{\Delta\Gamma_{l\alpha}}{\Gamma_{N_\alpha}}\,, \tag{40.3}$$

where Γ_{N_α} is the total decay width of the heavy Majorana neutrino N_α which, at tree-level, is given by $\Gamma_{N_\alpha} = \frac{m_{N_\alpha}}{8\pi}(h^\dagger h)_{\alpha\alpha}$. A non-zero CP asymmetry arises at one-loop level due to the interference between the tree-level graph with either the vertex or the self-energy graph. Following the terminology used for CP violation in neutral meson systems, we denote these two contributions as ε'-type and ε-type CP-violation respectively. In the two heavy-neutrino case ($\alpha, \beta = 1, 2; \alpha \neq \beta$), they can be expressed in a simple analytic form [8]:

$$\varepsilon'_{l\alpha} = \frac{{\rm Im}\left[(h^*_{l\alpha}h_{l\beta})(h^\dagger h)_{\alpha\beta}\right]}{8\pi\,(h^\dagger h)_{\alpha\alpha}}\frac{m_{N_\beta}}{m_{N_\alpha}}\left[1 - \left(1 + \frac{m^2_{N_\beta}}{m^2_{N_\alpha}}\right)\ln\left(1 + \frac{m^2_{N_\alpha}}{m^2_{N_\beta}}\right)\right], \tag{40.4}$$

$$\varepsilon_{l\alpha} = \frac{{\rm Im}\left[(h^*_{l\alpha}h_{l\beta})(h^\dagger h)_{\alpha\beta}\right] + \frac{m_\alpha}{m_\beta}\,{\rm Im}\left[h^*_{l\alpha}h_{l\beta}(h^\dagger h)_{\beta\alpha}\right]}{8\pi\,(h^\dagger h)_{\alpha\alpha}}\frac{(m^2_{N_\alpha} - m^2_{N_\beta})m_{N_\alpha}m_{N_\beta}}{(m^2_{N_\alpha} - m^2_{N_\beta})^2 + m^2_{N_\alpha}\Gamma^2_{N_\beta}}\,. \tag{40.5}$$

In the quasi-degenerate limit $|m_{N_\alpha} - m_{N_\beta}| \sim \frac{1}{2}\Gamma_{N_{\alpha,\beta}}$, the ε-type contribution becomes resonantly enhanced, as evident from (40.5).

Due to the Majorana nature of the heavy neutrinos, their decays to lepton and Higgs fields violate lepton number which, in presence of CP violation, leads to the generation of a lepton (or $B - L$) asymmetry. Part of this asymmetry is washed out due to the inverse decay processes $L\Phi \to N$, $L^c\Phi^c \to N$ and various $\Delta L = 1$ (e.g. $NL \to Qu^c$) and $\Delta L = 2$ (e.g. $L\Phi \to L^c\Phi^c$) scattering processes. In the flavor-diagonal limit, the total amount of $B - L$ asymmetry generated at a given temperature can be calculated by solving the following set of coupled Boltzmann equations [6]:

$$\frac{dN_{N_\alpha}}{dz} = -(D_\alpha + S_\alpha)(N_{N_\alpha} - N_{N_\alpha}^{\rm eq}) , \tag{40.6}$$

$$\frac{dN_{N_{B-L}}}{dz} = \sum_\alpha \varepsilon_\alpha D_\alpha (N_{N_\alpha} - N_{N_\alpha}^{\rm eq}) - N_{B-L} \sum_\alpha W_\alpha , \tag{40.7}$$

where $z \equiv m_{N_1}/T$, with N_1 being the lightest heavy neutrino, N_X denotes the number density in a portion of comoving volume containing one heavy-neutrino in ultra-relativistic limit, so that $N_{N_\alpha}^{\rm eq}(T \gg m_{N_\alpha}) = 1$, $\varepsilon_\alpha \equiv \sum_l (\varepsilon_{l\alpha} + \varepsilon'_{l\alpha})$ is the total CP asymmetry due to the decay of N_α, and D_α, S_α, W_α denote the decay, scattering and washout rates, respectively. Given the Hubble expansion rate $H(T) \simeq 1.66 g_* \frac{T^2}{M_{\rm Pl}}$, where g_* is the total relativistic degrees of freedom and $M_{\rm Pl} = 1.22 \times 10^{19}$ GeV is the Planck mass, we define the decay parameters $K_\alpha \equiv \frac{\Gamma_{D_\alpha}(T=0)}{H(T=m_{N_\alpha})}$, where $\Gamma_{D_\alpha} \equiv \sum_l (\Gamma_{l\alpha} + \Gamma^c_{l\alpha})$. For $K_\alpha \gg 1$, the system is in the *strong washout* regime, where the final lepton asymmetry is insensitive to any initial asymmetry present. The decay rates are given by $D_\alpha \equiv \frac{\Gamma_{D_\alpha}}{Hz} = K_\alpha x_\alpha z \frac{\mathcal{K}_1(z)}{\mathcal{K}_2(z)}$, where $x_\alpha \equiv \frac{m^2_{N_\alpha}}{m^2_{N_1}}$ and $\mathcal{K}_n(z)$ is the nth-order modified Bessel function of the second kind. Similarly, the washout rates induced by inverse decays are given by $W_\alpha^{\rm ID} = \frac{1}{4} K_\alpha \sqrt{x_\alpha} \mathcal{K}_1(z_\alpha) z_\alpha^3$, where $z_\alpha \equiv z\sqrt{x_\alpha}$. Other washout terms due to scattering can be similarly calculated [6]. The final $B - L$ asymmetry is given by $N^{\rm f}_{B-L} = \sum_\alpha \varepsilon_\alpha \kappa_\alpha (z \to \infty)$, where $\kappa_\alpha(z)$'s are the efficiency factors, obtained from (40.6) and (40.7):

$$\kappa_\alpha(z) = -\int_{z_{\rm in}}^{z} dz' \frac{D_\alpha(z')}{D_\alpha(z') + S_\alpha(z')} \frac{dN_{N_\alpha}}{dz'} \exp\left[-\int_{z'}^{z} dz'' \sum_\alpha W_\alpha(z'') \right] . \tag{40.8}$$

At temperatures $T \gg 100$ GeV, when the $(B + L)$-violating electroweak sphalerons are in thermal equilibrium, a fraction $a_{\rm sph} = \frac{28}{79}$ of the $B - L$ asymmetry is reprocessed to a baryon asymmetry. There is an additional entropy dilution factor $f = \frac{2387}{86}$ due to the standard photon production from the onset of leptogenesis till recombination. Combining all these effects, the predicted baryon asymmetry is

$$\eta_B = \frac{a_{\rm sph}}{f} N^{\rm f}_{B-L} \simeq 10^{-2} \sum_\alpha \varepsilon_\alpha \kappa_\alpha (z \to \infty) , \tag{40.9}$$

which has to be compared with the observed BAU given by (40.1).

40.3 Flavor Effects

Flavor effects in both heavy-neutrino and charged-lepton sectors, as well as the interplay between them, can play an important role in determining the final lepton asymmetry, especially in low-scale leptogenesis models [14]. These intrinsically-quantum effects can, in principle, be accounted for by extending the flavor-diagonal Boltzmann equations (40.6) and (40.7) for the number densities of individual flavor species to a semi-classical evolution equation for a *matrix of number densities*, analogous to the formalism presented in [15] for light neutrinos. This so-called 'density matrix' formalism has been adopted to describe flavor effects in various leptogenesis scenarios [16, 17]. It was recently shown [10, 18], in a semi-classical approach, that a consistent treatment of *all* pertinent flavor effects, including flavor mixing, oscillations and off-diagonal (de)coherences, necessitates a *fully* flavor-covariant formalism. It was further shown that there are two *distinct* physical phenomena, namely, resonant mixing of different heavy-neutrino flavors and coherent oscillations between them, which contribute to the CP-asymmetry in the resonant regime. Note that this is analogous to the experimentally-distinguishable phenomena of mixing and oscillations in the neutral meson systems.

One can go beyond the semi-classical 'density-matrix' approach to leptogenesis by means of a quantum field-theoretic analogue of the Boltzmann equations, known as the Kadanoff-Baym (KB) equations [19]. Such 'first-principles' approaches to leptogenesis are, in principle, capable of accounting consistently for all flavor effects, in addition to off-shell and finite-width effects, including thermal corrections. The source term for the lepton asymmetry obtained, at leading order, in this KB approach [20] was found to be exactly the same as that obtained in the semi-classical flavor-covariant approach of [10], confirming that flavor mixing and oscillations are indeed two *physically-distinct* phenomena. The proper treatment of these flavor effects can lead to a significant enhancement of the final lepton asymmetry, as compared to partially flavor-dependent or flavor-diagonal limits, thereby enlarging the viable parameter space for models of RL and enhancing the prospects of testing the leptogenesis mechanism.

40.4 A Concrete Model

Here we consider an example of a testable scenario of leptogenesis, with a flavor G_f and a CP symmetry [21] that strongly constrain lepton mixing angles and low as well as high energy CP phases [22]. G_f and CP are broken non-trivially in the neutrino and charged lepton sector. We choose in the following $G_f = \Delta(3\,n^2)$ [23] or $G_f = \Delta(6\,n^2)$ [24] (n even, $3 \nmid n$, $4 \nmid n$). LH lepton doublets L_l, $l = e, \mu, \tau$, transform in an irreducible faithful representation **3**, N_α are in an irreducible real representation **3′** and RH charged leptons l_R are assigned to **1** of G_f. The latter are distinguished by $Z_3^{(aux)}$, under which L_l and N_α are invariant. The CP symmetry is given by the CP transformation $X(s)(\mathrm{r})$ in the representation r and depends on the integer parameter s, $0 \leq s \leq n-1$ [25]. The form of the neutrino Dirac Yukawa coupling matrix h and the charged lepton mass matrix m_l are determined by the residual symmetries $G_\nu = Z_2 \times CP$ and $G_e = Z_3^{(D)}$ (the diagonal subgroup of Z_3 in G_f and $Z_3^{(aux)}$), respectively. Specifically [26],

$$h = \Omega(s)(\mathbf{3})\, R_{13}(\theta_L) \begin{pmatrix} h_1 & 0 & 0 \\ 0 & h_2 & 0 \\ 0 & 0 & h_3 \end{pmatrix} R_{13}(-\theta_R)\, \Omega(s)(\mathbf{3}')^\dagger\,, \tag{40.10}$$

where the unitary matrices $\Omega(s)(\mathbf{r})$ are determined by the CP transformation $X(s)(\mathrm{r})$ and $R_{13}(\theta)$ is a rotation in the (13)-plane through the angle θ. There are five real parameters, h_i, θ_L and θ_R, in (40.10). The charged lepton mass matrix m_l is diagonal with three undetermined entries corresponding to the charged lepton masses. The Majorana mass matrix M_N of RH neutrinos is invariant under G_f and CP, and is of the form

$$M_N = m_N \begin{pmatrix} 1 & 0 & 0 \\ 0 & 0 & 1 \\ 0 & 1 & 0 \end{pmatrix}. \tag{40.11}$$

We consider the two cases of strong normal (NO) and inverted ordering (IO): (*a*) strong NO arises for $h_1 = 0$ so that m_1 vanishes, $m_2 = h_2^2 v^2/m_N$ and $m_3 = h_3^2\,|\cos 2\,\theta_R|\, v^2/m_N$, while (*b*) strong IO arises for $h_3 = 0$ so that $m_3 = 0$, $m_1 = h_1^2\,|\cos 2\,\theta_R|\, v^2/m_N$ and $m_2 = h_2^2 v^2/m_N$, where $v \approx 174$ GeV is the SM Higgs vacuum expectation value. The two non-vanishing Yukawa couplings are fitted to the solar and the atmospheric mass squared differences $\Delta m^2_{\rm sol}$ and $\Delta m^2_{\rm atm}$ [27].

High Energy CP Phases and Leptogenesis. Higher-order corrections in general break the residual symmetries and thus change the forms of h, m_l and M_N given above. We focus on contributions to M_N that possess the residual symmetry G_e. These are proportional to κ that is a positive power of the symmetry breaking parameter, measured in units of m_N. As a result, a small splitting of the RH neutrino masses is induced:

$$m_1 = m_N\,(1 + 2\,\kappa) \;\; \text{and} \;\; m_2 = m_3 = m_N\,(1 - \kappa)\,. \tag{40.12}$$

The decay of N_α can thus generate η_B via resonant leptogenesis [8]. The CP asymmetries due to the decay of N_α are given in (40.5). We find that $\varepsilon_{l3} = 0$ for all l and $\varepsilon_{l\alpha}$ only has one contribution for $\alpha = 1, 2$ and we find

$$\varepsilon_{l1} \propto \begin{cases} \frac{h_2 h_3}{9}\,[-2\,h_2^2 + h_3^2\,(1 - \cos 2\,\theta_R)]\, \sin 3\,\phi_s\, \sin\theta_R\, \sin\theta_{L,l} & \text{(strong NO)} \\ \frac{h_1 h_2}{9}\,(-2\,h_2^2 + h_1^2\,(1 + \cos 2\,\theta_R))\, \sin 3\,\phi_s\, \cos\theta_R\, \cos\theta_{L,l} & \text{(strong IO)} \end{cases} \tag{40.13}$$

with $\phi_s = \frac{\pi s}{n}$, $\theta_{L,l} = \theta_L + \rho_l\, 4\pi/3$ and $\rho_e = 0$, $\rho_\mu = 1$, $\rho_\tau = -1$. The CP asymmetries ε_{l2} are the negative of ε_{l1}. We note that different values of s can lead to the same value of $\varepsilon_{l\alpha}$, in particular we find that

$$\varepsilon_{l\alpha}(s) = (-1)^s\, \varepsilon_{l\alpha}(n-s) = \varepsilon_{l\alpha}(n/2-s) = (-1)^{s+1}\, \varepsilon_{l\alpha}(n/2+s) \quad \text{for } s \le n/2\,. \tag{40.14}$$

Collider Signals. For the decay width of the RH neutrinos N_α we find

$$\begin{aligned} \Gamma_{N_1} &\approx \frac{m_N}{24\,\pi}\left(2\,h_1^2\cos^2\theta_R + h_2^2 + 2\,h_3^2\sin^2\theta_R\right)\,, \\ \Gamma_{N_2} &\approx \frac{m_N}{24\,\pi}\left(h_1^2\cos^2\theta_R + 2\,h_2^2 + h_3^2\sin^2\theta_R\right)\,, \\ \Gamma_{N_3} &\approx \frac{m_N}{8\,\pi}\left(h_1^2\sin^2\theta_R + h_3^2\cos^2\theta_R\right)\,. \end{aligned} \tag{40.15}$$

For m_N in the few hundred GeV range, we expect $h_i \sim 10^{-7}$ and thus decay lengths of a few mm to cm in the laboratory frame (for a typical boost factor of $\mathcal{O}(10)$). The interesting point is that for $\theta_R \approx \pi/2,\ 3\pi/2$ (for strong NO) or $\theta_R \approx 0,\ \pi$ (for strong IO) the RH neutrino N_3 can have a very long lifetime, since Γ_{N_3} tends to zero. The decay widths of N_1 and N_2 remain unsuppressed in these limits. Thus, $N)1, 2$ lead to either prompt or displaced vertex signatures at LHC [28], while N_3 can be long-lived enough to be probed at dedicated facilities like MATHUSLA [21].

In general, any observation of lepton number violation (LNV) at the LHC will yield a lower bound on the washout factor for the lepton asymmetry and could falsify *high*-scale leptogenesis as a viable mechanism behind the observed BAU [29]. However, one should keep in mind possible exceptions to this general argument, e.g. scenarios where LNV is confined to a specific flavor sector, models with new symmetries and/or conserved charges which could stabilize the baryon asymmetry against LNV washout, and models where lepton asymmetry can be generated below the observed LNV scale. An important related question is whether *low*-scale leptogenesis models could be ruled out from experiments. This has been investigated [30, 31] in the context of Left-Right symmetric models and it was shown that the minimum value of the RH gauge boson mass for successful leptogenesis, while satisfying all experimental constraints in the low-energy sector, is about 10 TeV. Thus, any positive signal for an RH gauge boson at the LHC might provide a litmus test for the mechanism of leptogenesis.

40.5 Conclusion

Leptogenesis is an attractive mechanism for dynamically generating the observed baryon asymmetry of the Universe, while relating it to the origin of neutrino mass. Resonant leptogenesis allows the relevant energy scale to be as low as the electroweak scale, thus offering a unique opportunity to test this idea in laboratory experiments. Flavor effects play an important role in the predictions for the lepton asymmetry, and hence, for the testability of the low-scale leptogenesis models. We have illustrated that models of resonant leptogenesis could lead to observable effects in current and future experiments, and may even be falsified in certain cases.

Acknowledgements This work was partially supported by the US Department of Energy under Grant No. DE-SC0017987.

References

1. P.A.R. Ade et al., Planck collaboration. Astron. Astrophys. **594**, A13 (2016)
2. J.M. Cline, *Les Houches Lectures*. hep-ph/0609145
3. M. Fukugita, T. Yanagida, Phys. Lett. B **174**, 45 (1986)
4. B. Garbrecht, E. Molinaro (eds.), Int. J. Mod. Phys. A **33**(5–6) (2018)
5. S. Davidson, A. Ibarra, Phys. Lett. B **535**, 25 (2002); W. Buchmüller, P. Di Bari, M. Plümacher, Nucl. Phys. B **643**, 367 (2002)
6. W. Buchmüller, P. Di Bari, M. Plümacher, Ann. Phys. **315**, 305 (2005)
7. J.R. Ellis, J.E. Kim, D.V. Nanopoulos, Phys. Lett. B **145**, 181 (1984); M. Kawasaki, K. Kohri, T. Moroi, A. Yotsuyanagi, Phys. Rev. D **78**, 065011 (2008)
8. A. Pilaftsis, Nucl. Phys. B **504**, 61 (1997); A. Pilaftsis, T.E.J. Underwood, Phys. Rev. D **56**, 5431 (1997); Nucl. Phys. B **692**, 303 (2004); Phys. Rev. D **72**, 113001 (2005)
9. A. Pilaftsis, Phys. Rev. Lett. **95**, 081602 (2005); F.F. Deppisch, A. Pilaftsis, Phys. Rev. D **83**, 076007 (2011)
10. P.S.B. Dev, P. Millington, A. Pilaftsis, D. Teresi, Nucl. Phys. B **886**, 569 (2014); Nucl. Phys. B **897**, 749 (2015)
11. F.F. Deppisch, P.S.B. Dev, A. Pilaftsis, New J. Phys. **17**(7), 075019 (2015)
12. S. Alekhin et al., Rept. Prog. Phys. **79**(12), 124201 (2016)
13. T. Hambye, New J. Phys. **14**, 125014 (2012)
14. S. Blanchet, P. Di Bari, New J. Phys. **14**, 125012 (2012)
15. G. Sigl, G. Raffelt, Nucl. Phys. B **406**, 423 (1993)
16. A. Abada et al., JCAP **0604**, 004 (2006); JHEP **0609**, 010 (2006)
17. E.K. Akhmedov, V.A. Rubakov, A.Y. Smirnov, Phys. Rev. Lett. **81**, 1359 (1998)
18. P.S.B. Dev, P. Millington, A. Pilaftsis, D. Teresi, Nucl. Part. Phys. Proc. **273**, 268 (2016); J. Phys. Conf. Ser. **631**, 012087 (2015); P.S.B. Dev, Springer Proc. Phys. **174**, 245 (2016)
19. L. Kadanoff, G. Baym, *Quantum Statistical Mechanics* (Benjamin, New York, 1962)
20. P.S.B. Dev, P. Millington, A. Pilaftsis, D. Teresi, Nucl. Phys. B **891**, 128 (2015)
21. D. Curtin et al., arXiv:1806.07396 [hep-ph]
22. F. Feruglio, C. Hagedorn, R. Ziegler, JHEP **1307**, 027 (2013). [arXiv:1211.5560 [hep-ph]]
23. C. Luhn, S. Nasri, P. Ramond, J. Math. Phys. **48**, 073501 (2007). [hep-th/0701188]
24. J.A. Escobar, C. Luhn, J. Math. Phys. **50**, 013524 (2009). [arXiv:0809.0639 [hep-th]]
25. C. Hagedorn, A. Meroni, E. Molinaro, Nucl. Phys. B **891**, 499 (2015). [arXiv:1408.7118 [hep-ph]]
26. C. Hagedorn, E. Molinaro, Nucl. Phys. B **919**, 404 (2017). [arXiv:1602.04206 [hep-ph]]
27. I. Esteban, M.C. Gonzalez-Garcia, M. Maltoni, I. Martinez-Soler, T. Schwetz, JHEP **1701**, 087 (2017). [arXiv:1611.01514 [hep-ph]]
28. J.C. Helo, M. Hirsch, S. Kovalenko, Phys. Rev. D **89**, 073005 (2014) [Erratum: Phys. Rev. D **93**, 099902 (2016)] [arXiv:1312.2900 [hep-ph]]
29. F.F. Deppisch, J. Harz, M. Hirsch, Phys. Rev. Lett. **112**, 221601 (2014)
30. J.M. Frere, T. Hambye, G. Vertongen, JHEP **0901**, 051 (2009)
31. P.S.B. Dev, C.H. Lee, R.N. Mohapatra, Phys. Rev. D **90**, 095012 (2014); J. Phys. Conf. Ser. **631**(1), 012007 (2015)

Chapter 41
India-Based Neutrino Observatory (INO): Physics and Status Report

D. Indumathi

Abstract We discuss the physics reach and current status of the India-based Neutrino Observatory (INO) project. We set this in the context of the proposed magnetised iron calorimeter (ICAL) detector at INO, whose main goal is the determination of the neutrino mass hierarchy using atmospheric neutrinos. We also discuss various possible synergies with other planned and upcoming experiments. We also mention the status of the mini-ICAL prototype that has been set up at IICHEP in Madurai.

41.1 Introduction

The India-based Neutrino Observatory (INO) project was mooted several years ago and is presently awaiting several clearances, as well as the outcome of litigation claiming that construction of the underground lab will pollute the ground water, affect dams and such structures that are about 40 km away, and adversely affect the local people living there apart from the local flora and fauna. A strong outreach programme has been launched to clear such misapprehensions. In the meanwhile, the detector simulations have progressed very well and the construction of the 85 ton mini-ICAL prototype is also complete. The entire physics programme is focussed on determining various unknown parameters in neutrino oscillation physics and is very strongly identified. Details have been published elsewhere [1]. In this talk we discuss the physics potential of the proposed magnetised iron calorimeter (ICAL) detector at INO as well as the status of the prototype and the project.

D. Indumathi (✉)
The Institute of Mathematical Sciences, CIT Campus, Chennai 600113, India
e-mail: indu@imsc.res.in
URL: http://imsc.res.in/~indu/

A. Giri and R. Mohanta (eds.), *16th Conference on Flavor Physics and CP Violation*, Springer Proceedings in Physics 234,
https://doi.org/10.1007/978-3-030-29622-3_41

41.2 INO and Atmospheric Neutrinos

Cosmic rays interact with C, O nuclei in Earth's atmosphere to give pions and kaons. These decay to muons and both neutrinos and anti-neutrinos:

$$\pi^{\pm}(K) \rightarrow \mu^{\pm}\, \nu_{\mu}\ ; \qquad \mu^{\pm} \rightarrow \nu_{\mu}\, e^{\pm}\, \nu_{e}\ . \tag{41.1}$$

ICAL has limited sensitivity to hadrons and electrons and is primarily sensitive to muons produced by charged current (CC) interactions of ν_{μ} with the detector. Hence it is specially suited to study neutrino oscillation parameters in the 2–3 sector, where the mixing angle θ_{23} is known to be large and near maximal, although its octant is unknown, and the modulus of the mass squared difference, $\Delta m^2_{32} \equiv m_3^2 - m_2^2$, is known to be $\Delta m^2_{32} \sim 2.4 \times 10^{-3}\ \text{eV}^2$.

Since atmospheric neutrinos have a large range in energy E and path-length L, atmospheric neutrino detectors such as ICAL must also have nearly 4π solid angle coverage, sensitivity to a large range of energies (the most interesting region for observing matter effects in the 2–3 sector of neutrino oscillations is 3–10 GeV), sensitivity to direction (up/down, since neutrinos that are formed in the atmosphere on the other side of the Earth also reach the detector), and good charge resolution so that the discrimination between μ^- and μ^+ produced in CC ν_{μ} interactions with the detector will point back to the original interaction being caused by a neutrino or anti-neutrino respectively. The remarkable feature of ICAL is that it is completely insensitive to the CP phase δ_{CP}. Hence it will be able to determine the 2–3 neutrino mass ordering without ambiguity. However, it is still sensitive to the value of θ_{23}.

41.3 Simulations Studies with ICAL

A detailed study of the physics potential of ICAL can be found in the ICAL white paper [1]. The magnetised ICAL detector will be composed of 151 layers of 5.6 cm thick iron plates, magnetised to about 1.4 T in the central region and with greater than 1 T over a substantial fraction of the detector. The iron layers are interspersed with the active detectors, the resistive plate chambers (RPCs) with 2 cm strip width in both x and y directions. The RPCs are gas detectors placed at a high voltage of about 9 KV; when charged particles pass through, they produce a discharge that is registered as a "hit" in the detector.

The NUANCE neutrino generator [2] was used to generate atmospheric neutrino interactions with the iron in ICAL. A GEANT-based software has been used to propagate the final state particles through the simulated detector; the hits are registered and stored with their (x, y, z) as well as timing information. When muons propagate through ICAL, they leave tracks that can be fitted using a Kalman filter algorithm to obtain the charge and momentum (magnitude and direction) of the muon. Hadron hits can be analysed to get the total energy alone.

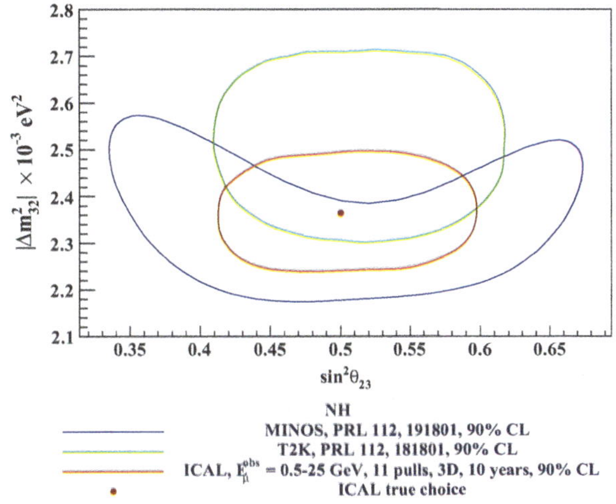

Fig. 41.1 Sensitivity to precision measurements of oscillation parameters in the 2–3 sector for the normal mass ordering [3]

41.3.1 *Precision Studies with ICAL*

Figure 41.1 shows the reach of ICAL with 10 years of data taking. It is seen that, like MINOS, it is very sensitive to $|\Delta m^2_{32}|$. In this and all other cases, the results have been marginalised over θ_{23} and θ_{13}; suitable pulls have been included for flux and cross section uncertainties, zenith angle dependence and energy tilt; significant improvement is seen when the constraints on the *ratios* of neutrino and anti-neutrino fluxes are implemented [3].

Figure 41.2 shows the sensitivity of ICAL with 10 years of data to the neutrino mass ordering in the 2–3 sector. It can be seen that the wrong hierarchy can be ruled out with 2, 3σ confidence with about 4,9 years' running. The improvement when hadron energy information [4] is included (so-called 3D analysis) as opposed to using muon energy and direction information alone (2D analysis) is very sharp.

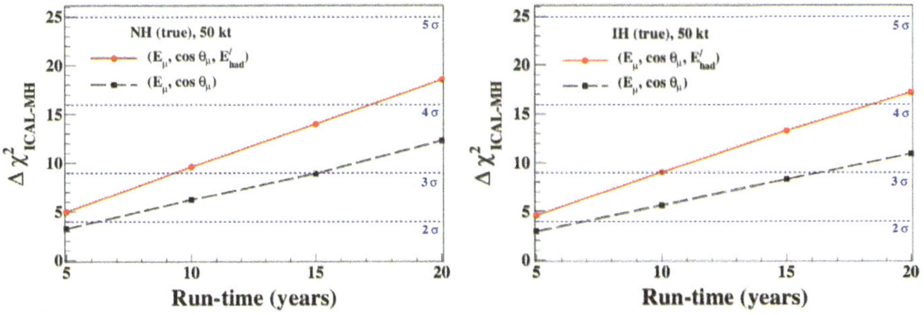

Fig. 41.2 Sensitivity to neutrino mass ordering in the 2–3 sector; notice the improvement on using hadron energy information [4]

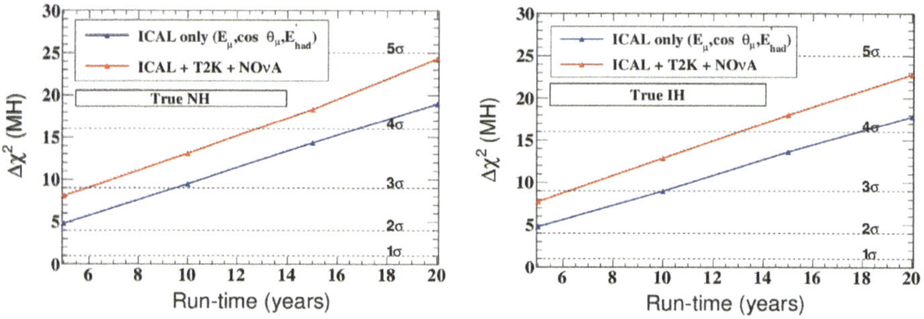

Fig. 41.3 Synergies with T2K and NOνA and subsequent improvement in determining the mass ordering

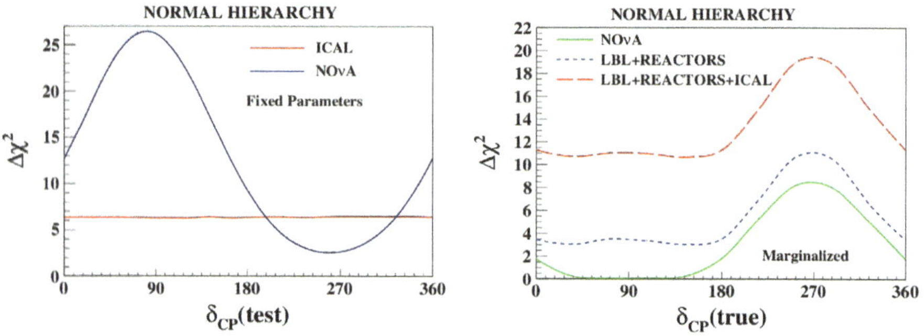

Fig. 41.4 Improvement in discrimination of the CP phase when ICAL data is combined with other experiments

Matter effect/mass hierarchy is the centrepiece of ICAL physics. It has a major role to play in understanding models of neutrino mass and mixing. It also impacts the determination of whether neutrinos are Majorana or Dirac type of fermions.

Apart from INO, MINOS [5], T2K [6], NOνA [7], PINGU/Icecube [8], DUNE [9], JUNO, [10], Hyper-K [11], all will/are probing mass hierarchy. Each is an amazing experiment. Most have to disentangle effects of CP phase from the hierarchy measurement; typically, they can accomplish this only for a fraction of possible δ_{CP} from $-\pi$ to π.

Significant improvements are obtained due to synergies with other experiments when determining the mass hierarchy, as can be seen in Fig. 41.3. Here a total luminosity of 8×10^{21} pot for T2K, and $3 + 3$ years' running with neutrino and anti-neutrino beam for NOνA are assumed. Remember, however, that these other experiments are already taking data while ICAL is yet to be built!

In fact, the complete independence of ICAL physics to the CP phase at these energies can be used to improve the determination of this parameter: see Fig. 41.4 where the earlier analysis of ICAL (without hadron information) was used.

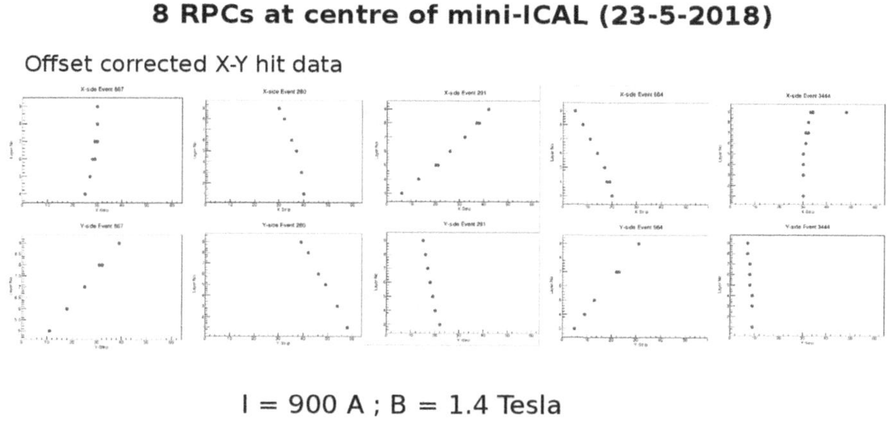

Fig. 41.5 Cosmic muon tracks in mini-ICAL on 23-5-2018

Several other physics possibilities exist: sterile neutrino searches, neutrino decay and limits on life-time, probing non-standard interactions, signatures of Lorentz invariance violation, as well as non-oscillation, non-neutrino physics apart from studying cosmic muons themselves. Many of these are discussed in [1].

41.4 Status of ICAL Prototype

A mini-ICAL with 12 layers of iron stacked in a 2 m × 2 m geometry has been built and the magnetic field configuration tested. The obtained field is close to the simulations, giving confidence in the simulations efforts. Typical hits (due to cosmic ray muons) in the detector are seen in Fig. 41.5.

Many studies on cosmic muons are on-going; in addition, the possibility of a near-surface ICAL is also being probed [12]. In conclusion, the physics simulations, RPC and detector R & D, and prototype studies are on in full swing. A larger engineering module is also in the pipe-line. Many outreach meetings are being held to counter the misinformation campaign against INO. Scientists and especially students of INO are working hard to convince locals, environmental activists, and politicians that the project poses no danger and will in fact be a boon for students in TN and India. We look forward to support from the entire community to take this project forward.

Acknowledgements Thanks to all the members of the ICAL collaboration.

References

1. ICAL Collaboration, Shakeel Ahmed et al., Physics Potential of the ICAL detector at the India-based Neutrino Observatory (INO), Pramana **88**, 79 (2017). arXiv:1505.07380 [physics.ins-det]
2. NUANCE neutrino generator, D. Casper, Nucl. Phys. Proc. Suppl. **112**, 161 (2002). http://www.ps.uci.edu/~nuint/nuance/default.htm
3. L.S. Mohan, D. Indumathi, Pinning down neutrino oscillation parameters in the 23 sector with a magnetised atmospheric neutrino detector: a new study, Eur. Phys. J. **C 77**, 54 (2017). arXiv:1605.04185 [hep-ph]
4. M.M. Devi, T. Thakore, S.K. Agarwalla, A. Dighe, Enhancing sensitivity to neutrino parameters at INO combining muon and hadron information. JHEP **1410**, 189 (2014). [arXiv:1406.3689]
5. MINOS Collaboration, P. Adamson et al., Combined analysis of ν_μ disappearance and $\nu_\mu \rightarrow \nu_e$ appearance in MINOS using accelerator and atmospheric neutrinos. Phys. Rev. Lett. (2014). [arXiv:1403.0867]
6. T2K Collaboration, K. Abe et al., Precise measurement of the neutrino mixing parameter θ_{23} from muon neutrino disappearance in an off-axis beam. arXiv:1403.1532
7. NOvA Collaboration, D. Ayres et al., The NOvA technical design report, technical report. FERMILAB-DESIGN-2007-01
8. IceCube-PINGU Collaboration, M. Aartsen et al., Letter of intent: the precision IceCube next generation upgrade (PINGU). arXiv:1401.2046
9. DUNE Collaboration, R. Acciarri et al., Long-Baseline Neutrino Facility (LBNF) and Deep Underground Neutrino Experiment (DUNE) conceptual design report vol. 2, The Physics Program for DUNE at LBNF. arXiv:1512.0614
10. JUNO Collaboration, F. An et al., Neutrino physics with JUNO. arXiv:1507.0561
11. K. Abe, T. Abe, H. Aihara, Y. Fukuda, Y. Hayato, et al., Letter of intent: the hyper-kamiokande experiment detector design and physics potential. arXiv:1109.3262
12. N. Panchal et al., A compact cosmic muon veto detector and possible use with the Iron Calorimeter detector for neutrinos. J. Instrum. **12**, T11002 (2017); N. Panchal, G. Majumder, V.M. Datar, Simulation of muon-induced neutral particle background for a shallow depth Iron Calorimeter detector (2018). arXiv:1809.08834 [physics.ins-det]

Chapter 42
Majorana Dark Matter, Massless Goldstone and Neutrino Mass in a New $B - L$ Model

Shivaramakrishna Singirala, Rukmani Mohanta, Sudhanwa Patra and Soumya Rao

Abstract We present a comprehensive study of Majorana dark matter in a $U(1)_{B-L}$ gauge extension of the standard model, where three exotic fermions with $B - L$ charges as -4, -4, $+5$ are added to make the model free from the triangle gauge anomalies. We compute the dark matter observables in scalar and gauge portal context and make a parameter scan consistent with the current experimental limits. A massless physical Goldstone boson plays a key role in the scalar portal relic density. Finally, we briefly discuss the neutrino mass generation at one-loop level.

42.1 Brief Introduction and Model Description

$B - L$ gauge extended scenarios are simple in built, rich in phenomenological perspective. These frameworks are free from triangle gauge anomalies with the inclusion of three right-handed neutrinos each with a $B - L$ charge -1. When it comes to fermionic dark matter (DM) study, an additional Z_2 symmetry needs to be imposed to avoid the DM decay. However, adding three exotic fermions charged -4, -4, $+5$ under new $U(1)$ can also make model anomaly free, first proposed in [1]. Moreover, it naturally provides a stable dark matter candidate without imposing any ad-hoc discrete symmetry.

S. Singirala (✉) · R. Mohanta
School of Physics, University of Hyderabad, Hyderabad 500046, India
e-mail: krishnas542@gmail.com

R. Mohanta
e-mail: rmsp@uohyd.ernet.in

S. Patra
Indian Institute of Technology Bhilai, GEC Campus, Sejbahar, Raipur 492015, Chhattisgarh, India
e-mail: sudhanwa@iitbhilai.ac.in

S. Patra
National Centre for Nuclear Research, Hoża 69, 00-681 Warsaw, Poland
e-mail: somu.jrf@gmail.com

A. Giri and R. Mohanta (eds.), *16th Conference on Flavor Physics and CP Violation*, Springer Proceedings in Physics 234,
https://doi.org/10.1007/978-3-030-29622-3_42

Table 42.1 Fields and their charges of the proposed $U(1)_{B-L}$ model

	Field	$SU(2)_L \times U(1)_Y$	$U(1)_{B-L}$
Fermions	$Q_L \equiv (u, d)_L^T$	$(\mathbf{2},\ 1/6)$	1/3
	u_R	$(\mathbf{1},\ 2/3)$	1/3
	d_R	$(\mathbf{1},\ -1/3)$	1/3
	$\ell_L \equiv (\nu,\ e)_L^T$	$(\mathbf{2},\ -1/2)$	−1
	e_R	$(\mathbf{1},\ -1)$	−1
	N_{1R}	$(\mathbf{1},\ 0)$	−4
	N_{2R}	$(\mathbf{1},\ 0)$	−4
	N_{3R}	$(\mathbf{1},\ 0)$	5
Scalars	H	$(\mathbf{2},\ 1/2)$	0
	ϕ_1	$(\mathbf{1},\ 0)$	−1
	ϕ_8	$(\mathbf{1},\ 0)$	8

In this work, we consider the above mentioned scenario where standard model is added with three exotic neutral fermions with $B - L$ charges $-4, -4, +5$. Furthermore, two scalar singlets ϕ_1 and ϕ_8 charged -1 and $+8$ under $U(1)_{B-L}$ are introduced to generate the mass terms for the exotic neutral fermions after the spontaneous breaking of $B - L$ gauge symmetry. Singlet dark matter in this variant has been explored recently in [2]. Detailed version of the present work can be found in [3].

Using the particle content listed in Table 42.1, one can write the following invariant Lagrangian

$$\begin{aligned}
\mathcal{L}_{\rm BL} &= -\frac{1}{3}\, g_{\rm BL}\overline{Q}_L\, Z'_\mu\gamma^\mu Q_L - \frac{1}{3}\, g_{\rm BL}\overline{u}_R\, Z'_\mu\gamma^\mu u_R - \frac{1}{3}\, g_{\rm BL}\overline{d}_R\, Z'_\mu\gamma^\mu d_R \\
&+ g_{\rm BL}\overline{\ell}_L\, Z'_\mu\gamma^\mu \ell_L + g_{\rm BL}\overline{e}_R\, Z'_\mu\gamma^\mu e_R + i\,\overline{N}_{1R}\left(\partial\!\!\!/ + 4i\, g_{\rm BL}\, Z'_\mu\gamma^\mu\right) N_{1R} \\
&+ i\,\overline{N}_{2R}\left(\partial\!\!\!/ + 4i\, g_{\rm BL}\, Z'_\mu\gamma^\mu\right) N_{2R} + i\,\overline{N}_{3R}\left(\partial\!\!\!/ - 5i\, g_{\rm BL}\, Z'_\mu\gamma^\mu\right) N_{3R} \\
&- \frac{y_{\alpha\beta}}{2}\left(\sum_{\alpha,\beta=1,2} \overline{N^c_{\alpha R}}\, N_{\beta R}\,\phi_8 + h.c.\right) - \frac{y_{\alpha 3}}{2}\left(\sum_{\alpha=1,2} \overline{N^c_{\alpha R}}\, N_{3R}\,\phi_1 + h.c.\right) \\
&+ \left|\left(\partial_\mu + i\, g_{\rm BL}\, Z'_\mu\right)\phi_1\right|^2 + \left|\left(\partial_\mu - 8\, i\, g_{\rm BL}\, Z'_\mu\right)\phi_8\right|^2 \\
&- \frac{1}{4} F^{\mu\nu}_{Z'} F^{Z'}_{\mu\nu} - V\left(H, \phi_1, \phi_8\right) + \mathcal{L}_{\rm SM}\,, \qquad (42.1)
\end{aligned}$$

where Z'_μ is the new gauge boson associated with $B - L$ gauge symmetry. Also $F^{Z'}_{\mu\nu} = \partial_\mu Z'_\nu - \partial_\nu Z'_\mu$ is the corresponding field strength tensor for $U(1)_{B-L}$. The scalar potential of the model is given by

$$
\begin{aligned}
V(H,\phi_1,\phi_8) = & \mu_H^2 H^\dagger H + \lambda_H (H^\dagger H)^2 + \mu_1^2 \phi_1^\dagger \phi_1 + \lambda_1 (\phi_1^\dagger \phi_1)^2 + \mu_8^2 \phi_8^\dagger \phi_8 \\
& + \lambda_8 (\phi_8^\dagger \phi_8)^2 + \lambda_{H1} (H^\dagger H)(\phi_1^\dagger \phi_1) + \lambda_{H8} (H^\dagger H)(\phi_8^\dagger \phi_8) \\
& + \lambda_{18} (\phi_1^\dagger \phi_1)(\phi_8^\dagger \phi_8). \qquad (42.2)
\end{aligned}
$$

The spontaneous symmetry breaking of $SU(2)_L \times U(1)_Y \times U(1)_{B-L}$ down to SM gauge group $SU(2)_L \times U(1)_Y$ is implemented with the scalars ϕ_1 and ϕ_8, denoted by $\phi_{1,8} = \frac{1}{\sqrt{2}}\left(v_{1,8} + h_{1,8} + iA_{1,8}\right)$. Then the SM gauge group is spontaneously broken to low energy theory by assigning a VEV v to SM Higgs doublet.

42.2 Mass Spectrum

The fermion and scalar mass matrices take the form

$$
M_N = \begin{pmatrix} y_{11}\langle\phi_8\rangle & y_{12}\langle\phi_8\rangle & y_{13}\langle\phi_1\rangle \\ y_{12}\langle\phi_8\rangle & y_{22}\langle\phi_8\rangle & y_{23}\langle\phi_1\rangle \\ y_{13}\langle\phi_1\rangle & y_{23}\langle\phi_1\rangle & 0 \end{pmatrix}, \quad M_S = \begin{pmatrix} 2\lambda_H v^2 & \lambda_{H1} v v_1 & \lambda_{H8} v v_8 \\ \lambda_{H1} v v_1 & 2\lambda_1 v_1^2 & \lambda_{18} v_1 v_8 \\ \lambda_{H8} v v_8 & \lambda_{18} v_1 v_8 & 2\lambda_8 v_8^2 \end{pmatrix}.
$$

One can diagonalize the above mass matrices using 3×3 unitary matrix U by $U^T_{\alpha(\beta)} M_{N(S)} U_{\alpha(\beta)} = \text{diag}\,[M_{N_{\text{Di}}(H_i)}]$, where $i = 1, 2, 3$. We denote the lightest physical scalar H_1 to be observed Higgs at LHC with $M_{H_1} = 125.09$ GeV and $v = 246$ GeV. In this work, the lightest Majorana mass eigenstate (N_{D1}) is considered to be the dark matter candidate.

Spontaneous breaking of $B - L$ gauge symmetry gives rise to two massless Goldstone modes such that one linear combination A_{G} of them will be eaten up by the neutral gauge boson corresponding to $U(1)_{B-L}$ gauge group and gives mass to Z' ($M_{Z'} = g_{\text{BL}}\sqrt{v_1^2 + 64 v_8^2}$) and the other orthogonal combination A_{NG} remains as massless Goldstone boson.

$$
A_{\text{G}} = \frac{-8v_8 A_8 + v_1 A_1}{\sqrt{v_1^2 + 64 v_8^2}}, \quad A_{\text{NG}} = \frac{v_1 A_8 + 8 v_8 A_1}{\sqrt{v_1^2 + 64 v_8^2}}. \qquad (42.3)
$$

This left over massless mode A_{NG} can couple to the scalar sector and the new fermion spectrum. The fermionic dark matter study can be made in the scalar and gauge portals as the DM interacts with the scalar sector ($H_{1,2,3}$ and A_{NG}) and the vector gauge boson Z'.

42.3 Dark Matter Phenomenology

42.3.1 Scalar Portal

In scalar portal, the DM annihilation channels with $f\bar{f}$, W^+W^-, ZZ, H_iH_j, $A_{\rm NG}A_{\rm NG}$ in final state mediated by $H_{1,2,3}$ and $N_{\rm D1}N_{\rm D1} \rightarrow H_iA_{\rm NG}$ mediated by $A_{\rm NG}$, can contribute to relic density. These s-channel contributions are featured to provide relic density consistent with Planck [4] near the resonance in propagator i.e., $M_{D1} \simeq \frac{M_{\rm prop}}{2}$ as displayed in the left panel of Fig. 42.1.

In direct detection (DD), a spin-independent (SI) contribution can arise by a t-channel process mediated by the massive scalars. Choosing three specific set of values for $M_{H_2} = \{1, 1.5, 2\}$ (in TeV) and varying M_{H_3} in the range $(M_{H_2}, 3)$ TeV, we show in right panel of Fig. 42.1, the DM-nucleon cross section computed for the parameter space consistent with 3σ region of Planck relic limit [4]. Here vertical (diagonal) data points represent resonance near $H_2(H_3)$ propagator. It is clear that the SI contribution is always within the experimental upper limit set by well known LUX [5], XENON1T [6] and PandaX-II [7] collaborations.

42.3.2 Gauge Portal

In gauge portal, the s-channel processes $N_{\rm D1}N_{\rm D1} \rightarrow f\bar{f}, Z'H_{1,2,3}$ can contribute to relic density. As similar to scalar portal, here the required relic density is met near the resonance in Z' propagator. The relevant gauge parameters, $M_{Z'}$ and $g_{\rm BL}$ are well constrained from collider searches.

ATLAS [8] results from the study of dilepton signals for the Z' boson provide the most stringent limits on $M_{Z'}$ and $g_{\rm BL}$. For the present $B-L$ model, we use CalcHEP [9] to compute the production cross section of Z'. We show in the left panel of

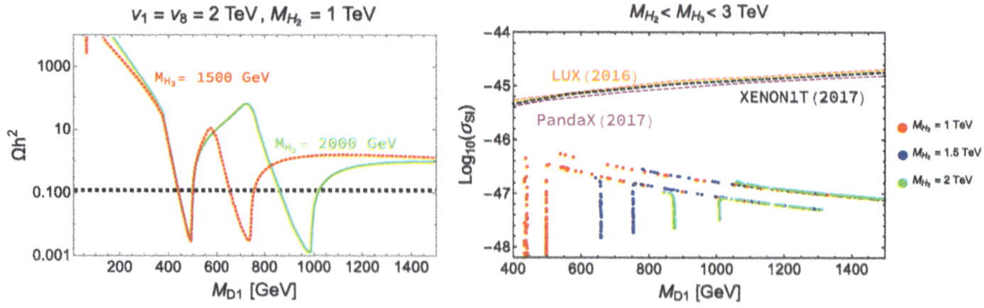

Fig. 42.1 Left panel depicts the scalar-portal relic density as a function of DM mass. Right panel shows SI DM-nucleon cross section for the parameter space that satisfies 3σ in current relic density. Dashed lines represent Planck limit at 3σ (left panel), upper limit on SI contribution set by LUX [5], XENON1T [6] and PandaX-II [7] (right panel)

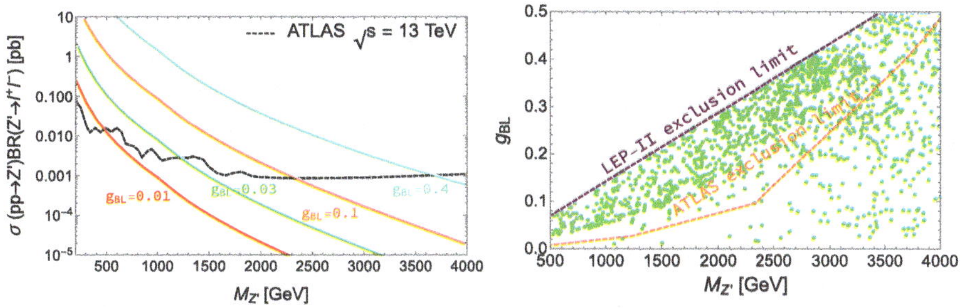

Fig. 42.2 Left panel shows ATLAS dilepton limits on the gauge parameters. The black dashed line represents the exclusion limit from ATLAS [8]. Right panel shows the $M_{Z'} - g_{\mathrm{BL}}$ parameter space that satisfies 3σ in current relic density. Magenta and orange dashed lines represent limits set by LEP-II and ATLAS

Fig. 42.2, dilepton $(ee, \mu\mu)$ signal in Z' production as a function of $M_{Z'}$. It can be seen that for $g_{\mathrm{BL}} = 0.4$, the region below $M_{Z'} \simeq 3.7$ TeV is excluded while for $g_{\mathrm{BL}} = 0.1$, $M_{Z'} < 2.3$ TeV is excluded and so on. Moreover, LEP-II [10] sets an upper limit on the ratio i.e., $\frac{M_{Z'}}{g_{\mathrm{BL}}} > 7$ TeV. Taking into account both the above discussed bounds, we show in the right panel of Fig. 42.2, the $M_{Z'} - g_{\mathrm{BL}}$ parameter space consistent with 3σ region of Planck [4]. Dashed magenta (orange) line denotes the LEP-II [10] (ATLAS [8]) limit on the gauge parameters. The region below the ATLAS limit survives all the experimental limits. Note that, as Z' coupling to Majorana fermion (quarks) is axial vector (vector) type, the contribution by WIMP-nucleon interaction is insensitive to DD experiments [11].

42.4 Light Neutrino Mass

In the present model, the neutrino mass can be generated at one-loop level through radiative mechanism. For this purpose, we introduce an additional inert doublet $\eta = \begin{pmatrix} \eta^+ \\ \frac{S+iA}{\sqrt{2}} \end{pmatrix}$ with the $B-L$ charge -3. Now, one can write the interaction term to generate light neutrino mass at one-loop level as

$$\sum_{\alpha=1,2} Y_{i\alpha} \overline{(\ell_L)_i} \tilde{\eta} N_{\alpha R}. \tag{42.4}$$

The scalar potential gets modified with the inclusion of additional terms as

$$\begin{aligned} V' &= V(H, \phi_1, \phi_8) + \mu_\eta(\eta^\dagger\eta) + \lambda_\eta(\eta^\dagger\eta)^2 + \lambda'_{H\eta}(H^\dagger\eta)(\eta^\dagger H) + \frac{\lambda_{\eta 18}}{2\Lambda^3}\left[(H^\dagger\eta)^2\phi_8\phi_1^2 + \text{h.c.}\right] \\ &+ (\eta^\dagger\eta)\left[\lambda_{H\eta}(H^\dagger H) + \lambda_{\eta 1}(\phi_1^\dagger\phi_1) + \lambda_{\eta 8}(\phi_8^\dagger\phi_8)\right], \end{aligned} \tag{42.5}$$

where Λ is the cut-off parameter. Assuming $m_0^2 = (M_S^2 + M_A^2)/2$ is much greater than $M_S^2 - M_A^2 = \frac{\lambda_{\eta 18}}{2\sqrt{2}\Lambda^3} v^2 v_1^2 v_8$, the expression for the radiatively generated neutrino mass is given by [12]

$$(\mathcal{M}_\nu)_{ij} = \frac{\lambda_{\eta 18} v^2 v_1^2 v_8}{32\sqrt{2}\pi^2\Lambda^3} \sum_{\alpha=1}^{2} \frac{Y_{i\alpha} Y_{j\alpha} M_{D\alpha}}{m_0^2 - M_{D\alpha}^2} \left[1 - \frac{M_{D\alpha}^2}{m_0^2 - M_{D\alpha}^2} \ln \frac{m_0^2}{M_{D\alpha}^2}\right]. \tag{42.6}$$

Here $M_{D\alpha} = (U^T M_N U)_\alpha$ and $N_{D\alpha} = U_{\alpha\beta}^\dagger N_\beta$, with M_N being the Majorana mass matrix. With a sample parameter space, $(Y, \lambda_{\eta 18}) \sim (10^{-2}, 10^{-3})$ and $(v_1, v_8, m_0, M_{D\alpha}, \Lambda) \sim (2, 2, 2, 0.5, 10)$ TeV, one can have $m_\nu \sim 10^{-11}$ GeV.

42.5 Conclusion

In this article, we made a detailed study of Majorana dark matter in a new $U(1)_{B-L}$ gauge extension of standard model, where three exotic fermions with $B - L$ charges $-4, -4, +5$ are added to avoid the triangle gauge anomalies. The scalar sector is equipped with two additional scalar singlets ϕ_1 and ϕ_8 with $B - L$ charges $-1, +8$ to break the $U(1)_{B-L}$ gauge symmetry, thereby giving mass to the exotic fermions and the heavy gauge boson Z'. We have explored the dark matter phenomenology in presence of scalar and vector boson as mediators. The scalar portal relic abundance has been studied with all possible annihilation channels and the effect of massless physical Goldstone boson is suitably addressed. The SI cross section has been calculated and investigated with the current limits from LUX (2016), XENON1T (2017) and PandaX-II (2017). Similar strategy is repeated for Z'-portal channels. LEP-II and ATLAS bounds on the Z' mass and $g_{\rm BL}$ impose strong constraints. However, we still have a viable parameter space satisfying the current experimental limits. We have also addressed the generation of light neutrino mass by adding an additional inert doublet η with $B - L$ charge assigned as -3. To conclude, we have made a complete systematic study of Majorana dark matter in a new variant of $B - L$ gauge extended model. This simple model survives the current collider limits while satisfies dark matter constraints and can be probed in future high luminosity data from LHC.

SS would like to acknowledge DST—Inspire Fellowship division, Govt of India for the financial support through ID No. IF130927. RM would like to thank Science and Engineering Research Board (SERB), Government of India for financial support through grant No. SB/S2/HEP-017/2013.

References

1. J.C. Montero, V. Pleitez, Gauging U(1) symmetries and the number of right-handed neutrinos. Phys. Lett. B **675**, 6468 (2009). https://doi.org/10.1016/j.physletb.2009.03.065
2. S. Singirala, et al., Singlet scalar Dark matter in $U(1)_{B-L}$ models without right-handed neutrinos. arXiv:1704.01107
3. S. Singirala, et al., Majorana dark matter in new $B-L$ model. arXiv:1710.05775
4. Planck Collaboration, P.A.R. Ade et al., Planck 2015 results-XIII. cosmological parameters. Astron. Astrophys. **594**, A13 (2016). https://doi.org/10.1051/0004-6361/201525830
5. LUX Collaboration, D.S. Akerib, et al., Results from a search for dark matter in the complete LUX exposure. Phys. Rev. Lett. **118**(2), 021303 (2017). https://doi.org/10.1103/PhysRevLett.118.021303
6. XENON Collaboration, E. Aprile, et al., First dark matter search results from the XENON1T experiment. Phys. Rev. Lett. **119**(18), 181301 (2017). https://doi.org/10.1103/PhysRevLett.119.181301
7. PandaX-II Collaboration, X. Cui, et al., Dark matter results from 54-ton-day exposure of PandaX-II experiment. Phys. Rev. Lett. **119**(18), 181302 (2017). https://doi.org/10.1103/PhysRevLett.119.181302
8. The ATLAS collaboration, Search for new phenomena in the dilepton final state using proton-proton collisions at $\sqrt{s}=13$ TeV with the ATLAS detector. ATLAS-CONF-2015-070 (2015)
9. A. Belyaev, et al., CalcHEP 3.4 for collider physics within and beyond the standard model. Comput. Phys. Commun. **184**, 1729–1769 (2013). https://doi.org/10.1016/j.cpc.2013.01.014
10. ALEPH and DELPHI and L3 and OPAL and LEP Electroweak Collaborations, S. Schael, et al., Electroweak measurements in electron-positron collisions at W-Boson-pair energies at LEP. Phys. Rep. **532**, 119–244 (2013). https://doi.org/10.1016/j.physrep.2013.07.004
11. P. Agrawal, et al., A classification of dark matter candidates with primarily spin-dependent interactions with matter. UMD-PP-10-004, RUNHETC-2010-07 (2010)
12. E. Ma, Verifiable radiative seesaw mechanism of neutrino mass and dark matter. Phys. Rev. D **73**, 077301 (2006). https://doi.org/10.1103/PhysRevD.73.077301

Chapter 43
T-Violation in Electron- and Antineutrino-Induced Hyperon Production

A. Fatima, M. Sajjad Athar and S. K. Singh

Abstract We study the effect of T-violation on the total scattering cross section and the polarization observables in the electron- and antineutrino-induced hyperon production off the free proton target. The assumption of T-violation requires the second class current form factors to be imaginary, which implies the transverse component of the hyperon polarization perpendicular to the plane of production to be nonzero. The observation of hyperons in the final state through their decay products, i.e., $\Lambda/\Sigma \longrightarrow N\pi$ and the structure of the angular distribution of pion will give information about the polarization of hyperons. The experimental observation of the transverse component of the polarization can be used to study the underlying physics of T-violation.

43.1 Introdution

Discrete symmetries like charge conjugation (C), parity (P) and time reversal (T) play an important role in particle physics. Earlier it was assumed that these symmetries are conserved in all the interactions. However, in 1956, Wu et al. [1] showed that parity is not conserved in weak interactions. Later, though a combined symmetry of C and P was observed to be conserved in pion decay, in 1964, it was found that CP is also not conserved in weak interactions in the neutral kaon sector. Furthermore, a combined operation of these three symmetries known as CPT is found to be conserved in all the interactions. Now, if CPT is conserved in weak interactions and CP is violated then T must be violated in order to conserve CPT.

In this work, we present an alternative way to study T-violation in weak interactions through the polarization of hyperons in their antineutrino induced production. The hyperons decay predominantly into pions i.e. $\Lambda(\Sigma^0) \rightarrow p\pi^-$ or $n\pi^0$, and $\Sigma^- \rightarrow n\pi^-$ and the asymmetry in the angular distribution of pions with respect to a given direction (which can be chosen to be the direction of the momentum of

A. Fatima (✉) · M. Sajjad Athar · S. K. Singh
Department of Physics, Aligarh Muslim University, Aligarh 202002, India
e-mail: atikafatima1706@gmail.com

A. Giri and R. Mohanta (eds.), *16th Conference on Flavor Physics and CP Violation*, Springer Proceedings in Physics 234,
https://doi.org/10.1007/978-3-030-29622-3_43

the hyperon, perpendicular to it or transverse to the reaction plane) determines the polarization component of the hyperon in that direction. The measurement of the transverse component of the polarization, gives information about the weak electric form factor which is forbidden by the T-invariance. Such a measurement will, thus, provide an opportunity to study the underlying physics of T-violation in neutrino interactions. The details of these calculations are given in Refs. [2, 3]. The present and future neutrino experiments like MicroBooNE and DUNE, which use liquid argon time projection chamber (LArTPC) detectors, may be able to study the T-violating effect through the measurements of the polarization of the hyperon. The LArTPC is itself a target as well as a detector and gives the three dimensional track of the interaction, therefore, the asymmetry in the angular distribution of pions may be studied by LArTPC and determines the polarization of hyperon.

43.2 Quasielastic Production of Hyperons

The quasielastic hyperon production processes are

$$\bar{\nu}_\mu(k) + p(p) \rightarrow \mu^+(k') + Y(p') \tag{43.1}$$

$$e^-(k) + p(p) \rightarrow \nu_e(k') + Y(p'), \quad Y = \Lambda, \Sigma^0, \tag{43.2}$$

where the quantities in the parentheses represent the four momenta of the corresponding particles. The transition matrix element $\mathcal{M}$ for the above processes is given by

$$\mathcal{M} = \frac{G_F}{\sqrt{2}} \sin\theta_c \, \ell^\mu \, J_\mu, \tag{43.3}$$

where the leptonic (ℓ^μ) and the hadronic (J_μ) currents are defined as

$$\ell^\mu = \bar{u}(k')\gamma^\mu(1 \pm \gamma_5)u(k), \tag{43.4}$$

$$\begin{aligned} J_\mu = \bar{u}(p') \Bigg[& \gamma_\mu f_1(Q^2) + i\sigma_{\mu\nu}\frac{q^\nu}{M + M_Y} f_2(Q^2) + \frac{2q_\mu}{M + M_Y} f_3(Q^2) \\ & - \left(\gamma_\mu\gamma_5 g_1(Q^2) + i\sigma_{\mu\nu}\gamma_5 \frac{q^\nu}{M + M_Y} g_2(Q^2) + \frac{2q_\mu\gamma_5}{M + M_Y} g_3(Q^2) \right) \Bigg] u(p). \end{aligned} \tag{43.5}$$

with $+(-)$ representing the antineutrino (electron) induced processes. The weak form factors $f_i(q^2)$ and $g_i(q^2)$, $(i = 1 - 3)$ are determined with various symmetry properties like T-invariance, G-invariance and SU(3) symmetry while assuming the hypothesis of conserved vector current and partially conserved axial vector current. The details are given in Refs. [2, 3].

Using the above definitions, the Q^2 distribution is written as

$$\frac{d\sigma}{dQ^2} = \frac{G_F^2 \sin^2\theta_c}{8\pi M^2 E_{\bar{\nu}_\mu}^2} \mathcal{L}_{\mu\nu}\mathcal{J}^{\mu\nu} \tag{43.6}$$

where the leptonic and hadronic tensors are obtained as

$$\mathcal{L}_{\mu\nu} = a \sum \ell_\mu \ell_\nu^\dagger, \qquad \mathcal{J}_{\mu\nu} = \sum J_\mu J_\nu^\dagger, \tag{43.7}$$

and $a = 1$ $\left(\frac{1}{2}\right)$ for the antineutrino (electron) induced process.

The vector form factors are obtained in terms of nucleon electromagnetic form factors for which various parameterizations are available in the literature; in our numerical calculations, we have used the parameterization given by Bradford et al. [4]. The axial vector and weak electric form factors $g_{1,2}(Q^2)$ are determined in terms of $g_{A,2}(Q^2)$, which are parameterized in the dipole form as

$$g_i(Q^2) = g_i(0)\left[1 + \frac{Q^2}{M_i^2}\right]^{-2}; \quad i = A, 2 \tag{43.8}$$

with $g_A(0) = 1.267$, $g_2(0) = i g_2^I(0)$ and $M_A = M_2 = 1.026$ GeV.

43.3 Polarization of the Hyperon

The polarization 4-vector (ξ^τ) of the hyperon produced in the quasielastic reaction is written as:

$$\xi^\tau = \left(g^{\tau\sigma} - \frac{p'^\tau p'^\sigma}{M_Y^2}\right) \frac{\mathcal{L}^{\alpha\beta}\mathrm{Tr}\left[\gamma_\sigma\gamma_5 \Lambda(p') J_\alpha \Lambda(p) \tilde{J}_\beta\right]}{\mathcal{L}^{\alpha\beta}\mathrm{Tr}\left[\Lambda(p') J_\alpha \Lambda(p) \tilde{J}_\beta\right]}. \tag{43.9}$$

One may write the polarization vector as

$$\boldsymbol{\xi} = \xi_P \mathbf{e}_P + \xi_L \mathbf{e}_L + +\xi_T \mathbf{e}_T,$$

where $\mathbf{e}_P$, $\mathbf{e}_L$, and $\mathbf{e}_T$ are the unit vectors corresponding to the perpendicular, longitudinal, and transverse directions along the momentum of the hyperon and are given as

$$\mathbf{e}_L = \frac{\boldsymbol{p}'}{|\boldsymbol{p}'|}, \qquad \mathbf{e}_P = \mathbf{e}_L \times \mathbf{e}_T, \qquad \mathbf{e}_T = \frac{\boldsymbol{p}' \times \boldsymbol{k}}{|\boldsymbol{p}' \times \boldsymbol{k}|};$$

and $\xi_{P,L,T}(Q^2)$ are described as

$$\xi_{P,L,T}(Q^2) = \boldsymbol{\xi} \cdot \mathbf{e}_{P,L,T}. \tag{43.10}$$

The longitudinal $P_L(Q^2)$, perpendicular $P_P(Q^2)$, and transverse $P_T(Q^2)$ components of the polarization vector are:

$$\begin{aligned}
P_L(Q^2) &= \frac{M_Y}{E_{p'}}\xi_L(Q^2) = \frac{M_Y}{E_{p'}}\frac{A(Q^2)\boldsymbol{k}\cdot\boldsymbol{p}' + B(Q^2)|\boldsymbol{p}'|^2}{N(Q^2)\,|\boldsymbol{p}'|},\\
P_P(Q^2) &= \xi_P(Q^2) = \frac{A(Q^2)[(\boldsymbol{k}\cdot\boldsymbol{p}')^2 - |\boldsymbol{k}|^2|\boldsymbol{p}'|^2]}{N(Q^2)\,|\boldsymbol{p}'|\,|\boldsymbol{p}'\times\boldsymbol{k}|},\\
P_T(Q^2) &= \xi_T(Q^2) = \frac{C(Q^2)M[(\boldsymbol{k}\cdot\boldsymbol{p}')^2 - |\boldsymbol{k}|^2|\boldsymbol{p}'|^2]}{N(Q^2)\,|\boldsymbol{p}'\times\boldsymbol{k}|}.
\end{aligned}$$

The expressions of $A(Q^2)$, $B(Q^2)$, $C(Q^2)$ and $N(Q^2)$ are given in Refs. [2, 3].

43.4 Results and Discussion

In Fig. 43.1, we present the results for $\sigma(E_e)$, $P_L(E_e)$, $P_P(E_e)$ and $P_T(E_e)$ for the process $e^- + p \rightarrow \nu_e + \Lambda$ obtained using $g_2(0) = 0$, 1 and 3 and $M_2 (= M_A) = 1.026$ GeV. We find that there is little dependence of $\sigma(E_e)$ and $P_L(E_e)$ on $g_2(0)$ while $P_P(E_e)$ and $P_T(E_e)$ have significant dependence on $g_2(0)$ especially for $g_2(0) \geq 1$.

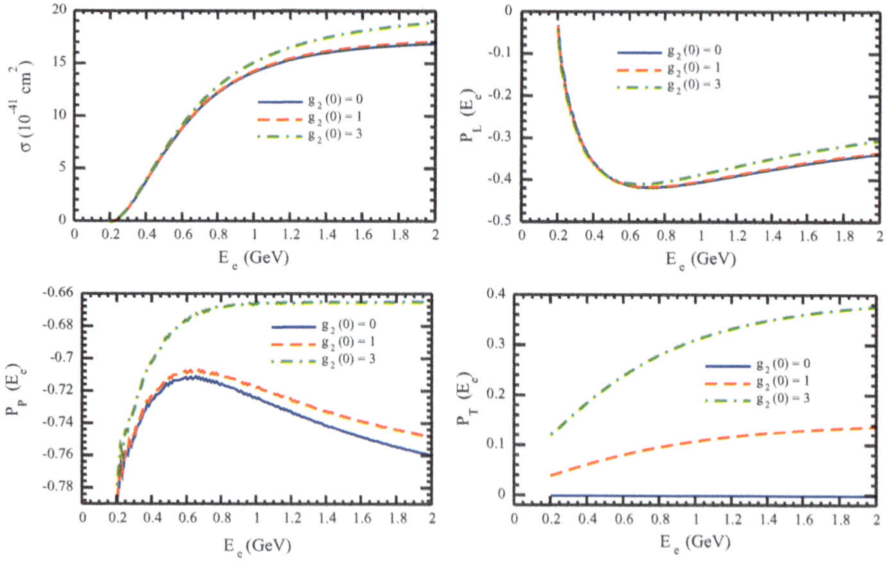

Fig. 43.1 Cross section and average polarization components for the e^- induced Λ production

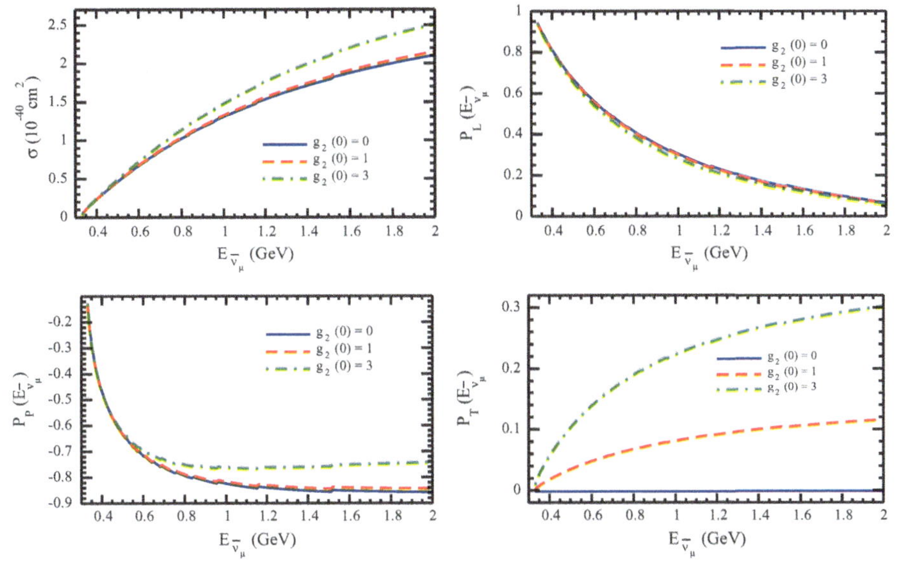

Fig. 43.2 Cross section and average polarization components for the $\bar{\nu}_\mu$ induced Λ production

In Fig. 43.2, the results for the total cross section $\sigma(E_{\bar{\nu}_\mu})$, and average polarizations $P_L(E_{\bar{\nu}_\mu})$, $P_P(E_{\bar{\nu}_\mu})$ and $P_T(E_{\bar{\nu}_\mu})$ are presented for the different values of $g_2(0) =$ 0, 1 and 3 for the process $\bar{\nu}_\mu + p \rightarrow \mu^+ + \Lambda$. We find that the value of $\sigma(E_{\nu_\mu})$ increases with energy for all values of $g_2(0)$. The effect of $g_2(0) \neq 0$ is to further increase the cross section and this increase is significant only for $|g_2(0)| \geq 1$. $P_L(E_{\nu_\mu})$ and $P_P(E_{\nu_\mu})$ components of the Λ polarization are less sensitive to the variation in $g_2(0)$. The transverse component of the Λ polarization increases with energy. Even for $g_2(0) = 1$, it could be $\sim 8\%$ for $E_{\nu_\mu} \sim 1$ GeV and becomes larger for $g_2(0) \neq 0$. Therefore, it may be concluded that in the hyperon production reactions induced by the electron as well as by the antineutrino, i.e., $e^- + p \longrightarrow \nu_e + \Lambda$ and $\bar{\nu}_\mu + p \longrightarrow \mu^+ + \Lambda$, respectively, it is feasible to study the effect of T-violation in the hyperon, provided the T-violating form factor $|g_2(0)| \geq 1$.

It is well known that with the increase in the axial dipole mass M_A, the cross section increases and the (anti)neutrino scattering experiments in the higher energy region suggest a higher value of M_A (like MiniBooNE, K2K, T2K, SciBooNE experiments) from the world average value i.e. $M_A = 1.026$ GeV. It is, therefore, possible that a nonzero value of $g_2(0)$ may lead to a smaller value of M_A. In Fig. 43.3, we show the dependence of cross section on M_A with the presence of $g_2(0)$. In the case of electron induced process i.e. $e^- + p \rightarrow \nu_e + \Lambda$, the results obtained by taking $M_A = 1.1$ (1.2) GeV and $g_2(0) = 0$ are comparable to that obtained with $M_A = 1.026$ GeV and $g_2(0) = 2$ (2). While in the case of antineutrino induced process i.e. $\bar{\nu}_\mu + p \rightarrow \mu^+ + n$, the results obtained by taking $M_A = 1.1$ (1.2) GeV and

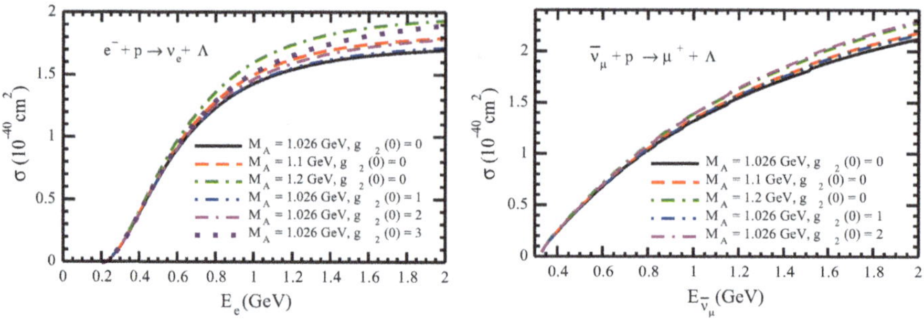

Fig. 43.3 Cross section with different values of M_A and $g_2(0)$ for the e^- and $\bar{\nu}_\mu$ induced Λ production

$g_2(0) = 0$ are comparable to that obtained with $M_A = 1.026$ GeV and $g_2(0) = 1$ (2). Therefore, a higher value of the total cross section may be obtained by either taking a nonzero value of $g_2(0)$ or increasing the value of M_A.

References

1. C.S. Wu, E. Ambler, R.W. Hayward, D.D. Hoppes, R.P. Hudson, Phys. Rev. **105**, 1413 (1957)
2. A. Fatima, M. Sajjad Athar, S.K. Singh, Eur. Phys. J. A **54**, 95 (2018)
3. A. Fatima, M. Sajjad Athar, S.K. Singh, Phys. Rev. D **98**, 033005 (2018)
4. R. Bradford et al., Nucl. Phys. Proc. Suppl. **159**, 127 (2006)

Chapter 44
Measurement of the Inclusive Muon Neutrino Charged-Current Cross-Section in the NOvA Near Detector

Biswaranjan Behera

Abstract The NOvA near detector, located off-axis in the NuMI beam, records a high rate of neutrino interactions with energies ranging from 1 to 4 GeV. These interactions are used to predict the neutrino energy spectrum at the NOvA far detector, 14 mrad off-axis and 810 km away, for neutrino oscillation measurements. We present a flux-averaged double-differential inclusive cross section measurement of muon-neutrino charged-current interactions in the NOvA near detector as a function of the muon energy and angle. We also present the measurement of the cross section as a function of neutrino energy in a restricted phase space of the muon kinematics.

44.1 Introduction

NOvA [1] is a two detector experiment. The far detector (FD) is 810 km from the NuMI production target and positioned 14 mrad off-axis from the NuMI beam, resulting in a narrow-band neutrino flux peaked around 2 GeV. The NOvA near detector (ND) is located approximately 1 km from the NuMI production target, off-axis such that the peak of the neutrino flux matches that of the far detector. However, the proximity of the ND to the production target means that it is exposed to a broader neutrino energy flux, and has a longer tail of neutrino interactions at higher energies than the FD. The high rate of interactions in the ND provides an opportunity for a rich program of neutrino-nucleus cross section measurements.

Neutrino oscillation experiments do not measure the oscillation probability, but rather the event rate. So we need to measure the number of neutrinos as a function of neutrino energy at near and far detectors. The number of neutrinos at the near detector is convolution of flux, cross-section and efficiency of near detector.

Biswaranjan Behera—For the NOvA collaboration.

B. Behera (✉)
IIT Hyderabad / Fermilab, Hyderabad, India
e-mail: bbehera@fnal.gov

A. Giri and R. Mohanta (eds.), *16th Conference on Flavor Physics and CP Violation*, Springer Proceedings in Physics 234,
https://doi.org/10.1007/978-3-030-29622-3_44

$$N_{\nu_\alpha}^{\text{ND}}(E_\nu) \propto \phi(E_\nu) \times \sigma(E_\nu) \times \varepsilon^{\text{ND}}(E_\nu, T_\alpha, \Theta_\alpha ...) \quad (44.1)$$

Flux ($\phi(E_\nu)$) is defined as number of neutrinos produced by the accelerator per cm^2, per energy, for a given number of protons on target. The probability of interaction of the neutrinos in the material of the detector is known as cross-section ($\sigma(E_\nu)$) and efficiency (ε) corrects for events which we lose during selection (threshold, acceptance, containment...). Due to convolution of flux and cross section, even if high correlation between flux, cross-section, and detector efficiency between near and far detector it will not remove all dependence.

The event rate approximately can be written as:

$$\frac{N_{\nu_\beta}^{\text{FD}}}{N_{\nu_\alpha}^{\text{ND}}} \approx \frac{\phi_{\nu_\beta}^{\text{FD}}(E_\nu)}{\phi_{\nu_\alpha}^{\text{ND}}(E_\nu)} \times \frac{\sigma_{\nu_\beta}^{\text{FD}}(E_\nu)}{\sigma_{\nu_\alpha}^{\text{ND}}(E_\nu)} \times \frac{\varepsilon_{\nu_\beta}^{\text{FD}}}{\varepsilon_{\nu_\alpha}^{\text{ND}}} \times P_{\nu_\alpha \to \nu_\beta}(E_\nu) \quad (44.2)$$

The oscillated flux at the far detector is not same as the un-oscillated flux measured at near detector as far detector sees a point source of neutrino beam, whereas near detector sees a broadband of neutrino source. Also, the acceptance of particles is different at both the near and far detectors, efficiency calculation depends on the simulated particles from an event generator which is based on the neutrino-nucleus interaction model. The dependence of efficiency on the kinematics of the final state which is driven by cross-section and the different model predicts different particle multiplicity in the final state which gives birth to uncertainty. The uncertainty does not cancel exactly due to the difference in the near and far detectors.

44.2 Event Selection

In the event selection for this analysis, we begin with some basic event quality criteria to reject obvious backgrounds and ensure that there is a reconstructed Kalman track [2] in the event. We next apply some loose fiducial and containment criteria based on reconstructed showers. With these criteria in place, we implement a new muon identification algorithm based on energy loss and scattering of reconstructed Kalman tracks, and optimize a cut on a score provided by that algorithm. Finally, we optimize the fiducial and containment selection criteria with the muon selection applied.

A general strategy in this analysis is to determine selection criteria that minimize the systematic uncertainty of the total cross section [3]:

$$\sigma = \frac{N_{\text{sel}} - N_{\text{bkg}}}{N_{\text{target}} \Phi \varepsilon} \quad (44.3)$$

where N_{sel} is the number of selected events, N_{bkg} is the estimated number of selected background events, Φ is the total flux, N_{target} is the number of targets and ε is the efficiency of the selection. Furthermore,

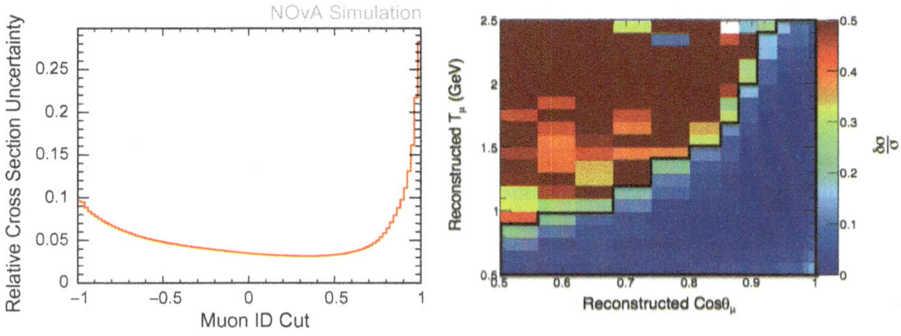

Fig. 44.1 Each bin is filled with total uncertainty if the chosen cut was the left edge of that bin. Muon ID > 0.34 are retained as candidate of ν_μ CC events (left). The fractional uncertainty on total cross-section as a function of kinematic phase space of muon (right)

$$\varepsilon = \frac{N_{\text{sel,signal}}}{N_{\text{signal}}} \tag{44.4}$$

Ignoring uncertainties on the flux and the number of targets (which depend very weakly on the selection criteria), and correlations between background estimates and efficiency, the fractional uncertainty on the cross section is:

$$\frac{\delta\sigma}{\sigma} = \sqrt{\frac{(\delta N_{\text{sel}}^{\text{stat}})^2 + (\delta N_{\text{bkg}}^{\text{stat}})^2 + (\delta N_{\text{bkg}}^{\text{syst}})^2}{(N_{\text{sel}} - N_{\text{bkg}})^2} + \left(\frac{\delta\varepsilon}{\varepsilon}\right)^2} \tag{44.5}$$

In the studies that follow, the statistical uncertainty on the number of selected events is negligible and therefore ignored, and the fractional uncertainty on the total cross section is plotted as either a function of a variable (e.g., vertex position) or as a function of a cut value (Fig. 44.1).

44.3 Neutrino Energy Reconstruction

Simulated neutrino interactions are used to relate reconstructed track and slice observables to the energy of the neutrino. The general procedure is to first determine the muon energy by relating the track length to the true energy of the muon. Next, the calorimetric energy in the slice not associated with the muon is associated to the hadronic energy in the event. This is done by requiring the summed reconstructed muon and hadronic energies to be that of the true neutrino energy.

44.4 Muon Kinematic Cuts

Because of the limited acceptance of the near detector, we constrain this measurement to the region of kinematic phase space of the final-state muon that has reasonable uncertainties on the measurement. Figure 44.1 shows the relative uncertainty in the differential cross section measurement in each bin. We limit our measurement to fractional uncertainties less than approximately 30%.

44.5 Efficiency Correction

The efficiency correction accounts for the loss of true signal to detector acceptance and the signal selection criteria. It is applied to the background subtracted, unfolded distribution, to recover the estimated true signal events as a function of a quantity of concern. In the measurement of the cross-section in the double-differential muon kinematic space, the efficiency correction is applied in the 2D space and naturally accounts for the observed differences in muon kinematics in the data and simulation, provided that those discrepancies arise due to different underlying true distributions. To capture this dependence in the cross-section measurement as a function of neutrino energy, our strategy for both, the energy-dependent and the double-differential measurements, the efficiency correction is done in the 3D space of neutrino energy, muon kinetic energy and muon $\cos\theta$ with respect to the mean beam direction. The selection efficiency depends on hadronic energy too, but this dependence is appropriately captured in the neutrino and muon energy space, since $E_{\text{had}} = E_\nu - E_\mu$. In this scheme, the reconstructed spectrum from data is generated, unfolded and efficiency corrected in 3D. Once the corrected 3D "true" data distribution is available, the distribution is projected to an appropriate axis/axes of the measurement.

44.6 Conclusion

NOvA has an excellent sensitivity to measure the cross-section in addition to neutrino oscillation study. The event and particle id selection criteria based on the optimization of the total uncertainty on the measured ν_μ inclusive cross-section. Measurement will be systematics-limited. The dominant systematic uncertainty is 8–10% (mostly normalization) and all other uncertainties add to ~5 %.

References

1. D.S. Ayres et al., The NOvA Technical Design Report
2. B. Behera, G. Davies, F. Psihas, Event Reconstruction in the NOvA Experiment. arXiv:1710.03772
3. B. Behera, Status of a Deep Learning Based Measurement of the Inclusive Muon Neutrino Charged-current Cross Section in the NOvA Near Detector. arXiv:1710.03766

Chapter 45
Baryogenesis via Leptogenesis from Asymmetric Dark Matter Using Higher Dimension Operator

Nimmala Narendra, Sudhanwa Patra, Narendra Sahu and Sujay Shil

Abstract We propose an extension of the standard model (SM) by including a dark sector comprising of three generations of heavy right-handed neutrinos, three singlet scalars and a singlet Dirac fermion. With this additional particle content we discuss the simultaneous solution to the baryon asymmetry and the Dark Matter(DM) relic density, in presence of higher dimensional operator.

45.1 Introduction

The galaxy rotation curve, gravitational lensing and large scale structure of the Universe undoubtedly confirmed the existence of Dark Matter (DM) in a large scale ($\gtrsim$ a few kpc) [1]. The relic abundance of DM is precisely measured by the WMAP [2] and PLANCK [3] experiments to be $\Omega_{\rm DM}h^2 = 0.1199 \pm 0.0027$. An interesting observation is that the ratio of the DM relic density and the baryon relic density is approximately equal to 5, i.e., $\Omega_{\rm DM} \approx 5\Omega_{\rm B}$. From these observations one can conclude that both of these relic densities may have a common origin [4]. The observed baryon asymmetry, in terms of the baryon to photon ratio, $\eta = n_B/n_\gamma$ is given

N. Narendra (✉) · N. Sahu
Indian Institute of Technology Hyderabad, Sangareddy 502285, Telangana, India
e-mail: ph14resch01002@iith.ac.in

N. Sahu
e-mail: nsahu@iith.ac.in

S. Patra
Indian Institute of Technology Bhilai, Chhattisgarh, Raipur, India
e-mail: sudha.astro@gmail.com

S. Shil
Institute of Physics, Sachivalaya Marg, Bhubaneswar 751005, Odisha, India
e-mail: sujay@iopb.res.in

Homi Bhabha National Institute Training School Complex,
Anushakti Nagar, Mumbai 400085, India

A. Giri and R. Mohanta (eds.), *16th Conference on Flavor Physics and CP Violation*, Springer Proceedings in Physics 234,
https://doi.org/10.1007/978-3-030-29622-3_45

as [5], $5.8 \times 10^{-10} \leq \eta \leq 6.6 \times 10^{-10}$ (BBN) $(95\%CL)$, where $\eta = 7.04 Y_B$ with $Y_B \equiv n_B/s$. Similarly the observed DM abundance can be expressed as

$$Y_{\rm DM} \equiv \frac{n_{\rm DM}}{s} = 4 \times 10^{-10} \left(\frac{1\text{GeV}}{M_{\rm DM}}\right) \left(\frac{\Omega_{\rm DM} h^2}{0.11}\right). \quad (45.1)$$

The standard model (SM), which is based on the gauge group $SU(3)_C \times SU(2)_L \times U(1)_Y$, provides the best understanding of fundamental particles and their interactions. However, it does not explain neither the DM abundance nor the baryon asymmetry of the Universe. In this article we made an attempt to solve these problems in a beyond SM framework.

45.2 The Model

We extend the SM symmetry with an additional symmetry, $U(1)_{B-L} \times U(1)_D \times Z_2$ and with a dark sector particles comprising of three generations of heavy right-handed neutrinos, a singlet scalar ϕ' and a singlet Dirac fermion χ. These particles are charged under these additional symmetry, while remain inert with respect to the SM gauge group. The $U(1)_{\rm B-L}$ is a gauge symmetry, which is broken spontaneously by the vacuum expectation value(vev) of an additional scalar $\phi_{\rm B-L}$ at a high scale and give Majorana masses to right-handed neutrinos as well as neutral gauge boson $Z_{\rm B-L}$, while $U(1)_D$ is a global symmetry and is allowed to break softly by the higher dimension operator. Moreover, the $U(1)_D$ symmetry provides a distinction between the dark sector fermions N_R and χ, which are having same charge under $U(1)_{\rm B-L} \times Z_2$ symmetry. The χ and ϕ' are stable and viable candidates of DM. The CP-violating out-of-equilibrium decay of lightest of the heavy right handed neutrinos, i.e., N_1, creates an asymmetry in χ and ϕ particles. In order to realize an asymmetric DM scenario, the symmetric component of χ and ϕ' should be annihilated away. To achieve this we introduce an additional singlet scalar ϕ, which mixes with the SM Higgs H. The particle content of the model can be seen in the Table. 45.1. The asymmetry in χ is transfered to the visible sector through a dimention-8 operator, which conserves $U(1)_{\rm B-L}$ symmetry. Thus there is a proportionality arises between

Table 45.1 Particles of the dark sector and their quantum numbers under the imposed symmetry

Fields	$SU(3)_C$	$SU(2)_L$	$U(1)_Y$	$U(1)_{B-L}$	$U(1)_D$	Z_2
N_R	1	1	0	–1	1	–
χ	1	1	0	–1	1/3	–
ϕ	1	1	0	0	0	+
ϕ'	1	1	0	0	2/3	+
ϕ'_{B-L}	1	1	0	+2	-2	+

the DM and the baryon relic densities. Note that $U(1)_D$ symmetry breaks softly by the higher dimentional operator.

The corresponding Lagrangian can be written as [6],

$$\mathcal{L} \supset M_\chi \overline{\chi}\chi + \lambda_{B-L}\phi_{B-L}\overline{(N_{Ri})^c}N_{Rj} + \lambda_{\rm DM}\overline{\chi}\chi\phi \\ + y_i\overline{N_{Ri}}\chi\phi' + h.c. - V(H,\phi,\phi') \quad (45.2)$$

where

$$V(H,\phi,\phi') = -\mu_H^2 H^\dagger H + \lambda_H(H^\dagger H)^2 + \frac{1}{2}M_\phi^2\phi^2 + \frac{1}{4}\lambda_\phi\phi^4 \\ + M_{\phi'}^2\phi'^\dagger\phi' + \lambda_{\phi'}(\phi'^\dagger\phi')^2 + \mu_\phi\phi(H^\dagger H) + \mu'_\phi\phi(\phi'^\dagger\phi') \\ + \frac{1}{2}\lambda_{H\phi}(H^\dagger H)\phi^2 + \lambda_{H\phi'}(H^\dagger H)(\phi'^\dagger\phi') + \frac{\lambda_{\phi\phi'}}{2}\phi^2(\phi'^\dagger\phi'). \quad (45.3)$$

We assume that the vev of $\langle\phi_{\rm B-L}\rangle = v_{\rm B-L} \sim 10^{10}$ GeV (say). Hence, in the low energy electroweak phenomenology $\phi_{\rm B-L}$ does not play any role. The electroweak phase transition occurs as the SM Higgs acquires a vev $v = \langle H\rangle$. After electroweak phase transition the scalar ϕ and SM-Higgs mix with each other and diagonalising the resulting scalar mass matrix we get new mass eigenstates, h_1 and h_2. The vev of SM-Higgs induces a non-zero vev to ϕ due to the trilinear term $\mu_\phi\phi(H^\dagger H)$ as given in (45.3). We assume that $\langle\phi\rangle = u << v$. In addition to that we also assume $< \phi' >= 0$. This implies ϕ' is stable and due to its charge under $U(1)_D$ symmetry, it acts as a viable candidate of DM.

45.3 Generation of Asymmetry in Dark Sector and Its Transfer from Dark Sector to Visible Sector

The CP-violating out-of-equilibrium decay of heavy right handed neutrino, N_1, generates a net $B - L$ asymmetry in χ and an equivalent amount of asymmetry generates in ϕ' as well. The CP asymmetry arises in the decay of N_1 due to the interference of tree level diagram with one loop vertex and self energy diagrams as shown in Fig. 45.1.

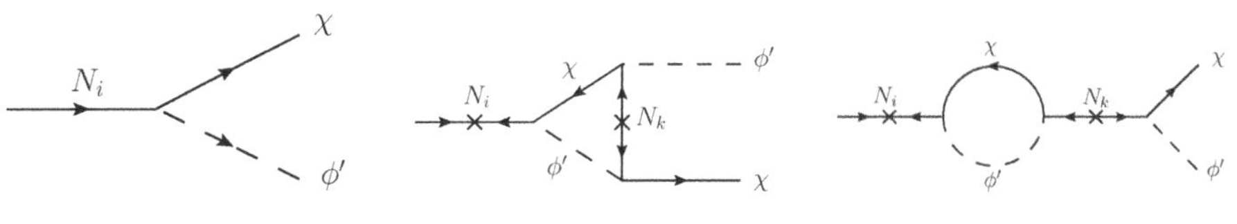

Fig. 45.1 CP-violation in the decay of N_1 through the interference of tree level diagram with vertex and self energy diagrams

The asymmetry ϵ_χ can be estimated as [7]

$$\epsilon_\chi = \frac{\Gamma(N_1 \to \chi\phi') - \Gamma(N_1 \to \bar{\chi}\phi')}{\Gamma_{N_1}} \simeq -\frac{1}{8\pi}\left(\frac{M_1}{M_2}\right)\frac{\mathrm{Im}\left[(y^\dagger y)^2\right]_{12}}{(y^\dagger y)_{11}}. \quad (45.4)$$

where M_i, $i = 1, 2, 3$ are the masses of heavy right-handed neutrinos and we assume $M_1 << M_2 << M_3$. Thus From N_1 decays, we get a net $B - L$ asymmetry [8]:

$$(n_{\mathrm{B-L}})_{\mathrm{total}} = \epsilon_\chi \kappa\, s \times \frac{n^{eq}_{N_1}(T \to \infty)}{s}. \quad (45.5)$$

where $(n^{eq}_{N_1}/s)(T \to \infty) = 135\zeta(3)/(4\pi^4 g_*)$ is the relativistic equilibrium abundance of N_1. and $s = (2\pi^2/45)g_* T^3$ is the entropy density. The κ arises via inverse decay and scattering processes and is a washout factor. The generated $B - L$ asymmetry in dark sector will be transfer to the visible sector via a higher dimensional operator [9]:

$$\mathcal{O}_8 = \frac{1}{M^4_{\mathrm{asy}}}\overline{\chi}^2(LH)^2. \quad (45.6)$$

Depending on the value of M_{asy}, the transfer operator will decouple from thermal plasma at different temperatures. We assume that, the dark matter χ is also in thermal equilibrium with the visible sector via the dimension eight operator $\mathcal{O}_8$ until the sphaleron decoupling temperature $T_{\mathrm{sph}} > M_W$. Using chemical equilibrium conditions we can get the total baryon asymmetry and the χ asymmetry. The χ asymmetry which is also the $B - L$ number density in the dark sector:

$$n_{\mathrm{B}} = \frac{30}{97}(n_{\mathrm{B-L}})_{\mathrm{v}} \quad , \quad n_\chi = (n_{\mathrm{B-L}})_{\mathrm{d}} = \frac{58}{291}(n_{B-L})_{\mathrm{v}} \,. \quad (45.7)$$

The total n_{B-L} of the Universe generated by N_1, is the sum of n_{B-L} in the visible and dark sectors. Therefore, we get

$$(n_{B-L})_{\mathrm{total}} = (n_{B-L})_{\mathrm{v}} + (n_{B-L})_{\mathrm{d}} = (n_{B-L})_{\mathrm{v}} + \frac{58}{291}(n_{B-L})_{\mathrm{v}} = \frac{349}{291}(n_{B-L})_{\mathrm{v}}. \quad (45.8)$$

Comparing (45.8) with (45.5) and using (45.7) we get the required CP asymmetry for observed DM abundance $\epsilon_\chi = 141.23(\eta/\kappa)(s/n^{eq}_{N_1}(T \to \infty))$. For $\kappa \sim 0.01$ we get $\epsilon_\chi \sim 10^{-6}$. Keeping (45.8) in (45.7), we can get,

$$n_{\mathrm{B}} = \frac{90}{349}(n_{\mathrm{B-L}})_{\mathrm{total}} \quad , \quad n_\chi = \frac{58}{349}(n_{B-L})_{\mathrm{total}} \quad (45.9)$$

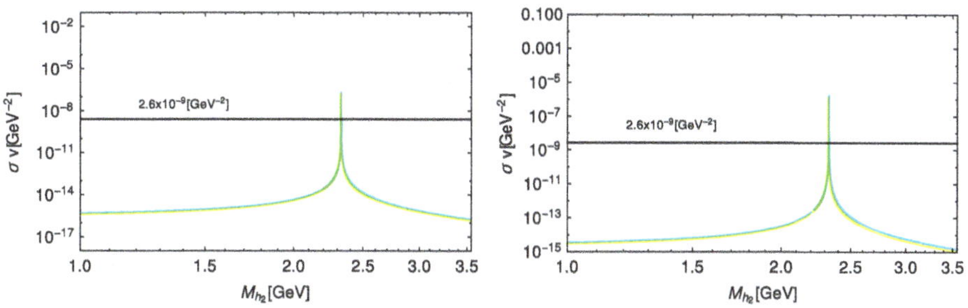

Fig. 45.2 The annihilation cross section of, ($\bar{\chi}\chi \rightarrow \bar{f}f$)(left panel), ($\phi'^{\dagger}\phi' \rightarrow \bar{f}f$)(right panel) as a function of M_{h_2} for a typical value of $\lambda_{\text{DM}} = 1 \times 10^{-2}$, $\lambda_{H\phi'} = 1 \times 10^{-3}$, $\mu_{\phi'} = 1 \times 10^{-3}$ and $\sin\gamma = 0.1$

The symmetry generated in χ and ϕ' is same. Therefore we can write

$$n_{\phi'} = n_{\chi} = (n_{B-L})_{\text{total}} \tag{45.10}$$

From the WMAP and PLANCK data observations, we have the ratio of the abundances of DM and baryons as, $\Omega_{\text{DM}} \approx 5\Omega_{\text{B}}$. Using this relation we can estimate the DM mass from (45.9) and (45.10) as,

$$M_{\chi} = \frac{450M_p - 349M_{\phi'}}{58} \tag{45.11}$$

where M_p is proton mass. From these observations we stick to the mass of DM to be $M_{\chi} = M_{\phi'} \approx 1$ GeV in below Sect. 45.4.

45.4 Depletion of Symmetric Component of DM

The symmetric component of DM can be depleted via the annihilation of DM particles($\bar{\chi}\chi \rightarrow \bar{f}f$ and $\phi'^{\dagger}\phi' \rightarrow \bar{f}f$) to the SM particles through the mediation of scalar ($\phi - H$ mixing) portal. We get the efficient depletion of the symmetric component of the DM at the resonance of the cross section, where actually the mass of h_2 is twice of the DM mass as shown in below Fig. 45.2.

45.5 Conclusion

We considered an extension of SM with additional symmetries $U(1)_{B-L} \times U(1)_D \times Z_2$ by including the three generations of heavy right handed neutrinos, $N_{iR}, i = 1, 2, 3$, a singlet Dirac fermion χ and three singlet scalars ϕ_{B-L}, ϕ' and ϕ. The additional symmetries allows two of the dark sector particles χ and ϕ' as the candidates of DM. The out-of-equilibrium decay of lightest heavy right handed neutrino creates an asymmetry in dark sector. Latter this asymmetry in dark sector transfer to the visible sector via higher dimensional operator. The symmetric component of the DM can be annihilate away by the mediation of ϕ-SM Higgs mixing portal. We found an efficient depletion of symmetric component of the DM, when extra scalar mass is equal to twice of the DM mass, irrespective of all other parameters in the model. Since the observed relic density ratio of the DM to baryons gives the DM mass to be $M_\chi = M_{\phi'} \approx 1$ GeV, so we get the extra scalar mass ≈ 2 GeV, which can be searched at the colliders and via an indirect gamma ray search.

References

1. G. Jungman, M. Kamionkowski, K. Griest, Phys. Rept. **267**, 195 (1996) [hep-ph/9506380]; G. Bertone, D. Hooper, J. Silk, Phys. Rept. **405**, 279 (2005) [hep-ph/0404175]
2. G. Hinshaw et al., WMAP collaboration. Astrophys. J. Suppl. **208**, 19 (2013). [arXiv:1212.5226 [astro-ph.CO]]
3. P.A.R. Ade et al., Planck collaboration. Astron. Astrophys. **594**, A13 (2016). [arXiv:1502.01589 [astro-ph.CO]]
4. S. Nussinov, Phys. Lett. **165B**, 55 (1985); K. Griest, D. Seckel, Nucl. Phys. B **283**, 681 (1987) Erratum: [Nucl. Phys. B **296**, 1034 (1988)]
5. C. Patrignani et al., [Particle Data Group], Chin. Phys. C **40**(10), 100001 (2016)
6. N. Narendra, S. Patra, N. Sahu, S. Shil, arXiv:1805.04860 [hep-ph]
7. W. Buchmuller, M. Plumacher, Phys. Rept. **320**, 329 (1999)
8. W. Buchmuller, P. Di Bari, M. Plumacher, Annals Phys. **315**, 305 (2005) https://doi.org/10.1016/j.aop.2004.02.003 [hep-ph/0401240]
9. W.Z. Feng, P. Nath, G. Peim, Phys. Rev. D **85**, 115016 (2012). [arXiv:1204.5752 [hep-ph]]

Chapter 46
Impact of Active-Sterile Neutrino Mixing at Currently Running Long-Baseline Experiments

Rudra Majhi, C. Soumya and Rukmani Mohanta

Abstract In this work, we explore the effect of active-sterile neutrino mixing on oscillation probability and on mass hierarchy sensitivity of currently running long-baseline neutrino experiments. We find that the existence of sterile neutrino can lead to new kind of degeneracies among the oscillation parameters which deteriorate the mass hierarchy sensitivity of NOνA experiment. We also notice that the addition of data from T2K experiment helps in resolving the degeneracies.

46.1 Introduction

In the three flavor framework, neutrino oscillation can be characterized by three mixing angles ($\theta_{12}, \theta_{13}, \theta_{23}$) and two mass squared differences Δm^2_{21}, Δm^2_{31} and one Dirac type CP phase δ_{CP}. All of these oscillation parameters are measured very precisely except δ_{CP}. However, we do not have sufficient knowledge about the octant of atmospheric mixing angle (θ_{23}), the sign on atmospheric mass splitting $|\Delta m^2_{31}|$, and the value of CP-violating phase δ_{CP}. The long baseline experiments play a major role in the determination of unknowns due to the presence of enhanced matter effect [1]. However, the existence of four fold degeneracies among the oscillation parameters greatly affect the sensitivities of these experiments. Therefore, the resolution of degeneracies among the oscillation parameters are the primary concern in neutrino oscillation studies.

The short baseline anomalies hint towards the existence of at least one sterile neutrino. Moreover, latest result from the MiniBooNE is also in good agreement with

R. Majhi (✉) · R. Mohanta
School of Physics, University of Hyderabad, Hyderabad 500 046, India
e-mail: rudra.majhi95@gmail.com

R. Mohanta
e-mail: rukmani98@gmail.com

C. Soumya
Institute of Physics, Sachivalaya Marg, Bhubaneswar 751005, India
e-mail: soumyac20@gmail.com

A. Giri and R. Mohanta (eds.), *16th Conference on Flavor Physics and CP Violation*, Springer Proceedings in Physics 234,
https://doi.org/10.1007/978-3-030-29622-3_46

the excess of events reported by LSND and hence provides strong hint for existence of eV-scale sterile neutrino [2]. Though sterile neutrinos are blind to weak interaction, they can mix with active neutrinos. Therefore, this mixing can have effect on the determination of mass ordering of neutrinos. In presence of one sterile neutrino, the parametrization of neutrino mixing requires additional oscillation parameters: three mixing angles ($\theta_{14}, \theta_{24}, \theta_{34}$), two phases ($\delta_{14}, \delta_{34}$), and one mass squared difference (Δm^2_{41}). With an effective 4×4 mixing matrix, the vacuum oscillation probability in 3+1 framework is given by

$$\begin{aligned} P \approx\, & 4 \sin^2\theta_{23} \sin^2\theta_{13} \sin^2\Delta + 8 \sin\theta_{13} \sin\theta_{12} \cos\theta_{12} \sin\theta_{23} \cos\theta_{23} (\alpha\Delta) \sin\Delta \cos(\Delta + \delta_{13}) \\ & + 4 \sin\theta_{14} \sin\theta_{24} \sin\theta_{13} \sin\theta_{23} \sin\Delta \sin(\Delta + \delta_{13} - \delta_{14}), \end{aligned} \tag{46.1}$$

where $\Delta \equiv \Delta_{31} L/4E$, $\alpha \equiv \Delta_{21}/\Delta_{31}$, $\Delta_{ij} = m_i^2 - m_j^2$ and L, E are respectively baseline and energy of the neutrino beam.

46.2 Simulation details

As we focus on currently running long baseline experiments NOνA and T2K, we simulate these experiments using GLoBES software package along with snu plugin [3, 4]. The auxiliary files and experimental specification of these experiments that we use for the analysis is taken from [5]. In our analysis, we use the values of oscillation parameters as given in Table 46.1.

Table 46.1 The values of oscillation parameters that we consider in analysis [6]

Parameters	True values	Test value range	Parameters	True values	Test value range
$\sin^2\theta_{12}$	0.304	NA	$\sin^2\theta_{14}$	0.025	$\theta_{14}(0°, 15°)$
$\sin^2 2\theta_{13}$	0.085	NA	$\sin^2\theta_{24}$	0.025	$\theta_{14}(0°, 15°)$
$\sin^2\theta_{23}$	0.5	$0.4 \longrightarrow 0.6$			
	(LO 0.44)	$0.4 \longrightarrow 0.5$			
	(HO 0.56)	$0.5 \longrightarrow 0.6$			
δ_{CP}	$-90°$	$-180° \longrightarrow 180°$	$\sin^2\theta_{34}$	0	NA
$\frac{\Delta m^2_{12}}{10^{-5}\text{eV}^2}$	7.4	NA	$\frac{\Delta m^2_{14}}{1\text{ eV}^2}$	1	NA
$\frac{\Delta m^2_{31}}{10^{-3}\text{eV}^2}$	2.5(NH)	$2.36 \longrightarrow 2.64$	δ_{14}	$-90°$	$-180° \longrightarrow 180°$
	-2.5(IH)	$-2.64 \longrightarrow -2.36$	δ_{34}	$0°$	$-180° \longrightarrow 180°$

46.3 Degeneracies and MH sensitivity

In order to analyse degeneracies among the oscillation parameters at probability level, we show ν_e appearance oscillation probability as a function of δ_{CP} in the left (middle) panel of Fig. 46.1. From the left panel of figure, it can be seen that the bands NH-HO and IH-LO bands are very well separated in neutrino channel, whereas the bands for NH-LO and IH-HO are overlapped each other which results in the degeneracies among the oscillation parameters. It can be shown that in the anti-neutrino channel the case is just opposite. Therefore, a combined analysis of neutrino and anti-neutrino data helps in the resolution of degeneracies and improves the sensitivity of LBL experiment to the unknowns in standard paradigm. From the middle panel, it can be seen that there emerged new types of degeneracies among the oscillation parameters in presence of sterile neutrino for $\delta_{14}(= -90°)$ which can worsen the sensitivity to the unknowns. Another way of representing these degeneracies among oscillation parameters is by using the bi-probability plot as given in right panel of Fig. 46.1. The ellipses in the figure correspond to 3 flavor case. Whereas the bands represent the oscillation probabilities in presence of sterile neutrino with all possible values of new phase δ_{14}. From the figure, it can be seen that the ellipses for LO and HO are very well separated for both hierarchies, whereas the ellipses for NH and IH for both LO and HO are overlapped with each other and give rise to degeneracies. Therefore, NOνA experiment is more sensitive to octant of θ_{23} than that of mass hierarchy. While in $3 + 1$ paradigm, the bands are overlapped each other for all combinations and give rise to new degeneracies. More degeneracies between lower and higher octant along with standard case, indicate that experiment is loosing its sensitivity in presence of sterile neutrino. Next, we show the allowed parameter space in $\theta_{23} - \delta_{CP}$ plane at 90% C.L. for each hierarchy-octant combination as given in Fig. 46.2. From the figure, it can be seen that the allowed parameter space in presence of sterile neutrino is enlarged, which indicates that the degeneracy resolution capability is decreased significantly. However, the synergy of T2K and NOνA improves the degeneracy resolution capability. Finally, we discuss how MH sensitivity of NOνA gets modified in presence of sterile neutrino. From Fig. 46.3, one

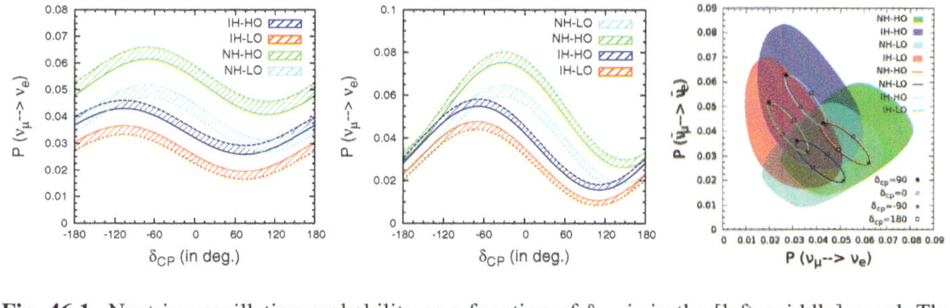

Fig. 46.1 Neutrino oscillation probability as a function of δ_{CP} is in the [left, middle] panel. The left (middle) panel is for 3 flavor (3 + 1) case with $\delta_{14} = -90°$. [Right panel] Bi-probability plots for NOνA in 3 years in ν and 3 years in $\bar{\nu}$ mode

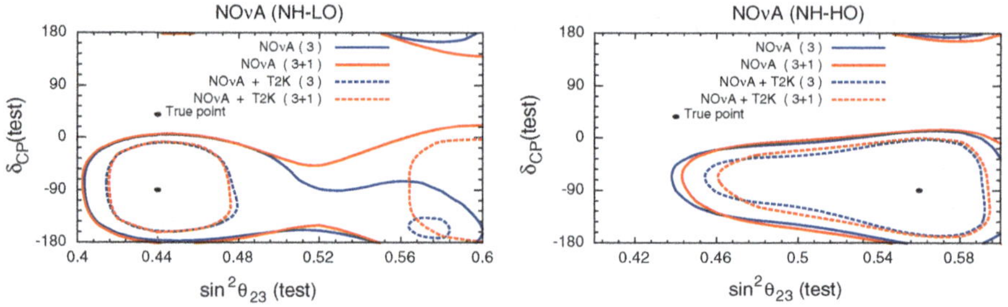

Fig. 46.2 The allowed parameter space in $\theta_{23} - \delta_{CP}$ plane

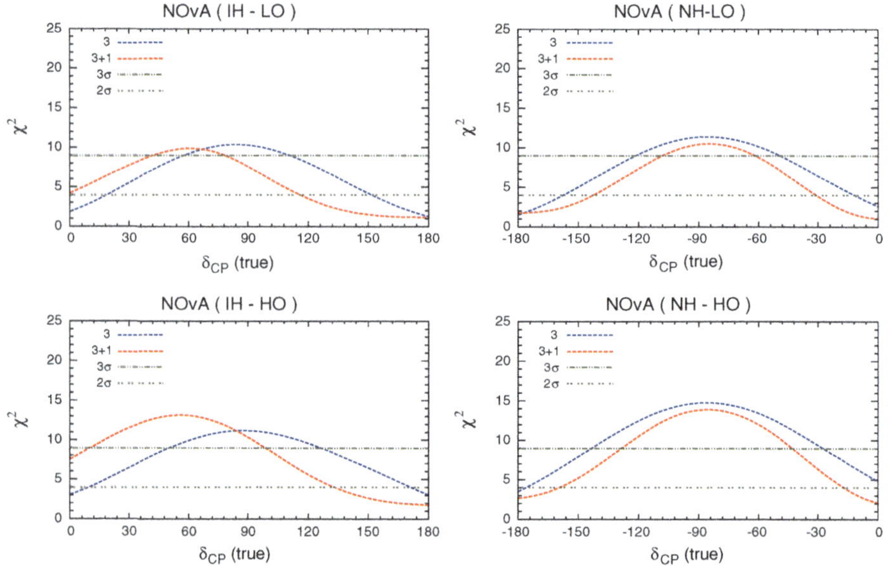

Fig. 46.3 MH sensitivity as a function of true values of δ_{CP}

can see that the wrong mass hierarchy can be ruled out significantly above 2σ in the favourable regions, i.e., lower-half plane (upper-half plane) for NH (IH) in the standard paradigm. Whereas, in presence of sterile neutrino the δ_{CP} coverage for the mass hierarchy sensitivity is significantly reduced.

46.4 Conclusion

In this paper, we discussed the effect of active-sterile mixing on the degeneracy resolution capability and MH sensitivity of NOνA experiment. We found that introduction of sterile neutrino gives rise to new kind of degeneracies among the oscillation parameters which results in reduction of δ_{CP} coverage for MH sensitivity of

NOνA experiment. We also found that addition of T2K data helps in resolving the degeneracies among the oscillation parameters.

Acknowledgements We thank SERB and INSPIRE for financial support.

References

1. M.V. Diwan, et al., Long-baseline neutrino experiments. Ann. Rev. Nucl. Part. Sci. **66**, 47–71 (2016). https://doi.org/10.1146/annurev-nucl-102014-021939
2. A.A. Aguilar-Arevalo et al., Observation of a significant excess of electron-like events in the MiniBooNE short-baseline neutrino experiment. FERMILAB-PUB-18-219, LA-UR-18-24586 (2018)
3. P. Huber et al., Simulation of long-baseline neutrino oscillation experiments with GLoBES. Comput. Phys. Commun. **167**, 195 (2005). https://doi.org/10.1016/j.cpc.2005.01.003
4. P. Huber et al., New features in the simulation of neutrino oscillation experiments with GLoBES 3.0. Comput. Phys. Commun. **177**, 432438 (2007). https://doi.org/10.1016/j.cpc.2007.05.004
5. C. Soumya et al., A Comprehensive study of the discovery potential of NOA, T2K, and T2HK experiments. Adv. High Energy Phys. 9139402 (2016). https://doi.org/10.1155/2016/9139402
6. D. Dutta et al., Capabilities of long-baseline experiments in the presence of a sterile neutrino. JHEP **1611**, 122 (2016). https://doi.org/10.1007/JHEP11(2016)122

Chapter 47
S_3 Extended Standard Model and Scalar Triplet Leptogenesis

Subhasmita Mishra and Anjan Giri

Abstract The standard model is extended with the S_3 symmetry in which non zero θ_{13} is obtained in the type I+II seesaw framework of neutrino mixing consistent with the current observation. Non zero contribution of CP asymmetry is discussed from the decay of scalar triplet in S_3 extended standard model. Presence of right-handed neutrinos could explain both neutrino oscillation and non zero lepton asymmetry from the vertex correction of triplet decay, without including any other heavy triplet.

47.1 Introduction

Despite huge success, Neutrino mass, leptogenesis, dark matter etc. are not explained by the standard model (SM) and therefore, it seems logical to extend the same. In this work, we extend the SM with the simplest permutation symmetry S_3 along with 3 right handed neutrinos. Moreover, we also add two Higgs doublets and one Higgs triplet to explain neutrino phenomenology and leptogenesis. The type II seesaw framework isn't enough to explain neutrino masses and mixing in this model, and therefore, we consider type I+II seesaw mechanism to explain neutrino sector in consistent with the current experimental data. Lepton asymmetry from the out of equilibrium decay of right handed neutrinos has already been discussed in the literature. But there are only few studies focused on the generation of lepton asymmetry from scalar or fermion triplets. It is known that the lepton asymmetry is the most preferable way to generate baryon asymmetry through sphaleron transition. In this work, due to the heavy mass, right handed neutrinos decouple much earlier than the triplet scalars. Hence the total asymmetry due to the decay of scalar triplets survives. It has also been discussed in the literature that one scalar triplet isn't enough to generate non zero CP asymmetry in one loop level, but the presence of heavy right handed neutrinos can

S. Mishra (✉) · A. Giri
IIT, Hyderabad, India
e-mail: subhasmita.mishra92@gmail.com

A. Giri
e-mail: giria@iith.ac.in

A. Giri and R. Mohanta (eds.), *16th Conference on Flavor Physics and CP Violation*, Springer Proceedings in Physics 234,
https://doi.org/10.1007/978-3-030-29622-3_47

generate a large CP asymmetry from scalar triplet decay and we considered the one flavor regime to explain the generation of nonzero lepton asymmetry [1, 2].

47.2 Neutrino Masses and Mixing

In this model, adding one scalar triplet couldn't explain neutrino mixing, hence we added three extra right handed neutrinos. Presence of right handed neutrinos not only solve the neutrino mixing, but also contribute to the non zero CP asymmetry through vertex correction in scalar triplet decay (Table 47.1). In the Higgs sector, we redefined the two Higgs doublets as $H_{\pm} = \frac{H_1 \pm H_2}{\sqrt{2}}$, where we considered the mixing between H_+ and H_3 as $H_L = H_3 \cos\beta + H_+ \sin\beta$, $H_H = -H_3 \sin\beta + H_+ \cos\beta$, and $v = \sqrt{{v_+}^2 + {v_3}^2} = 246\,\text{GeV}$. Since we have, more than one Higgs in the model, flavor changing neutral current in the tree level can be suppressed by considering other Higgs, except the standard model Higgs, to be very heavy, say order of TeV. Finetuning between the soft symmetry breaking parameters is required to achieve TeV order mass which is not possible by electroweak symmetry breaking. The $SM \otimes S_3 \otimes Z_2$ invariant Lagrangian for type I+II Yukawa interaction in the charged and neutral lepton sector in the isospin basis of scalar triplet is given by [3]

$$
\begin{aligned}
L_m = & -h_1\left[\overline{L_e}\tilde{H}_2 N_{1R} + \overline{L_\mu}\tilde{H}_1 N_{1R} + \overline{L_e}\tilde{H}_1 N_{2R} - \overline{L_\mu}\tilde{H}_2 N_{2R}\right] - h_3\left[\overline{L_\tau}\ \tilde{H}_1 N_{1R} + \overline{L_\tau}\ \tilde{H}_2 N_{2R}\right] \\
& -h_4\left[\overline{L_\tau}\ \tilde{H}_3 N_{3R}\right] - \frac{1}{2}\overline{N^{c}_{\mathrm{iR}}} M_{\mathrm{iR}} N_{\mathrm{iR}} - \frac{1}{2}\overline{N^{c}_{\mathrm{3R}}} M_{\mathrm{3R}} N_{\mathrm{3R}} - y_{l2}\left[\overline{L_e} H_2 E_{1R} + \overline{L_\mu} H_1 E_{1R} + \right. \\
& \left.\overline{L_e} H_1 E_{2R} - \overline{L_\mu} H_2 E_{2R}\right] - y_{l4}\left[\overline{L_\tau}\ H_1 E_{1R} + \overline{L_\tau}\ H_2 E_{2R}\right] - y_{l5}[\overline{L_e} H_1 E_{3R} + \overline{L_\mu} H_2 E_{3R}] \\
& + y_t\left[\overline{L^c}_e i\tau_2 \Delta L_e + \overline{L^c}_\mu i\tau_2 \Delta L_\mu\right] + y_t'[\overline{L^c}_\tau i\tau_2 \Delta L_\tau] + \mu[\tilde{H}_1{}^\dagger \Delta^\dagger H_1 + \tilde{H}_2{}^\dagger \Delta^\dagger H_2] \\
& + \mu'[\tilde{H}_3{}^\dagger \Delta_i{}^\dagger H_3] + \ h.c \quad (i = 1, 2).
\end{aligned}
$$

Since in this model, type II seesaw alone can't explain the neutrino masses and mixing, we considered type $I + II$ seesaw mass matrix. Which are given by ${m_\nu}^I = M_{\mathrm{D}} M_R^{-1} (M_{\mathrm{D}})^T$ and ${m_\nu}^{II} = \frac{2\mu_L Y_\Delta v^2}{{M_\Delta}^2}$, where Y_Δ and μ_L are the triplet-lepton and triplet-SM like Higgs coupling. Considering the Majorana neutrino mass in the diagonal basis, for simplicity, the small neutrino mass matrix M_ν^{I+II} is written below,

Table 47.1 Particle contents and quantum numbers under SM and S_3

Particles	SM - Group	S_3	Z_2
$(L_e, L_\mu), L_\tau$	$(1, 2, -1)$	2, 1	+1
$(E_{1R}, E_{2R}), E_{3R}$	$(1, 1, -2)$	2, 1	+1
$(N_{1R}, N_{2R}), N_{3R}$	$(1, 1, 0)$	2, 1	+1, −1
$(H_1, H_2), H_3$	$(0, 2, 1)$	2, 1	+1, −1
Δ	$(0, 3, 2)$	1	+1

where, $x_1 = \frac{2(\mu \sin^2\beta + \mu' \cos^2\beta)v^2}{{M_\Delta}^2}$, $\eta_1 = \frac{\sqrt{2}h_1 v \sin\beta}{\sqrt{M_{1R}}}$, $\eta_3 = \frac{\sqrt{2}h_3 v \sin\beta}{\sqrt{M_{1R}}}$ and $\eta_4 = \frac{\sqrt{2}h_4 v \cos\beta}{\sqrt{M_{3R}}}$. After redefining the phases, the remnant phases are ϕ and ϕ_Δ in type I and type II Yukawa respectively in the mass matrix.

$$
{M_\nu}^{I+II} = \begin{pmatrix} {\eta_1}^2 + y_t x_1 & 0 & \eta_1\eta_3 \\ 0 & {\eta_1}^2 + y_t x_1 & 0 \\ \eta_1\eta_3 & 0 & {\eta_3}^2 + {\eta_4}^2 e^{2i\phi} + |y_t' x_1| e^{i\phi_\Delta} \end{pmatrix}.
$$

The simple block diagonal form of mass matrix is easy to diagonalize, where the mixing matrices for both neutrino and charged leptons are given as following [3, 4]

$$
U_\nu = \begin{pmatrix} \cos\theta & \sin\theta e^{-i\phi_\nu} & 0 \\ 0 & 0 & 1 \\ -\sin\theta e^{i\phi_\nu} & \cos\theta & 0 \end{pmatrix}, \quad U_{el} = \begin{pmatrix} \frac{x}{\sqrt{2(1-x^2)}} & \frac{1}{\sqrt{2(1+x^2)}} & \frac{1}{\sqrt{2(1+\sqrt{z})}} \\ \frac{-x}{\sqrt{2(1-x^2)}} & \frac{-1}{\sqrt{2(1+x^2)}} & \frac{1}{\sqrt{2(1+\sqrt{z})}} \\ \frac{\sqrt{1-2x^2}}{\sqrt{1-x^2}} & \frac{x}{\sqrt{1+x^2}} & \frac{\sqrt{z}}{\sqrt{(1+\sqrt{z})}} \end{pmatrix}.
$$

Where, $x = \frac{m_e}{m_\mu}$ and $z = \frac{m_e^2 m_\mu^2}{m_\tau^4}$, ${U_\nu}^T M_\nu U_\nu = \text{Diag}(m_{\nu_1} e^{i\phi_1}, m_{\nu_2} e^{i\phi_2}, m_{\nu_3})$ and $U_{el} M_l {M_l}^\dagger {U_{el}}^\dagger = \text{Diag}({m_e}^2, {m_\mu}^2, {m_\tau}^2)$. From the above parameterization, we found $\sin\theta_{13} \approx 0.004$, $\tan\theta_{12} \approx 0.56$ for $\theta = \frac{\pi}{6}$ and $\sin\theta_{23} \approx 0.707$. And we can also get the neutrino masses in terms of S_3 parameters, so that the allowed parameter space can be obtained. Considering $r_1 = {\eta_1}^2 + y_t x_1$ and $r_2 = |{\eta_3}^2 + {\eta_4}^2 + y_t' x_1|$ and ϕ_{eff} as the relative phase between ϕ and ϕ_Δ in the mass matrix, we can write the neutrino masses after diagonalization in terms of S_3 parameters as,

$$
m_{\nu_3} = r_1, \quad m_{\nu_2} = | \frac{(r_1 + r_2 e^{i\phi_{\text{eff}}})}{2} + \frac{1}{2}\sqrt{(r_1 - r_2 e^{i\phi_{\text{eff}}})^2 + (2\eta_3\eta_4)^2} \,|\, e^{i\phi_1}, \tag{47.1}
$$

$$
m_{\nu_1} = | \frac{(r_1 + r_2 e^{i\phi_{\text{eff}}})}{2} - \frac{1}{2}\sqrt{(r_1 - r_2 e^{i\phi_{\text{eff}}})^2 + (2\eta_3\eta_4)^2} | e^{i\phi_2}, \tag{47.2}
$$

$$
\tan\phi_\nu = \frac{r_2 \sin\phi_{\text{eff}}}{[r_1 + r_2 \cos^2\phi_{\text{eff}}]}. \tag{47.3}
$$

Which prefers inverted hierarchical mass spectrum of active neutrinos. Here ϕ_1, ϕ_2 are the Majorana like phases and the phase ϕ_ν appearing in the parameterization of UPMNS matrix is the dirac like phase.

47.3 Leptogenesis

Leptogenesis from scalar triplet decay is studied in the literature [5, 6]. Where the lepton number and flavor violating decay of scalar triplet generates non zero lepton asymmetry. Unlike the right handed neutrinos, triplet scalars aren't Majorana particles, hence, there will be an asymmetry in decay of particle and antiparticle.

And due to the gauge interactions, scalar triplet will thermalize and will freeze out once the mass exceeds the temperature of thermal bath. In this model, we consider the lightest right-handed neutrino mass to be ($O(10^{11})$ GeV) being heavier than the mass of scalar triplet ($O(10^{10})$ GeV), so that the asymmetry generated by the decay of scalar triplet only survives. As the scalar triplet has two decay modes, any of the decay modes need to be out of equilibrium to satisfy the Sakharov's conditions. Depending upon the masses of the heavy particles, there could be other possibilities of generating lepton asymmetry solely from scalar triplet or right handed neutrino or from both. But in this model we considered only the first case, where the non zero CP asymmetry contribution comes only from the scalar triplet being lighter in mass.

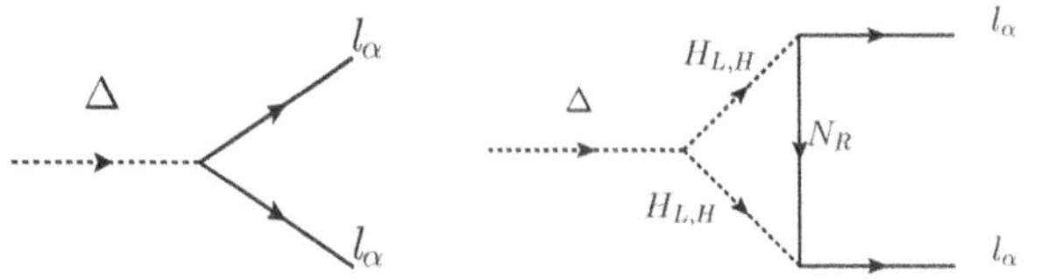

The tree level branching ratios and total decay rate for all possible decay modes of the scalar triplet are given by:

$$B_l = \frac{\frac{M_\Delta}{8\pi} Tr(Y_\Delta Y_\Delta^\dagger)}{\Gamma_{tot}}, \quad B_h = \frac{{\mu_L}^2 + {\mu_H}^2 + {\mu_-}^2}{8\pi M_\Delta \Gamma_{tot}}, \quad \Gamma_{tot} = \frac{M_\Delta}{8\pi}\left(|Y_\Delta|^2_{ii} + |\lambda|^2\right),$$

where $\lambda = \frac{({\mu_L}^2 + {\mu_H}^2 + {\mu_-}^2)^{0.5}}{M_\Delta}$ and μ_L, μ_H, μ_- are the coupling of triplet with H_H, H_L, H_- respectively. CP asymmetry from the interference of tree level and right handed neutrino vertex correction diagram is given by

$$\begin{aligned}
\varepsilon^N &= \frac{-1}{4\pi} \sum_i M_{iR} \frac{Im\left[\mu_\Delta (Y_\Delta)_{ii} (h_\nu^\dagger)_{ii}\right]}{{M_\Delta}^2 Tr[Y_\Delta {Y_\Delta}^\dagger] + {\mu_\Delta}^2} ln\left(1 + \frac{{M_\Delta}^2}{{M_{iR}}^2}\right) \\
&\approx \frac{-1}{4\pi}\left(\frac{{M_\Delta}^2}{M_3}\right) \frac{\mu_\Delta y_t' \tilde{h}_3^2 \sin(\phi_\Delta + 2\phi)}{{M_\Delta}^2 Tr({Y_\Delta}^\dagger Y_\Delta) + |\mu_\Delta|^2} \quad (M_R > M_\Delta).
\end{aligned}$$

Due to the diagonal structure of the triplet-Yukawa coupling matrix, flavor consideration isn't possible in this model. Hence the CP asymmetry of order 10^{-7} to 10^{-8}, inconsistent with the observed baryon asymmetry, comes from the lepton number violating right handed neutrino vertex diagram. And both of the decay modes of the scalar triplet need to be out of equilibrium. Here, leptogenesis is done in the mass basis of charged leptons and hence the neutrino Yukawa coupling (h_ν) is modified as $\tilde{h}_\nu = h_\nu . U_{el}$ (Fig. 47.1 and Table 47.2).

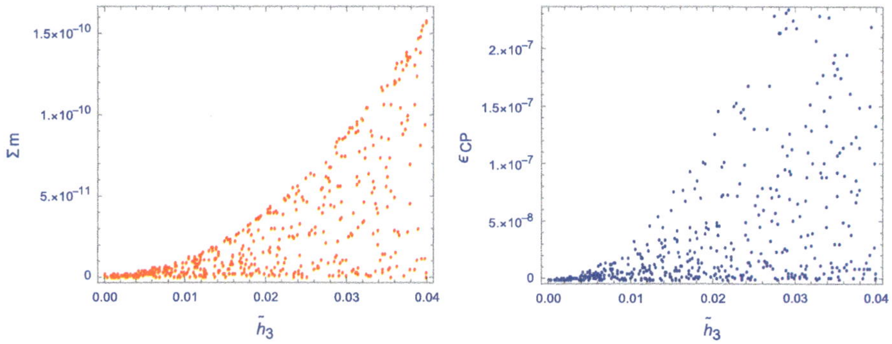

Fig. 47.1 1st figure shows the allowed parameter space for the modified Yukawa coupling from neutrino mass (GeV) constraint and the 2nd figure shows the constraints on the same Yukawa coupling from CP asymmetry to achieve the observed baryon asymmetry of order of 10^{-11}

Table 47.2 Bench mark points for the parameters satisfying the constraints from neutrino mass and observed baryon asymmetry

Parameters	$M_{\Delta 1}$ (GeV)	M_N (GeV)	$\tilde{h}_3$	$y_{t1}'\mu_L$	$\sin\phi_{CP}$	$\sum m_\nu$ (eV)	ε_{CP}
BP1	10^{10}	2×10^{11}	0.05	2.1×10^3	10^{-2}	0.016	3.1×10^{-8}
BP2	10^{10}	5.1×10^{11}	0.03	1.1×10^4	10^{-2}	0.015	3.7×10^{-8}

47.4 Conclusion

In this model, imposition of S_3 symmetry leads to a simple flavor structure of neutrino mass matrix, which helped us to study neutrino mixing in consistent with the current observation. We discussed the non zero lepton asymmetry from the decay of scalar triplet through right handed neutrino loop in one flavor regime.The detail calculation is discussed in [7].

SM would like to acknowledge DST for the financial support.

References

1. E. Ma, U. Sarkar, Phys. Rev. Lett. **80**, 5716 (1998). https://doi.org/10.1103/PhysRevLett.80.5716 [hep-ph/9802445]
2. N. Haba, K. Yoshioka, Nucl. Phys. B **739**, 254 (2006). https://doi.org/10.1016/j.nuclphysb.2006.01.027 [hep-ph/0511108]
3. T. Araki, J. Kubo, E.A. Paschos, Eur. Phys. J. C **45**, 465 (2006). https://doi.org/10.1140/epjc/s2005-02434-3 [hep-ph/0502164]
4. A.A. Cruz, M. Mondragn, arXiv:1701.07929 [hep-ph]
5. D. Aristizabal Sierra, M. Dhen, T. Hambye, JCAP **1408**, 003 (2014). https://doi.org/10.1088/1475-7516/2014/08/003, [arXiv:1401.4347 [hep-ph]]

6. R. Gonzalez Felipe, F.R. Joaquim, H. Serodio, Int. J. Mod. Phys. A **28**, 1350165 (2013). https://doi.org/10.1142/S0217751X13501650, [arXiv:1301.0288 [hep-ph]]
7. S. Mishra, A. Giri (Submitted)

Chapter 48
Measurement of Atmospheric Neutrino Mass-Squared Splittings at INO-ICAL Experiment

Zubair Ahmad Dar, Daljeet Kaur, Sanjeev Kumar and Md. Naimuddin

Abstract In the present work we explore the ability of the magnetized Iron Calorimeter detector (ICal) at the India-based Neutrino Observatory (INO) to identify the neutrinos and antineutrinos on an event by event basis. Consequently, the experimental observation of the difference of mass-squared splittings in ν_μ and $\bar{\nu}_\mu$ oscillations at INO has been studied. The detector sensitivity is estimated so as to confirm a non-zero difference in the splittings for neutrinos and antineutrinos using charged current ν_μ and $\bar{\nu}_\mu$ interactions with the detector under the influence of earth matter effect for ten years exposure of the detector. A χ^2 is minimized with respect to observed parameters to find out the oscillation parameters for ν_μ and $\bar{\nu}_\mu$ independently.

48.1 Introduction

The neutrino comes always with surprise since its induction in the Standard Model (SM) of Particle Physics. It was assumed to be massless in the SM but later many experiments provided evidence for the small neutrino mass through the phenomenon of neutrino oscillation. This phenomenon is well established by several experiments involving solar [1], atmospheric [2, 3], accelerator [4], and reactor neutrinos [5], and the parameters governing the oscillations are being measured to a high precision. In this phenomenon, the neutrino flavor eigenstates $(\nu_e, \nu_\mu, \nu_\tau)$ are considered to be the quantum superpositions of mass eigenstates (ν_1, ν_2, ν_3) with definite masses (m_1, m_2, m_3) and represented mathematically by

$$|\nu_i\rangle = \sum_\alpha U_{\alpha i} \, |\nu_\alpha\rangle \tag{48.1}$$

Z. A. Dar (✉)
Department of Physics, Aligarh Muslim University, Aligarh, India
e-mail: zubairnazir4766@gmail.com

D. Kaur · S. Kumar · Md. Naimuddin
Department of Physics, Delhi University, Delhi, India

A. Giri and R. Mohanta (eds.), *16th Conference on Flavor Physics and CP Violation*, Springer Proceedings in Physics 234,
https://doi.org/10.1007/978-3-030-29622-3_48

where, $|\nu_i\rangle$ represents a neutrino with a definite mass m_i ($i = 1, 2, 3$), $|\nu_\alpha\rangle$ denotes a neutrino of a specific flavor, and $U_{\alpha i}$ is the famous Pontecorvo-Maki-Nakagawa-Sakata (PMNS) [6, 7] lepton mixing matrix. This matrix is parametrized in terms of six variables i.e., three mixing angles, $\theta_{12}, \theta_{13}, \theta_{23}$; two mass differences, $\Delta m^2_{21} = m^2_2 - m^2_1$, and $\Delta m^2_{31} = m^2_3 - m^2_1$; and a CP phase δ_{CP}. The task of any oscillation experiment is to measure these values and obtain the probabilities for oscillation between different flavors. There are many physics problems yet to be answered like the neutrino mass hierarchy, leptonic Dirac CP-violating phase etc., which can be determined from the oscillation experiments.

A consequence of the CPT-theorem is that a particle and the corresponding antiparticle have equal mass and their couplings are closely related. Therefore, the parameters governing neutrino and antineutrino oscillations are considered to be identical. Comparing the oscillation parameters of neutrinos and antineutrinos one could test the hypothesis of CPT-conservation as any difference between them may indicate a sign of new physics.

The probability for muon neutrinos with the energy E_ν and propagation length L is given by

$$P(\nu_\mu \to \nu_\mu) \simeq 1 - 4\cos^2\theta_{13}\sin^2\theta_{23} \times [1 - \cos^2\theta_{13}\sin^2\theta_{23}]\sin^2\left(\frac{1.267|\Delta m^2_{32}|L}{E_\nu}\right). \tag{48.2}$$

Now, the probability of muon antineutrinos can be obtained simply by replacing the oscillation parameters in Eq. 48.2 with that of antineutrino, denoted mathematically by placing bar on them.

In the present paper, we have obtained the sensitivity of the iron calorimeter (ICAL) for the measurement of neutrino and antineutrino oscillation parameters with an assumption that $|\Delta m^2_{32}|$ and $\sin^2\theta_{23}$ are different for neutrinos and antineutrinos. In the first case, we consider the scenario when the difference ($\delta_m = |\Delta m^2_{32}| - |\overline{\Delta m^2}_{32}|$) is nonzero while keeping ($\delta_\theta = \sin^2\theta_{23} - \sin^2\overline{\theta}_{23}$) as zero. In another case, with the realistic detector resolutions and efficiencies of the ICAL incorporated, we calculate the sensitivity for the ICAL detector by varying all the atmospheric oscillation parameters ($|\Delta m^2_{32}|$, $|\overline{\Delta m^2}_{32}|$, $\sin^2\theta_{23}$, $\sin^2\overline{\theta}_{23}$). Using these results, we obtain the potential of the ICAL detector to observe the difference between mass-squared splittings of the neutrinos and antineutrinos i.e. ($|\overline{\Delta m^2}_{32}| - |\Delta m^2_{32}|$) and its sensitivity for ruling out the identical oscillation parameter hypothesis.

48.2 The INO-ICAL Experiment

The India-based Neutrino Observatory (INO) is a multi-institutional science project with a huge detector (around 50 kT) designed to study the atmospheric oscillations, located in Tamil Nadu, India. An important feature about the location is 1 km rock overburden above the site acting as a natural shield to minimize the cosmic ray background. The Iron Calorimeter (ICAL) detector will be magnetized to about 1.5 T, which will enable to distinguish ν_μ and $\overline{\nu}_\mu$ with its good charge identification capability. The aim is to operate the detector in the few GeV energy region with the focus on to observe the neutrino and anti-neutrino oscillations separately, using the ν_μ ($\overline{\nu}_\mu$) disappearance channel with good precision. One of the primary goals of the ICAL experiment is to study the nature of neutrino mass hierarchy. The ICAL detector will consist of three modules with each having a dimension of 16 m × 16 m × 14.5 m. Resistive Plate Chambers (RPCs) of dimension 2m × 2m having gas thickness of 3 mm will be used as active detector elements to detect the charged particles produced inside the detector and a large number of them i.e. almost 30000 RPCs are going to be used for the INO-ICAL detector.

The good tracking capability and energy resolution of the detector makes it well suited for the study of neutrino physics. Apart from muons, the ICAL detector will also measure the energy of the hadronic showers, improving the overall sensitivity to neutrino oscillation parameters.

48.3 Methodology

The magnetic property of the ICAL detector enables one to perform the measurement separately for neutrinos and antineutrinos. In this analysis, earth matter effects have been incorporated, and a large set of unoscillated data is generated with the NUANCE event generator [9] using Honda atmospheric neutrino flux [10], with the exposure of 1000 years of the ICAL detector. The detector geometry is integrated through the GEANT4 simulation package. The oscillation parameters used are listed in Table 48.1 with $\delta_{CP} = 0$ as ICAL is not sensitive to this phase.

A χ^2 method based on Poisson probability distribution with the inclusion of systematic uncertainties is used to estimate the sensitivity of oscillation parameters. The χ^2 is separately calculated for neutrinos and antineutrinos as well as a combined measurement is performed.

Table 48.1 The oscillation parameters used in the analysis

Oscillation parameters	True values	Marginalization range
$\sin^2\theta_{13}$	0.0234	Fixed
$\lvert\Delta m^2_{\overline{32}}\rvert$ (eV^2)	2.4×10^{-3}	(2.0–3.0) $\times 10^{-3}$
$\lvert\Delta \overline{m^2}_{32}\rvert$ (eV^2)	2.4×10^{-3}	(2.0–3.0) $\times 10^{-3}$
$\sin^2\theta_{23}$	0.5	0.3–0.7
$\sin^2\overline{\theta}_{23}$	0.5	0.3–0.7

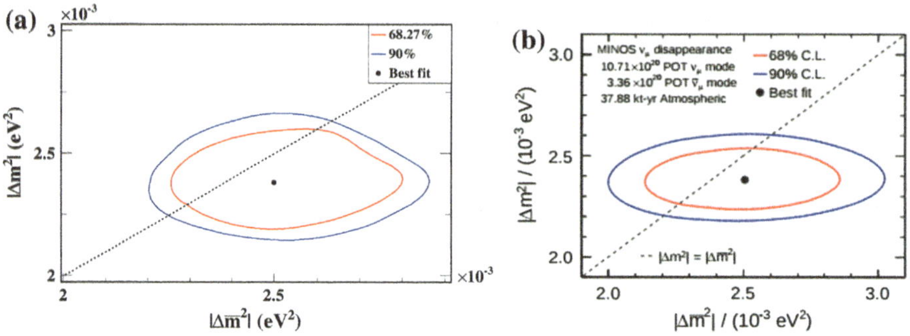

Fig. 48.1 The 68 and 90% CL contours on the $|\Delta m^2_{32}|$ and $|\Delta \overline{m^2}_{32}|$ parameter space (**a**) ICAL experiment using atmospheric data only (**b**) MINOS experiment using combined beamline and atmospheric data

48.4 Results and Discussion

48.4.1 Two Fit Extract from Four Fit Data

From the complete data set of four parameters a two parameter data set for $|\Delta m^2_{32}|$ and $|\Delta \overline{m^2}_{32}|$ is extracted while minimizing the other two parameters i.e. $\sin^2\theta_{23}$ and $\sin^2\overline{\theta}_{23}$. The ICAL sensitivity for atmospheric mass-squared splitting is measured, and it is observed that ICAL can measure $|\Delta m^2_{32}|$ and $|\Delta \overline{m^2}_{32}|$ with a precision of about 10.41% and 12.87% at 90% confidence levels (CL), respectively. The identical mass splittings and mixing angles for neutrinos and antineutrinos are shown using a diagonal dashed line in Figure 48.1. The MINOS collaboration using both beamline and atmospheric data [8] has also performed a similar kind of measurement. A comparison is drawn as shown in Fig. 48.1 using similar oscillation parameters, and it is observed that the sensitivity of ICAL for neutrinos is almost comparable to that of MINOS while the the sensitivity of ICAL for antineutrinos i.e. (12.87% at 90% CL) is relatively improved than MINOS (20% at 90% CL).

48.4.2 ICAL Sensitivity for $|\Delta M^2_{32}| - |\Delta \overline{m^2}_{32}| \neq 0$

In this section, the true values of the oscillation parameters i.e. $|\Delta m^2_{32}|$ and $|\Delta \overline{m^2}_{32}|$ have been varied independently in a range ($0.0021 - 0.0028\text{eV}^2$). We estimate the $\chi^2(\nu + \overline{\nu})$ only on the null hypothesis line where the $|\Delta m^2_{32}|$ and $|\Delta \overline{m^2}_{32}|$ values are equal and a minimum χ^2 value is chosen where this line tangentially coincides with the corresponding contour. A χ^2 is binned as a function $(|\Delta m^2_{32}| - |\Delta \overline{m^2}_{32}|)$. The result obtained will be several minimum points corresponding to a common $(|\Delta m^2_{32}| - |\Delta \overline{m^2}_{32}|)$. We choose only those points which have the smallest χ^2 and

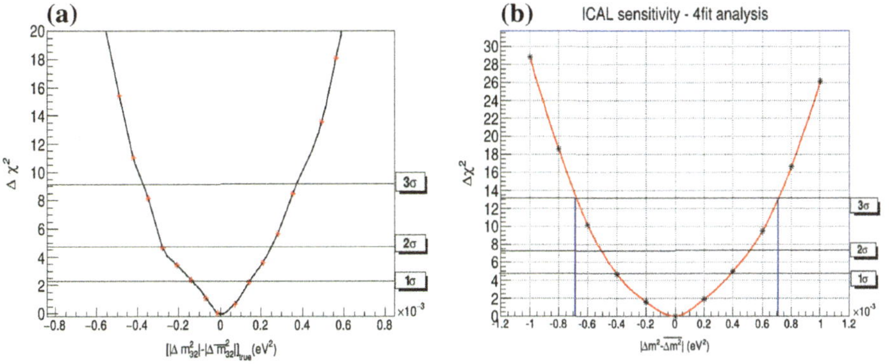

Fig. 48.2 The INO-ICAL sensitivity for $(|\Delta m^2_{32}| - |\overline{\Delta m^2}_{32}|)_{\text{True}}(\text{eV}^2)$ at 1σ, 2σ and 3σ confidence levels using **a** two parameter and **b** four parameter analysis

plotted as the function of $|\Delta m^2_{32}| - |\overline{\Delta m^2}_{32}|$, as shown in Fig. 48.2. The plot depicts the INO-ICAL potential for ruling out the null hypothesis $|\Delta m^2_{32}| = |\overline{\Delta m^2}_{32}|$ when analyzed in two parameter case. It is found that ICAL can rule out the null hypothesis of $|\Delta m^2_{32}| = |\overline{\Delta m^2}_{32}|$ at more than 3σ level if the difference of true values of $\delta_m \geq +0.4 \times 10^{-3}\text{eV}^2$ or $\delta_m \leq -0.4 \times 10^{-3}\text{eV}^2$[11].

48.4.3 ICAL Potential Using Four Parameter Analysis

All the four atmospheric oscillation parameters i.e. $|\Delta m^2_{32}|$, $|\overline{\Delta m^2}_{32}|$, $\sin^2\theta_{23}$ and $\sin^2\overline{\theta}_{23}$ are varied independently as given in Table 48.1. The true values of oscillation parameters are kept i.e. $\delta_m \neq 0$ and $\delta_\theta \neq 0$ nonidentical in order to obtain the complete sensitivity of ICAL detector. The true values are varied simultaneously in a grid of 6×5 for both the neutrino and antineutrino planes. A minimum χ^2 is calculated as a function of the difference in the true values of mass squared splittings while keeping marginalization over $[\sin^2\theta_{23}$ and $\sin^2\overline{\theta}_{23}]_{\text{True}}$. Figure 48.2 represents the INO-ICAL potential for ruling out the null hypothesis $|\Delta m^2_{32}| = |\overline{\Delta m^2}_{32}|$ and $\sin^2\theta_{23} = \sin^2\overline{\theta}_{23}$ at different CLs. It is found that the sensitivity with the four parameters is almost 30% less than that with the two parameters.

Acknowledgements One of the authors, Zubair Ahmad Dar, thanks Department of Science and Technology for the project EMR/2016/002285 under which this work was carried out.

References

1. Q.R. Ahmad et al., Phys. Rev. Lett. **89**, 011301 (2002)
2. Y. Fukuda et al., Phys. Rev. Lett. **81**, 1562 (1998)
3. Y. Fukuda et al., Phys. Rev. Lett. **82**, 2644 (1999)
4. S.H. Ahn et al., Phys. Lett. B **511**, 178 (2001)
5. K. Eguchi et al., Phys. Rev. Lett. **90**, 021802 (2003)
6. B. Pontecorvo, Measonium and anti-mesonium. Zh. Eksp. Theor. Fiz. **33**, 549 (1957)
7. B. Pontecorvo, Sov. Phys. JETP **26**, 984 (1968) [Zh. Eksp. Teor. Fiz 53, 1717 (1967)]
8. M.M. Medeiros et al., 33rd International Cosmic Ray Conference, Rio de Janeiro, July, 2013. http://www.cbpf.br/~icrc2013/papers/icrc2013-0958.pdf
9. D. Casper, Nucl. Phys. Proc. Suppl. **112** (2002)
10. M. Honda, T. Kajita et al., Phys. Rev. D **70** (2004)
11. Daljeet Kaur et al., Phys. Rev. D **95**, 093005 (2017)

Chapter 49
Impact of Non-unitary Neutrino Mixing on the Physics Potential of NOνA Experiment

C. Soumya and Rukmani Mohanta

Abstract The extended theories of Standard Model (SM) require heavy right-handed neutrinos to accommodate observed neutrino oscillation data. If the mass of such sterile neutrinos are of the order of TeV scale, then they can mix with active neutrinos, which leads to non-unitary neutrino mixing. In this paper, we investigate the effect of non-unitary neutrino mixing on the physics potential of the long baseline experiment. We find that non-unitary mixing can deteriorate the sensitivity of NOνA, to the current unknowns in neutrino sector due to the existence of new degeneracies among the oscillation parameters in presence of non-unitary mixing. We also find that the sensitivity to the unknowns enhanced, if we add oscillation data from T2K experiment as it enables to reduce the degeneracies among the oscillation parameters.

49.1 Introduction

The discovery of neutrino oscillation is the first confirmed evidence for new physics beyond the Standard Model (SM) as it implies that neutrino is no more a massless particle. As of now, the three flavor neutrino oscillation framework has become the successful theoretical framework to explain the observed oscillation data. This standard picture of neutrino oscillation is described by three mixing angles: $\theta_{12}, \theta_{13}, \theta_{23}$, one phase: $\delta_{\rm CP}$, and two mass squared differences: $\Delta m^2_{21}, \Delta m^2_{31}$. The determination of the reactor mixing angle by Daya Bay, RENO, and Double Chooze, T2K experiments, opens up a way to explore current unknowns in neutrino sector such as neutrino mass ordering, CP-violation in leptonic sector, and the octant of atmospheric mixing angle. The main objective of current and future neutrino oscillation experiments is to determine these unknowns.

C. Soumya (✉)
Institute of Physics, Sachivalaya Marg, Sainik School Post, Bhubaneswar 751005, India
e-mail: soumyac20@gmail.com

R. Mohanta
University of Hyderabad, Hyderabad 500046, India
e-mail: rukmani98@gmail.com

A. Giri and R. Mohanta (eds.), *16th Conference on Flavor Physics and CP Violation*, Springer Proceedings in Physics 234,
https://doi.org/10.1007/978-3-030-29622-3_49

In general, the determination of the oscillation parameters are done by taking the assumption that the neutrino mixing matrix is unitary [1]. However, many extensions of the Standard Model require additional fermion fields to incorporate massive neutrino and lead to active-sterile neutrino mixing, which gives rise to unitarity violation [2, 3] in active neutrino mixing. The low-scale seesaw models like inverse seesaw model permits significantly large mixing between the active and sterile neutrinos and gives rise to significant non-unitary lepton mixing. In this paper, we investigate the effect of non-unitary neutrino mixing on the physics potential of NOνA experiment.

49.2 Non-unitary Neutrino Mixing

In the presence of heavy sterile neutrino, the neutrino mixing matrix is of the form,

$$U_{eff} = \begin{pmatrix} N & \Theta \\ R & S \end{pmatrix},$$

where N is the non-unitary active neutrino mixing matrix.

$$N = (1 - \frac{1}{2}\Theta^{\dagger}\Theta)U_{\text{PMNS}} = (1 - \eta)U_{\text{PMNS}} \tag{49.1}$$

where $\eta(=\frac{1}{2}\Theta^{\dagger}\Theta)$ is called non-unitarity parameter. In presence of non-unitary lepton mixing, the charged current and neutral current interaction Lagrangian gets modified as [3]

$$-\mathcal{L}_{int} = V_{\text{CC}} \sum_{i,j} N^*_{ei} N_{ej} \bar{\nu}_i \gamma^0 \nu_j + V_{\text{NC}} \sum_{\alpha,i,j} N^*_{\alpha i} N_{\alpha j} \bar{\nu}_i \gamma^0 \nu_j \;, \tag{49.2}$$

which yields the effective Hamiltonian as

$$\mathcal{H}^N_m = \frac{1}{2E} \begin{pmatrix} 0 & 0 & 0 \\ 0 & \Delta m^2_{21} & 0 \\ 0 & 0 & \Delta m^2_{31} \end{pmatrix} + N^{\dagger} \begin{pmatrix} V_{\text{CC}} + V_{\text{NC}} & 0 & 0 \\ 0 & V_{\text{NC}} & 0 \\ 0 & 0 & V_{\text{NC}} \end{pmatrix} N,$$

with $\Delta m^2_{ij} = m^2_i - m^2_j$, charged current matter potential $V_{\text{CC}} = \sqrt{2} G_F n_e$ and neutral current matter potential $V_{\text{NC}} = -G_F n_n/\sqrt{2}$. Hence, the oscillation probability is given by

$$P_{\alpha\beta}(E, L) = |\langle \nu_\beta | \nu_\alpha(L) \rangle|^2 = \left| \left(N e^{-i\mathcal{H}^N_m L} N^{\dagger} \right)_{\beta\alpha} \right|^2 .$$

49.3 Simulation Details

The long baseline neutrino oscillation experiment NOνA, which uses NuMI beam with a power of 700 kW, has a 14 kt far detector situated almost 810 km away from the Fermilab. T2K is another currently running Japan based off-axis long baseline experiment with baseline 295 km and its oscillation peaks around 0.6 GeV. We use GLoBES [4, 5] package along with MonteCUBES [6] to simulate these experiments and the experimental specifications are explicitly given in [7].

49.4 Mass Hierarchy, Octant and CP-violation Sensitivities

In this section, we compare the sensitivity of NOνA to rule out the CP-conserving points, wrong octant, and wrong mass hierarchy in both standard and non-unitary mixing schemes. In order to obtain the sensitivities in terms of χ^2, we follow the method that is described in [7]. The sensitivity of NOνA is given in Fig. 49.1. From the figure, it can be seen that non-unitary lepton mixing significantly affects the sensitivities of current unknowns in neutrino sector. The sensitivities are deteriorated significantly in presence of non-unitary lepton mixing and they crucially depend up on the new CP-violating phase in the non-unitary mixing. The results for synergy of T2K and NOνA is given in Fig. 49.2. From the figure, it can be seen that while adding the T2K data, there is a significant enhancement in the sensitivities of the unknowns.

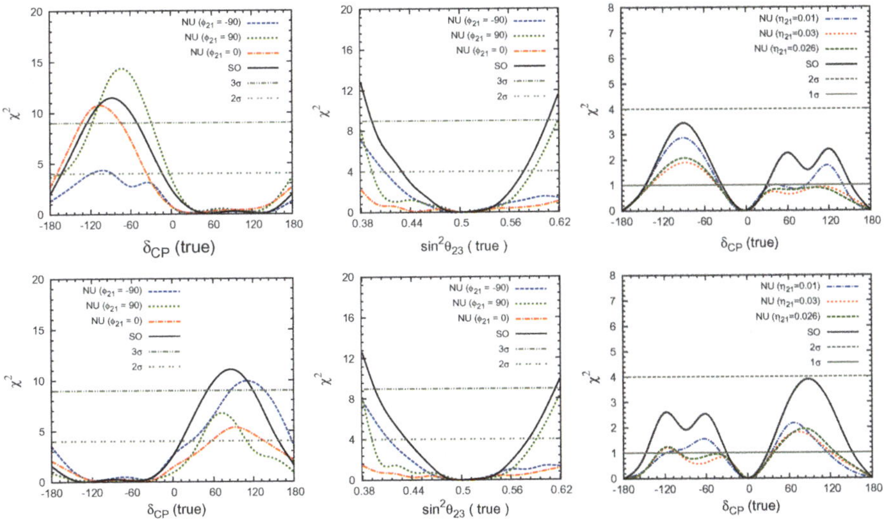

Fig. 49.1 The MH (left panel), octant (middle panel), and CPV (right panel) sensitivities of NOνA. The true hierarchy is assumed to be normal (inverted) in upper (lower) panel

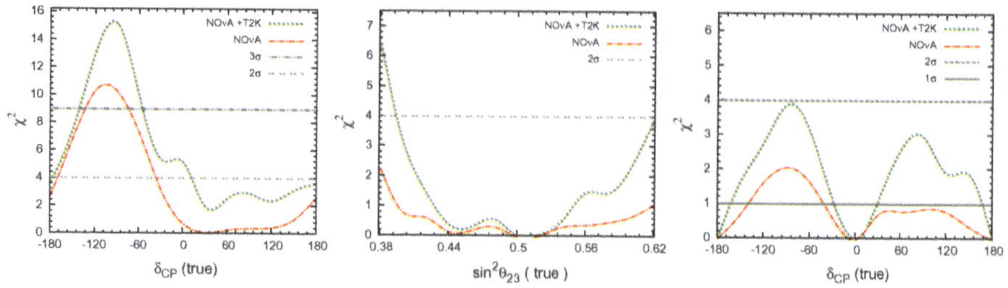

Fig. 49.2 The MH (left panel), octant (middle panel), and CPV (right panel) sensitivities of NOνA and T2K. The true hierarchy is assumed to be normal

49.5 Conclusions

We have studied the effect of non-unitary lepton mixing on the determination of neutrino mass hierarchy, octant of atmospheric mixing angle and CP violating phase. We found that non-unitary lepton mixing significantly affect the sensitivities of current unknowns in neutrino sector. In fact, the mass hierarchy sensitivity, octant sensitivity, and CPV sensitivity are deteriorated significantly in presence of non-unitary lepton mixing and the sensitivities crucially depend up on the new CP-violating phase in the non-unitary mixing. Moreover, the oscillation parameter degeneracy resolution capability of NOνA experiment is reduced in presence of non-unitarity parameters as they introduced new degeneracies among the oscillation parameters. However, we have seen that the synergy between the currently running experiments T2K and NOνA has improved degeneracy resolution capability.

Acknowledgements This work is supported by SERB, Govt. of India.

References

1. P.F. de Salas et al., Status of neutrino oscillations 2018: 3σ hint for normal mass ordering and improved CP sensitivity. Phys. Lett. B **782**, 633 (2018). https://www.sciencedirect.com/science/article/pii/S0370269318304672?via%3Dihub, https://doi.org/10.1016/j.physletb.2018.06.019, arXiv:1708.01186 [hep-ph]
2. R.N. Mohapatra et al., Neutrino masses and mixings in gauge models with spontaneous parity violation. Phys. Rev. D **23**, 165 (1981). https://journals.aps.org/prd/abstract/10.1103/PhysRevD.23.165, https://doi.org/10.1103/PhysRevD.23.165
3. S. Antusch et al., Unitarity of the leptonic mixing matrix. JHEP **0610**, 084 (2006). http://iopscience.iop.org/article/10.1088/1126-6708/2006/10/084/meta, https://doi.org/10.1088/1126-6708/2006/10/084, arXiv:hep-ph/0607020
4. P. Huber et al., Simulation of long-baseline neutrino oscillation experiments with GLoBES (General Long Baseline Experiment Simulator). Comput. Phys. Commun. **167**, 195 (2005). https://www.sciencedirect.com/science/article/pii/S0010465505000743?via%3Dihub, https://doi.org/10.1016/j.cpc.2005.01.003, arXiv:hep-ph/0407333

5. P. Huber et al., New features in the simulation of neutrino oscillation experiments with GLoBES 3.0: general long baseline experiment simulator. Comput. Phys. Commun. **177**, 432 (2007). https://www.sciencedirect.com/science/article/pii/S0010465507002421?via%3Dihub, https://doi.org/10.1016/j.cpc.2007.05.004, arXiv:hep-ph/0701187
6. M. Blennow et al., Neutrino oscillation parameter sampling with MonteCUBES. Comput. Phys. Commun. **181**, 227 (2010). https://www.sciencedirect.com/science/article/pii/S001046550900304X?via%3Dihub, https://doi.org/10.1016/j.cpc.2009.09.014, arXiv:0903.3985 [hep-ph]
7. C. Soumya, M. Rukmani, Non-unitary lepton mixing in an inverse seesaw and its impact on the physics potential of long-baseline experiments. J. Phys. G **45**(9), 095003 (2018). http://iopscience.iop.org/article/10.1088/1361-6471/aad2cc/meta, https://doi.org/10.1088/1361-6471/aad2cc, arXiv:1708.05372(hep-ph)

Chapter 50
A_4 Realization of Linear Seesaw and Neutrino Phenomenology

M. Sruthilaya, R. Mohanta and Sudhanwa Patra

Abstract We consider the A_4 realization of linear seesaw by extending the standard model (SM) particle content with two types of right-handed neutrinos along with the flavon fields, and the SM symmetry with $A_4 \times Z_4 \times Z_3$ and a global symmetry $U(1)_X$. We also discuss the scope of this model to explain the baryon asymmetry of the Universe through leptogenesis.

50.1 Introduction

The experimental observation of neutrino oscillation confirmed that neutrinos have tiny but non-zero mass. The smallness of neutrino mass is addressed by many seesaw mechanisms such as type-I, type-II type-III, Linear and Inverse seesaw etc. Predictions of type-I, type-II type-III seesaw mechanism are very heavy particles which are beyond the reach of present and future experiments. This makes them untestable. Whereas linear and inverse seesaw explain light neutrino mass with predictions of comparatively light particles. Hence, models employing linear and inverse seesaw mechanism to generate light neutrino mass can be tested in near future experiments. This makes them more interesting. Another important concern in neutrino physics is whether three generations of fermions are connected by any symmetry. The special values of neutrino mixing angles support well the idea of existence flavor symmetry across fermion generations. Moreover, it has been shown in may papers that the observed neutrino mixing can be explained if existence of A_4 flavor symmetry is assumed. In light of these facts we consider A_4 realization of linear seesaw to explain

M. Sruthilaya (✉) · R. Mohanta
University of Hyderabad, Hyderabad 500046, India
e-mail: msruthi28@gmail.com

R. Mohanta
e-mail: rmsp@uohyd.ernet.in

S. Patra
Indian Institute of Technology Bhilai, GEC Campus, Sejbahar, Raipur 492015, India
e-mail: sudha.astro@gmail.com

A. Giri and R. Mohanta (eds.), *16th Conference on Flavor Physics and CP Violation*, Springer Proceedings in Physics 234,
https://doi.org/10.1007/978-3-030-29622-3_50

Table 50.1 The particle content and their charge assignments for an A_4 realization of linear seesaw mechanism

Fields	e_R	μ_R	τ_R	L	H	N_R	S_R	ϕ_T	ϕ_S	ξ	ξ'	ρ	ρ'
A_4	1	$1''$	$1'$	3	1	3	3	3	3	1	$1'$	1	1
Z_4	$-i$	$-i$	$-i$	$-i$	1	i	1	1	i	i	i	$-i$	-1
Z_3	1	1	1	1	1	1	ω	1	ω	ω	ω	ω^2	1
X	-1	-1	-1	-1	0	-1	1	0	0	0	0	-2	0

active neutrino mass and mixing. To do this we extend the standard model symmetry with flavor symmetry $A_4 \times Z_4 \times Z_3$ and the particle content of standard model with three generations of two types of heavy Majorana neutrinos namely, right-handed neutrino and sterile neutrino. The existence of heavy Majorana neutrinos and Majorana phases in neutrino mixing matrix add up the source of baryon asymmetry of universe. Hence, we explore the scope of the model to explain the baryon asymmetry of Universe.

50.2 The Model

With the particle content that belongs to different representations under $A_4 \times Z_4 \times Z_3$ as given in Table 50.1 the Yukawa Lagrangian in charged lepton and neutrino sector take the forms [1]

$$\mathcal{L}_l = -\left\{\left[\frac{\lambda_e}{\Lambda}\left(\bar{L}\phi_T\right) He_R\right] + \left[\frac{\lambda_\mu}{\Lambda}\left(\bar{L}\phi_T\right)' H\mu_R\right] + \left[\frac{\lambda_\tau}{\Lambda}\left(\bar{L}\phi_T\right)'' H\tau_R\right]\right\},$$

$$-\mathcal{L}_{\succeq} = y_1\overline{L}\widetilde{H}N_R\frac{\rho'}{\Lambda} + y_2\overline{L}\widetilde{H}S_R\frac{\rho}{\Lambda} + \left(\lambda^{\phi}_{NS}\phi_s + \lambda^{\xi}_{NS}\xi + \lambda^{\xi'}_{NS}\xi'\right)\overline{N_R}S^c_R,$$

respectively. Then the full mass matrix for neutral leptons in the basis $N = (\nu_L, N^c_R, S^c_R)^T$ is given as

$$M = \begin{pmatrix} 0 & m_D & m_{LS} \\ m^T_D & 0 & m_{RS} \\ m^T_{LS} & m^T_{RS} & 0 \end{pmatrix} \tag{50.1}$$

The resulting mass formula for light neutrinos is governed by linear seesaw mechanism [2],

$$m_\nu = m_D m^{-1}_{RS} m^T_{LS} + \text{transpose}. \tag{50.2}$$

With the vacuum alignment for the scalar and flavon fields

$$\langle H\rangle = (0, v)^T,\ \langle \phi_S\rangle = v_T(1,0,0),\ \langle \phi_S\rangle = v_S(1,1,1),\ \langle \xi\rangle = v_\xi,\ \langle \xi'\rangle = v_{\xi'},\ \langle \rho\rangle = v_\rho,\ \langle \rho'\rangle = v_{\rho'},$$

the various mass matrices are found to be

$$M_l = v\frac{v_T}{\Lambda}\text{diag}\left(\lambda_e, \lambda_\mu, \lambda_\tau\right), \tag{50.3}$$

$$m_D = y_1 v\frac{v_{\rho'}}{\Lambda}\begin{pmatrix} 1\,0\,0 \\ 0\,0\,1 \\ 0\,1\,0 \end{pmatrix}, \quad m_{LS} = y_2 v\frac{v_\rho}{\Lambda}\begin{pmatrix} 1\,0\,0 \\ 0\,0\,1 \\ 0\,1\,0 \end{pmatrix},$$

$$m_{RS} = \frac{a}{3}\begin{pmatrix} 2 & -1 & -1 \\ -1 & 2 & -1 \\ -1 & -1 & 2 \end{pmatrix} + b\begin{pmatrix} 1\,0\,0 \\ 0\,0\,1 \\ 0\,1\,0 \end{pmatrix} + d\begin{pmatrix} 0\,0\,1 \\ 0\,1\,0 \\ 1\,0\,0 \end{pmatrix}, \tag{50.4}$$

where $a = \lambda^{\phi}_{NS} v_S$, $b = \lambda^{\xi}_{NS} v_\xi$ and $d = \lambda^{\xi'}_{NS} v_{\xi'}$.

50.3 Neutrino Mass and Mixing

The special form of Dirac neutrino mass matrix m_D gives rise to the relation

$$m_i = \frac{|k_1 k_2|}{M_i}, \tag{50.5}$$

between the light neutrino mass (m_i) and heavy neutrino mass (M_i). Also the matrix that diagonalizes light neutrino mass matrix is same as the matrix that diagonalizes m_{RS}^{-1}, which is $U_{TBM}.U_{13}.P$. Where U_{TBM} is the tri-bimaximal mixing matrix whereas the complex unitary matrix U_{13} is given by

$$U_{13} = \begin{pmatrix} \cos\theta & 0 & \sin\theta e^{-i\psi} \\ 0 & 1 & 0 \\ -\sin\theta e^{i\psi} & 0 & \cos\theta \end{pmatrix}, \tag{50.6}$$

and P is a diagonal matrix that contains the phases associated with the eigen values of light neutrino mass matrix.

Using the constrain, $r = \left(\frac{\Delta m^2_{21}}{|\Delta m^2_{31}|}\right) = 0.0291 \pm 0.00085$, we obtained the parametric space for which the model gives mixing angles and Dirac CP phase inline with recent experimental results as shown in Fig. 50.1. The correlation between Dirac CP phase and lightest neutrino mass and sum of active neutrino masses are also given Fig. 50.1.

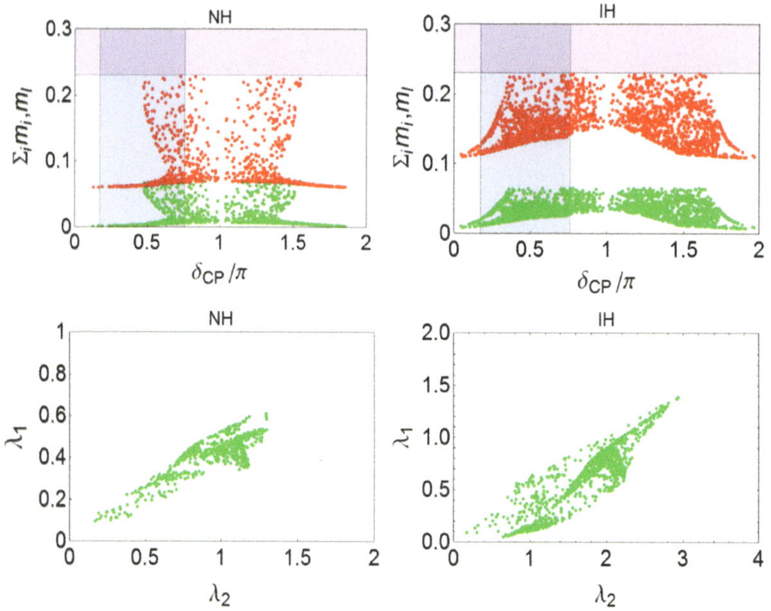

Fig. 50.1 The correlation plots between Dirac CP phase and lightest neutrino mass (green dots) and sum of active neutrino masses (red dots) for normal and inverted hierarchies are given respectively in the top left and right panels. The parametric space in $\lambda_2(|\frac{a}{b}|) - \lambda_1(|\frac{d}{b}|)$ plane for normal and inverted hierarchies are given in bottom left and right panels respectively

50.4 Leptogenesis

The dimensional six operators

$$-\left\{\left[\lambda_{N\phi}\phi_S + \lambda_{N\xi}\xi + \lambda_{N\xi'}\xi'\right]\frac{\rho\rho'}{\Lambda^2}\overline{N}_R N_R^c + \left[\lambda_{S\phi}\phi_S + \lambda_{S\xi}\xi + \lambda_{S\xi'}\xi'\right]\frac{\rho^\dagger\rho'}{\Lambda^2}\overline{S}_R S_R^c\right\}, \quad (50.7)$$

in the lepton sector generate baryon asymmetry through resonant leptogenesis [3], which in the case of normal hierarchy is related to model parameters as follows

$$\eta_B \leq 0.174\left(\frac{\left(\frac{m_3(\text{eV})}{10^{-2}}\right)^2}{|y_1|^2 K_{N_3}^3 (\ln K_{N_3})^{0.6}}\right). \quad (50.8)$$

where m_3 is the mass of heaviest light neutrino and $K_{N_3} \approx 0.234 \left[\frac{m_3\ (\mathrm{eV})}{10^{-2}}\right] \frac{v_{\rho'}}{v_\rho}$. Which intern gives lower limit on the ratio, $\frac{v_\rho}{v_{\rho'}}$ as 5×10^{-5} to generate baryon asymmetry of 6.9×10^{-10}, the baryon asymmetry of Universe.

50.5 Conclusions

We considered the A_4 realization of linear seesaw and found that the model can explain the experimental observations on neutrino mass an mixing. We also found that the model is fit enough to explain the baryon asymmetry of the universe.

Acknowledgements SM would like to thank the University Grants Commission for financial support. RM acknowledges the support from the Science and Engineering Research Board (SERB), Government of India, through Grant No. SB/S2/HEP-017/2013.

References

1. M. Sruthilaya, R. Mohanta, S. Patra, A_4 realization of linear seesaw and neutrino phenomenology. Eur. Phys. J. C **78**, 719 (2018). https://doi.org/10.1140/epjc/s10052-018-6181-6
2. M. Malinsky, J.C. Romao, J.W.F. Valle, Novel supersymmetric SO(10) seesaw mechanism. Phys. Rev. Lett. **95**, 161801 (2005). https://doi.org/10.1103/PhysRevLett.95.161801
3. P.-H. Gu, U. Sarkar, Leptogenesis with linear, inverse or double seesaw. Phys. Lett. B **694**, 226232 (2011). https://doi.org/10.1016/j.Physletb.2010.09.062

Chapter 51
Sterile Neutrino Discovery Potential of NOvA and DUNE Using Neutral-Current Interactions

Akshay Chatla and Bindu A. Bambah

Abstract Sterile neutrino mixing can be studied at long-baseline experiments using neutral-current (NC) data. A reduction in NC event rate is expected in presence of sterile neutrino due to the mixing between sterile and active neutrinos, thus giving direct sensitivity to sterile neutrino channel. In this work, we see that NC channel has good sensitivity to sterile mixing angles at long-baseline. We take current best fit values of NOvA as true values and try to find the extent to which the sterile neutrino parameters will be constrained by the NOvA and DUNE in their future runs.

51.1 Introduction

Precise measurement of neutrino oscillation parameters is an important objective of all neutrino oscillation experiments. Anomalous results of LSND and MiniBooNE could be attributed to a new neutrino introducing new $\Delta m^2 \sim 1$ eV^2. As the LEP experiment limits number of active neutrinos flavors to 3, the new neutrino must be a sterile (No weak interaction) neutrino (ν_s). Introducing sterile neutrino gives rise to new oscillation parameters. Short baseline (SBL) experiments are sensitive to the sterile mixing angles. These sterile mixing angles can also be measured at long baseline (LBL) using NC channel. An energy-dependent suppression of NC event rate will occur in presence of sterile neutrino due to mixing between sterile and active neutrinos. Standard oscillations do not effect NC channel since all the active flavors take part in NC events. Thus, the degeneracies and uncertainties present in active neutrino parameter space have a small effect on NC event rates in contrast to charge-current (CC) events. Using this direct sensitivity to sterile neutrino channel, we try to find the capacity of NOvA and DUNE experiments to measure and constrain sterile mixing angles in their future runs.

A. Chatla (✉) · B. A. Bambah
School of Physics, University of Hyderabad, Hyderabad 500046, India
e-mail: chatlaakshay@gmail.com

A. Giri and R. Mohanta (eds.), *16th Conference on Flavor Physics and CP Violation*, Springer Proceedings in Physics 234,
https://doi.org/10.1007/978-3-030-29622-3_51

51.2 Theoretical Framework

In this paper, we worked with one light sterile neutrino (3+1 model). In 3+1 neutrino model, the flavour and mass eigenstates are coupled with a 4×4 mixing matrix. A suitable parametrization of the mixing matrix is

$$\mathrm{U}_{\mathrm{PMNS}_{3+1}} = \mathrm{R}_{34}\tilde{\mathrm{R}}_{24}\tilde{\mathrm{R}}_{14}\mathbf{R}_{23}\tilde{\mathbf{R}}_{13}\mathbf{R}_{12}. \tag{51.1}$$

here R_{ij} and $\tilde{R_{ij}}$ represent real and complex 4×4 rotation in the plane containing the 2×2 sub-block in (i,j) sub-block

$$R_{ij}^{2\times2} = \begin{pmatrix} c_{ij} & s_{ij} \\ -s_{ij} & c_{ij} \end{pmatrix} \qquad \tilde{R_{ij}}^{2\times2} = \begin{pmatrix} c_{ij} & \tilde{s_{ij}} \\ -\tilde{s_{ij}}^{*} & c_{ij} \end{pmatrix} \tag{51.2}$$

where, $c_{ij} = \cos\theta_{ij}$, $s_{ij} = \sin\theta_{ij}$, $\tilde{s_{ij}} = s_{ij}e^{-i\delta_{ij}}$ and δ_{ij} are the CP phases.

The bold matrices in (51.1) represent standard 3 flavour model. We see that addition of one sterile neutrino introduces 3 new mixing angles and 2 new CP-phases. Measuring these new parameters is important for the study of sterile neutrinos. Experiments like NOvA and DUNE are useful to measure these parameters. Sterile neutrino search can be done at both SBL and LBL, each sensitive to different parameters. In this work, analysis is done for sterile neutrino search at far detectors (LBL) of NOvA and DUNE.

An approximate expression for NC disappearance probability [1] at LBL is

$$\begin{aligned} 1 - \mathrm{P}(\nu_\mu \rightarrow \nu_s) \approx 1 - \frac{1}{2}\cos^4\theta_{14}\cos^2\theta_{34}\sin^2 2\theta_{24} \\ + \mathrm{A}\sin^2\Delta_{31} - \mathrm{B}\sin 2\Delta_{31}, \end{aligned} \tag{51.3}$$

$$where,\ \Delta_{ij} \equiv \Delta m_{ij}^2 L/4E, \text{ L is baseline and E is neutrino energy} \tag{51.4}$$

$$A = \sin^2\theta_{34}\sin^2 2\theta_{23} \tag{51.5}$$

$$B = \frac{1}{2}\sin\delta_{24}\sin\theta_{24}\sin 2\theta_{34}\sin 2\theta_{23} \tag{51.6}$$

We see that the NC oscillations are sensitive to θ_{24}, θ_{34} and δ_{24}. This disappearance of NC events which is independent of standard oscillations is our probe for sterile search in this work.

51.3 Results

We used GLoBES (General Long Baseline Experiment simulator)[2, 3] to simulate the data for NOvA and DUNE. The simulation and experimental details we used are in [4, 5] and the references therein. In Fig. 51.1, we plot NC disappearance probability (P_{NC}) plot to show effects of sterile mixing angles on reduction of NC event rates. In the left (right) plot in Fig. 51.1, we show P_{NC} while varying θ_{24} (θ_{34}) and keeping other sterile mixing angles zero. We see that both θ_{24} and θ_{34} play a major role in NC oscillations as implied by Eqn 51.3.

In Fig. 51.2, we show how well the sterile mixing angles will be constrained by NOvA (3+$\bar{3}$) and DUNE (3.5+$\bar{3.5}$) in their future runs. This analysis is done by taking standard 3 flavour model as true hypothesis and checking the confidence level to which 3+1 model i.e. test hypothesis can be rejected. In this analysis, we did not simulate near detector explicitly. So, this analysis is valid for $0.05 < \Delta m^2 < 0.5\text{eV}^2$ as sterile oscillations are not developed at near detector in this mass range. We explore allowed regions in θ_{24} and θ_{34} plane from NOvA and DUNE simulation data for their full run times, considering latest NOvA results as true values. As seen in the Fig. 51.2, DUNE is twice as good as NOvA at constraining θ_{24} and θ_{34} which can be attributed to it's high expected event rates and detector efficiency. We see that the constraints are similar for the 2 best-fit values of NOvA, as expected since standard oscillation parameters have negligible effect on NC channel.

In conclusion, expected NC event disappearance in presence of sterile neutrino is calculated for future runs of NOvA and DUNE. We find that DUNE is better at constraining sterile mixing angles θ_{24} and θ_{34} than NOvA as expected due to high NC event rates in DUNE. Further analysis while including near detector oscillations need to be done to explore the mass ranges $\Delta m^2 > 0.5\text{eV}^2$.

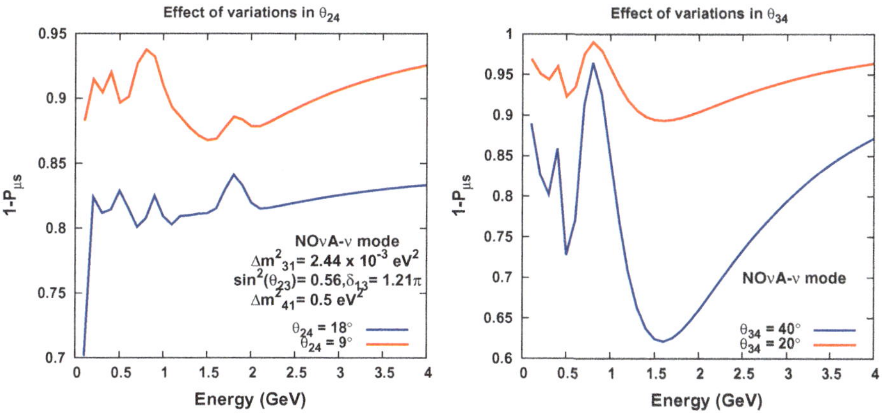

Fig. 51.1 NC disappearance probability plots for NOvA while changing values of θ_{24} (left) and θ_{34} (right). Here θ_{24} (θ_{34}) takes values 9° and 18° (20° and 40°) while other sterile mixing angles are kept zero

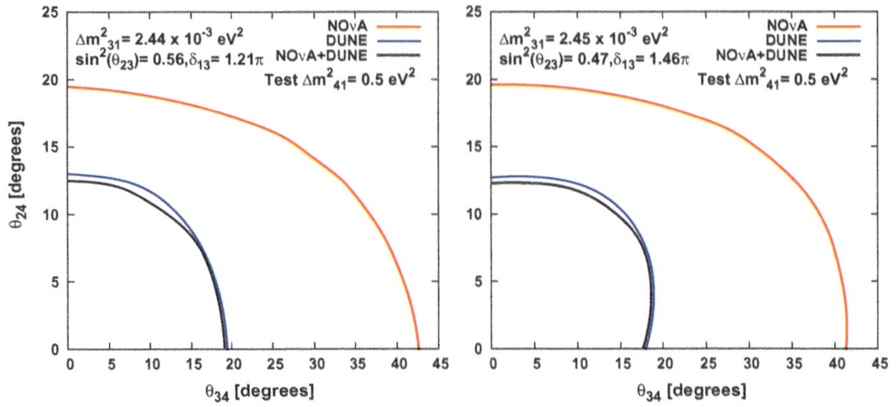

Fig. 51.2 Expected sensitivity at 90% C.L projected in the θ_{34} versus θ_{24} plane, for $\Delta m^2_{41} = 0.5 eV^2$ in future full-run of NOvA and DUNE for two best-fit values of NOvA

References

1. P. Adamson et al. NOvA Collaboration, Phys. Rev. D **96**(7), 072006 (2017)
2. P. Huber et al. Comput. Phys. Commun. **167**, 195 (2005), [hep-ph/0407333]
3. P. Huber et al. Comput. Phys. Commun. **177**, 432 (2007), [hep-ph/0701187]
4. R. Gandhi, B. Kayser, S. Prakash, S. Roy, JHEP **1711**, 202 (2017)
5. A. Chatla et al., Adv. High Energy Phys. **2018**, 2547358 (2018)

Part V
Flavour Anomalies and New Physics

Chapter 52
Lepton Universality Violating Anomalies in B Decays

Alakabha Datta

Hints of lepton universality violating (LUV) new physics have shown up in several B decays. Experimental measurements show deviation from the standard model (SM) predictions of lepton universal interactions. These measurement anomalies occur in charged current and neutral current transitions. I will review the present measurements and future measurements that can tell us more about these anomalies. I will focus on joint explanation of the anomalies and point out that LUV new physics can often lead to lepton flavor violation effects. Finally, I will discuss light mediator models to explain the anomalies.

52.1 Introduction

At present, there are several measurements of B decays that may indicate the presence of physics beyond the standard model (SM). In particular these measurements indicate lepton universality violation (LUV). In the SM the gauge bosons couple equally to all members of the quark and the lepton families. Hence measurement of LUV is evidence of physics beyond the SM.

The LUV are observed in two groups—in charged current (CC) processes and in the neutral current (NC) processes. We will discuss them separately first and then consider how the CC and NC LUV might have a common origin.

This research was supported by the U.S. NSF under Grant No. PHY-1414345.

A. Datta (✉)
Department of Physics and Astronomy, University of Mississippi,
108 Lewis Hall, 38677 Oxford, MS, USA
e-mail: datta@olemiss.edu

A. Giri and R. Mohanta (eds.), *16th Conference on Flavor Physics and CP Violation*, Springer Proceedings in Physics 234,
https://doi.org/10.1007/978-3-030-29622-3_52

52.1.1 CC LUV

The charged-current decays $B \to D^{(*)}\tau\nu_\tau$ have been measured by the BaBar [1], Belle [2] and LHCb [3] Collaborations. It is found that the values of the ratios $R(D^{(*)}) \equiv \mathcal{B}(\bar{B} \to D^{(*)}\tau^-\bar{\nu}_\tau)/\mathcal{B}(\bar{B} \to D^{(*)}\ell^-\bar{\nu}_\ell)$ ($\ell = e, \mu$) considerably exceed their SM predictions. These ratios of branching fractions have certain advantages over the absolute branching fraction measurements as they are relatively less sensitive to form factor variations and several systematic uncertainties such as those on the experimental efficiency as well as the dependence on the value of $|V_{cb}|$ cancel in the ratios. There are lattice determination of the ratio $R(D)_{SM}$ [4, 5] that are in general agreement with one another.

$$\begin{aligned} R(D)_{SM} &= 0.299 \pm 0.011, && \text{FNAL/MILC[4]} \\ R(D)_{SM} &= 0.300 \pm 0.008, && \text{HPQCD[5]} \\ R(D^*)_{SM} &= 0.252 \pm 0.003. && \end{aligned} \tag{52.1}$$

Calculation of $R(D^*)_{SM}$ is not available from lattice and so one has to use SM phenomenological prediction [6–8] where the form factors are obtained from fits to the angular distributions in $\bar{B} \to D^{(*)}\ell^-\bar{\nu}_\ell$.

By averaging the most recent measurements, the HFAG Collaboration has found [9]

$$\begin{aligned} R(D) &= 0.407 \pm 0.039 \pm 0.024, \\ R(D^*) &= 0.304 \pm 0.013 \pm 0.007, \end{aligned} \tag{52.2}$$

where the first uncertainty is statistical and the second is systematic. $R(D^*)$ and $R(D)$ exceed the SM predictions by 3.4σ and $2.3\ \sigma$, respectively. The combined analysis of $R(D^*)$ and $R(D)$, taking into account measurement correlations, leads to a deviation is 4.1σ from the SM prediction [9].

In general there have been many analyses of the $R(D^{(*)})$ puzzles both in model independent framework as well as in specific models (see, for example, [10–15]).

52.1.1.1 Distributions and CP Violation

The new physics proposed for the $R(D^{(*)})$ puzzled can be probed in distributions [10, 16, 17]. Some of the observables in the distributions have been measured. In the coming years more of these will be measured. These measurements will discover or constrain new physics and will provide important clues to the nature of new physics. An important observable is CP violation in the distribution [16, 18] as this is free of hadronic uncertainties. Measurement of non zero value of CP violation will be a clear sign of new physics. The complete three-angle distribution for the decay $\bar{B} \to D^*(\to D\pi)l^-\bar{\nu}_l$ in the presence of NP can be expressed in terms of four

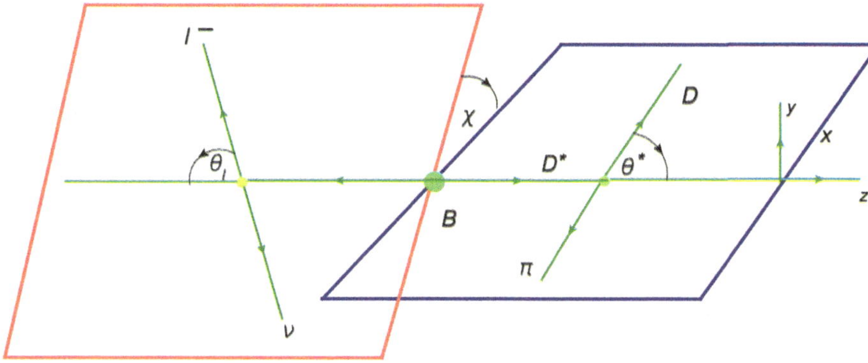

Fig. 52.1 The description of the angles θ_l, θ^* and χ in the angular distribution of $\bar{B} \to D^*(\to D\pi)l^-\nu_l$ decay

kinematic variables, q^2 (the momentum transfer squared), two polar angles θ_l, θ^{D^*}, and the azimuthal angle χ. The angle θ_l is the polar angle between the charged lepton and the direction opposite to the D^* meson in the $(l\nu_l)$ rest frame. The angle θ_{D^*} is the polar angle between the D meson and the direction of the D^* meson in the $(D\pi)$ rest frame. The angle χ is the azimuthal angle between the two decay planes spanned by the 3-momenta of the $(D\pi)$ and $(l\nu_l)$ systems. These angles are described in Fig. 52.1. The three-angle distribution can be obtained by using the helicity formalism.
We can write the angular distribution for $\bar{B} \to D^*(\to D\pi)l^-\bar{\nu}_l$ as [10, 19–22]

$$\frac{d^4\Gamma}{dq^2\, d\cos\theta_l\, d\cos\theta_{D^*}\, d\chi} = \frac{9}{32\pi} NF\Big(\sum_{i=1}^{8} I_i + \frac{m_l^2}{q^2}\sum_{j=1}^{8} J_i\Big), \qquad (52.3)$$

where the I_i and J_i are functions of the helicity amplitudes and the helicity angles [16].

The complex NP couplings lead to CP violation which is sensitive to the angular terms $\sin\chi$ and $\sin 2\chi$ in the distribution. The coefficients of these terms are triple products (TP) and have the structure $\sim Im[\mathcal{A}_i\mathcal{A}_j^*] \sim \sin(\phi_i - \phi_j)$, where $\mathcal{A}_{i,j} = |\mathcal{A}_{i,j}|e^{i\phi_{i,j}}$. In the SM these terms vanish, as there is only one dominant contribution to the decay and so all amplitudes have the same weak phase. Hence any non-zero measurements of TPs are clear signs of NP without any hadronic uncertainties. For the charged conjugate modes, the weak phases change sign and $\overline{\mathcal{A}_{i,j}} = |\mathcal{A}_{i,j}|e^{-i\phi_{i,j}}$ and the TPs change sign. Even though we focus on τ final states, we should point out that this distribution is applicable also for e and μ in the final state. Since experiments have already studied the distributions for e, μ final states it might be worth checking the $\sin\chi$ and $\sin 2\chi$ terms in the distributions for these decays for signals of non-SM physics. There are other b decays that can be used to constrain the models discussed to explain the $R(D^{(*)})$ measurements. It was pointed out that the underlying quark level transition $b \to c\tau^- \bar{\nu}_\tau$ in the $R(D^{(*)})$ puzzles can be probed in both B and Λ_b decays [23].

52.1.1.2 NC LUV

The LHCb and Belle Collaboration has made measurements of $B \to K^* \mu^+ \mu^-$ and $B_s^0 \to \phi \mu^+ \mu^-$ that deviate from the SM predictions.

The main discrepancy is in the angular observable P_5' [24] though the significance of the discrepancy depends on the assumptions about the theoretical hadronic uncertainties.

To find clear evidence of new physics one should consider observables largely free of hadronic uncertainties. One such observable is $R_K \equiv \mathcal{B}(B^+ \to K^+ \mu^+ \mu^-)/\mathcal{B}(B^+ \to K^+ e^+ e^-)$ [25, 26], which has been measured by LHCb [27]:

$$R_K^{\rm expt} = 0.745^{+0.090}_{-0.074}\,({\rm stat}) \pm 0.036\,({\rm syst})\,, \quad 1 \le q^2 \le 6.0\,{\rm GeV}^2\,. \qquad (52.4)$$

This differs from the SM prediction, $R_K^{\rm SM} = 1 \pm 0.01$ [28] by 2.6σ. Note, the observable R_K is a measure of lepton flavor universality and requires different new physics for the muons versus the electrons, while it is possible to explain the anomalies in the angular observables in $b \to s \mu^+ \mu^-$ in terms of lepton flavor universal new physics [29].

Recently, the LHCb Collaboration reported the measurement of the ratio $R_{K^*} \equiv \mathcal{B}(B^0 \to K^{*0} \mu^+ \mu^-)/\mathcal{B}(B^0 \to K^{*0} e^+ e^-)$ in two different ranges of the dilepton invariant mass-squared q^2 [30]:

$$R_{K^*}^{\rm expt} = \begin{array}{lll} 0.66^{+0.11}_{-0.07}\,({\rm stat}) \pm 0.03\,({\rm syst})\,, & 0.045 \le q^2 \le 1.1\,{\rm GeV}^2\,, & ({\rm low~q^2}) \\ 0.69^{+0.11}_{-0.07}\,({\rm stat}) \pm 0.05\,({\rm syst})\,, & 1.1 \le q^2 \le 6.0\,{\rm GeV}^2\,, & ({\rm central~q^2})\,. \end{array} \qquad (52.5)$$

These differ from the SM predictions by 2.2–2.4σ (low q^2) and 2.4-2.5σ (central q^2), which further strengthens the hint of lepton non-universality observed in R_K.

Fits to new physics with heavy mediators have been considered by several groups but it has generally been difficult to understand the low q^2 measurements with heavy mediators [31].

52.1.1.3 Lepton Universality Violation and Lepton Flavor Violation

Any interesting question is whether the CC LUV and the NC LUV are related. In [32, 33], it was pointed out that, assuming the scale of NP is much larger than the weak scale, operators contributing to R_K anomalies should be made invariant under the full $SU(3)_C \times SU(2)_L \times U(1)_Y$ gauge group. There are two possibilities if only left handed particles are involved:

$$\begin{aligned}\mathcal{O}_1^{NP} &= \frac{G_1}{\Lambda_{NP}^2}(\bar{Q}'_L\gamma_\mu Q'_L)(\bar{L}'_L\gamma^\mu L'_L)\,,\\ \mathcal{O}_2^{NP} &= \frac{G_2}{\Lambda_{NP}^2}(\bar{Q}'_L\gamma_\mu\sigma^I Q'_L)(\bar{L}'_L\gamma^\mu\sigma^I L'_L)\\ &= \frac{G_2}{\Lambda_{NP}^2}\left[2(\bar{Q}'^i_L\gamma_\mu Q'^j_L)(\bar{L}'^j_L\gamma^\mu L'^i_L)-(\bar{Q}'_L\gamma_\mu Q'_L)(\bar{L}'_L\gamma^\mu L'_L)\right]\,, \qquad (52.6)\end{aligned}$$

where G_1 and G_2 are both $O(1)$, and the σ^I are the Pauli matrices. Here $Q' \equiv (t', b')^T$ and $L' \equiv (\nu'_\tau, \tau')^T$. The key point is that $\mathcal{O}_2^{NP}$ contains both neutral-current (NC) and charged-current (CC) interactions. The NC and CC pieces can be used to respectively explain the R_K and $R_{D^{(*)}}$ puzzles. In this scenario new physics affects only the third generation but via mixing effect LUV and lepton flavor violation(LFV) effects involving lighter generations are generated [33, 34].

In [35] UV completions that can give rise to $\mathcal{O}_{1,2}^{NP}$ (52.6), were discussed. These include leptoquark models and vector boson (VB) models. Concrete models are discussed as in [36–39]. Within specific models there are new LUV processes as well as lepton flavor violating(LFV) processes. A fairly exhaustive analysis of specific models was carried out in the recent publication [40]. Several decay modes were discussed which in the future could distinguish among the different models. In particular it was shown that the observation of $\tau \to 3\mu$ would be a clear sign of the VB model while the observation of $\Upsilon \to \mu\tau$ would point towards the leptoquark model.

52.2 Light Mediators

As indicated earlier the low q^2 $R_{K^{(*)}}$ measurement cannot be understood in terms of heavy mediators. This motivates considering light mediator solution to the neutral current B anomalies and in [41, 42] it was proposed that the NC anomalies could explained by the exchange of a light mediator. In these models the flavor changing $b \to s$ vertex may arise from loops of hidden sector particles [42, 43]. These models may be tested not only in B experiments but also in neutrino interactions such as coherent neutrino scattering [43].

References

1. J.P. Lees et al. (BaBar Collaboration), Phys. Rev. Lett. **109**, 101802 (2012). https://doi.org/10.1103/PhysRevLett.109.101802. arXiv:1205.5442 [hep-ex]; J.P. Lees et al. (BaBar Collaboration), Measurement of an excess of $\bar{B} \to D^{(*)}\tau^-\bar{\nu}_\tau$ decays and implications for charged Higgs Bosons. Phys. Rev. D **88**(7), 072012 (2013). https://doi.org/10.1103/PhysRevD.88.072012. arXiv:1303.0571 [hep-ex]
2. M. Huschle et al. (Belle Collaboration), Measurement of the branching ratio of $\bar{B} \to D^{(*)}\tau^-\bar{\nu}_\tau$ relative to $\bar{B} \to D^{(*)}\ell^-\bar{\nu}_\ell$ decays with hadronic tagging at Belle. Phys. Rev. D **92**(7), 072014

(2015). https://doi.org/10.1103/PhysRevD.92.072014.arXiv:1507.03233 [hep-ex]; A. Abdesselam et al. (Belle Collaboration), arXiv:1603.06711 [hep-ex]; S. Hirose et al. (Belle Collaboration), Phys. Rev. Lett. **118**(21), 211801 (2017). https://doi.org/10.1103/PhysRevLett.118.211801. arXiv:1612.00529 [hep-ex]
3. R. Aaij et al. (LHCb Collaboration), Measurement of the ratio of branching fractions $\mathcal{B}(\bar{B}^0 \to D^{*+}\tau^-\bar{\nu}_\tau)/\mathcal{B}(\bar{B}^0 \to D^{*+}\mu^-\bar{\nu}_\mu)$. Phys. Rev. Lett. **115**(11), 111803 (2015). https://doi.org/10.1103/PhysRevLett.115.159901. arXiv:1506.08614 [hep-ex]; FPCP conference (2017)
4. J.A. Bailey et al., Phys. Rev. Lett. **109**, 071802 (2012). https://doi.org/10.1103/PhysRevD.92.034506. arXiv:1206.4992 [hep-ph]; J. A. Bailey et al. (MILC Collaboration), Phys. Rev. D **92**(3), 034506 (2015). arXiv:1503.07237 [hep-lat]. https://doi.org/10.1103/PhysRevLett.109.071802
5. H. Na et al. (HPQCD Collaboration), Phys. Rev. D **92**(5), 054510 (2015). Erratum: Phys. Rev. D **93**, no. 11, 119906 (2016). https://doi.org/10.1103/PhysRevD.93.119906, https://doi.org/10.1103/PhysRevD.92.054510. arXiv:1505.03925 [hep-lat]
6. M. Tanaka, R. Watanabe, Phys. Rev. D **82**, 034027 (2010). https://doi.org/10.1103/PhysRevD.82.034027. arXiv:1005.4306 [hep-ph]
7. S. Fajfer, J.F. Kamenik, I. Nisandzic, Phys. Rev. D **85**, 094025 (2012). https://doi.org/10.1103/PhysRevD.85.094025. arXiv:1203.2654 [hep-ph]
8. F.U. Bernlochner, Z. Ligeti, M. Papucci, D.J. Robinson, Phys. Rev. D **95**(11), 115008 (2017). https://doi.org/10.1103/PhysRevD.95.115008. arXiv:1703.05330 [hep-ph]
9. Y. Amhis et.al., Averages of b-hadron,c-hadron, and lepton properties as of winter 2016. HFLAV: Semileptonic B Decay Parameters
10. A. Datta, M. Duraisamy, D. Ghosh, Phys. Rev. D **86**, 034027 (2012). https://doi.org/10.1103/PhysRevD.86.034027. arXiv:1206.3760 [hep-ph]
11. A. Celis, M. Jung, X.Q. Li, A. Pich, JHEP **1301**, 054 (2013). https://doi.org/10.1007/JHEP01(2013). arXiv:1210.8443 [hep-ph]
12. A. Crivellin, A. Kokulu, C. Greub, Phys. Rev. D **87**(9), 094031 (2013). https://doi.org/10.1103/PhysRevD.87.094031. arXiv:1303.5877 [hep-ph]
13. I. Dorner, S. Fajfer, N. Konik, I. Niandi?, JHEP **1311**, 084 (2013). https://doi.org/10.1007/JHEP11(2013)084. arXiv:1306.6493 [hep-ph]
14. M. Freytsis, Z. Ligeti, J.T. Ruderman, Phys. Rev. D **92**(5), 054018 (2015). https://doi.org/10.1103/PhysRevD.92.054018. arXiv:1506.08896 [hep-ph]
15. N.G. Deshpande, X.G. He. arXiv:1608.04817 [hep-ph]
16. M. Duraisamy, A. Datta, JHEP **1309**, 059 (2013). https://doi.org/10.1007/JHEP09(2013). arXiv:1302.7031 [hep-ph]
17. M. Duraisamy, P. Sharma, A. Datta, Phys. Rev. D **90**(7), 074013 (2014). https://doi.org/10.1103/PhysRevD.90.074013. arXiv:1405.3719 [hep-ph]
18. K. Hagiwara, M.M. Nojiri, Y. Sakaki, Phys. Rev. D **89**(9), 094009 (2014). https://doi.org/10.1103/PhysRevD.89.094009. arXiv:1403.5892 [hep-ph]
19. W. Dungel et al. (Belle Collaboration), Phys. Rev. D **82**, 112007 (2010). arXiv:1010.5620 [hep-ex]
20. J.D. Richman, P.R. Burchat, Rev. Mod. Phys. **67**, 893 (1995). [hep-ph/9508250]
21. J.G. Korner, G.A. Schuler, Z. Phys. C **46**, 93 (1990)
22. J.G. Korner, G.A. Schuler, Phys. Lett. B **231**, 306 (1989)
23. A. Datta, S. Kamali, S. Meinel, A. Rashed, JHEP **1708**, 131 (2017). https://doi.org/10.1007/JHEP08(2017). arXiv:1702.02243 [hep-ph]
24. S. Descotes-Genon, T. Hurth, J. Matias, J. Virto, Optimizing the basis of $B \to K^*ll$ observables in the full kinematic range. JHEP **1305**, 137 (2013). https://doi.org/10.1007/JHEP05(2013)137. arXiv:1303.5794 [hep-ph]
25. G. Hiller, F. Kruger, More model independent analysis of $b \to s$ processes. Phys. Rev. D **69**, 074020 (2004). arXiv:hep-ph/0310219
26. C. Bobeth, G. Hiller, G. Piranishvili, Angular distributions of $\bar{B}_d \to Kl^+l^-$ decays. JHEP **0712**, 040 (2007). arXiv:0709.4174 [hep-ph]

27. R. Aaij et al. (LHCb Collaboration), Test of lepton universality using $B^+ \to K^+\ell^+\ell^-$ decays. Phys. Rev. Lett. **113**, 151601 (2014). arXiv:1406.6482 [hep-ex]
28. M. Bordone, G. Isidori, A. Pattori, Eur. Phys. J. C **76**(8), 440 (2016). arXiv:1605.07633 [hep-ph]
29. A. Datta, M. Duraisamy, D. Ghosh, Phys. Rev. D **89**(7), 071501 (2014). https://doi.org/10.1103/PhysRevD.89.071501. arXiv:1310.1937 [hep-ph]
30. R. Aaij et al. (LHCb Collaboration), arXiv:1705.05802 [hep-ex]
31. A.K. Alok, B. Bhattacharya, A. Datta, D. Kumar, J. Kumar, D. London. arXiv:1704.07397 [hep-ph]
32. R. Alonso, B. Grinstein, J.M. Camalich, Lepton universality violation and lepton flavor conservation in B-meson decays. JHEP **1510**, 184 (2015). https://doi.org/10.1007/JHEP10(2015)184. arXiv:1505.05164 [hep-ph]
33. B. Bhattacharya, A. Datta, D. London, S. Shivashankara, Simultaneous explanation of the R_K and $R(D^{(*)})$ Puzzles. Phys. Lett. B **742**, 370 (2015). https://doi.org/10.1016/j.physletb.2015.02.011. arXiv:1412.7164 [hep-ph]
34. S.L. Glashow, D. Guadagnoli, K. Lane, Lepton flavor violation in B Decays?. Phys. Rev. Lett. **114**, 091801 (2015). https://doi.org/10.1103/PhysRevLett.114.091801. arXiv:1411.0565 [hep-ph]
35. L. Calibbi, A. Crivellin, T. Ota, Effective field theory approach to $b \to s\ell\ell^{(\prime)}$, $B \to K^{(*)}\nu\bar{\nu}$ and $B \to D^{(*)}\tau\nu$ with third generation couplings. Phys. Rev. Lett. **115**, 181801 (2015). https://doi.org/10.1103/PhysRevLett.115.181801. arXiv:1506.02661 [hep-ph]
36. A. Greljo, G. Isidori, D. Marzocca, On the breaking of Lepton Flavor Universality in B decays. JHEP **1507**, 142 (2015). https://doi.org/10.1007/JHEP07(2015)142. arXiv:1506.01705 [hep-ph]
37. S.M. Boucenna, A. Celis, J. Fuentes-Martin, A. Vicente, J. Virto, Non-abelian gauge extensions for B-decay anomalies. arXiv:1604.03088 [hep-ph]
38. W. Altmannshofer, P.S. Bhupal Dev, A. Soni, Phys. Rev. D **96**(9), 095010 (2017). https://doi.org/10.1103/PhysRevD.96.095010. arXiv:1704.06659 [hep-ph]
39. S. Sahoo, R. Mohanta, A.K. Giri, Phys. Rev. D **95**(3), 035027 (2017). https://doi.org/10.1103/PhysRevD.95.035027. arXiv:1609.04367 [hep-ph]
40. B. Bhattacharya, A. Datta, J. P. Guvin, D. London, R. Watanabe, Simultaneous explanation of the R_K and $R_{D^{(*)}}$ Puzzles: a model analysis, arXiv:1609.09078 [hep-ph]
41. A. Datta, J. Liao, D. Marfatia, Phys. Lett. B **768**, 265 (2017). arXiv:1702.01099 [hep-ph]
42. A. Datta, J. Kumar, J. Liao, D. Marfatia, Phys. Rev. D **97**(11), 115038 (2018). https://doi.org/10.1103/PhysRevD.97.115038. arXiv:1705.08423 [hep-ph]
43. A. Datta, B. Dutta, S. Liao, D. Marfatia, L.E. Strigari, arXiv:1808.02611 [hep-ph]

Chapter 53
Probing Flavor Physics in BSM Scenarios with SM Null Tests

Jusak Tandean

Abstract As direct searches for new physics (NP) beyond the standard model (SM) at the LHC and other colliders have so far come up empty, increasingly attention has been paid to the intensity frontier, where high-precision measurements can probe rare processes which are possibly sensitive to indirect signals of NP. An important tool in the quest for NP effects on rare processes is the null tests of the SM. These are observables which vanish or are very small in the SM and which therefore could receive potentially large contributions from NP. Among such tests are, to be discussed here, asymmetries in the charm-meson decays $D \to P_1 P_2 \ell^+ \ell^-$ ($P = \pi, K$ and $\ell = e, \mu$), direct CP asymmetries in Cabibbo-favored nonleptonic charm-meson decays, muon asymmetries in the hyperon decay $\Sigma^+ \to p\mu^+\mu^-$, and searches for the rare transitions $b \to ss\bar{d}, dd\bar{s}$ and for nonleptonic B_s-meson decays which fully break isospin symmetry. These are under ongoing or near-future investigations in the LHCb and other experiments.

Observables which are predicted to be very small or vanish within the standard model (SM) serve, at least in principle, as its (approximate or exact) null tests [1]. They are potentially good places to look for indications of new physics (NP) beyond the SM (BSM), as any unambiguous nonnegligible result of the measurement of such an observable would be a compelling hint of NP BSM. If an experiment performing this kind of test produces no positive result, the null outcome will, on the flip side, translate into a constraint on the NP scenario under consideration.

Over the years there have already been numerous efforts to conduct SM null tests with various flavor-changing neutral current (FCNC) transitions. Examples are the searches for processes manifesting charged-lepton-flavor violation, such as $\mu \to e\gamma$ and $B^0 \to \mu^\pm\tau^\mp$, which have so far come up negative [2]. Since they are forbidden in the SM, their observation would be clear evidence for NP. Another instance of SM null test is the deviation of the ratio $R_{K^{(*)}}$ of the branching fractions of the decays

J. Tandean (✉)
Department of Physics, National Taiwan University, Taipei 106, Taiwan
e-mail: jtandean@yahoo.com

Physics Division, National Center for Theoretical Sciences, Hsinchu 300, Taiwan

A. Giri and R. Mohanta (eds.), *16th Conference on Flavor Physics and CP Violation*, Springer Proceedings in Physics 234,
https://doi.org/10.1007/978-3-030-29622-3_53

$B^{(*)} \to K^{(*)}\mu^+\mu^-$ and $B^{(*)} \to K^{(*)}e^+e^-$ from the SM lepton-flavor universality [3]. Thus the recent LHCb finding of $R_{K^{(*)}} \sim 0.7$ [4] would have a NP origin if conclusively confirmed by future data.

If the $R_{K^{(*)}}$ and related $b \to s\mu^+\mu^-$ anomalies [4] turn out be caused by BSM interactions, it will be important to explore the potential implications for other processes in order to gain a comprehensive picture about the nature of the NP. One way to help achieve this is to conduct as many additional SM null tests as is feasible, some of which are already being carried out by LHCb and will be discussed in what follows.

53.1 Asymmetries in Charm-Meson Decays $D \to P_1 P_2 \ell^+\ell^-$

The amplitudes for the rare charm-meson decays of interest consist of short distance (SD) and long distance (LD) components. The SD one arises from FCNCs which in the SM suffer from severe suppression due to loop diagrams involving down-type quarks leading to highly efficient Glashow-Iliopoulos-Maiani (GIM) mechanism. Since NP does not have to be subject to the same GIM cancelation, the SD component could get sizable corrections from NP. It turns out, however, that the amplitudes are dominated by LD physics, the calculation of which has significant uncertainties and which consequently makes probing the potential NP contribution quite challenging. Nevertheless, it is possible to devise SM null tests with these decays, especially asymmetries which in the SM are expected to be tiny [5, 6].

For $D \to P_1 P_2 \ell^+\ell^-$ ($P = \pi, K$ and $\ell = e, \mu$), one can start with the differential rate [6]

$$d^5\Gamma = \sum_i c_i(\theta_\ell, \phi)\, I_i(q^2, p^2, \cos\theta_{P_1})\, \frac{dq^2 dp^2 d\cos\theta_{P_1} d\cos\theta_\ell d\phi}{2\pi} \tag{53.1}$$

where θ_ℓ is the angle between ℓ^- and D in the dilepton rest-frame, θ_{P_1} is the angle between P_1 (the meson containing the quark emitted from the semileptonic weak $\bar{u}c\ell\ell$ vertex) and the direction opposite to the D motion in the $P_1 P_2$ rest-frame, ϕ is the angle between the normals of the $\ell^+\ell^-$ and $P_1 P_2$ planes in the D rest-frame, q^2 (p^2) is the invariant mass squared of the lepton ($P_1 P_2$) pair, $c_i(\theta_\ell, \phi)$ are simple functions of θ_ℓ and ϕ [6], and $I_i(q^2, p^2, \cos\theta_{P_1})$, $i = 1, 2, ..., 9$, are angular coefficients given in terms of so-called transversity amplitudes [7]. Subsequently, from I_i one can construct a number of SM null tests [6].

To list some examples, since in the SM photon-mediated contributions dominate $c \to u\ell^+\ell^-$, one can see that $I_5 = -\big[\mathrm{Re}\big(H_0^L H_\perp^{L*}\big) - (L \to R)\big]\sin\theta_{P_1}/4$, $I_6 = \big[\mathrm{Re}\big(H_\parallel^L H_\perp^{L*}\big) - (L \to R)\big]\sin^2\theta_{P_1}/4$, and $I_7 = -\big[\mathrm{Re}\big(H_0^L H_\parallel^{L*}\big) - (L \to R)\big]$ $\sin\theta_{P_1}/4$ are null tests, as they will be zero in the absence of axial-vector contributions. Moreover, as detailed in [6], one can utilize $I_{5,6,8,9}$ to form CP asymmetries as additional null tests, and the corresponding integrated asymmetries can reach a few percent if generated by NP BSM. Recently LHCb [8] has begun to measure some of the asymmetries in $D^0 \to (\pi^+\pi^-, K^+K^-)\mu^+\mu^-$ and the results, which are not yet precise, are in line with the SM predictions.

Similarly to $R_{K^{(*)}}$, the ratio $R^D_{P_1 P_2}$ of branching fractions of $D \to P_1 P_2 \ell^+\ell^-$ for $\ell = \mu, e$,

$$R^D_{P_1 P_2} = \frac{\int_{q^2_{\min}}^{q^2_{\max}} \mathrm{d}q^2\, \mathrm{d}\mathcal{B}(D \to P_1 P_2 \mu^+\mu^-)/\mathrm{d}q^2}{\int_{q^2_{\min}}^{q^2_{\max}} \mathrm{d}q^2\, \mathrm{d}\mathcal{B}(D \to P_1 P_2 e^+e^-)/\mathrm{d}q^2} \tag{53.2}$$

with the same $q^2_{\min,\max}$ cuts, can check if there is violation of lepton-flavor universality in the up-type-quark sector. The SM predictions are $R^D_{\pi\pi} = 1.00 \pm \mathcal{O}(\%)$ and $R^D_{KK} = 1.00 \pm \mathcal{O}(\%)$ for $q^2_{\min} = 4m^2_\mu$ and $q^2_{\max} = (m_D - m_{P_1} - m_{P_2})^2$, implying that $R^D_{\pi\pi,KK} - 1$ are SM null tests [6]. At present, the relevant empirical information is still incomplete [2, 9], but the situation will likely change once more data become available from LHCb, BESIII, and other facilities.

53.2 Direct CP Violation in Cabibbo-Favored Charm Decays

In the SM the leading-order effective Hamiltonian describing the Cabibbo-favored nonleptonic decays of charm mesons, such as $D^0 \to \pi^+K^-$ and $D_s^+ \to \eta^{(\prime)}\pi^+$, is of the form $\mathcal{H} = G_{\mathrm{F}} V^*_{cs} V_{ud} \big(C_1\, \overline{s}\Gamma^\kappa c\, \overline{u}\Gamma_\kappa d + C_2\, \overline{u}\Gamma^\kappa c\, \overline{s}\Gamma_\kappa d\big)/\sqrt{2}$ + H.c., where V_{kl} are elements of the Cabibbo-Kobayashi-Maskawa (CKM) matrix, $C_{1,2}$ are the Wilson coefficients, and $\Gamma^\mu = \gamma^\mu(1-\gamma_5)$. Accordingly, as is well known, their amplitudes at tree level do not have two or more terms with different strong and weak phases which are necessary for direct CP violation to occur [10]. Nevertheless, incorporating loop contributions, such as box and dipenguin diagrams, can lead in the SM to direct CP violation, albeit it is very small [11]. Hence direct CP asymmetries in these modes are SM null tests.

In the presence of NP, the decay amplitudes may acquire extra terms with new strong and weak phases which give rise to nonnegligible CP violation. This can happen in, for instance, gauged left-right models, where the direct CP asymmetries can reach the per-mill level or more [11, 12], which is potentially detectable at LHCb and Belle II [12]. This could even be realized in the not-too-distant future at LHCb, which has recently already reported an upper limit of order 1% for the asymmetry in $D_s^\pm \to \eta'\pi^\pm$ [13].

53.3 Muon Asymmetries in $\Sigma^+ \to p\mu^+\mu^-$

Recently LHCb has turned up evidence for the rare hyperon decay $\Sigma^+ \to p\mu^+\mu^-$ and reported its branching fraction to be $\left(2.2^{+1.8}_{-1.3}\right) \times 10^{-8}$ [14]. It is consistent with the earlier HyperCP finding $\left(8.6^{+6.6}_{-5.4} \pm 5.5\right) \times 10^{-8}$ [15] and with the SM prediction $\mathcal{B}_{\rm SM} \in [1.6, 9.0] \times 10^{-8}$ [16]. Despite the agreement, in view of the still sizable experimental errors, more data are clearly needed to test the theory more stringently. This will likely be achievable with the larger event sample that LHCb expects to collect in the near future [17].

As the amount of data on $\Sigma^+ \to p\mu^+\mu^-$ grows, it will be possible to test the SM predictions further via additional observables. Its dimuon invariant-mass distribution has already been examined in [16]. In this section we look at a few asymmetries pertaining to the muons in this decay which have recently been proposed in [18]. Some of these observables turn out to be tiny in the SM and therefore can serve as its null tests.

Including SM and potential NP contributions, one can express the amplitude for $\Sigma^+(p_\Sigma) \to p(p_p)\,\mu^+(p_+)\,\mu^-(p_-)$ as [18]

$$
\begin{aligned}
\mathcal{M} = {} & \big[iq_\kappa \bar{u}_p(\tilde{\rm A} + \gamma_5\tilde{\rm B})\sigma^{\nu\kappa}u_\Sigma - \bar{u}_p\gamma^\nu(\tilde{\rm C} + \gamma_5\tilde{\rm D})u_\Sigma\big]\bar{u}_\mu\gamma_\nu v_{\bar\mu} \\
& + \bar{u}_p\gamma^\nu(\tilde{\rm E} + \gamma_5\tilde{\rm F})u_\Sigma\,\bar{u}_\mu\gamma_\nu\gamma_5 v_{\bar\mu} + \bar{u}_p(\tilde{\rm G} + \gamma_5\tilde{\rm H})u_\Sigma\,\bar{u}_\mu v_{\bar\mu} \\
& + \bar{u}_p(\tilde{\rm J} + \gamma_5\tilde{\rm K})u_\Sigma\,\bar{u}_\mu\gamma_5 v_{\bar\mu}\,,
\end{aligned}
\tag{53.3}
$$

where $q = p_\Sigma - p_p = p_+ + p_-$ and $\tilde{\rm A}, \tilde{\rm B}, \ldots, \tilde{\rm K}$ are complex coefficients depending on q^2. In the SM this mode is virtually controlled by long-distance physics, most of which is due to photon-exchange diagrams [18, 19] and enters $\tilde{\rm A}$, $\tilde{\rm B}$, $\tilde{\rm C}$, and $\tilde{\rm D}$.

One can then construct the forward-backward asymmetry $\mathcal{A}_{\rm FB}$ defined by

$$
\mathcal{A}_{\rm FB} = \frac{\int_{-1}^{1} dc_\theta\,\Gamma''\,{\rm sgn}(c_\theta)}{\int_{-1}^{1} dc_\theta\,\Gamma''}\,, \qquad \Gamma'' \equiv \frac{d\Gamma'}{dc_\theta}\,, \qquad c_\theta \equiv \cos\theta\,,
\tag{53.4}
$$

where θ is the angle between the μ^- and p directions in the rest frame of the dimuon system and $\Gamma' \equiv d\Gamma(\Sigma^+ \to p\mu^+\mu^-)/dq^2$ is the differential decay rate. In terms of the aforesaid coefficients [18]

$$
\begin{aligned}
\mathcal{A}_{\rm FB} = \frac{\beta^2\,\bar\lambda}{64\pi^3\,\Gamma' m_\Sigma^3}\,{\rm Re}\Big\{ & \big[{\rm M}_+\tilde{\rm A}^*\tilde{\rm F} - {\rm M}_-\tilde{\rm B}^*\tilde{\rm E} - (\tilde{\rm A}^*\tilde{\rm G} + \tilde{\rm B}^*\tilde{\rm H})m_\mu\big]q^2 \\
& + (\tilde{\rm C}^*\tilde{\rm F} + \tilde{\rm D}^*\tilde{\rm E})q^2 - ({\rm M}_+\tilde{\rm C}^*\tilde{\rm G} - {\rm M}_-\tilde{\rm D}^*\tilde{\rm H})m_\mu\Big\}\,,
\end{aligned}
\tag{53.5}
$$

where $\beta = \left(1 - 4m_\mu^2/q^2\right)^{1/2}$, ${\rm M}_\pm = m_\Sigma \pm m_p$, and $\bar\lambda = \left({\rm M}_+^2 - q^2\right)\left({\rm M}_-^2 - q^2\right)$. Also of interest is the integrated forward-backward asymmetry $\tilde{A}_{\rm FB}$ given by

$$\tilde{A}_{\mathrm{FB}} = \frac{\int_{q^2_{\min}}^{q^2_{\max}} dq^2 \int_{-1}^{1} dc_\theta\, \Gamma''\, \mathrm{sgn}(c_\theta)}{\Gamma(\Sigma^+ \to p\mu^+\mu^-)}\,, \qquad q^2_{\min} = 4m^2_\mu, \quad q^2_{\max} = \left(m_\Sigma - m_p\right)^2. \tag{53.6}$$

In the dimuon system's rest frame, with enough data, one can additionally examine the $\mu^\pm$ polarization asymmetries. When μ^- is polarized, one can write the differential rate as $\Gamma'\left(1 + \mathcal{P}^-_{\mathrm{T}}\varsigma^-_x + \mathcal{P}^-_{\mathrm{N}}\varsigma^-_y + \mathcal{P}^-_{\mathrm{L}}\varsigma^-_z\right)/2$, where L, N, and T refer to the longitudinal, normal, and transverse polarizations of μ^- along the chosen z-, y-, and x-axes, respectively, as described in [18], and $\varsigma^-_{x,y,z}$ are the spin three-vector components satisfying $(\varsigma^-_x)^2 + (\varsigma^-_y)^2 + (\varsigma^-_z)^2 = 1$. The corresponding polarization asymmetries $\mathcal{P}^-_{\mathrm{L,N,T}}$ are [18]

$$\begin{aligned}
\mathcal{P}^-_{\mathrm{L}} &= \frac{\beta^2\sqrt{\bar\lambda}}{192\pi^3\,\Gamma'\,m^3_\Sigma}\,\mathrm{Re}\Big\{\Big[-3\big(2\mathrm{M}_+\tilde{\mathrm{A}}^*\tilde{\mathrm{E}} + \tilde{\mathrm{H}}^*\tilde{\mathrm{K}}\big)q^2 - 2\big(\hat{m}^2_+ + 3q^2\big)\tilde{\mathrm{C}}^*\tilde{\mathrm{E}}\Big]\hat{m}^2_- \\
&\qquad + \Big[3\big(2\mathrm{M}_-\tilde{\mathrm{B}}^*\tilde{\mathrm{F}} - \tilde{\mathrm{G}}^*\tilde{\mathrm{J}}\big)q^2 - 2\big(\hat{m}^2_- + 3q^2\big)\tilde{\mathrm{D}}^*\tilde{\mathrm{F}}\Big]\hat{m}^2_+ \\
&\qquad + 6\big(\mathrm{M}_+\tilde{\mathrm{F}}^*\tilde{\mathrm{H}}\,\hat{m}^2_- - \mathrm{M}_-\tilde{\mathrm{E}}^*\tilde{\mathrm{G}}\,\hat{m}^2_+\big)m_\mu\Big\}\,, \\
\mathcal{P}^-_{\mathrm{N}} &= \frac{\beta^2\,\bar\lambda\sqrt{q^2}}{256\pi^2\,\Gamma'\,m^3_\Sigma}\,\mathrm{Im}\Big\{2\Big[(\mathrm{M}_+\tilde{\mathrm{A}} + \tilde{\mathrm{C}})^*\tilde{\mathrm{F}} + (\tilde{\mathrm{D}} - \mathrm{M}_-\tilde{\mathrm{B}})^*\tilde{\mathrm{E}}\Big]m_\mu - \big(\tilde{\mathrm{A}}^*\tilde{\mathrm{G}} + \tilde{\mathrm{B}}^*\tilde{\mathrm{H}}\big)q^2 \\
&\qquad - \big(\tilde{\mathrm{C}}^*\tilde{\mathrm{G}} - \tilde{\mathrm{E}}^*\tilde{\mathrm{J}}\big)\mathrm{M}_+ + \big(\tilde{\mathrm{D}}^*\tilde{\mathrm{H}} - \tilde{\mathrm{F}}^*\tilde{\mathrm{K}}\big)\mathrm{M}_-\Big\}\,, \\
\mathcal{P}^-_{\mathrm{T}} &= \frac{\beta\,\bar\lambda\sqrt{q^2}}{256\pi^2\,\Gamma'\,m^3_\Sigma}\,\mathrm{Re}\Big\{2\Big[2\big(\mathrm{M}_+\tilde{\mathrm{A}} + \tilde{\mathrm{C}}\big)^*\big(\tilde{\mathrm{D}} - \mathrm{M}_-\tilde{\mathrm{B}}\big) - \mathrm{M}_-\tilde{\mathrm{A}}^*\tilde{\mathrm{E}} + \mathrm{M}_+\tilde{\mathrm{B}}^*\tilde{\mathrm{F}}\Big]m_\mu \\
&\qquad - \mathrm{M}_+\tilde{\mathrm{C}}^*\tilde{\mathrm{J}} + \mathrm{M}_-\tilde{\mathrm{D}}^*\tilde{\mathrm{K}} + \beta^2\big(\mathrm{M}_+\tilde{\mathrm{E}}^*\tilde{\mathrm{G}} - \mathrm{M}_-\tilde{\mathrm{F}}^*\tilde{\mathrm{H}}\big)\Big\} \\
&\quad - \frac{\beta\,\bar\lambda\,\mathrm{Re}\Big[\big(\tilde{\mathrm{A}}^*\tilde{\mathrm{J}} + \tilde{\mathrm{B}}^*\tilde{\mathrm{K}}\big)q^4 + 2\big(\tilde{\mathrm{C}}^*\tilde{\mathrm{E}} + \tilde{\mathrm{D}}^*\tilde{\mathrm{F}}\big)\mathrm{M}_+\mathrm{M}_-m_\mu\Big]}{256\pi^2\,\Gamma'\,m^3_\Sigma\sqrt{q^2}}\,,
\end{aligned} \tag{53.7}$$

where $\hat{m}^2_\pm = \mathrm{M}^2_\pm - q^2$. Like $\tilde{A}_{\mathrm{FB}}$, one can define the integrated asymmetries

$$\tilde{P}^-_{\mathrm{L,N,T}} = \frac{\int_{q^2_{\min}}^{q^2_{\max}} dq^2\, \Gamma'\, \mathcal{P}^-_{\mathrm{L,N,T}}}{\Gamma(\Sigma^+ \to p\mu^+\mu^-)}\,. \tag{53.8}$$

Analogously, one could write down [18] the μ^+ polarization asymmetries $\mathcal{P}^+_{\mathrm{L,N,T}}$ and $\tilde{P}^+_{\mathrm{L,N,T}}$.

Numerically, $|\tilde{A}_{\mathrm{FB}}|$ and $|\tilde{P}^\pm_{\mathrm{L,N}}|$ are found to be under 10^{-4} in the SM [18]. This is because these asymmetries involve no more than one factor from the dominant LD component, as may be seen in (53.5) and (53.7). In contrast, $\tilde{P}^\pm_{\mathrm{T}}$ can be large, up to $\sim 60\%$ in size [18], because they contain interference terms between two LD factors. The smallness of $\tilde{A}_{\mathrm{FB}}$ and $\tilde{P}^\pm_{\mathrm{L,N}}$ within the SM implies that they can potentially provide access to NP. A clear measurement of any one of these quantities exceeding the percent level would be a strong hint of NP.

53.4 Rare Decays $b \to ss\bar{D}, dd\bar{S}$

In the SM the transition $b \to ss\bar{d}$ ($dd\bar{s}$) arises mainly from box diagrams involving up-type quarks and the W boson and, unlike the familiar $b \to s\bar{s}d$ ($d\bar{d}s$), is predicted to have a tiny branching fraction, below 10^{-11} [20], which is brought about by strong GIM suppression and a small CKM factor. Quests for the corresponding exclusive modes, such as $B^- \to K^-K^-\pi^+, K^-\bar{K}^{*0}$ $(\pi^-\pi^-K^+, \pi^-K^{*0})$, can then be regarded as SM null tests.

There are models BSM which can hugely amplify the branching fractions of these decays relative to their SM expectations, by up to several orders of magnitude [21]. The predictions are thus close to potentially detectable levels at ongoing and near-future searches. For instance, LHCb has recently set the upper limits $\mathcal{B}(B^+ \to K^+K^+\pi^-) < 1.1 \times 10^{-8}$ and $\mathcal{B}(B^+ \to \pi^+\pi^+K^-) < 4.6 \times 10^{-8}$ at 90% confidence level [22].

53.5 Rare Nonleptonic B_s Decays

There are nonleptonic decay modes of the B_s meson which are predicted to have relatively small rates in the SM because the effects of the tree four-quark operators are suppressed and, as a consequence, their penguin counterparts have sizable impact. These processes can offer approximate null tests of the SM, as NP may modify the penguin contributions and boost the rates significantly. Here we focus on $\bar{B}_s \to (\eta, \eta', \phi)(\pi^0, \rho^0)$, the only $\bar{B}_s$ decays into two charmless mesons which are both strangeness changing and purely isospin-violating. Hence their amplitudes have no QCD-penguin terms and proceed instead from charmless tree and electroweak-penguin operators [23], the tree ones being CKM-factor suppressed. Among the six modes, only $\bar{B}_s \to \phi\rho^0$ has been seen with a branching fraction of $(2.7 \pm 0.8) \times 10^{-7}$ [2], compatible with its estimate in the SM albeit with a sizable error [24].

The interest in these decays stems from the possibility that the anomalies recently found in $b \to s\mu^+\mu^-$ data [4] have a NP origin and that the same NP can enhance one or more of the above B_s decays substantially. This has been explored in [25] within a particular BSM scenario where an electrically neutral and uncolored spin-one particle, the Z' boson, is responsible for the NP effects on these two sets of $b \to s$ transitions. It is assumed to couple nonuniversally to SM fermions, have no mixing with SM gauge bosons, and possess mass $m_{Z'}$ higher than the electroweak scale, but the Z' is not necessarily a gauge boson.

The pertinent Z' interactions are described by

$$\mathcal{L}_{Z'} \supset -\left[\bar{s}\gamma^\kappa\left(\Delta_L^{sb}P_L + \Delta_R^{sb}P_R\right)b\,Z'_\kappa + \text{H.c.}\right] - \Delta_V^{\mu\mu}\,\bar{\mu}\gamma^\kappa\mu\,Z'_\kappa$$
$$-\left[\bar{u}\gamma^\kappa\left(\Delta_L^{uu}P_L + \Delta_R^{uu}P_R\right)u + \bar{d}\gamma^\kappa\left(\Delta_L^{dd}P_L + \Delta_R^{dd}P_R\right)d\right]Z'_\kappa, \quad (53.9)$$

where $\Delta_{L,R}^{sb}$ are complex constants, whereas $\Delta_V^{\mu\mu}$ and $\Delta_{L,R}^{uu,dd}$ are real due to the hermiticity of $\mathcal{L}_{Z'}$. Other Z' couplings to SM fermions are taken to be absent. For simplicity, the choices $\Delta_{L,R}^{sb} = \rho_{L,R} V_{ts}^* V_{tb}$ with real $\rho_{L,R}$ are made [25].

There are constraints on these couplings that need to be taken into account. Obviously, $\Delta_{L,R}^{sb}\Delta_V^{\mu\mu}$ must have values in line with the observed $b \to s\mu^+\mu^-$ anomalies. Since the $\Delta_{L,R}^{sb}$ contribute at tree level to B_s-$\bar{B}_s$ mixing, restrictions from its data are also to be respected. Interestingly, these two sets of conditions together lead to the requirement $\Delta_L^{sb} \sim 10\Delta_R^{sb}$ [25], and so $\rho_L \sim 10\rho_R$. The aforementioned $\bar{B}_s \to \phi\rho^0$ measurement restrains $\Delta_{L,R}^{sb}\Delta_{L,R}^{uu,dd}$. In addition, all the Z' couplings have to satisfy the bounds inferred from collider data.

For definiteness, numerically one can set $\rho_R = 0.1\,\rho_L$ and $m_{Z'} = 1\,\text{TeV}$ and impose 2σ ranges of the various empirical constraints. With these choices, it is demonstrated in [25] that within the allowed parameter space the Z' impact on $\bar{B}_s^0 \to (\eta, \phi)\pi^0$ can cause their rates to increase by about an order of magnitude with respect to their SM expectations [26]. Such possible large enhancement implies that the Z' effect on these modes could be easily distinguishable from the SM predictions which have sizable uncertainties. On the other hand, the enhancement factors for $\bar{B}_s^0 \to \eta'\pi^0, (\eta, \eta')\rho^0$ are at most a few, partly because of the $\bar{B}_s \to \phi\rho^0$ constraint. One or more of these predictions may be testable in the near future at LHCb.

53.6 Conclusions

If the recently observed $b \to s\mu^+\mu^-$ anomalies are due to new physics, its impact can be expected to appear in some other flavor sectors. Null tests of the SM may probe such NP to help reveal its properties. The foregoing presents a number of SM null tests which are among the subjects of ongoing and near-future flavor experiments and which could offer important complementary information concerning the potential NP responsible for these anomalies.

Acknowledgements This work was supported in part by the MOE Academic Excellence Program (Grant No. 105R891505).

References

1. T.E. Browder et al., Rev. Mod. Phys. **81**, 1887 (2009)
2. M. Tanabashi et al. [Particle Data Group], Phys. Rev. D **98**, no. 3, 030001 (2018)
3. G. Hiller, F. Kruger, Phys. Rev. D **69**, 074020 (2004), C. Bouchard et al. [HPQCD Collaboration], Phys. Rev. Lett. **111**(16), 162002 (2013), Erratum, Phys. Rev. Lett. **112**(14), 149902 (2014), M. Bordone, G. Isidori, A. Pattori, Eur. Phys. J. C **76**(8), 440 (2016)
4. I. Bediaga et al. [LHCb Collaboration], arXiv:1808.08865
5. L. Cappiello, O. Cata, G. D'Ambrosio, JHEP **04**, 135 (2013)
6. S. De Boer, G. Hiller, Phys. Rev. D **98**(3), 035041 (2018)

7. F. Kruger, L.M. Sehgal, N. Sinha, R. Sinha, Phys. Rev. D **61**, 114028 (2000), Erratum, Phys. Rev. D **63**, 019901 (2001)
8. R. Aaij *et al.* [LHCb Collaboration], Phys. Rev. Lett. **121**(9), 091801 (2018)
9. M. Ablikim et al. [BESIII Collaboration], Phys. Rev. D **97**(7), 072015 (2018)
10. M. Artuso, B. Meadows, A.A. Petrov, Ann. Rev. Nucl. Part. Sci. **58**, 249 (2008)
11. D. Delepine, G. Faisel, C.A. Ramirez, Phys. Rev. D **87**(7), 075017 (2013)
12. N. Haba, H. Umeeda, T. Yamada. arXiv:1806.03424 [hep-ph]
13. R. Aaij et al., LHCb Collaboration. Phys. Lett. B **771**, 21 (2017)
14. R. Aaij et al. [LHCb Collaboration], Phys. Rev. Lett. **120**(22), 221803 (2018)
15. H. Park et al., HyperCP collaboration. Phys. Rev. Lett. **94**, 021801 (2005)
16. X.G. He, J. Tandean, G. Valencia, Phys. Rev. D **72**, 074003 (2005)
17. D.M. Santos [LHCb Collaboration], EPJ Web Conf. **179**, 01013 (2018)
18. X.G. He, J. Tandean, G. Valencia,. arXiv:1806.08350
19. I.V. Lyagin, E.K. Ginzburg, Sov. Phys. JETP **14**, 653 (1962), L. Bergstrom, R. Safadi, P. Singer, Z. Phys. C **37**, 281 (1988)
20. K. Huitu, D.X. Zhang, C.D. Lu, P. Singer, Phys. Rev. Lett. **81**, 4313 (1998)
21. K. Huitu, C.D. Lu, P. Singer, D.X. Zhang, Phys. Lett. B **445**, 394 (1999), S. Fajfer, J.F. Kamenik, N. Kosnik, Phys. Rev. D **74**, 034027 (2006), D. Pirjol, J. Zupan, JHEP **1002**, 028 (2010), F.M. Bhutta, Y. Li, C.D. Lu, Y.H. Xie, arXiv:1807.05350
22. R. Aaij et al., LHCb Collaboration. Phys. Lett. B **765**, 307 (2017)
23. R. Fleischer, Phys. Lett. B **332**, 419 (1994), N.G. Deshpande, X.G. He, J. Trampetic, Phys. Lett. B **345**, 547 (1995)
24. A. Ali, et al., Phys. Rev. D **76**, 074018 (2007), H.Y. Cheng, C.K. Chua, Phys. Rev. D **80**, 114026 (2009), L. Hofer, D. Scherer, L. Vernazza, JHEP **1102**, 080 (2011), G. Faisel, JHEP **1208**, 031 (2012), Phys. Lett. B **731**, 279 (2014), C. Wang, S.H. Zhou, Y. Li, C.D. Lu, Phys. Rev. D **96**, no. 7, 073004 (2017)
25. G. Faisel, J. Tandean, JHEP **02**, 074 (2018)
26. A.R. Williamson, J. Zupan, Phys. Rev. D **74**, 014003 (2006), Erratum, Phys. Rev. D **74**, 03901 (2006), W. Wang, Y.M. Wang, D.S. Yang, C.D. Lu, Phys. Rev. D **78**, 034011 (2008), G. Faisel, Eur. Phys. J. C **77**(6), 380 (2017)

Chapter 54
Flavor and Dark Matter Connection

Avelino Vicente

Abstract In recent years, the LHCb collaboration has published results on the measurement of several observables associated to semileptonic $b \to s$ transitions. Interestingly, various deviations from their expected values in the Standard Model have been found, including some tantalizing hints pointing towards the violation of Lepton Flavor Universality. We discuss New Physics models that address these anomalies and explore their possible connection to the dark matter of the Universe.

54.1 Introduction

One of the most relevant open questions in current physics is the nature of the dark matter (DM) that makes up 27% of the energy density of the Universe [1]. Many ideas and proposals have been put forward, including the possibility that the DM is composed of particles. This popular scenario requires an extension of the Standard Model (SM) with new states and dynamics, since the SM particles do not have the required properties to be a good DM candidate.

Lepton Flavor Universality (LFU) is a central feature in the SM. The fact that gauge bosons couple with the same strength to the three generations of leptons is well rooted in the SM construction and has a strong experimental support. Nevertheless, this expectation is broken in some New Physics (NP) scenarios, and this can lead to clear signatures of physics beyond the SM. In fact, these signatures might have been observed already. Since 2013, the LHCb collaboration has reported on the measurement of several observables associated to semileptonic $b \to s$ transitions, finding some tensions with the SM expectations, including possible hints of the violation of LFU.

A. Vicente (✉)
Instituto de Física Corpuscular (CSIC-Universitat de València), Apdo. 22085, 46071 Valencia, Spain
e-mail: avelino.vicente@ific.uv.es

A. Giri and R. Mohanta (eds.), *16th Conference on Flavor Physics and CP Violation*, Springer Proceedings in Physics 234,
https://doi.org/10.1007/978-3-030-29622-3_54

We are interested in NP models that aim at an explanation of the so-called $b \to s$ *anomalies* while introducing a dark sector with the ingredients to accommodate the astrophysical and cosmological indications of the existence of DM [2]. Several authors have explored this direction [3–22]. The rest of the manuscript is organized as follows. We review the $b \to s$ anomalies and provide a model independent interpretation in Sect. 54.2. Section 54.3 classifies the proposed New Physics explanations to these anomalies with a link to the dark matter problem and presnts two example models that illustrate this connection. We finally summarize in Sect. 54.4.

54.2 The $b \to s$ Anomalies

There are two types of $b \to s$ anomalies: (1) branching ratios and angular observables, and (2) lepton flavor universality violating (LFUV) anomalies.

Branching ratios and angular observables: In 2013, the LHCb collaboration reported on the measurement of several observables in the decay $B \to K^* \mu^+ \mu^-$ with 1 fb^{-1} of integrated luminosity. Interestingly, several deviations with respect to the SM expectations were found. The most popular one was a 3.7σ discrepancy in one of the dimuon invariant mass bins in the P_5' angular observable [23]. Moreover, LHCb also found a systematic deficit in several branching ratios, mainly $\text{BR}(B_s \to \phi \mu^+ \mu^-)$ [24]. These anomalies were later confirmed with the addition of further data with the presentation by LHCb of new results in 2015, using in this case their full Run 1 dataset with 3 fb^{-1} [25, 26].

LFUV anomalies: several observables have been proposed in order to test LFU experimentally. In particular, one can consider the $R_{K^{(*)}}$ ratios, given by [27]

$$R_{K^{(*)}} = \frac{\Gamma(B \to K^{(*)} \mu^+ \mu^-)}{\Gamma(B \to K^{(*)} e^+ e^-)} . \tag{54.1}$$

These observables are measured in specific dilepton invariant mass squared ranges $q^2 \in [q^2_{\text{min}}, q^2_{\text{max}}]$. These ratios are very close to one in the SM, but this prediction can be altered by NP violating LFU. Moreover, hadronic uncertainties are expected to cancel to a high degree. Therefore, a deviation in $R_{K^{(*)}}$ would be regarded as a very clear sign of LFUV. Interestingly, several measurements by the LHCb collaboration point in this direction. The measurement of R_K in the region $[1, 6]$ GeV^2 was reported in 2014 [28], whereas the measurement of the R_{K^*} ratio in two q^2 bins was announced in 2017 [29]. These were the results:

$$\begin{aligned} R_K &= 0.745^{+0.090}_{-0.074} \pm 0.036 , \quad q^2 \in [1, 6] \text{ GeV}^2 , \\ R_{K^*} &= 0.660^{+0.110}_{-0.070} \pm 0.024 , \quad q^2 \in [0.045, 1.1] \text{ GeV}^2 , \\ R_{K^*} &= 0.685^{+0.113}_{-0.069} \pm 0.047 , \quad q^2 \in [1.1, 6.0] \text{ GeV}^2 . \end{aligned} \tag{54.2}$$

When these are compared to the SM predictions [30],

$$
\begin{aligned}
R_K^{\rm SM} &= 1.00 \pm 0.01\,, \quad q^2 \in [1,6]\,\mathrm{GeV}^2\,,\\
R_{K^*}^{\rm SM} &= 0.92 \pm 0.02\,, \quad q^2 \in [0.045,1.1]\,\mathrm{GeV}^2\,,\\
R_{K^*}^{\rm SM} &= 1.00 \pm 0.01\,, \quad q^2 \in [1.1,6.0]\,\mathrm{GeV}^2\,,
\end{aligned}
\tag{54.3}
$$

one finds deviations from the SM at the 2.6 σ level in the case of R_K, 2.2 σ for R_{K^*} in the low-q^2 region, and 2.4 σ for R_{K^*} in the central-q^2 region.

While the first category, angular observables and branching ratios, might be affected by hadronic uncertainties and the possibility of uncontrolled QCD effects cannot be discarded, the second one, composed by the $R_{K^{(*)}}$ ratios, is clean from this issue and can only be explained by NP violating LFU.

In order to interpret these experimental results it is convenient to adopt a language based on effective field theory. The effective Hamiltonian for $b \to s$ transitions can be written as

$$
\mathcal{H}_{\rm eff} = -\frac{4G_F}{\sqrt{2}} V_{tb} V_{ts}^* \frac{e^2}{16\pi^2} \sum_i \left(C_i\, \mathcal{O}_i + C_i'\, \mathcal{O}_i' \right) + \text{h.c.}. \tag{54.4}
$$

Here G_F is the Fermi constant, e the electric charge and V the Cabibbo-Kobayashi-Maskawa matrix. The effective operators contributing to $b \to s$ transitions are denoted by $\mathcal{O}_i$ and $\mathcal{O}_i'$, while C_i and C_i' denote their Wilson coefficients. The operators that turn out to be relevant for the interpretation of the $b \to s$ anomalies are

$$
\mathcal{O}_9 = \left(\bar{s} \gamma_\mu P_L b\right) \left(\bar{\ell} \gamma^\mu \ell\right)\,, \qquad \mathcal{O}_9' = \left(\bar{s} \gamma_\mu P_R b\right) \left(\bar{\ell} \gamma^\mu \ell\right)\,, \tag{54.5}
$$

$$
\mathcal{O}_{10} = \left(\bar{s} \gamma_\mu P_L b\right) \left(\bar{\ell} \gamma^\mu \gamma_5 \ell\right)\,, \qquad \mathcal{O}_{10}' = \left(\bar{s} \gamma_\mu P_R b\right) \left(\bar{\ell} \gamma^\mu \gamma_5 \ell\right)\,. \tag{54.6}
$$

Here $\ell = e, \mu, \tau$. Unless necessary, we will omit flavor indices in the Wilson coefficients in order to simplify the notation. It proves convenient to split the Wilson coefficients into their SM and NP pieces, defining

$$
C_9 = C_9^{\rm SM} + C_9^{\rm NP}\,, \tag{54.7}
$$

$$
C_{10} = C_{10}^{\rm SM} + C_{10}^{\rm NP}\,. \tag{54.8}
$$

Several independent global fits [31–38] have compared a large set of experimental measurements of observables associated to $b \to s$ transitions to their expected values in the SM, finding a remarkable tension, only alleviated by the introduction of NP contributions. In particular, there is a general agreement on the qualitative fact that global fits improve substantially with a negative contribution in $C_9^{\mu,\rm NP}$, with $C_9^{\mu,\rm NP} \sim -25\% \times C_9^{\mu,\rm SM}$. NP contributions in other muonic Wilson coefficients can affect the global fit, but they are sub-dominant compared to $C_9^{\mu,\rm NP}$. For instance, the addition of NP in the one-dimensional direction given by $C_9^{\mu,\rm NP} = -C_{10}^{\mu,\rm NP}$ also serves to

Table 54.1 New scalars and fermions in the model of [3]

Field	Spin	$SU(3)_c \times SU(2)_L \times U(1)_Y \times U(1)_X$
ϕ	0	$(\mathbf{1}, \mathbf{1}, 0, 2)$
χ	0	$(\mathbf{1}, \mathbf{1}, 0, -1)$
$Q_{L,R}$	$\frac{1}{2}$	$(\mathbf{3}, \mathbf{2}, \frac{1}{6}, 2)$
$L_{L,R}$	$\frac{1}{2}$	$(\mathbf{1}, \mathbf{2}, -\frac{1}{2}, 2)$

improve the fit, and this can be regarded as a hint in favor of purely left-handed NP interactions. Moreover, no hint for NP is found in contributions involving electrons or tau leptons.[1]

54.3 Model Classification

In general, the models explaining the $b \to s$ anomalies with a link to the dark matter problem can be classified into two categories:

- **Portal models**: in these models the mediator responsible for the NP contributions to $b \to s$ transitions is also the mediator for the production of DM in the early Universe.
- **Loop models**: in these models the required NP contributions to $b \to s$ transitions are induced via loops containing the DM particle.

There are also some *hybrid models* that share some properties with both categories.

54.3.1 A Portal Model

We will first discuss the portal model introduced in [3]. An extension of this model that also accounts for neutrino masses has been recently discussed in [40].

The model adds a new $U(1)_X$ factor to the SM gauge symmetry, with its gauge boson denoted as Z' and its gauge coupling as g_X. All the SM particles are singlets under this new symmetry, while the new states beyond the SM, the vector-like (VL) fermions Q and L and the complex scalar fields ϕ and χ, are charged. Table 54.1 shows the charges of the new scalars and fermions in the model. In addition to the usual canonical kinetic terms, the new VL fermions Q and L have gauge-invariant mass terms,

$$\mathcal{L}_m = m_Q \, \overline{Q} Q + m_L \, \overline{L} L \, . \tag{54.9}$$

[1]See also [39] for a recent analysis of the $b \to s$ anomalies based on gauge invariant effective operators.

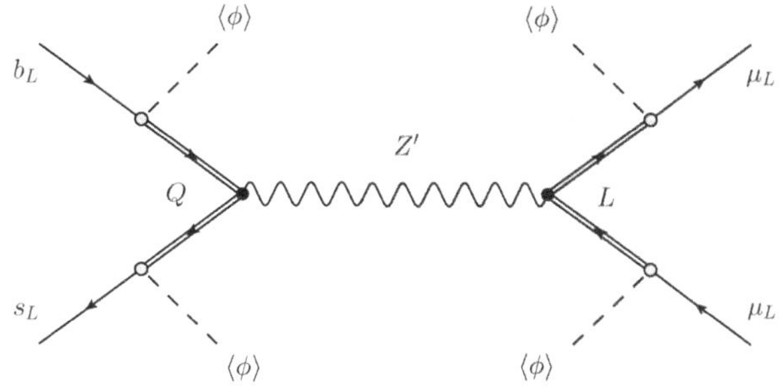

Fig. 54.1 Generation of $\mathcal{O}_9$ and $\mathcal{O}_{10}$ in the model of [3]

They also have Yukawa couplings with the SM doublets q and ℓ and the scalar ϕ,

$$\mathcal{L}_Y = \lambda_Q \, \overline{Q_R} \, \phi \, q_L + \lambda_L \, \overline{L_R} \, \phi \, \ell_L + \text{h.c.}. \tag{54.10}$$

Here λ_Q and λ_L are 3 component vectors. We will consider that the scalar potential of the model leads to a vacuum expectation value (VEV) for the ϕ scalar, $\langle\phi\rangle = \frac{v_\phi}{\sqrt{2}}$, breaking $U(1)_X$ spontaneously and inducing a mass for the Z' boson, $m_{Z'} = 2 g_X v_\phi$. In contrast, the scalar χ does not get a VEV. This leads to the existence of a remnant $\mathbb{Z}_2$ parity, under which χ is odd and all the other particles are even. This mechanism [41–43] stabilizes χ without the need of any additional symmetry.

The solution to the $\boldsymbol{b \to s}$ **anomalies** in this model is diagrammatically shown in Fig. 54.1. The Yukawa couplings in (54.10) induce mixings between the VL and SM fermions after $U(1)_X$ breaking. This mixing results in Z' effective couplings to the SM fermions. Since the SM fermions participating in the Yukawa interactions are purely left-handed, the model predicts $C_9^{\mu,\text{NP}} = -C_{10}^{\mu,\text{NP}}$. It is possible to show that by using $|\lambda_L^\mu| \sim 1 \gg |\lambda_Q^{b,s}|$ one can accommodate the required values for $C_9^{\mu,\text{NP}}$ and $C_{10}^{\mu,\text{NP}}$ determined by the global fits to $b \to s$ data and, at the same time, be compatible with all constraints.

In what concerns to the **Dark Matter** predictions of the model, we already pointed out that χ is automatically stable due to the remnant $\mathbb{Z}_2$ symmetry that is left after symmetry breaking. Therefore, this is the DM candidate. Its production in the early Universe takes place via $2 \leftrightarrow 2$ processes mediated by the massive Z' boson. This establishes a link with the $b \to s$ anomalies and justifies the choice of name *portal models* for the category represented by this model.

54.3.2 A Loop Model

We now consider the model introduced in [14], a simple illustration of the category of loop models. In this case, the SM symmetry is extended with a global (not gauge) $U(1)_X$ symmetry. As in the previous model, all SM fields are assumed to be singlets

Table 54.2 New scalars and fermions in the model of [14]

Field	Spin	$SU(3)_c \times SU(2)_L \times U(1)_Y$	$U(1)_X$
X	0	$(\mathbf{1}, \mathbf{1}, 0)$	-1
$Q_{L,R}$	$\frac{1}{2}$	$(\mathbf{3}, \mathbf{2}, \frac{1}{6})$	1
$L_{L,R}$	$\frac{1}{2}$	$(\mathbf{1}, \mathbf{2}, -\frac{1}{2})$	1

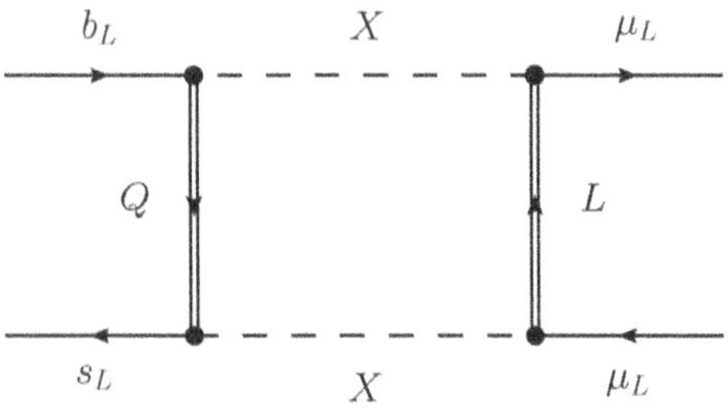

Fig. 54.2 Generation of $\mathcal{O}_9$ and $\mathcal{O}_{10}$ in the model of [14]

under this symmetry. In contrast, the new fields, the VL fermions Q and L and the complex scalar X, are charged. Table 54.2 details the new scalar and fermionic fields and their charges under the symmetries of the model.

The Lagrangian of the model contains the same Dirac mass terms as in (54.9), as well as the Yukawa couplings

$$\mathcal{L}_Y = \lambda_Q \, \overline{Q_R} \, X \, q_L + \lambda_L \, \overline{L_R} \, X \, \ell_L + \text{h.c.}. \tag{54.11}$$

Here λ_Q and λ_L are 3 component vectors. We consider a vacuum with $\langle X \rangle = 0$. This preserves the global $U(1)_X$ symmetry and stabilizes the lightest state with a non-vanishing charge under this symmetry. Furthermore, the conservation of $U(1)_X$ forbids the mixing between SM and VL fermions.

The solution of the $\boldsymbol{b \to s}$ **anomalies** comes now at the 1-loop level, as shown in Fig. 54.2. No NP contributions to $b \to s$ transitions are generated at tree-level in this model, as can be easily checked. As in the previous case, the left-handed chirality of the fermions involved in the new Yukawa interactions leads to $C_9^{\mu,\text{NP}} = -C_{10}^{\mu,\text{NP}}$, and one can obtain the required ranges for these Wilson coefficients by properly adjusting the parameters of the model.

Finally, we move on to the **Dark Matter** phenomenology of the model. Assuming that the lighest state charged under $U(1)_X$ is the neutral scalar X, it constitutes the DM candidate in the model. As discussed in detail in [14], the most relevant DM annihilation channels for the determination of the DM relic density are $XX^* \leftrightarrow \mu^+\mu^-, \nu\nu$, and this is due to the fact that one requires a large $|\lambda_L^\mu|$ in order to account for the $b \to s$ anomalies. Interestingly, the model is also testable in direct DM detection experiments, such as XENON1T.

54.4 Summary

Flavor and Dark Matter may seem two completely independent issues, but they might be connected to the same fundamental physics. In these proceedings we have discussed models that link the solution to the $b \to s$ anomalies, a subject of great interest in current flavor physics, to the existence of a dark sector. In doing this, these models extend the SM with new ingredients, opening new model building directions that would not be explored in the absence of this connection. It would definitely be fascinating to find a deep bond between these two areas of physics.

Acknowledgements I am very grateful to the organizers of the FPCP 2018 conference for their hospitality and congratulate them for the exciting discussions and friendly atmosphere during the meeting. This work was supported by the Spanish grants SEV-2014-0398 and FPA2017-85216-P (AEI/FEDER, UE), PROMETEO/2018/165 and SEJI/2018/033 (Generalitat Valenciana) and the Spanish Red Consolider MultiDark FPA201790566REDC.

References

1. P.A.R. Ade et al., Planck collaboration. Astron. Astrophys. **594**, A13 (2016). https://doi.org/10.1051/0004-6361/201525830. [arXiv:1502.01589 [astro-ph.CO]]
2. A. Vicente, Adv. High Energy Phys. **2018**, 3905848 (2018). https://doi.org/10.1155/2018/3905848. [arXiv:1803.04703 [hep-ph]]
3. D. Aristizabal Sierra, F. Staub, A. Vicente, Phys. Rev. D **92** (2015) no. 1, 015001. https://doi.org/10.1103/PhysRevD.92.015001[arXiv:1503.06077 [hep-ph]]
4. G. Bélanger, C. Delaunay, S. Westhoff, Phys. Rev. D **92**, 055021 (2015). https://doi.org/10.1103/PhysRevD.92.055021. [arXiv:1507.06660 [hep-ph]]
5. B. Allanach, F.S. Queiroz, A. Strumia, S. Sun, Phys. Rev. D **93** (2016) no. 5, 055045 Erratum: [Phys. Rev. D **95** (2017) no. 11, 119902] https://doi.org/10.1103/PhysRevD.93.055045. https://doi.org/10.1103/PhysRevD.95.119902. arXiv:1511.07447 [hep-ph]
6. M. Bauer, M. Neubert, Phys. Rev. D **93** (2016) no. 11, 115030 https://doi.org/10.1103/PhysRevD.93.115030. arXiv:1512.06828 [hep-ph]
7. A. Celis, W.Z. Feng, M. Vollmann, Phys. Rev. D **95** (2017) no. 3, 035018 https://doi.org/10.1103/PhysRevD.95.035018. arXiv:1608.03894 [hep-ph]
8. W. Altmannshofer, S. Gori, S. Profumo, F.S. Queiroz, JHEP **1612** (2016) 106.https://doi.org/10.1007/JHEP12(2016)106. arXiv:1609.04026 [hep-ph]
9. P. Ko, T. Nomura, H. Okada, Phys. Lett. B **772**, 547 (2017). https://doi.org/10.1016/j.physletb.2017.07.021. [arXiv:1701.05788 [hep-ph]]
10. P. Ko, T. Nomura, H. Okada, Phys. Rev. D **95** (2017) no. 11, 111701 https://doi.org/10.1103/PhysRevD.95.111701 arXiv:1702.02699 [hep-ph]
11. J.M. Cline, J.M. Cornell, D. London, R. Watanabe, Phys. Rev. D **95** (2017) no.9, 095015. https://doi.org/10.1103/PhysRevD.95.095015 arXiv:1702.00395 [hep-ph]
12. F. Sala, D.M. Straub, Phys. Lett. B **774**, 205 (2017). https://doi.org/10.1016/j.physletb.2017.09.072. [arXiv:1704.06188 [hep-ph]]
13. J. Ellis, M. Fairbairn and P. Tunney, Eur. Phys. J. C **78** (2018) (3), 238. https://doi.org/10.1140/epjc/s10052-018-5725-0 arXiv:1705.03447 [hep-ph]
14. J. Kawamura, S. Okawa, Y. Omura, Phys. Rev. D **96** (2017) no.7, 075041 https://doi.org/10.1103/PhysRevD.96.075041 arXiv:1706.04344 [hep-ph]
15. S. Baek, Phys. Lett. B **781**, 376 (2018). https://doi.org/10.1016/j.physletb.2018.04.012. arXiv:1707.04573 [hep-ph]

16. J.M. Cline, Phys. Rev. D **97** (2018) no.1, 015013. https://doi.org/10.1103/PhysRevD.97.015013 arXiv:1710.02140 [hep-ph]
17. J.M. Cline, J.M. Cornell, Phys. Lett. B **782**, 232 (2018). https://doi.org/10.1016/j.physletb.2018.05.034. arXiv:1711.10770 [hep-ph]
18. L. Dhargyal, arXiv:1711.09772 [hep-ph]
19. C.W. Chiang, H. Okada,. arXiv:1711.07365 [hep-ph]
20. A. Falkowski, S. F. King, E. Perdomo and M. Pierre, JHEP **1808** (2018) 061. https://doi.org/10.1007/JHEP08(2018)061 arXiv:1803.04430 [hep-ph]
21. G. Arcadi, T. Hugle, F.S. Queiroz, Phys. Lett. B **784**, 151 (2018). https://doi.org/10.1016/j.physletb.2018.07.028. arXiv:1803.05723 [hep-ph]
22. S. Baek, C. Yu,. arXiv:1806.05967 [hep-ph]
23. R. Aaij et al., LHCb Collaboration. Phys. Rev. Lett. **111**, 191801 (2013). https://doi.org/10.1103/PhysRevLett. [arXiv:1308.1707 [hep-ex]]. 111.191801
24. R. Aaij et al., LHCb collaboration. JHEP **1307**, 084 (2013). https://doi.org/10.1007/JHEP07(2013). [arXiv:1305.2168 [hep-ex]]. 084
25. R. Aaij et al., LHCb Collaboration. JHEP **1602**, 104 (2016). https://doi.org/10.1007/JHEP02(2016). [arXiv:1512.04442 [hep-ex]]. 104
26. R. Aaij et al., LHCb Collaboration. JHEP **1509**, 179 (2015). https://doi.org/10.1007/JHEP09(2015). [arXiv:1506.08777 [hep-ex]]. 179
27. G. Hiller, F. Kruger, Phys. Rev. D **69** (2004) 074020 https://doi.org/10.1103/PhysRevD.69.074020[hep-ph/0310219]
28. R. Aaij et al., LHCb collaboration. Phys. Rev. Lett. **113**, 151601 (2014). https://doi.org/10.1103/PhysRevLett. [arXiv:1406.6482 [hep-ex]]. 113.151601
29. R. Aaij et al., LHCb collaboration. JHEP **1708**, 055 (2017). https://doi.org/10.1007/JHEP08(2017). [arXiv:1705.05802 [hep-ex]]. 055
30. M. Bordone, G. Isidori, A. Pattori, Eur. Phys. J. C **76** (2016) no.8, 440 https://doi.org/10.1140/epjc/s10052-016-4274-7 arXiv:1605.07633 [hep-ph]
31. B. Capdevila, A. Crivellin, S. Descotes-Genon, J. Matias and J. Virto, JHEP **1801** (2018) 093 https://doi.org/10.1007/JHEP01(2018)093 arXiv:1704.05340 [hep-ph]
32. W. Altmannshofer, P. Stangl, D. M. Straub, Phys. Rev. D **96** (2017) no.5, 055008 https://doi.org/10.1103/PhysRevD.96.055008[arXiv:1704.05435 [hep-ph]]
33. G. D'Amico, M. Nardecchia, P. Panci, F. Sannino, A. Strumia, R. Torre and A. Urbano, JHEP **1709** (2017) 010. https://doi.org/10.1007/JHEP09(2017)010[arXiv:1704.05438 [hep-ph]]
34. G. Hiller, I. Nisandzic, Phys. Rev. D **96** (2017) no.3, 035003. https://doi.org/10.1103/PhysRevD.96.035003 arXiv:1704.05444 [hep-ph]
35. L.S. Geng, B. Grinstein, S. Jger, J. Martin Camalich, X. L. Ren and R. X. Shi, Phys. Rev. D **96** (2017) no.9, 093006 https://doi.org/10.1103/PhysRevD.96.093006 arXiv:1704.05446 [hep-ph]
36. M. Ciuchini, A. M. Coutinho, M. Fedele, E. Franco, A. Paul, L. Silvestrini, M. Valli, Eur. Phys. J. C **77** (2017) no.10, 688 https://doi.org/10.1140/epjc/s10052-017-5270-2 arXiv:1704.05447 [hep-ph]
37. A. K. Alok, B. Bhattacharya, A. Datta, D. Kumar, J. Kumar and D. London, Phys. Rev. D **96** (2017) no.9, 095009. https://doi.org/10.1103/PhysRevD.96.095009 arXiv:1704.07397 [hep-ph]
38. T. Hurth, F. Mahmoudi, D. Martinez Santos and S. Neshatpour, Phys. Rev. D **96** (2017) no. 9, 095034 https://doi.org/10.1103/PhysRevD.96.095034 arXiv:1705.06274 [hep-ph]
39. A. Celis, J. Fuentes-Martin, A. Vicente and J. Virto, Phys. Rev. D **96** (2017) no.3, 035026 doi:https://doi.org/10.1103/PhysRevD.96.035026[arXiv:1704.05672 [hep-ph]]
40. P. Rocha-Moran, A. Vicente,. arXiv:1810.02135 [hep-ph]
41. L.M. Krauss, F. Wilczek, Phys. Rev. Lett. **62**, 1221 (1989).https://doi.org/10.1103/PhysRevLett.62.1221
42. B. Petersen, M. Ratz, R. Schieren, JHEP **0908** (2009) 111. https://doi.org/10.1088/1126-6708/2009/08/111[arXiv:0907.4049 [hep-ph]]
43. D. Aristizabal Sierra, M. Dhen, C.S. Fong, A. Vicente, Phys. Rev. D **91** (2015) no. 9, 096004. https://doi.org/10.1103/PhysRevD.91.096004. arXiv:1412.5600 [hep-ph]

Chapter 55
B-Sector Anomalies—The Top Connection

David London, Pratishruti Saha and Ryoutaro Watanabe

Abstract Measurements of B-meson decays have revealed several departures from the Standard Model. In contrast, measurements of top-quark processes seem to adhere strictly to the Standard Model. This presents a conundrum given that the bottom quark and the top quark belong to the same $SU(2)$ doublet. Several models have been proposed in the literature to explain the B-sector anomalies. We confront some such models with top-sector observables to examine whether they can be tested by such measurements.

55.1 Introduction

The study of meson decays and oscillations has provided a wealth of information about the nature of fundamental particles and interactions. Flavor experiments provided crucial inputs for the construction of the Standard Model (SM). At present, while all the components of the minimal Standard Model have been discovered, explorations continue for physics beyond the Standard Model. Once again, flavor observables are expected to play an important role in this quest.

In recent times, a multitude of measurements related to B-meson decays have exhibited departures from Standard Model predictions [1–5]. The ratio $\mathcal{B}(B^+ \to K^+\mu^+\mu^-) / \mathcal{B}(B^+ \to K^+e^+e^-)$, labelled as R_K, was found to be approximately 25% smaller than the SM prediction [1]. Similarly, the measured value of $R_{K^*} \equiv \mathcal{B}(B^0 \to K^{*0}\mu^+\mu^-) / \mathcal{B}(B^0 \to K^{*0}e^+e^-)$ is 2.1–2.5 standard deviations away from the SM prediction [2]. The angular distributions of $B \to K^*\mu^+\mu^-$ also show a discrepancy [3]. Deviations are seen in both the branching fraction and the angular distributions of $B_s^0 \to \phi\mu^+\mu^-$ [4]. Finally, the ratio $R_{D^{(*)}} \equiv \mathcal{B}(\bar{B} \to D^{(*)}\tau^-\bar{\nu}_\tau) / \mathcal{B}(\bar{B} \to$

D. London · R. Watanabe
Physique des Particules, Université de Montréal, C.P. 6128, succ. centre-ville, Montréal H3C 3J7, QC, Canada

P. Saha (✉)
Harish-Chandra Research Institute, Chhatnag Road, Jhunsi, Allahabad 211019, India
e-mail: pratishrutisaha@hri.res.in

A. Giri and R. Mohanta (eds.), *16th Conference on Flavor Physics and CP Violation*, Springer Proceedings in Physics 234,
https://doi.org/10.1007/978-3-030-29622-3_55

$D^{(*)}\ell^-\bar{\nu}_\ell$) is also observed to be different from the SM prediction [5]. The individual deviations are not very large and lie in the 2-σ to 3-σ range. Nonetheless, they display certain intriguing patterns that are more suggestive of some hitherto unknown dynamics than of statistical fluctuations. Specifically,

(i) the observables relate to processes involving 2nd and 3rd generation quarks and leptons;
(ii) they indicate non-universal couplings in the lepton sector.

The first four observables described above involve the transition $b \to s\mu^+\mu^-$. Several new physics (NP) scenarios have been proposed as explanations [6]. It has been found that, in order to explain the data, the NP must essentially couple to left-chiral bottom quarks (b_L). This allows for a negative interference between the SM and NP amplitudes and lowers the prediction for R_K which is ~ 1 in the SM. NP couplings to right-chiral bottom quarks (b_R) are not ruled out, but they are, by themselves, insufficient for explaining the observed data. This immediately brings up an interesting prospect. The Standard Model relates left-chiral bottom quarks to left-chiral top quarks (t_L) via $SU(2)$ symmetry. Hence any NP that couples to b_L must necessarily couple to t_L. Curiously enough, all measurements of top processes agree well with the SM [7]. Therefore, the questions being addressed here are : (a) How are top-sector observables affected by NP models constructed to explain the B anomalies? (b) Can top-sector observables be used to confirm or reject certain models?

55.2 The Generic Z' Model

As mentioned earlier, several NP models have been proposed as explanations for the anomalies in the B-sector. These can be broadly categorized into extended-Higgs models, leptoquark models and Z' models. In this study, only Z' models are considered.

A typical Z' model propounds a $U(1)$ symmetry which is spontaneously broken at a high scale resulting in a massive, neutral gauge boson at the electroweak scale. Here the Z' couples to left-chiral quarks and leptons. The mass eigenstates are related to the gauge eigenstates via

$$u'_{Li} = U_{ij}\, u_{Lj} \tag{55.1}$$

$$d'_{Li} = D_{ij}\, d_{Lj} \tag{55.2}$$

where i is the generation index and the gauge eigenstates are shown on the LHS. The CKM matrix is then given by

$$V_{CKM} = U^\dagger D \tag{55.3}$$

Thus, once the Z' couplings to *down-type* quarks are fixed, the couplings to *up-type* quarks are completely determined in the terms of the *down-type* couplings and the elements of V_{CKM}.

As in [8], we denote the flavor-changing coupling of the Z' to b_L and s_L as g_{sb}. The existence of g_{sb} mandates the existence of g_{ss} and g_{bb}. However, couplings to d_L are assumed to be negligible as would be required by the constraints from $K^0 - \bar{K}^0$ and $B^0 - \bar{B}^0$ mixing. Similarly, in the lepton sector, couplings to electrons are assumed to be negligible, and only $g_{\mu\mu}$, $g_{\tau\tau}$ and $g_{\mu\tau}$ are taken into consideration.

55.3 Top Processes

Z' mediated processes would contribute to top quark production at the LHC through both flavor-changing as well as flavor-conserving interactions. Table 55.1 shows the processes that are affected as well as the respective SM predictions. Prima facie, one might expect s-channel single top production and like-sign ditop production to be the most promising channels as these have the lowest SM backgrounds. However, there are large experimental uncertainties in s-channel single top production. As for like-sign ditop production, note that this final state can be identified only when both top quarks decay to $b\,\ell^+\,\nu_\ell$. As a result, the effective cross section in this channel is suppressed.

55.4 Numerical Results

The existing constraints on g_{sb} and $g_{\mu\mu}$ are summarized in Table 55.2 [8]. In order to remain perturbative, g_{ss} and g_{bb} can, at most, be $O(1)$. Therefore, given that, $g_{u_i u_j} \equiv f(g_{ss}, g_{bb}, g_{sb}, V_{CKM})$, we have $g_{cc}, g_{tt} \lesssim O(1)$; $g_{uc} \lesssim O(10^{-1})$; $g_{uu}, g_{ct} \lesssim O(10^{-2})$ and $g_{ut} \lesssim O(10^{-3})$. This would appear to suggest that enhancements are more likely in $t\bar{t}$ cross section than in single top or like-sign ditop cross sections. However, it is important to note that such an enhancement would only occur in the sub-process $q\bar{q} \to t\bar{t}$ and the dominant $gg \to t\bar{t}$ sub-process would remain unaffected.

Table 55.1 Top processes affected by Z'-mediation

	Process	σ_{SM} for LHC@13TeV
Top pair production	$pp \to t\bar{t}$	~ 850 pb
t-channel single top production	$pp \to tj$	~ 150 pb
s-channel single top production	$pp \to t\bar{b}$	~ 7 pb
Like-sign ditop production	$pp \to tt$	~ 0

Table 55.2 Existing constraints on Z' couplings

Process	Limits
$b \to s\,\mu^+\mu^-$	$\frac{g_{sb}\,g_{\mu\mu}}{M_{Z'}^2} = -0.0011 \pm 0.0002\ \text{TeV}^{-2}$
$B_s^0 - \bar{B}_s^0$ mixing	$\frac{g_{sb}}{M_{Z'}} = \pm(1.0^{+2.0}_{-3.9}) \times 10^{-3}\ \text{TeV}^{-1}$
$B \to K^* \nu\bar{\nu}$	$\frac{g_{sb}\,(g_{\mu\mu} + g_{\tau\tau})}{M_{Z'}^2} \in [-0.014, 0.034]\ \text{TeV}^{-2}$
$\nu N \to \nu N \mu^+\mu^-$	$\frac{g_{\mu\mu}}{M_{Z'}} = 0 \pm 1.13\ \text{TeV}^{-1}$

Furthermore, it turns out that $D^0 - \bar{D}^0$ mixing imposes a far more stringent limit on g_{uc}, and, in turn, on g_{ss} and g_{bb}. Consistency with measurements of $D^0 - \bar{D}^0$ mixing requires $g_{ss}/M_{Z'} < 0.003\ \text{TeV}^{-1}$ and $g_{bb}/M_{Z'} < 4.5\ \text{TeV}^{-1}$. There are also constraints on g_{bb} from Υ decays. However these are weaker than those mentioned above.

With these considerations in mind, we proceed to calculate the top production cross section for all the channels discussed in the previous section. We explore $M_{Z'} \sim$ O(10 GeV) to O(100 GeV). In all cases, the enhancement in the cross section due to Z' contributions remains within a few percent of the SM prediction and the net resultant cross section is consistent with the LHC measurements within experimental uncertainties.

55.5 Conclusions

The symmetry structure of the Standard Model ensures that any NP that couples to b_L also couples to t_L. Deviation from SM predictions in B-meson decays point to the presence of NP that couples to b_L. We analyze a generic Z' model as an instance of such NP. We do not make any assumptions about the UV completion of the model. Instead, we place limits on the Z' mass and couplings purely on the basis of existing data. We find that the Z' parameter space that survives these constraints is automatically consistent with all measurements in the top sector. This apparent cul-de-sac has important consequences. At present, all top observables are in good agreement with SM predictions. Should any discrepancies arise in the future, the scope of Z' models that couple with t_L would already be severely restricted.

Acknowledgements P.S. would like to thank George Wei-Shu Hou and Alakabha Datta for useful comments. This work was financially supported in part by NSERC of Canada (D.L. and R.W.). P.S. is supported by the Dept. of Science and Technology, India.

References

1. R. Aaij et al., LHCb Collaboration. Phys. Rev. Lett. **113**, 151601 (2014)
2. R. Aaij et al., LHCb Collaboration. JHEP **1708**, 055 (2017)
3. R. Aaij et al., LHCb Collaboration. JHEP **1602**, 104 (2016); A. Abdesselam et al. [Belle Collaboration], BELLE-CONF-1603, arXiv:1604.04042
4. R. Aaij et al., LHCb Collaboration. JHEP **1509**, 179 (2015)
5. J.P. Lees et al., BaBar Collaboration. Phys. Rev. D **88**, 072012 (2013); R. Aaij et al. [LHCb Collaboration] Phys. Rev. Lett. **115**, 111803 (2015), Erratum: Phys. Rev. Lett. 115, 159901 (2015)
6. *See, for example,* J.M. Cline, J.M. Cornell, D. London, R. Watanabe, Phys. Rev. D **95**, 095015 (2017) and references therein
7. LHC Top Physics Working Group. https://twiki.cern.ch/twiki/bin/view/LHCPhysics/LHCTopWGSummaryPlots
8. J. Kumar, D. London, R. Watanabe, Phys. Rev. D **99**, 015007 (2019)

Chapter 56
Anatomy of $b \to c$ Anomalies

Debjyoti Bardhan

56.1 Experimental Status

One of the very few signs of any deviation from the Standard Model (SM) is a long-standing anomaly in the charged current decay $B \to D^{(*)}$. The measured quantities are the two ratios defined as

$$R_{D^{(*)}} = \frac{\mathcal{B}(B \to D^{(*)}\tau\nu_\tau)}{\mathcal{B}(B \to D^{(*)}\ell\nu_\ell)} \tag{56.1}$$

where $\ell = e, \mu$. This suggests a violation of lepton flavour universality.

The experimental values are all greater than the SM and deviation is 2.3σ in R_D and 3.0σ in R_{D^*}, which gives a combined deviation of 3.62σ overall [1]. In light of this rather large deviation, we have decided to take a detailed look at the decay, first in [2] and then recently in [3].

56.2 Basis of Effective Operators and the Effective Scale

In the Standard Model, the decay at the quark level is mediated by the W-boson. The effective Lagrangian for this process, after integrating out the heavy W-boson is

$$\mathcal{L}_{\text{eff}}^{b\to cl\nu}|_{\text{SM}} = -\sqrt{2}G_F V_{cb}\left[\bar{c}\gamma^\mu(1-\gamma_5)b\right]\left[\bar{l}\gamma_\mu P_L \nu\right] \tag{56.2}$$

D. Bardhan (✉)
Ben Gurion University of the Negev, Be'er Sheva, Israel
e-mail: debjyotiarr@gmail.com

A. Giri and R. Mohanta (eds.), *16th Conference on Flavor Physics and CP Violation*, Springer Proceedings in Physics 234,
https://doi.org/10.1007/978-3-030-29622-3_56

Naively, one would think that the effective scale for this process in the SM is simply $\Lambda_{\rm SM}^2 = \left(\sqrt{2}G_F\right)^{-1}$. However, the presence of the CKM matrix element raises the effective scale to

$$\Lambda_{\rm SM}^2 = \left(\sqrt{2}G_F V_{cb}\right)^{-1} \approx (1.2\ {\rm TeV})^2 \tag{56.3}$$

The list of effective operators that we shall use in our analysis is

$$\begin{aligned}
\mathcal{O}_{\rm VL}^{cb\ell\nu} &= [\bar{c}\,\gamma^{\mu}\,b][\bar{\ell}\,\gamma_{\mu}\,P_L\,\nu] & \mathcal{O}_{\rm PL}^{cb\ell\nu} &= [\bar{c}\,\gamma_5\,b][\bar{\ell}\,P_L\,\nu]\\
\mathcal{O}_{\rm AL}^{cb\ell\nu} &= [\bar{c}\,\gamma^{\mu}\gamma_5\,b][\bar{\ell}\,\gamma_{\mu}\,P_L\,\nu] & \mathcal{O}_{\rm TL}^{cb\ell\nu} &= [\bar{c}\,\sigma^{\mu\nu}\,b][\bar{\ell}\,\sigma_{\mu\nu}P_L\,\nu]\\
\mathcal{O}_{\rm SL}^{cb\ell\nu} &= [\bar{c}\,b][\bar{\ell}\,P_L\,\nu] & &
\end{aligned}$$

where the subscript 'L' refers to the left handed neutrinos and the first letter refers to the type of operator (V = Vector, A = Axial-vector etc). One can easily construct another list with right handed neutrinos, and it has been studied in [4]; we shall not consider them here. We can tune the Wilson Coefficients (WCs) of each of these operators (denoted by C_X, where X is the same subscript as the operator) in order to explain the anomalies in a model-independent fashion. Any deviation from the SM values implies the existence of New Physics (NP). Furthermore, we shall assume that NP is present only in the tauonic WCs and the electronic or muonic WCs are at the SM value. In the SM, $C_{\rm VL}^{\tau} = -C_{\rm AL}^{\tau} = 1$, and all the other WCs are zero. All WCs considered in the work and in this note are assumed to be real. All the explanations and constraints in this note are taken at a 1σ level, unless otherwise mentioned.

56.3 Model-Independent Explanations

Our choice of operator combinations would be
(a) Vector-Axial Vector (b) Scalar-Pseudoscalar (c) Tensor-Scalar

56.3.1 Vector-Axial Vector

In the left plot of Fig. 56.1, we plot the different regions which satisfy the experimental values of R_D and R_{D^*}. The red band for the R_D is vertical because R_D is independent of $C_{\rm AL}^{\tau}$. The region where the green band (which is for R_{D^*}) and the red band overlap is the region which explains the two anomalies together. For this proceeding, we will ignore the blue band which shows the region explaining $R_{J/\psi}$, since that measurement has very large error bars and need higher precision before consideration. The slanted brown dashed line represents the $C_{\rm VL}^{\tau} = -C_{\rm AL}^{\tau}$ condition, which ensured by the $SU(2)_L \times U(1)_Y$ gauge invariance. The overlap region

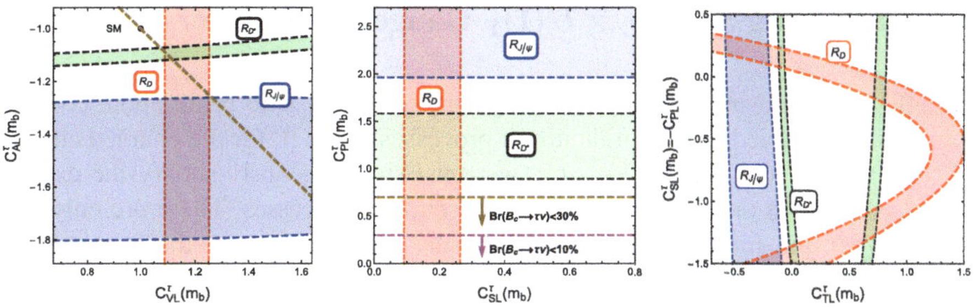

Fig. 56.1 Plots in the plane of certain Wilson Coefficient combinations. Refer to the text for more details

roughly corresponds to $C^{\tau}_{\text{VL}} = -C^{\tau}_{\text{AL}} = 1.1$, which means that the change in each from the SM is about 10%, since the SM value is $C^{\tau}_{\text{VL}}|_{\text{SM}} = -C^{\tau}_{\text{AL}}|_{\text{SM}} = 1.0$.

56.3.2 Scalar-Pseudoscalar

The explanation of the anomalies by the scalar and pseudoscalar operators runs into direct conflict with the branching ratio (BR) $\mathcal{B}(B_c \to \tau\nu)$. The authors of [5] put a strong bound of 30% on this BR, derived from the B_c lifetime. The authors of [6] put a stronger bound on the BR of 10%, derived from Z-pole measurements in LEP. These limits put a bound on the range of possible C^{τ}_{PL}, since the BR also depends on the same pseudoscalar operator that contributes to R_{D^*}. This situation is shown in the middle plot in Fig. 56.1. From this figure, it is clear that a simultaneous explanation of the anomalies in the two ratios is impossible, even allowing for the relaxed bound of 30% on the BR.

56.3.3 Tensor-Scalar

An explanation may be attempted in the $C^{\tau}_{\text{SL}} - C^{\tau}_{\text{TL}}$ plane, with the additional constraint that $C^{\tau}_{\text{SL}} = -C^{\tau}_{\text{PL}}$, as is shown in the right plot in Fig. 56.1. This gives us four regions of overlap. The two regions on the bottom half of the plot are ruled out by the $\mathcal{B}(B_c \to \tau\nu)$ constraint, while the region on the top right is basically the region in which $C^{\tau}_{\text{SL}} = 0$ and thus it is an explanation using only the tensor operator.

56.4 Imposing $SU(2)_L \times U(1)_Y$ Gauge Invariance

The effect of imposing the electroweak gauge symmetry is to bring about correlation between charged and neutral current processes. Thus if, for the charged current process, we are looking at $b \to c\tau\nu$, we must simultaneously satisfy the experimental constraints on any correlated neutral current processes. There are only two operators—a four-fermion operator and a scalar-lepton operator—that involve left-handed fields and also lead to lepton flavour non-universality. The $SU(2)_L \times U(1)_Y$ invariant six-dimensional operators relevant for this process, taken from the list in [7], are

$$\mathcal{L}^{\text{dim6}} = -\frac{1}{\Lambda^2} \sum_{p'r's't'} \Big\{ [C_{lq}^{(3)}]'_{p'r's't'} \left(\bar{l}'_{p'}\gamma_\mu \sigma^I l'_{r'}\right)\left(\bar{q}'_{s'}\gamma^\mu \sigma^I q'_{t'}\right) + \\ [C_{\phi l}^{(3)}]'_{p'r'} \left(\phi^\dagger i \overleftrightarrow{D}^I_\mu \phi\right)\left(\bar{l}'_{p'}\sigma^I \gamma^\mu l'_{r'}\right) + \text{H.c.} \Big\} \tag{56.4}$$

(The primes denote the fact that all the quantities are in the gauge basis).

56.4.1 The Scalar-Fermion Operator

Let's take the second operator in (56.4) first. The operator can be decomposed into

$$\left(\phi^\dagger i \overleftrightarrow{D}^I_\mu \phi\right)\left(\bar{l}'_{p'}\sigma^I \gamma^\mu l'_{r'}\right) = \Big[-\frac{1}{2}\frac{g_2}{\cos\theta_W} Z_\mu \left(\overline{\nu'}_{p'}\gamma^\mu P_L \nu'_{r'}\right) + \frac{1}{2}\frac{g_2}{\cos\theta_W} Z_\mu \left(\overline{e'}_{p'}\gamma^\mu P_L e'_{r'}\right) \\ - \frac{g_2}{\sqrt{2}} W^+_\mu \left(\overline{\nu'}_{p'}\gamma^\mu P_L e'_{r'}\right) - \frac{g_2}{\sqrt{2}} W^-_\mu \left(\overline{e'}_{p'}\gamma^\mu P_L \nu'_{r'}\right)\Big]\left(v^2 + 2vh + h^2\right) \tag{56.5}$$

where D_μ is the usual covariant derivative. In order to go from the gauge basis (as shown by the primed subscripts) to the mass (unprimed) basis, we need to introduce the mixing matrices.

$$(e_{L,R})_{r'} = (V^e_{L,R})_{r'r}(e_{L,R})_r, \quad (\nu_{L,R})_{r'} = (V^\nu_{L,R})_{r'r}(\nu_{L,R})_r, \\ (u_{L,R})_{r'} = (V^u_{L,R})_{r'r}(u_{L,R})_r, \quad (d_{L,R})_{r'} = (V^d_{L,R})_{r'r}(d_{L,R})_r \tag{56.6}$$

The CKM and the PMNS matrices are then defined to be

$$V_{\text{CKM}} = (V^u_L)^\dagger V^d_L, \quad V_{\text{PMNS}} = (V^\nu_L)^\dagger V^e_L$$

Without any loss of generality, we can assume that the left-chiral down-type quarks and the left-chiral charged leptons are in the mass basis. Thus

$$V^e_L = I_{3\times3} \quad , \qquad V^d_L = I_{3\times3} \tag{56.7}$$

Using this choice of basis, we have

$$V_{\text{CKM}} = V_L^{u\dagger} \quad , \qquad V_{\text{PMNS}} = V_L^{\nu\dagger} \tag{56.8}$$

Using (56.6), (56.7) and (56.8), we have

$$\begin{aligned}
&[C_{\phi l}^{(3)}]'_{p'r'} \left(\phi^\dagger i \overleftrightarrow{D}^I_\mu \phi\right) \left(\bar{l}'_{p'}\, \sigma^I \gamma^\mu\, l'_{r'}\right) \\
&= \sum_{p,r} \Bigg[\sum_{p',r'} [C_{\phi l}^{(3)}]'_{p'r'} \Bigg[-\frac{1}{2}\frac{g_2}{\cos\theta_W} (V_{\text{PMNS}})_{pp'} (V_{\text{PMNS}})^\dagger_{r'r} Z_\mu \left(\bar{\nu}_p \gamma^\mu P_L \nu_r\right) \\
&\quad + \frac{1}{2}\frac{g_2}{\cos\theta_W} Z_\mu \left(\bar{e}_p \gamma^\mu P_L e_r\right) - \frac{g_2}{\sqrt{2}} (V_{\text{PMNS}})_{pp'} W_\mu^+ \left(\bar{\nu}_p \gamma^\mu P_L e_r\right) \\
&\quad - \frac{g_2}{\sqrt{2}} (V_{\text{PMNS}})^\dagger_{r'r} W_\mu^- \left(\bar{e}_p \gamma^\mu P_L \nu_r\right) \Bigg]\Bigg] \left(v^2 + 2vh + h^2\right)
\end{aligned} \tag{56.9}$$

One might be tempted to think that a direct contribution to the charged current vertex will help explain $R_{D^{(*)}}$. The bound on the following ratio from LEP is [8]:

$$\frac{\mathcal{B}(W \to \tau\nu)}{0.5(\mathcal{B}(W \to e\nu) + \mathcal{B}(W \to \mu\nu))} = 1.077 \pm 0.026 \tag{56.10}$$

Thus, we might get an overall increment of 10% on the ratio at 1σ. Assuming that this is due to NP only in the tau final state, the deviation in the Wilson Coefficient is only 5% each, i.e. $\Delta C^\tau_{\text{VL}} = -\Delta C^\tau_{\text{AL}} \leq 0.05$. Since the necessary deviation is 10% in each, this clearly doesn't produce enough deviation from SM physics. Additionally, the operator contributes to neutral current vertices like $Z\nu\nu$ and $Z\tau\tau$. Both processes have bounds from LEP. The Lagrangian for the former process is

$$\mathcal{L}_{Z\nu\nu} = -\frac{g}{\cos\theta_W}\left(1 + \Delta g_L^\nu\right) Z_\mu \bar{\nu}_\tau \gamma^\mu P_L \nu_\tau + \text{H.c} \tag{56.11}$$

The amount of NP is measured by Δg_L^ν. This quantity is related to the WC of the operator as

$$\Delta g_L^\nu = -\frac{v^2}{2\Lambda^2}\left[V_{\text{PMNS}}\left([\tilde{C}_{\phi l}^{(3)e\nu}] + [\tilde{C}_{\phi l}^{(3)\nu e}]^\dagger\right) V^\dagger_{\text{PMNS}}\right]_{33} \tag{56.12}$$

The bound on this quantity from LEP [9] puts bounds on the WCs.

$$\left|\Delta g_L^\nu\right| \lesssim 1.2 \times 10^{-3} \implies \left|\Delta C^\tau_{\text{VL}}\right| < 0.022; \;\; \left|\Delta C^\tau_{\text{VL}}\right| = \left|\Delta C^\tau_{\text{AL}}\right| \tag{56.13}$$

which is again much less than the 10% deviation that is needed. The constraints with the $Z\tau\tau$ vertex due to the LEP data is even stronger. The Lagrangian is

$$\mathcal{L}_{Z\nu\nu} = -\frac{g}{\cos\theta_W} Z_\mu \left(\left(g_L^\tau + \Delta g_L^\tau\right) \bar{\tau}\gamma^\mu P_L \tau + \left(g_R^\tau + \Delta g_R^\tau\right) \bar{\tau}\gamma^\mu P_R \tau \right) \tag{56.14}$$

From the Lagrangian, we have

$$\Delta g_L^\tau = \frac{1}{2} \left[\left([\tilde{C}_{\phi l}^{(3)e\nu}] + [\tilde{C}_{\phi l}^{(3)\nu e}]^\dagger \right) V_{\mathrm{PMNS}} \right]_{33} \frac{v^2}{\Lambda^2}, \quad \Delta g_R^\tau = 0 \tag{56.15}$$

The LEP bounds then force

$$|\Delta g_L^\tau| \lesssim 6 \times 10^{-4} \quad \Rightarrow \Delta C_{\mathrm{VL}}^\tau \lesssim 0.001 \tag{56.16}$$

which is again extremely small.

56.4.2 The Four-Fermion Operator

We can perform the same analysis with the four-fermion operator. It can be decomposed as

$$\begin{aligned}
&\left(\bar{l}'_{p'}\gamma_\mu\sigma^I l'_{r'}\right)\left(\bar{q}'_{s'}\gamma^\mu\sigma^I q'_{t'}\right) \\
&\quad = \left(\bar{\nu}'_{p'}\gamma^\mu P_L \nu'_{r'}\right)\left(\bar{u}'_{s'}\gamma_\mu P_L u'_{t'}\right) + \left(\bar{e}'_{p'}\gamma^\mu P_L e'_{r'}\right)\left(\bar{d}'_{s'}\gamma_\mu P_L d'_{t'}\right) - \left(\bar{e}'_{p'}\gamma^\mu P_L e'_{r'}\right)\left(\bar{u}'_{s'}\gamma_\mu P_L u'_{t'}\right) \\
&\quad - \left(\bar{\nu}'_{p'}\gamma^\mu P_L \nu'_{r'}\right)\left(\bar{d}'_{s'}\gamma_\mu P_L d'_{t'}\right) + 2\left(\bar{\nu}'_{p'}\gamma^\mu P_L e'_{r'}\right)\left(\bar{d}'_{s'}\gamma_\mu P_L u'_{t'}\right) + 2\left(\bar{e}'_{p'}\gamma^\mu P_L \nu'_{r'}\right)\left(\bar{u}'_{s'}\gamma_\mu P_L d'_{t'}\right)
\end{aligned} \tag{56.17}$$

Note that the first four terms contribute to neutral current processes, while the last two terms contribute to charged current processes. The contribution to the process of interest $(\bar{\tau}\gamma_\mu P_L \nu_\tau)(\bar{c}\gamma^\mu P_L b)$ from this operator, in order to explain the anomaly at 1σ level, is

$$\begin{aligned}
&-\left(([C_{lq}^{(3)}]'_{3313} + [C_{lq}^{(3)}]'^*_{3331})V_{cd} + ([C_{lq}^{(3)}]'_{3323} + [C_{lq}^{(3)}]'^*_{3332})V_{cs} + ([C_{lq}^{(3)}]'_{3333} + [C_{lq}^{(3)}]'^*_{3333})V_{cb}\right) \\
&\quad \geq 0.015\left(\frac{\Lambda_{\mathrm{SM}^2}}{\mathrm{TeV}^2}\right)
\end{aligned} \tag{56.18}$$

Each of the factors in the parenthesis can be constrained by considering certain processes which the $b \to c\tau\nu$ process is correlated with. The term within the first parenthesis in (56.18) is to be found upon considering $B^0 \to \pi^0\bar{\nu}\nu$ decays, since that involves the decay $(\bar{\nu}\gamma^\nu P_L \nu)(\bar{d}\gamma_\nu P_L b)$. From LEP bounds [10] and theoretical calculations [11], we have

$$-0.018\left(\frac{\Lambda^2}{\mathrm{TeV}^2}\right) \lesssim [C_{lq}^{(3)}]'_{3313} + [C_{lq}^{(3)}]'^*_{3331} \lesssim 0.023\left(\frac{\Lambda^2}{\mathrm{TeV}^2}\right) \tag{56.19}$$

Similarly, by considering the branching ratio of the decay $B^0 \to K^{*0}\bar{\nu}\nu$, from [10] and [12], one finds,

$$-0.005\left(\frac{\Lambda^2}{\text{TeV}^2}\right) \lesssim [C_{lq}^{(3)}]'_{3323} + [C_{lq}^{(3)}]'^{*}_{3332} \leq 0.025\left(\frac{\Lambda^2}{\text{TeV}^2}\right) \tag{56.20}$$

These two bounds taken together give us, from (56.18):

$$([C_{lq}^{(3)}]'_{3333} + [C_{lq}^{(3)}]'^{*}_{3333})V_{cb} \gtrsim 0.03\left(\frac{\Lambda^2}{\text{TeV}^2}\right) \tag{56.21}$$

However, the same operator also contributes to the $(\bar{\tau}\gamma^{\mu}P_L\tau)\left(\bar{t}\gamma_{\mu}P_Lt\right)$ vertex. The top quarks can be then thought to form a loop and attach to a Z-boson, effectively contributing to a $Z\tau\tau$ vertex. From LEP bounds on this vertex, it can be calculated that

$$\left|[C_{lq}^{(3)}]'_{3333} + [C_{lq}^{(3)}]'^{*}_{3333}\right| \lesssim \frac{0.017}{V_{cb}}\left(\frac{\Lambda}{\text{TeV}}\right)^2 \frac{1}{1+0.6\log\frac{\Lambda}{\text{TeV}}} \tag{56.22}$$

which is in clear contradiction with (56.21).

Note that this means explaining the anomaly with the four-fermion operator alone will not be possible. It might so happen that a singlet operator, viz. $\left(\bar{l}'_{p'}\gamma_{\mu}l'_{r'}\right)\left(\bar{q}'_{s'}\gamma^{\mu}q'_{t'}\right)$, also contributes and cancels the contribution of this operator to the neutral current processes, thereby helping us evade the severe constraints outlined here.

56.5 The $\Delta F = 2$ Decays

The imposition of gauge symmetry doesn't allow us to relate $\Delta F = 1$ processes to $\Delta F = 2$ processes. In order to do this, we must assume a power counting rule, in the form of either a model with an $SU(2)$ triplet or a leptoquark. The relevant operator for the $\Delta F = 2$ process, expanded in the mass basis, is:

$$[C_{qq}^{(3)}]'_{p'r's't'}\left(\bar{q}'_{p'}\gamma_{\mu}\tau^I q'_{r'}\right)\left(\bar{q}'_{s'}\gamma^{\mu}\tau^I q'_{t'}\right) = \left(\frac{1}{4}[\tilde{C}_{qq}^{(3)dddd}]_{prst}\left(\bar{d}_p\gamma^{\mu}P_Ld_r\right)\left(\bar{d}_s\gamma_{\mu}P_Ld_t\right) + \ldots\right) \tag{56.23}$$

The contribution to $(\bar{s}\gamma^{\mu}P_Lb)(\bar{s}\gamma_{\mu}P_Lb)$ is then given by

$$\frac{1}{4}[\tilde{C}_{qq}^{(3)dddd}]_{2323} + \frac{1}{4}([\tilde{C}_{qq}^{(3)dddd}]_{3232})^* = \frac{1}{4}[C_{qq}^{(3)}]'_{2323} + \frac{1}{4}([C_{qq}^{(3)}]'_{3232})^* \tag{56.24}$$

The coefficient contributing to the $\Delta F = 1$ may be written as a product of quark and leptonic couplings.

$$[C_{lq}^{(3)}]'_{3323} = \frac{[C_{lq;l}^{(3)}]'_{33}}{\Lambda}\frac{[C_{lq;q}^{(3)}]'_{23}}{\Lambda}, \ [C_{lq}^{(3)}]'_{3332} = \frac{[C_{lq;l}^{(3)}]'_{33}}{\Lambda}\frac{[C_{lq;q}^{(3)}]'_{32}}{\Lambda} \tag{56.25}$$

where the subscripts l and q refer to the leptonic and quark parts respectively. Following the same procedure for the $\Delta F = 2$ decays, we have

$$(\bar{s}\gamma^{\mu}P_L b)\left(\bar{s}\gamma_{\mu}P_L b\right) : \frac{1}{2}\frac{([C_{lq;q}^{(3)}]'_{23} + [C_{lq;q}^{(3)}]'_{32})^2}{\Lambda^2} \tag{56.26}$$

From the experimental data for $\bar{B}_s - B_s$ mixing [13], we have

$$\frac{1}{2}\frac{([C_{lq;q}^{(3)}]'_{23} + [C_{lq;q}^{(3)}]'_{32})^2}{\Lambda^2} < 5.6 \times 10^{-11}\,\mathrm{GeV}^{-2} \implies ([C_{lq;q}^{(3)}]'_{23} + [C_{lq;q}^{(3)}]'_{32}) < 0.011\left(\frac{\Lambda}{\mathrm{TeV}}\right) \tag{56.27}$$

However, this sets a very strong bound for the coefficient of the $\Delta F = 1$ decay processes

$$\frac{C_{\mathrm{VL}}^{\tau} - C_{\mathrm{AL}}^{\tau}}{\Lambda_{\mathrm{SM}}^2} = \frac{1}{2}V_{cs}\frac{[C_{lq;l}^{(3)}]'_{33}}{\Lambda}\frac{([C_{lq;q}^{(3)}]'_{23} + [C_{lq;q}^{(3)}]'_{32})}{\Lambda} \implies [C_{lq;l}^{(3)}]'_{33} < 0.008\left(\frac{\mathrm{TeV}}{\Lambda}\right)[C_{lq;l}^{(3)}]'_{33} \tag{56.28}$$

If $[C_{lq;l}^{(3)}]'_{33} \equiv \mathcal{O}(1)$, then there is strong tension between the $R_{D^{(*)}}$ anomalies and $\bar{B}_s - B_s$ mixing, since, to satisfy the anomalies, one needs $(C_{\mathrm{VL}}^{\tau} - C_{\mathrm{AL}}^{\tau}) = 0.20$. This constraint is avoided if the SU(2) triplet contribution is replaced by a leptoquark, since then, the contribution, will occur at the loop level, rather than at the tree-level as in the triplet case.

Summary

In this short note, we reported on an in-depth study of R_{D^*}. In this context, we consider different $\Delta F = 1$ and $\Delta F = 2$ processes. Taking $SU(2)_L \times U(1)_Y$ symmetric six-dimensional effective operators in account, we find that certain $\Delta F = 1$ processes like $B \to K\nu\nu$, $B \to \pi\nu\nu$, $B \to \tau\nu$ etc. and vertices such as $W\tau\nu$, $Z\nu\nu$ and $Z\tau\tau$ are correlated and experimental data for them pose severe difficulties in explaining $R_{D^{(*)}}$. In the case of $\Delta F = 2$, we have to resort to some power-counting rules and, in this case, neutral meson mixing, such as $K - \bar{K}$, $B_d - \bar{B}_d$ and $B_s - \bar{B}_s$, are very constraining. In this case, instead of an $SU(2)$ triplet, a leptoquark may help in explaining $R_{D^{(*)}}$.

References

1. https://hflav-eos.web.cern.ch/hflav-eos/semi/summer18/RDRDs.html
2. D. Bardhan, P. Byakti, D. Ghosh, JHEP **1701**, 125 (2017). https://doi.org/10.1007/JHEP01(2017)125, arXiv:1610.03038 [hep-ph]
3. A. Azatov, D. Bardhan, D. Ghosh, F. Sgarlata, E. Venturini, arXiv:1805.03209 [hep-ph]
4. A. Azatov, D. Barducci, D. Ghosh, D. Marzocca, L. Ubaldi, arXiv:1807.10745 [hep-ph]

5. R. Alonso, B. Grinstein, J. Martin Camalich, Phys. Rev. Lett. **118**(8), 081802 (2017). https://doi.org/10.1103/PhysRevLett.118.081802, arXiv:1611.06676 [hep-ph]
6. A.G. Akeroyd, C.H. Chen, Phys. Rev. D **96**(7), 075011 (2017). https://doi.org/10.1103/PhysRevD.96.075011, arXiv:1708.04072 [hep-ph]
7. B. Grzadkowski, M. Iskrzynski, M. Misiak, J. Rosiek, JHEP **1010**, 085 (2010). https://doi.org/10.1007/JHEP10(2010)085, arXiv:1008.4884 [hep-ph]
8. [ALEPH, DELPHI, L3 and OPAL Collaborations and LEP Electroweak Working Group], hep-ex/0511027
9. S. Schael et al., [ALEPH, DELPHI, L3, OPAL and SLD Collaborations and LEP Electroweak Working Group and SLD Electroweak Group and SLD Heavy Flavour Group], Phys. Rept. **427**, 257 (2006). https://doi.org/10.1016/j.physrep.2005.12.006 [hep-ex/0509008]
10. J. Grygier et al., [Belle Collaboration], Phys. Rev. D **96**(9), 091101 (2017) Addendum: [Phys. Rev. D **97**(9), 099902 (2018)]. https://doi.org/10.1103/PhysRevD.97.099902, https://doi.org/10.1103/PhysRevD.96.091101, arXiv:1702.03224 [hep-ex]
11. C. Hambrock, A. Khodjamirian, A. Rusov, Phys. Rev. D **92**(7), 074020 (2015). https://doi.org/10.1103/PhysRevD.92.074020, arXiv:1506.07760 [hep-ph]
12. A.J. Buras, J. Girrbach-Noe, C. Niehoff, D.M. Straub, JHEP **1502**, 184 (2015). https://doi.org/10.1007/JHEP02(2015)184, arXiv:1409.4557 [hep-ph]
13. N. Carrasco et al., [ETM Collaboration], JHEP **1403**, 016 (2014). https://doi.org/10.1007/JHEP03(2014)016, arXiv:1308.1851 [hep-lat]

Chapter 57
Reconciling B-meson Anomalies, Neutrino Masses and Dark Matter

Girish Kumar, Chandan Hati, Jean Orloff and Ana M. Teixeira

Abstract We explore the connection of the leptoquark solution to the recently reported B-meson anomalies with a mechanism of neutrino mass generation and a viable dark matter candidate. We consider a model consisting of two scalar leptoquarks and three generations of triplet fermions: neutrino masses are radiatively generated at the 3-loop level and, by imposing a discrete Z_2 symmetry, one can obtain a viable dark matter candidate. We discuss the constraints on the flavour structure of this model arising from numerous flavour observables. The rare decay $K \to \pi^+ \nu\bar{\nu}$ and charged lepton flavour violating $\mu - e$ conversion in nuclei are found to provide the most stringent constraint on this class of models.

57.1 Introduction

Following the observation of the Higgs boson at the LHC, which was the last missing piece of the standard model (SM) to be discovered, a strong effort is being made to carry out tests of the SM, and unveil the presence of new physics (NP). The observation of neutrino oscillations and the cosmological evidence for dark matter in the Universe are among some of the most important reasons to believe that SM is a low-energy limit of a more fundamental theory, realised at some unknown high scale. Recently, the observed lepton flavour universality violation (LFUV) in B-decays has provided new hints of NP. In particular, measurements of the following ratios show significant deviations from the SM predictions

---

G. Kumar (✉)
Tata Institute of Fundamental Research, Homi Bhabha Rd., Mumbai 400005, India
e-mail: girishk@theory.tifr.res.in

C. Hati · J. Orloff · A. M. Teixeira
Laboratoire de Physique de Clermont, CNRS/IN2P3 - UMR 6533,
4 Avenue Blaise Pascal, 63178 Aubière, France

A. Giri and R. Mohanta (eds.), *16th Conference on Flavor Physics and CP Violation*, Springer Proceedings in Physics 234,
https://doi.org/10.1007/978-3-030-29622-3_57

$$R_{K^{(*)}} = \frac{\mathrm{BR}(B \to K^{(*)} \mu^+ \mu^-)}{\mathrm{BR}(B \to K^{(*)} e^+ e^-)} \text{ and } R_{D^{(*)}}^{\tau/\ell} = \frac{\Gamma(B \to D^{(*)} \tau^- \bar{\nu})}{\Gamma(B \to D^{(*)} \ell^- \bar{\nu})}; \; \ell = e, \mu . \tag{57.1}$$

The measured values of $R_{K^{(*)}}$, and the corresponding predictions in the SM in low dilepton invariant mass squared q^2 bins are as follows [1–3]

$$\begin{aligned} R_{K[1,6]} &= 0.745 \pm^{0.090}_{0.074} \pm 0.036, \quad R_K^{\mathrm{SM}} = 1.00, \pm 0.01 \\ R_{K^*[0.045,1.1]} &= 0.66^{+0.11}_{-0.07} \pm 0.03 , \quad R^{\mathrm{SM}}_{K^*[0.045,1.1]} \sim 0.92 \pm 0.02 \\ R_{K^*[1.1,6]} &= 0.69^{+0.11}_{-0.07} \pm 0.05 , \quad R^{\mathrm{SM}}_{K^*[1.1,6]} \sim 1.00 \pm 0.01 , \end{aligned} \tag{57.2}$$

which respectively correspond to 2.6σ, 2.4σ and 2.5σ deviations from the theoretical expectations for $R_{K[1,6]}$, $R_{K^*[0.045,1.1]}$ and $R_{K^*[1.1,6]}$. On the other hand, experimental values for LFUV ratios $R_{D^{(*)}}^{\tau/\ell}$ are larger than the predictions in the SM. The current experimental world averages are $R_D^{\tau/\ell} = 0.407 \pm 0.039 \pm 0.024$ and $R_{D^*}^{\tau/\ell} = 0.306 \pm 0.013 \pm 0.007$, corresponding to a 2.3σ and 3.0σ excess over the respective SM values [4]. If combined together, the discrepancy is at the level of 3.8σ. Interestingly, measurements of $R_{D^{(*)}}^{e/\mu}$ (related to e/μ modes) do not show any sign of LFUV, and are consistent with the SM predictions. Here, we discuss an extension of the SM by two scalar triplet leptoquarks h_1, h_2 and three generation of triplet fermions Σ_R^i. The corresponding quantum charges under the SM gauge group $\mathrm{SU(3)}_C \times \mathrm{SU(2)}_L \times \mathrm{U(1)}_Y$ and an additional discrete symmetry Z_2 (which ensures the stability of the dark matter candidate) are listed in Table 57.1. As we proceed to discuss, this model has the potential to explain the neutral B-meson decay anomalies, neutrino oscillation data, and also provide a suitable dark matter candidate. We also discuss possible contributions to several flavour observables such as neutral meson mixing, rare meson decays, and charged lepton flavour violating (cLFV) decays, and study the constraints arising from these processes on the model. For a detailed analysis we refer to the original work [5].

The relevant interactions in the Lagrangian are

$$\begin{aligned} \mathcal{L} = \mathcal{L}_{\mathrm{SM}} &+ y_{ij} \, \bar{Q}_L^{C\,i} \, i\tau^2 \, (\boldsymbol{\tau}.\mathbf{h}_1) \, L_L^j + \tilde{y}_{ij} \, \overline{(\boldsymbol{\tau}.\boldsymbol{\Sigma})_R}^{C\,i,ab} [i\tau^2 \, (\boldsymbol{\tau}.\mathbf{h}_2) \, \epsilon^T]^{ba} \, d_R^j \\ &- \frac{1}{2} \overline{\Sigma^C}^i \, M_{ij}^{\Sigma} \, \Sigma^j - V_{\mathrm{scalar}}^{H,h} + \mathrm{H.c.} , \end{aligned} \tag{57.3}$$

Table 57.1 New fields (in addition to the usual SM fields) and their transformation under the SM gauge group and discrete symmetry Z_2. All the SM fields are even under Z_2

	Field	$\mathrm{SU(3)}_C \times \mathrm{SU(2)}_L \times \mathrm{U(1)}_Y$	Z_2
Fermions	Σ_R	$(\mathbf{1}, \mathbf{3}, 0)$	-1
Scalars	H	$(\mathbf{1}, \mathbf{2}, 1/2)$	1
	h_1	$(\bar{\mathbf{3}}, \mathbf{3}, 1/3)$	1
	h_2	$(\bar{\mathbf{3}}, \mathbf{3}, 1/3)$	-1

where Q_L is the left-handed quark doublet, L_L is the left-handed lepton doublet and d is the right-handed singlet down-type quark. Here i, j are labels for generations, a, b denote SU(2) indices and τ^c ($c = 1, 2, 3$) are the Pauli matrices. In writing the above Lagrangian, we have forbidden the di-quark interaction of h_1 to be consistent with proton decay bounds. The first term in the second line is the mass term for fermion triplets Σ_R^i, and $V_{\rm scalar}^{H,h}$ is the scalar potential. The full expression of the scalar potential can be found in [5]. In particular, it contains a quadratic term $(\lambda_h/4)\,{\rm Tr}(h_1^\dagger\, h_2\, h_1^\dagger\, h_2)$ which is instrumental for generating neutrino masses in this model.

57.2 Radiative Neutrino Masses and Dark Matter

In this model, the Z_2 symmetry forbids any tree level realisation of the conventional seesaw mechanism. However, neutrino masses are generated radiatively and the leading contribution arises at three loop level. After calculating all the relevant loop diagrams, one obtains

$$(m_\nu)_{\alpha\beta} \;=\; -30\,\frac{\lambda_h}{\left(4\pi^2\right)^3\, m_{h_2}}\; y_{\alpha i}^T\, m_{D_i}\, \tilde{y}_{ij}^T\, G\left(\frac{m_{\Sigma_j}^2}{m_{h_2}^2},\, \frac{m_{h_1}^2}{m_{h_2}^2}\right)\, \tilde{y}_{jk}\, m_{D_k}\, y_{k\beta}, \qquad (57.4)$$

where y and $\tilde{y}$ are Yukawa couplings which have been defined in (57.3), m_D denotes the diagonal down-quark mass matrix and $G(a, b)$ is a loop function given in [5]. The charged leptons are assumed to be in the physical basis and the Pontecorvo-Maki-Nakagawa-Sakata (PMNS) matrix, U_{ij}, parametrises the misalignment between flavour and mass eigenstates of the three SM neutrinos. The diagonal neutrino mass matrix is obtained using the relation $m_\nu^{\rm diag} = U_{i\alpha}^T\,(m_\nu)_{\alpha\beta}\,U_{\beta i}$.

A numerical analysis of neutrino oscillation data can be performed using a modified Casas-Ibarra parametrisation [5] to obtain one of the Yukawa couplings in terms of other Yukawa coupling as

$$\tilde{y} \;=\; [F(\lambda_h, m_\Sigma, m_{h_{1,2}})]^{-1/2}\, \mathcal{R}\, \sqrt{m_\nu^{\rm diag}}\; U^\dagger y^{-1} m_d^{-1}\,. \qquad (57.5)$$

Here $\mathcal{R}$ is an arbitrary complex orthogonal matrix and the function $F(\lambda_h, m_\Sigma, m_{h_{1,2}})$ can be found in [5].

Since all the SM fields are Z_2-even, the lightest Z_2-odd particle is stable in this model. Noting that the electroweak radiative corrections render the charged component slightly heavier than the neutral ones, the neutral component (Σ^0) of the lightest lepton triplet (Σ_R^1) emerges as a dark matter candidate. The relic density, predominantly governed by the gauge interactions, can be obtained as

$$\Omega\, h^2 \;\simeq\; \frac{1.07\times 10^9\, x_f}{g_*^{1/2}\, M_{\rm Pl}({\rm GeV})\, \langle\sigma_{\rm eff}|\bar{v}|\rangle}, \qquad (57.6)$$

where $x_f \equiv m_\Sigma/T_f$ is the freeze-out temperature and $\langle\sigma_{\rm eff}|\bar{v}|\rangle$ is obtained by including the gauge coannihilation of the charged and neutral components of Σ_R^1 as described in [5]. The numerical analysis of the relic abundance yields a rough benchmark limit for the Σ_R^1 mass given by 2.425 TeV $< m_\Sigma <$ 2.465 TeV to obtain the correct $\Omega h^2 = 0.1186 \pm 0.0020$. This can in turn be used as a first constraint on the parameter space of the model.

57.3 *B*-meson Decay Anomalies

The $B \to K^{(*)}\ell^+\ell^-$ decays are described by the low-energy effective Hamiltonian [6]

$$\mathcal{H}_{\rm eff} = -\frac{4G_F}{\sqrt{2}}\frac{\alpha_{em}}{4\pi}V_{tb}V_{ts}^*\left[C_9\,(\bar{s}_L\gamma^\mu b_L)(\bar{\ell}\gamma_\mu\ell) + C_{10}\,(\bar{s}_L\gamma^\mu b_L)(\bar{\ell}\gamma_\mu\gamma_5\ell)\right] + \text{H.c.}\,, \tag{57.7}$$

where only effective operators relevant for addressing LFUV in B-decays have been kept. New contributions to Wilson coefficients (WC) C_9 and C_{10} are given by

$$C_9^{\ell\ell'} = -C_{10}^{\ell\ell'} = \frac{\pi\, v^2}{\alpha_e\, V_{tb}\, V_{ts}^*}\,\frac{y_{b\ell'}\, y_{s\ell}^*}{m_{h_1}^2}\,. \tag{57.8}$$

where v is the Higgs vacuum expectation value. Model-independent studies to explain LFUV in $b \to s\ell^+\ell^-$ decays advocate NP in left-handed currents, modifying C_9 and C_{10} [6]. Adapting the NP solutions obtained in [6] to our case, we obtain $-0.7 \lesssim \text{Re}[C_{9,\rm NP}^{\mu\mu} - C_{9,\rm NP}^{ee}] \lesssim -0.4$, which translates to the following constraint (at 1σ)

$$0.64\times 10^{-3} \lesssim \frac{\text{Re}[y_{b\mu}\, y_{s\mu}^* - y_{be}\, y_{se}^*]}{(m_{h_1}/1\text{TeV})^2} \lesssim 1.12\times 10^{-3}\,. \tag{57.9}$$

In Fig. 57.2 (left plot) we show the parameter space in the plane of the Yukawa couplings $y_{b\mu}\, y_{s\mu}^*$ and $y_{be}\, y_{se}^*$ satisfying the above condition to explain the R_K and R_{K^*} anomalies. We note that in this model the h_1 leptoquark also contributes to $b \to c\ell^-\bar{\nu}$ at tree level. After integrating out the h_1 leptoquark, we obtain the following effective NP Hamiltonian

$$\mathcal{H}_{\rm eff}(b \to c\ell\bar{\nu}_i) = -\frac{v^2\,(yU)_{3i}\,(Vy^*)_{2\ell}}{4\,V_{cb}\,m_{h_1}^2}\,(\bar{c}_L\gamma^\mu b_L)\left(\bar{\ell}_L\gamma_\mu\nu_{Li}\right) + \text{H.c.} \tag{57.10}$$

Once various flavour constraints are taken into account, one finds SM-like R_D and R_{D^*} in this model. The LFUV ratios $R_{D^{(*)}}^{e/\mu}$, reflecting μ/e universality, are also SM-like, and consistent with current data.

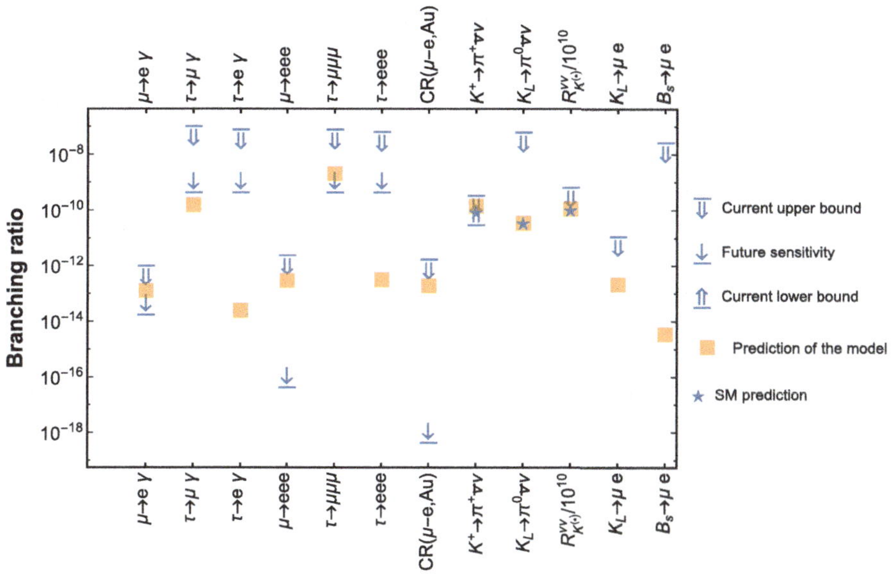

Fig. 57.1 Predictions of various processes for the flavour texture of the h_1 leptoquark, as discussed in the text. The relevant information on the corresponding SM prediction and experimental data is also shown. For details see [5]

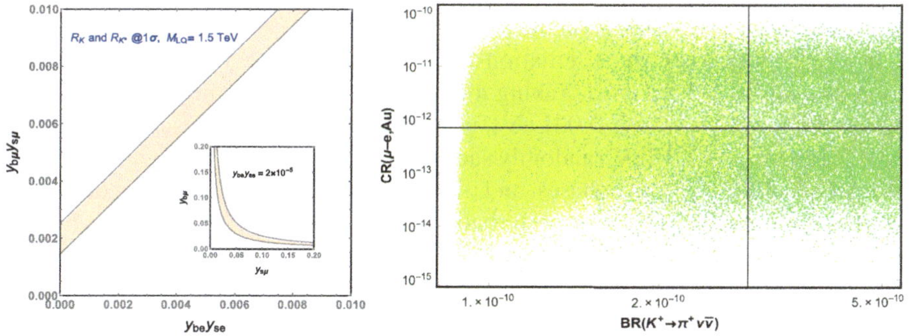

Fig. 57.2 The left plot shows the parameter space explaining R_K and R_{K^*} anomalies. In the right plot we show the allowed solutions from the neutrino oscillation data in the plane of BR($K^+ \to \pi^+ \nu\bar{\nu}$) and $\mu - e$ conversion rate for gold nuclei (for textures similar to (57.11)). The yellow(green) points correspond to non-perturbative (perturbative) Yukawa couplings $\tilde{y}$

57.4 Constraints from Flavour Data

We note that the simplest explanation of $b \to s\mu^+\mu^-$ anomalies requires nonzero values of the muon couplings $y_{b\mu} y^*_{s\mu}$ complying with (57.9). However, as evident from the relation in (57.5), accommodating neutrino oscillation data implies a non-trivial flavour structure of the leptoquark h_1 Yukawa couplings, thereby inducing new contributions to a plethora of flavour processes. One can consider generic parametrisations of y in terms of powers of a small parameter ϵ, with each entry weighed by an $\mathcal{O}(1)$ real coefficient a_{ij}: $y_{ij} = a_{ij} \odot \epsilon^{n_{ij}}$, with $\odot$ denoting that there is no summation implied over i, j. After incorporating constraints from all the flavour violating processes, one can identify the allowed flavour textures which simultaneously explain $R_{K^{(*)}}$ while being consistent with all the constraints from flavour violating processes. Among the several textures analysed in [5], one example of an allowed texture for the Yukawa couplings is given by

$$y \simeq \begin{pmatrix} \epsilon^4 & \epsilon^5 & \epsilon^2 \\ \epsilon^3 & \epsilon^3 & \epsilon^4 \\ \epsilon^4 & \epsilon & \epsilon \end{pmatrix}, \tag{57.11}$$

with $\epsilon \sim 0.215$ (for a leptoquark mass $m_{h_1} = 1.5$ TeV). The value of the parameter ϵ is obtained from the $R_{K^{(*)}}$ anomaly constraint given in (57.9), by setting the product $y_{b\mu} y^*_{s\mu} \sim \epsilon^4$ (with $\epsilon \sim \mathcal{O}(1)$). In Fig. 57.1 we show the predictions for some of the most important flavour violating processes for the texture given in (57.11). A complete analysis suggests that $K^+ \to \pi^+ \nu\bar{\nu}$ and $\mu - e$ conversion in nuclei provide the most stringent constraints on the textures.

In what concerns neutrino oscillation data, the best-fit values from the global oscillation analysis of [7] are used. Taking a normal ordering with the lightest neutrino mass in the range $[10^{-8}$ eV, 0.001 eV$]$, using the benchmark values $m_\Sigma = 2.45$, 3.5 and 4.5 TeV, $m_{h_2} = 2.7$ TeV, randomly sampling the three complex angles in $\mathcal{R}$ from the intervals: $[0, 2\pi]$ for the phases, and $[-4\pi, 4\pi]$ for the angles, one can obtain the couplings $\tilde{y}$ for different y (for perturbative regimes $\lambda_h \lesssim 4\pi$). Finally, each entry of the couplings must comply with perturbativity requirements, $|y(\tilde{y})| \lesssim 4\pi$ and the various constraints from flavour violating processes must be satisfied. In Fig. 57.2 (right plot) we show the results of the scan, for the texture given in (57.11) in the plane of neutrinoless conversion in nuclei and the $K^+ \to \pi^+ \nu\bar{\nu}$ decay; the colour code distinguishes between perturbative and non-perturbative entries of the y and $\tilde{y}$ couplings.

57.5 Summary

We discussed a model containing two leptoquarks h_1 and h_2 and three Majorana triplets in addition to the SM fields. Neutrino masses are radiatively generated at the 3-loop level, and the neutral component of the lightest triplet is a viable dark matter candidate. The model allows to accommodate the observed anomalies in $B \to K^{(*)}\ell^+\ell^-$ decays, R_K and R_{K^*}. We discuss contributions to low energy flavour observables and cLFV processes and identify processes which provide the tightest constraints on this class of models: these turn out to be $\mu - e$ conversion in nuclei and $K^+ \to \pi^+\nu\bar{\nu}$ decays.

References

1. R. Aaij et al., [LHCb Collaboration], Phys. Rev. Lett. **113**, 151601 (2014). https://doi.org/10.1103/PhysRevLett.113.151601, arXiv:1406.6482 [hep-ex]
2. R. Aaij et al., [LHCb Collaboration], JHEP **1708**, 055 (2017). https://doi.org/10.1007/JHEP08(2017)055, arXiv:1705.05802 [hep-ex]
3. M. Bordone, G. Isidori, A. Pattori, Eur. Phys. J. C **76**(8), 440 (2016). https://doi.org/10.1140/epjc/s10052-016-4274-7, arXiv:1605.07633 [hep-ph]
4. Y. Amhis et al., [HFLAV Collaboration], Eur. Phys. J. C **77**(12), 895 (2017). https://doi.org/10.1140/epjc/s10052-017-5058-4, arXiv:1612.07233 [hep-ex]; online updates are available at the webpage : https://hflav.web.cern.ch/
5. C. Hati, G. Kumar, J. Orloff, A.M. Teixeira, arXiv:1806.10146 [hep-ph]
6. G. Hiller, I. Nisandzic, Phys. Rev. D **96**(3), 035003 (2017). https://doi.org/10.1103/PhysRevD.96.035003, arXiv:1704.05444 [hep-ph]
7. P.F. de Salas, D.V. Forero, C.A. Ternes, M. Tortola, J.W.F. Valle, Phys. Lett. B **782**, 633 (2018). https://doi.org/10.1016/j.physletb.2018.06.019, arXiv:1708.01186 [hep-ph]

Chapter 58
Ultraviolet Complete Leptoquark Scenario Addressing the *B* Physics Anomalies

Nejc Košnik, Damir Bečirević, Ilja Doršner, Svjetlana Fajfer, Darius A. Faroughy and Olcyr Sumensari

Abstract We present the model with two light scalar leptoquarks which are able to explain the currently observed deviations from the SM predictions in LFU measurements in B meson decays. The two leptoquarks, $S_3(\bar{3}, 3, 1/3)$ and $R_2(3, 2, 7/6)$, accommodate both $R_{K^{(*)}}$ and $R_{D^{(*)}}$ and are in accord with other flavor observables, as well as with Z-pole measurements and direct searches at the LHC. The two LQ states are part of a realistic Grand Unified Theory that remains perturbative to the scale of unification. The predictions of this scenario are enhanced lepton flavor violation signals, $B \to K^{(*)}\nu\nu$ decays as well as observable signals in leptoquark production at the LHC.

58.1 Introduction

Lepton flavor universality (LFU) ratios are among the most useful observables to test the flavor sector of the Standard Model (SM). Recently, several experiments measured the following LFU ratios $R_{D^{(*)}} = \mathcal{B}(B \to D^{(*)}\tau\bar{\nu}_\tau)/\mathcal{B}(B \to D^{(*)}l\bar{\nu}_l), l = e, \mu$, whose values are larger than $R^{\rm SM}_{D^{(*)}}$. The value of R_D [1–3] is $\sim 2\sigma$ above the

N. Košnik (✉) · S. Fajfer · D. A. Faroughy
Jožef Stefan Institute, Jamova 39, 1001 Ljubljana, Slovenia
e-mail: nejc.kosnik@ijs.si

N. Košnik · S. Fajfer
Faculty of Mathematics and Physics, University of Ljubljana, Jadranska 19, 1000 Ljubljana, Slovenia

D. Bečirević
Laboratoire de Physique Théorique (Bât. 210), Université Paris-Sud and CNRS, 91405 Orsay-Cedex, France

I. Doršner
University of Split, Faculty of Electrical Engineering, Mechanical Engineering and Naval Architecture in Split (FESB), Ruđera Boškovića 32, 21 000 Split, Croatia

O. Sumensari
Dipartimento di Fisica e Astronomia "G. Galilei", Università di Padova, Padova, Italy

Istituto Nazionale Fisica Nucleare, Sezione di Padova, 35131 Padova, Italy

A. Giri and R. Mohanta (eds.), *16th Conference on Flavor Physics and CP Violation*, Springer Proceedings in Physics 234,
https://doi.org/10.1007/978-3-030-29622-3_58

SM prediction, while R_{D^*} is $\sim 3\sigma$ larger than SM prediction [4]. On the other hand, the LHCb experiment has measured LFU ratios $R_{K^{(*)}} = \mathcal{B}(B \to K^{(*)}\mu\mu)/\mathcal{B}(B \to K^{(*)}ee)$ driven by the neutral-current process $b \to sll$ and found them to be below the expected values in the SM [5, 6]. The significances of $R_{K^{(*)}}$ deviations from the SM predictions are at $\sim$2.5 σ level [7, 8].

The effective theory approach shows that one or more TeV scale mediators which couple to left-handed currents with predominantly third generation fit well with the LFU data. Among the most prominent New Physics (NP) mediators affecting the LFU observables are leptoquarks (LQs). In the effective theory approach, it has been shown that particular vector LQ can generate suitable $V - A$ operators for the LFU anomalies that satisfy both low-energy and high-p_T constraints. An alternative approach is to introduce several mediators, where the low energy $V-A$ structure can be generated by integrating out two scalar LQs. One can also explore two scalar LQs scenario even when they are known to generate operators with Lorentz structures other than $V-A$ for $R_{D^{(*)}}$ (see [4] and references therein).

We propose an ultraviolet complete model that is based on the $SU(5)$ Grand Unified Theory (GUT) and contains two light scalar LQs that address both LFU anomalies. The light LQs are $R_2(\mathbf{3}, \mathbf{2}, 7/6)$ and $S_3(\overline{\mathbf{3}}, \mathbf{3}, 1/3)$, where the numbers in brackets denote their representation under the SM gauge group. At low energies, R_2 generates a combination of scalar and tensor effective operators that accommodate $R_{D^{(*)}}$, while S_3 generates a $V-A$ operator which fits well the observed $R_{K^{(*)}}$. The Yukawa couplings of the leptoquarks are not independent since their origin is from a common $SU(5)$ contraction. We take into account all relevant flavor constraints and find that the preferred region of the parameter space resolves both LFU anomalies and is compatible with direct searches at the LHC. In the following we present our model outline, discuss the low-energy phenomenology, present the results and predictions, and finally we conclude.

58.2 Model Outline

The interactions of R_2 and S_3 with the SM fermions in the mass eigenstate basis are [4]

$$\begin{aligned}\mathcal{L} \supset &+ (VY_R E_R^\dagger)^{ij} \bar{u}_{Li} \ell_{Rj} R_2^{\frac{5}{3}} + (Y_R E_R^\dagger)^{ij} \bar{d}_{Li} \ell_{Rj} R_2^{\frac{2}{3}} + (U_R Y_L U)^{ij} \bar{u}_{Ri} \nu_{Lj} R_2^{\frac{2}{3}} \\ &- (U_R Y_L)^{ij} \bar{u}_{Ri} \ell_{Lj} R_2^{\frac{5}{3}} + (Y_L U)^{ij} \bar{d}^C_{Li} \nu_{Lj} S_3^{\frac{1}{3}} - 2^{\frac{1}{2}} (V^* Y_L U)^{ij} \bar{u}^C_{Li} \nu_{Lj} S_3^{-\frac{2}{3}} \\ &+ 2^{\frac{1}{2}} Y_L^{ij} \bar{d}^C_{Li} \ell_{Lj} S_3^{\frac{4}{3}} + (V^* Y_L)^{ij} \bar{u}^C_{Li} \ell_{Lj} S_3^{\frac{1}{3}}.\end{aligned} \quad (58.1)$$

Here Y_L, Y_R are Yukawa matrices, $R_2^{(Q)}$ and $S_3^{(Q)}$ are charge-Q eigenstates of leptoquarks. The unitary matrices $U_{L,R}$, $D_{L,R}$, $E_{L,R}$, and N_L rotate between mass and gauge basis of up-type quarks, down-type quarks, charged leptons, and neutrinos, respectively. $V \equiv U_L D_L^\dagger = U_L$ is the CKM matrix, $U \equiv E_L N_L^\dagger = N_L^\dagger$ is the PMNS matrix. We assume the following structure for the Yukawa matrices

$$Y_R E_R^\dagger = \begin{pmatrix} 0 & 0 & 0 \\ 0 & 0 & 0 \\ 0 & 0 & y_R^{b\tau} \end{pmatrix}, \quad U_R Y_L = \begin{pmatrix} 0 & 0 & 0 \\ 0 & y_L^{c\mu} & y_L^{c\tau} \\ 0 & 0 & 0 \end{pmatrix}, \tag{58.2}$$

where U_R is a rotation by angle θ between second and third generation. Free parameters of the model are thus m_{R_2}, m_{S_3}, $y_R^{b\tau}$, $y_L^{c\mu}$, $y_L^{c\tau}$, and θ.

Without relying on a particular UV completion of the model or some underlying symmetry, there is no reason for the Yukawa couplings of S_3 to be related to the ones of R_2. In our case we consider two leptoquarks within the $SU(5)$ based unification model where the scalar sector contains representations of dimensionality $\underline{\mathbf{45}}$ and $\underline{\mathbf{50}}$ whereas the SM fermions reside in $\overline{\mathbf{5}}_i$ and $\mathbf{10}_i$, with $i(=1,2,3)$ counting generations. All the low-energy operators of (58.1) can be generated with $SU(5)$ contractions $a^{ij}\mathbf{10}_i\overline{\mathbf{5}}_j\overline{\underline{\mathbf{45}}}$, and $b^{ij}\mathbf{10}_i\mathbf{10}_j\underline{\mathbf{50}}$, where a and $b(=b^T)$ are matrices in the flavor space. The former contraction couples $R_2 \in \underline{\mathbf{45}}$ ($S_3 \in \underline{\mathbf{45}}$) with the right-handed up-type quarks (quark doublets) and leptonic doublets, while the latter generates couplings of $R_2 \in \underline{\mathbf{50}}$ with the quark doublets and right-handed charged leptons. To break $SU(5)$ down to the SM we use scalar representation $\underline{\mathbf{24}}$ and write a term in the scalar potential $m\,\underline{\mathbf{45}}\,\overline{\underline{\mathbf{50}}}\,\underline{\mathbf{24}}$. The two R_2 leptoquarks that reside in $\underline{\mathbf{45}}$ and $\underline{\mathbf{50}}$ then mix and allow us to have one light R_2 and one heavy R_2 in the spectrum, where the latter state completely decouples from the low-energy spectrum for large values of m [4].

LQs are potentially dangerous for proton decay, in particular if they couple to diquarks. The S_3 leptoquark would not couple to the diquark if $SU(5)$ contraction $c^{ij}\mathbf{10}_i\mathbf{10}_j\underline{\mathbf{45}}$ was forbidden or suppressed. Furthermore, S_3 must not mix with any other LQ with diquark couplings. Both conditions can be met in a generic $SU(5)$ framework [9].

58.3 LFU Puzzles and Related Flavor Constraints

In the presented model the R_2 LQ accommodates the $R_{D^{(*)}}$ anomaly by inducing scalar and tensor Wilson coefficients, whereas contribution of S_3 to the vector operator to the effective Lagrangian is negligible:

$$\begin{aligned} \mathcal{L}_{\text{eff}}^{d\to u\ell\bar{\nu}} = -\frac{4\,G_F}{\sqrt{2}} V_{ud}\big[& (1+g_{V_L})(\bar{u}_L\gamma_\mu d_L)(\bar{\ell}_L\gamma^\mu\nu_L) \\ & + g_{S_L}(\mu)\,(\bar{u}_R d_L)(\bar{\ell}_R\nu_L) \\ & + g_T(\mu)\,(\bar{u}_R\sigma_{\mu\nu}d_L)(\bar{\ell}_R\sigma^{\mu\nu}\nu_L)\big]. \end{aligned} \tag{58.3}$$

At the matching scale the Wilson coefficients are related to the LQ couplings:

$$g_{S_L}(\Lambda) = 4\,g_T(\Lambda) = \frac{y_L^{u\ell'}\,y_R^{d\ell *}}{4\sqrt{2}\,m_{R_2}^2\,G_F V_{ud}}. \tag{58.4}$$

From the above equations we learn that the only charged current transition affected by R_2 in our scenario is $b \to c\tau\bar{\nu}_\ell$. To compute R_D we have employed lattice QCD results for the $B \to D$ form factors, whereas for R_{D^*} we consider the form factors extracted from the $B \to D^* l\bar{\nu}$ $(l = e, \mu)$.

The $R_{K^{(*)}}$ anomaly is accounted for through purely left-handed tree-level contributions of S_3 to the vector and axial-vector Wilson coefficients [10]

$$\delta C_9^{\mu\mu} = -\delta C_{10}^{\mu\mu} = \frac{\pi v^2}{\lambda_t \alpha_{\rm em}} \frac{\sin 2\theta \, (y_L^{c\mu})^2}{2m_{S_3}^2} \,. \tag{58.5}$$

The mixing factor $\sin 2\theta$, originating from the matrix U_R, plays an important role in suppressing effect in $R_{K^{(*)}}$ relative to the one in $R_{D^{(*)}}$. The 1σ interval $C_9^{\mu\mu} = -C_{10}^{\mu\mu} \in (-0.85, -0.50)$ has been obtained by performing a fit to R_K, R_{K^*}, and $\mathcal{B}(B_s \to \mu\mu)$. The left-handed (weak triplet) nature of the S_3 LQ imply contributions to both neutral and charged current semileptonic processes. Among the charged current observables the LFU ratios $R_{D^{(*)}}^{\mu/e} = \mathcal{B}(B \to D^{(*)}\mu\bar{\nu})/\mathcal{B}(B \to D^{(*)}e\bar{\nu})$ impose severe constraints on S_3 couplings. We have also considered $\mathcal{B}(B \to \tau\bar{\nu})$ and the kaon LFU ratio $R_{e/\mu}^K = \Gamma(K^- \to e^-\bar{\nu})/\Gamma(K^- \to \mu^-\bar{\nu})$. In Sect. 58.4 we confront the predictions of $R_{\nu\nu}^{(*)}$ with experimental bounds, $R_{\nu\nu} < 3.9$ and $R_{\nu\nu}^* < 2.7$ [11]. The loop-induced neutral-current constraints affect both LQ's couplings. We have taken into account the $B_s - \bar{B}_s$ mixing frequency, which is modified by the S_3 box-diagram, proportional to $\sin^2 2\theta \left[(y_L^{c\mu})^2 + (y_L^{c\tau})^2\right]^2 / m_{S_3}^2$, the upper limit on lepton flavor violating τ decays $\mathcal{B}(\tau \to \mu\phi)$ and $\mathcal{B}(\tau \to \mu\gamma)$. The Z-boson couplings to leptons measured at LEP [12] are also modified at loop level by both LQs.

58.4 Fit to Low Energy Observables

Taking into account the aforementioned flavor observables we have performed a fit for parameters $y_R^{b\tau}$, $y_L^{c\mu}$, $y_L^{c\tau}$ and θ, while fixing the masses to $m_{R_2} = 0.8$ TeV and $m_{S_3} = 2$ TeV. The opposite sign of interference terms in R_D and R_{D^*} require complex Wilson coefficient g_{S_L} (Fig. 58.1), where we have put the complex phase in $y_{b\tau}^R$. The SM is excluded at 3.8σ, the best fit point provides a good agreement with $R_{D^{(*)}}$ and $R_{K^{(*)}}$. Note that the required large imaginary part in $y_R^{b\tau}$ could be in principle tested in the future experiments measuring neutron EDM [13]. The best fit point is consistent with the LHC constraints [4] superimposed in gray on the same plot.

The consistency of model with low energy data requires that $\mathcal{B}(B \to K\mu\tau)$ is bounded and at 1σ we obtain $1.1 \times 10^{-7} \lesssim \mathcal{B}(B \to K\mu^\pm\tau^\mp) \lesssim 6.5 \times 10^{-7}$, whereas related decay modes are predicted to be $\mathcal{B}(B \to K^*\mu\tau) \approx 1.9 \times \mathcal{B}(B \to K\mu\tau)$ and $\mathcal{B}(B_s \to \mu\tau) \approx 0.9 \times \mathcal{B}(B \to K\mu\tau)$. Another important prediction is a $\gtrsim 50\%$ enhancement of $\mathcal{B}(B \to K^{(*)}\nu\nu)$, which will be further tested at Belle-II. Remarkably, these two observables are highly correlated [4]. Furthermore, we derive a lower bound on $\mathcal{B}(\tau \to \mu\gamma)$, just below the current experimental limit, $\mathcal{B}(\tau \to \mu\gamma) \gtrsim 1.5 \times 10^{-8}$.

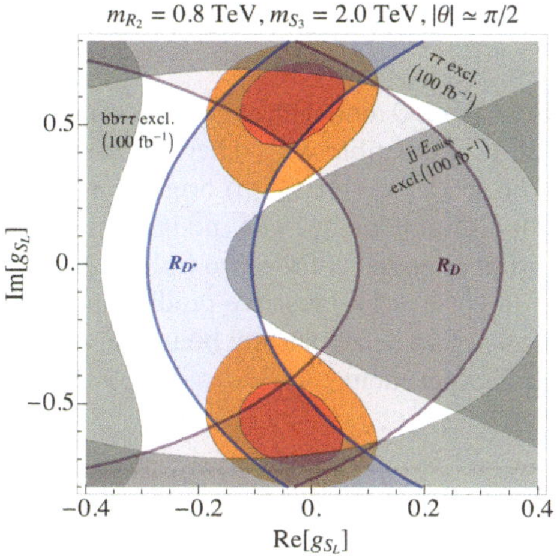

Fig. 58.1 Results of the flavor fit in the g_{S_L} plane, scalar coupling entering the transition $b \to c\tau\bar{\nu}_\tau$. The $1\,\sigma\,(2\,\sigma)$ fitted regions are rendered in red (orange). Separate constraints from R_D and R_{D^*} to $2\,\sigma$ accuracy are shown by the blue and purple regions, respectively. The LHC exclusions are depicted by the gray regions

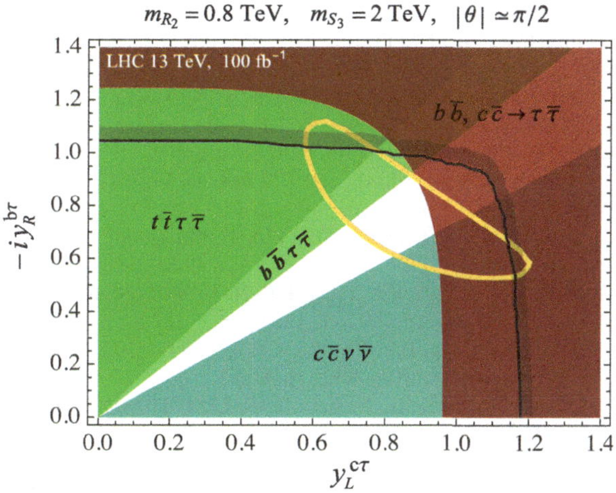

Fig. 58.2 Most important LHC limits for each LQ process at a projected luminosity of $100\,\text{fb}^{-1}$. The red region corresponds is excluded from the high-p_T di-tau search by ATLAS [14], the green and turquoise exclusion regions come from LQ pair production searches by CMS [15–17]. The region of Yukawa couplings above the black line is excluded due to their non-perturbative values below the GUT scale. The yellow contour denotes the $1\,\sigma$ region of the fit to the low-energy observables

Finally, our description of $R_{D^{(*)}}$ anomaly requires relatively light LQ states, not far from the TeV scale, and these LQs are necessarily accessible at the LHC, as we discuss in [4] (see also Fig. 58.2).

58.5 Conclusions

We have proposed a two scalar LQ model that accommodates the observed LFU ratios in B-meson decays, and is compatible with other low energy constraints and direct searches at the LHC. The model has an $SU(5)$ origin that relates Yukawa couplings of the two LQs through a mixing angle. The model remains perturbative up to the unification scale. We propose signals of the two light LQs at the LHC and spell out predictions for several flavor observables. We predict and correlate $\mathcal{B}(B \to K\mu\tau)$ with $\mathcal{B}(B \to K^{(*)}\nu\nu)$, as well as derive a lower bound for $\mathcal{B}(\tau \to \mu\gamma)$, which should be in reach of the Belle 2 experiment.

References

1. J.P. Lees et al., [BaBar Collaboration], Phys. Rev. D **88**(7), 072012 (2013). https://doi.org/10.1103/PhysRevD.88.072012, arXiv:1303.0571 [hep-ex]
2. S. Hirose et al., [Belle Collaboration], Phys. Rev. Lett. **118**(21), 211801 (2017). https://doi.org/10.1103/PhysRevLett.118.211801, arXiv:1612.00529 [hep-ex]
3. R. Aaij et al., [LHCb Collaboration], Phys. Rev. Lett. **115**(11), 111803 (2015) Erratum: [Phys. Rev. Lett. **115**(15), 159901 (2015)]. https://doi.org/10.1103/PhysRevLett.115.159901, https://doi.org/10.1103/PhysRevLett.115.111803, arXiv:1506.08614 [hep-ex]
4. D. Bečirević, I. Doršner, S. Fajfer, N. Košnik, D.A. Faroughy, O. Sumensari, Phys. Rev. D **98**(5), 055003 (2018). https://doi.org/10.1103/PhysRevD.98.055003, arXiv:1806.05689 [hep-ph]
5. R. Aaij et al., [LHCb Collaboration], Phys. Rev. Lett. **113**, 151601 (2014). https://doi.org/10.1103/PhysRevLett.113.151601, arXiv:1406.6482 [hep-ex]
6. R. Aaij et al., [LHCb Collaboration], JHEP **1708**, 055 (2017). https://doi.org/10.1007/JHEP08(2017)055, arXiv:1705.05802 [hep-ex]
7. G. Hiller, F. Kruger, Phys. Rev. D **69**, 074020 (2004). https://doi.org/10.1103/PhysRevD.69.074020 [hep-ph/0310219]
8. M. Bordone, G. Isidori, A. Pattori, Eur. Phys. J. C **76**(8), 440 (2016). https://doi.org/10.1140/epjc/s10052-016-4274-7, arXiv:1605.07633 [hep-ph]
9. I. Doršner, S. Fajfer, N. Košnik, Eur. Phys. J. C **77**(6), 417 (2017). https://doi.org/10.1140/epjc/s10052-017-4987-2, arXiv:1701.08322 [hep-ph]
10. I. Doršner, S. Fajfer, D.A. Faroughy, N. Košnik, JHEP **1710**, 188 (2017). https://doi.org/10.1007/JHEP10(2017)188, arXiv:1706.07779 [hep-ph]
11. J. Grygier et al., [Belle Collaboration], Phys. Rev. D **96**(9), 091101 (2017) Addendum: [Phys. Rev. D **97**(9), 099902 (2018)]. https://doi.org/10.1103/PhysRevD.97.099902, https://doi.org/10.1103/PhysRevD.96.091101, arXiv:1702.03224 [hep-ex]
12. S. Schael et al., [ALEPH and DELPHI and L3 and OPAL and SLD Collaborations and LEP Electroweak Working Group and SLD Electroweak Group and SLD Heavy Flavour Group], Phys. Rept. **427**, 257 (2006). https://doi.org/10.1016/j.physrep.2005.12.006 [hep-ex/0509008]
13. W. Dekens, J. De Vries, M. Jung, K.K. Vos, arXiv:1809.09114 [hep-ph]
14. M. Aaboud et al., [ATLAS Collaboration], JHEP **1801**, 055 (2018). https://doi.org/10.1007/JHEP01(2018)055, arXiv:1709.07242 [hep-ex]
15. A.M. Sirunyan et al., [CMS Collaboration], JHEP **1707**, 121 (2017). https://doi.org/10.1007/JHEP07(2017)121, arXiv:1703.03995 [hep-ex]
16. CMS Collaboration [CMS Collaboration], CMS-PAS-SUS-16-036
17. A.M. Sirunyan et al., [CMS Collaboration], Eur. Phys. J. C **78**, 707 (2018). https://doi.org/10.1140/epjc/s10052-018-6143-z, arXiv:1803.02864 [hep-ex]

Chapter 59
Interplay of $R_{D^{(*)}}$ and $Z\ell\ell$ in the Scalar Leptoquark Scenario

Monika Blanke, Marta Moscati and Ulrich Nierste

Abstract The deviations of the experimental values of $R_{D^{(*)}}$ from the Standard Model predictions hint at the presence of new physics in the B sector, and the existence of a scalar SU(2)-singlet leptoquark with hypercharge $-1/3$ has often been considered as explanation of this tension. The particle modifies the coupling of the charged leptons to the Z boson, and comparison with data from electroweak precision experiments severely constrains its parameter space. We evaluate the contribution of such a leptoquark to the charged leptons Z couplings including corrections of order $\mathcal{O}(M_Z^2/m_t^2)$ and electroweak renormalisation effects.

59.1 Introduction

In the recent years the charged current decay of the B meson $B \to D^{(*)}\ell\bar{\nu}_\ell$ showed tension with the Standard Model (SM) predictions in quantities testing lepton flavour universality. In the ratios of interest

$$R_{D^{(*)}} = \frac{\mathcal{B}(B \to D^{(*)}\tau\bar{\nu}_\tau)}{\mathcal{B}(B \to D^{(*)}\ell\bar{\nu}_\ell)}, \quad [\ell = e, \mu] \tag{59.1}$$

M. Blanke
Institut für Kernphysik, Karlsruhe Institute of Technology, Hermann-von-Helmholtz-Platz 1, 76344 Eggenstein-Leopoldshafen, Germany
e-mail: monika.blanke@kit.edu

M. Moscati (✉) · U. Nierste
Institut für Theoretische Teilchenphysik, Karlsruhe Institute of Technology, Engesserstraße 7, 76128 Karlsruhe, Germany
e-mail: marta.moscati@kit.edu

U. Nierste
e-mail: ulrich.nierste@kit.edu

A. Giri and R. Mohanta (eds.), *16th Conference on Flavor Physics and CP Violation*, Springer Proceedings in Physics 234,
https://doi.org/10.1007/978-3-030-29622-3_59

the dependence on the squared CKM element $|V_{cb}|^2$ and the corresponding uncertainty (see for instance [1–3] for recent reviews) drop out. In addition, the uncertainties originating from the form factors are significantly reduced [4–7].

Experimentally, the BaBar [8, 9] and Belle [10–13] collaborations measured R_D and R_{D^*} while the LHCb collaboration [14–16] measured R_{D^*}. Combining the experimental results and comparing to the theory predictions [17–20][1] the HFLAV collaboration [26] finds a tension at the level of $3.8\,\sigma$.

A scalar SU(2)-singlet leptoquark (LQ) with hypercharge $-1/3$, that contributes at tree-level to $B \to D^{(*)}\ell\bar{\nu}_\ell$, has often been discussed as candidate for explaining the anomaly [27–29]. In [27, 29] the authors already pointed out that the additional particle contributes at loop level to the coupling of charged leptons to the Z gauge boson. Consequently, requiring agreement with the electroweak (EW) precision measurements of these couplings severely constrains the parameter space, in some cases ruling out the LQ as explanation of the anomaly [29]. We reevaluate these contributions including terms up to order $\mathcal{O}(M_Z^2/m_t^2)$ and including EW renormalisation effects.

The rest of the proceeding is organised as follows. In Sect. 59.2 we introduce the simplified model extending the SM with the scalar LQ. In Sect. 59.3 we study the additional contributions to the Z couplings and analyse their implication on the explanation of the $R_{D^{(*)}}$ anomaly.

59.2 The Model

In analysing the model, we follow the notation of [27]. We introduce a scalar LQ transforming under the SM gauge group as $\phi \sim (3, 1, -1/3)$.[2] Following [27], we set to 0 all the ϕ-quark-quark couplings in order not to destabilise the proton.[3]

The part of the Lagrangian involving ϕ then reads

$$\mathcal{L}_\phi = (D_\mu\phi)^\dagger D^\mu\phi - M_\phi^2|\phi|^2 - g_{h\phi}|H|^2|\phi|^2 + \overline{Q}^c\lambda^L i\tau_2 L\phi^* + \overline{u}_R^c\lambda^R e_R\phi^* + \text{h.c.}, \tag{59.2}$$

where H is the Higgs doublet, $\lambda^{L,R}$ are matrices in flavour space, $\Psi^c = C\overline{\Psi}^T$ are charge-conjugate spinors, and τ_2 is the second Pauli matrix. This Lagrangian is written in the weak interaction basis. Switching to the mass basis for quarks and charged leptons, the couplings to the fermions take the form

---

[1]The predictions are consistent with the previous evaluations of R_D in [4, 5, 21–23] and of R_{D^*} in [6]. For a recent analysis of new physics contributions to $b \to c\tau\nu$ transitions in the effective theory, see [24], while recent discussions of long-distance electromagnetic effects in R_D can be found in [21, 25].

[2]The particle corresponds to the S_1 of [30].

[3]This assumption can be motivated with the introduction of an additional symmetry in the UV-complete model.

$$\mathscr{L}_\phi \supset \overline{u}_L^c \lambda_{ue}^L e_L \phi^* - \overline{d}^c \lambda_{d\nu}^L \nu_L \phi^* + \overline{u}_R^c \lambda_{ue}^R e_R \phi^* + \text{h.c.}, \tag{59.3}$$

where

$$\lambda_{ue}^L = U_u^T \lambda^L U_e, \quad \lambda_{d\nu}^L = U_d^T \lambda^L, \quad \lambda_{ue}^R = V_u^T \lambda^R V_e \tag{59.4}$$

and U_q (V_q) denote the rotations of the left-handed (right-handed) fermion fields. Since $V_{\text{CKM}} = U_u^\dagger U_d$, we have

$$V_{\text{CKM}}^T \lambda_{ue}^L = \lambda_{d\nu}^L U_e \tag{59.5}$$

We neglected the rotation matrix for the neutrino sector (and hence the PMNS matrix), therefore the neutrino fields are always expressed in the interaction basis. In this way, the fields $\nu_{e,\mu,\tau}$ will always correspond to the same admixture of mass eigenstates that arises from the W coupling in the SM, and hence interferences between SM and NP are possible. For instance, the matrix element $\lambda_{d\nu 33}^L$ corresponds to the interaction of a b quark (in the mass basis) with the ν_τ neutrino (in the interaction basis).

59.3 ϕ Contribution to $Z\ell\ell$ Coupling and Implications on $R_{D^{(*)}}$

The LQ contributes to the $Z\ell\ell$ coupling at one-loop level through the diagrams in Fig. 59.1. Additional contributions come from EW renormalisation effects, relating the Z coupling to the electromagnetic and W couplings. The latter receives the additional contribution from the diagram shown in Fig. 59.2. For what concerns the electromagnetic coupling, contributions coming from the diagrams in Fig. 59.1 as well as modifications of the vacuum polarisation of the photon, originating from the $A\phi\phi$ coupling as shown in Fig. 59.3, need to be taken into account.

Furthermore, in obtaining our result we kept corrections up to order $\mathscr{O}(M_Z^2/m_t^2)$.

The reduction of the tensor structure and the evaluation of the loop integrals have been performed with the Mathematica packages FeynCalc and PackageX [31–34].

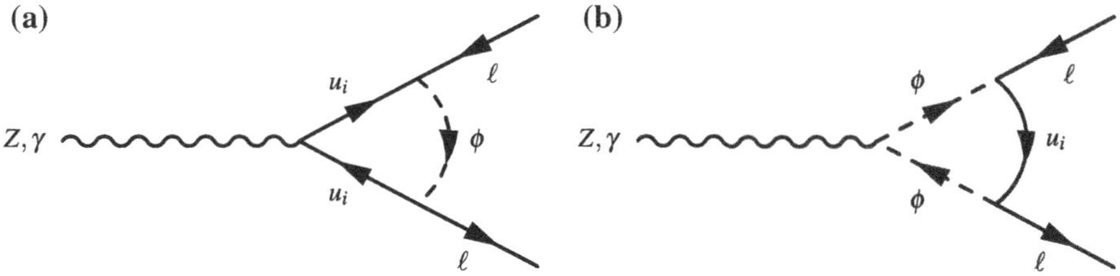

Fig. 59.1 Triangle diagrams contributing to the Z and EM coupling

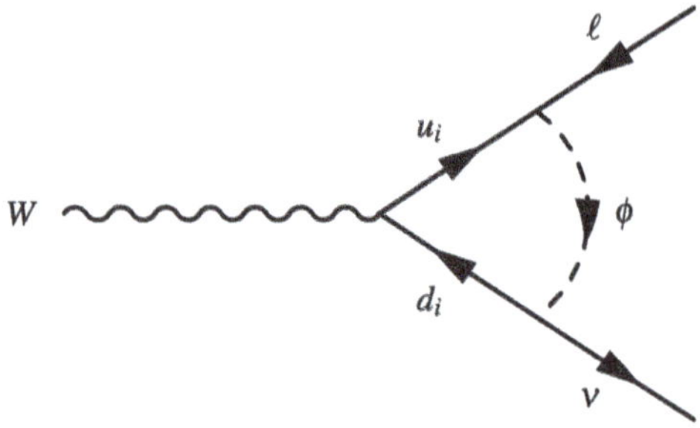

Fig. 59.2 Triangle diagram contributing to the W coupling

Fig. 59.3 Leptoquark contribution to the vacuum polarisation of the photon

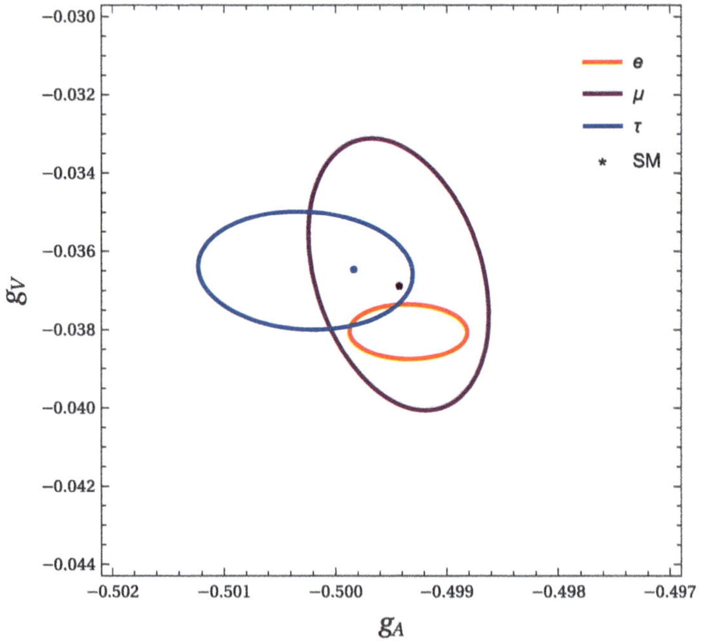

Fig. 59.4 Comparison of the modified $Z\ell\ell$ coupling with the measured couplings. The ellipses represent the 68% C.L. experimental regions while the black star represent the SM value (LFU). The coloured dots represent the prediction within the LQ model

A pedagogical explanation of how to treat vertices involving charge conjugation can be found in [35].

To analyse the interplay between the $Z\ell\ell$ vertex corrections and the explanation of the $R_{D^{(*)}}$ anomaly, we took as benchmark a LQ of mass $M_\phi = 1$ TeV having as nonvanishing couplings[4] $\lambda^L_{b\nu_\tau} = -1.3$ and $\lambda^R_{c\tau} = 0.3$. This point corresponds to the scenario $(x_{33}, y_{32}) = (1.3, 0.3)$ of [29], which is claimed to reduce the tension for the $R_{D^{(*)}}$ anomaly to within 1σ while passing the Z-couplings constraint for the τ lepton. We analyse here the effects on all the three charged lepton generations.

In Fig. 59.4 we plot the 68% C.L. experimental regions [36] in the plane $g_A - g_V$, together with the SM point obtained by fitting all the EW precision measurements, marked in black. The coloured dots represent the prediction of the model including the scalar LQ.

We observe that, due to the choice of the couplings, the τ receives the largest contribution, but the corresponding point still lies in the 68% experimental region. The μ and e couplings receive nonzero contributions from the photon vacuum polarisation through EW renormalisation, but these effects do not significantly change the consistency with experimental values.

59.4 Summary and Perspectives

In this work we improved the study of the effects of a scalar SU(2)-singlet leptoquark with hypercharge $-1/3$ on the coupling of charged leptons to the Z boson including corrections of order $\mathcal{O}(M_Z^2/m_t^2)$ and EW renormalisation effects. The role of the latter is particularly interesting since it implies a modification of all the three charged lepton couplings even when the structure of the couplings of the LQ to fermions suppresses its direct one-loop contribution through diagrams like the one shown in Fig. 59.1.

This observation could be extended to other LQs as well, for which all the three Z couplings would then provide a constraint even in scenarios in which some of the couplings are suppressed.

Acknowledgements M.M. acknowledges the support by the DFG-funded Doctoral School KSETA.

References

1. G. Ricciardi, Semileptonic and leptonic B decays, circa 2016. Mod. Phys. Lett. **A32**(05), 1730005 (2017)
2. B. Grinstein, Determination of V_{ub} and V_{cb} from semileptonic decays. PoS **HQL2016**, 061 (2017)
3. F. De Fazio, Theory overview of tree-level B decays. PoS, **EPS-HEP2017**, 210 (2017)

[4] A non-zero coupling matrix λ^L_{ue} is generated by SU(2) invariance and CKM rotations.

4. U. Nierste, S. Trine, S. Westhoff, Charged-Higgs effects in a new B → D tau nu differential decay distribution. Phys. Rev. **D78**, 015006 (2008)
5. J.F. Kamenik, F. Mescia, $B \to D\tau\nu$ branching ratios: opportunity for lattice QCD and hadron colliders. Phys. Rev. **D78**, 014003 (2008)
6. S. Fajfer, J.F. Kamenik, I. Nisandzic, On the $B \to D^*\tau\bar{\nu}_\tau$ sensitivity to new physics. Phys. Rev. **D85**, 094025 (2012)
7. R. Alonso, J. Martin Camalich, S. Westhoff, Tau properties in $B \to D\tau\nu$ from visible final-state kinematics. Phys. Rev. **D95**(9), 093006 (2017)
8. J.P. Lees et al., Evidence for an excess of $\bar{B} \to D^{(*)}\tau^-\bar{\nu}_\tau$ decays. Phys. Rev. Lett. **109**, 101802 (2012)
9. J.P. Lees et al., Measurement of an excess of $\bar{B} \to D^{(*)}\tau^-\bar{\nu}_\tau$ decays and implications for charged Higgs bosons. Phys. Rev. **D88**(7), 072012 (2013)
10. M. Huschle et al., Measurement of the branching ratio of $\bar{B} \to D^{(*)}\tau^-\bar{\nu}_\tau$ relative to $\bar{B} \to D^{(*)}\ell^-\bar{\nu}_\ell$ decays with hadronic tagging at Belle. Phys. Rev. **D92**(7), 072014 (2015)
11. Y. Sato et al., Measurement of the branching ratio of $\bar{B}^0 \to D^{*+}\tau^-\bar{\nu}_\tau$ relative to $\bar{B}^0 \to D^{*+}\ell^-\bar{\nu}_\ell$ decays with a semileptonic tagging method. Phys. Rev. **D94**(7), 072007 (2016)
12. S. Hirose et al., Measurement of the τ lepton polarization and $R(D^*)$ in the decay $\bar{B} \to D^*\tau^-\bar{\nu}_\tau$. Phys. Rev. Lett. **118**(21), 211801 (2017)
13. S. Hirose et al., Measurement of the τ lepton polarization and $R(D^*)$ in the decay $\bar{B} \to D^*\tau^-\bar{\nu}_\tau$ with one-prong hadronic τ decays at Belle. Phys. Rev. **D97**(1), 012004 (2018)
14. R. Aaij et al., Measurement of the ratio of branching fractions $\mathcal{B}(\bar{B}^0 \to D^{*+}\tau^-\bar{\nu}_\tau)/\mathcal{B}(\bar{B}^0 \to D^{*+}\mu^-\bar{\nu}_\mu)$. Phys. Rev. Lett. **115**(11), 111803 (2015). [Erratum: Phys. Rev. Lett. **115**(15), 159901 (2015)]
15. R. Aaij et al., Measurement of the ratio of the $B^0 \to D^{*-}\tau^+\nu_\tau$ and $B^0 \to D^{*-}\mu^+\nu_\mu$ branching fractions using three-prong τ-lepton decays. Phys. Rev. Lett. **120**(17), 171802 (2018)
16. R. Aaij et al., Test of lepton flavor universality by the measurement of the $B^0 \to D^{*-}\tau^+\nu_\tau$ branching fraction using three-prong τ decays. Phys. Rev. **D97**(7), 072013 (2018)
17. D. Bigi, P. Gambino, Revisiting $B \to D\ell\nu$. Phys. Rev. **D94**(9), 094008 (2016)
18. F.U. Bernlochner, Z. Ligeti, M. Papucci, D.J. Robinson, Combined analysis of semileptonic B decays to D and D^*: $R(D^{(*)})$, $|V_{cb}|$, and new physics. Phys. Rev. **D95**(11), 115008 (2017). [Erratum: Phys. Rev. **D97**(5), 059902 (2018)]
19. D. Bigi, P. Gambino, S. Schacht, $R(D^*)$, $|V_{cb}|$, and the heavy quark symmetry relations between form factors. JHEP **11**, 061 (2017)
20. S. Jaiswal, S. Nandi, S.K. Patra, Extraction of $|V_{cb}|$ from $B \to D^{(*)}\ell\nu_\ell$ and the standard model predictions of $R(D^{(*)})$. JHEP **12**, 060 (2017)
21. D. Becirevic, N. Kosnik, Soft photons in semileptonic B → D decays. Acta Phys. Polon. Supp. **3**, 207–214 (2010)
22. J.A. Bailey et al., $B \to D\ell\nu$ form factors at nonzero recoil and $|V_{cb}|$ from 2+1-flavor lattice QCD. Phys. Rev. **D92**(3), 034506 (2015)
23. H. Na, C.M. Bouchard, G. Peter Lepage, C. Monahan, J. Shigemitsu, $B \to Dl\nu$ form factors at nonzero recoil and extraction of $|V_{cb}|$. Phys. Rev. **D92**(5), 054510 (2015). [Erratum: Phys. Rev. **D93**(11), 119906 (2016)]
24. M. Blanke, A. Crivellin, S. de Boer, M. Moscati, U. Nierste, I. Niandi, T. Kitahara, Impact of polarization observables and $B_c \to \tau\nu$ on new physics explanations of the $b \to c\tau\nu$ anomaly (2018)
25. S. de Boer, T. Kitahara, I. Nisandzic, Soft-photon corrections to $\bar{B} \to D\tau^-\bar{\nu}_\tau$ relative to $\bar{B} \to D\mu^-\bar{\nu}_\mu$. Phys. Rev. Lett. **120**(26), 261804 (2018)
26. Y. Amhis et al., Averages of b-hadron, c-hadron, and τ-lepton properties as of summer 2016. Eur. Phys. J. **C77**(12), 895 (2017). Updated average of $R(D)$ and $R(D^*)$ for Summer 2018 at https://hflav-eos.web.cern.ch/hflav-eos/semi/summer18/RDRDs.html
27. M. Bauer, M. Neubert, Minimal leptoquark explanation for the $\mathrm{R}_{D^{(*)}}$, R_K, and $(g-2)_g$ anomalies. Phys. Rev. Lett. **116**(14), 141802 (2016)
28. D. Beirevi, N. Konik, O. Sumensari, R.Z. Funchal, Palatable leptoquark scenarios for lepton flavor violation in exclusive $b \to s\ell_1\ell_2$ modes. JHEP **11**, 035 (2016)

29. Y. Cai, J. Gargalionis, M.A. Schmidt, R.R. Volkas, Reconsidering the one leptoquark solution: flavor anomalies and neutrino mass. JHEP **10**, 047 (2017)
30. I. Dorner, S. Fajfer, A. Greljo, J.F. Kamenik, N. Konik, Physics of leptoquarks in precision experiments and at particle colliders. Phys. Rept. **641**, 1–68 (2016)
31. V. Shtabovenko, R. Mertig, F. Orellana, New developments in FeynCalc 9.0. Comput. Phys. Commun. **207**, 432–444 (2016)
32. R. Mertig, M. Bohm, A. Denner, FEYN CALC: computer algebraic calculation of Feynman amplitudes. Comput. Phys. Commun. **64**, 345–359 (1991)
33. V. Shtabovenko, FeynHelpers: connecting FeynCalc to FIRE and Package-X. Comput. Phys. Commun. **218**, 48–65 (2017)
34. H.H. Patel, Package-X 2.0: a mathematica package for the analytic calculation of one-loop integrals. Comput. Phys. Commun. **218**, 66–70 (2017)
35. A. Denner, H. Eck, O. Hahn, J. Kublbeck, Feynman rules for fermion number violating interactions. Nucl. Phys. **B387**, 467–481 (1992)
36. C. Patrignani et al., Review of particle physics. Chin. Phys. **C40**(10), 100001 (2016)

Chapter 60
Simultaneous Explanation of $R_{K^{(*)}}$ and $R_{D^{(*)}}$ in Various New Physics Models

Dinesh Kumar, Ashutosh Kumar Alok, Jacky Kumar and Ruchi Sharma

Abstract We explore the possibility of simultaneous explanation of $R_{K^{(*)}}$ and $R_{D^{(*)}}$ measurements in a model dependent approach. We reanalysed the four new physics models: Standard model like vector boson (VB), $SU(2)_L$-singlet vector leptoquark (U_1), $SU(2)_L$-triplet scalar leptoquark (S_3) and $SU(2)_L$ triplet vector leptoquark (U_3) models. The new physics couplings are assumed only to the third generation in the weak basis. We found that vector boson model is not consistent with the present data as the branching ratio of $\mathcal{B}(\tau \to 3\mu)$ evades the upper bound on the best fit point. The three leptoquark models cannot simultaneously accomodate the $R_{K^{(*)}}$ and $R_{D^{(*)}}$ measurements.

60.1 Introduction

The measurements of R_K, $R_{D^{(*)}}$ alongwith the recent measurement of R_{K^*} show the discrepancies between the experimental value and Standard Model (SM) prediction. These observables are very clean as hadronic uncertainties cancels at large extent. In this sense, these observables can be powerful tool to test the Lepton Flavour Universality (LFU) and New Physics (NP) beyond the SM. The measurement of $R_{D^{(*)}} \equiv \Gamma(B \to D^{(*)} \tau \bar{\nu})/\Gamma(B \to D^{(*)} l \bar{\nu})$ $(l = e, \mu)$ [1] disagrees with the SM

D. Kumar (✉)
Department of Physics, University of Rajasthan, Jaipur 302004, India
e-mail: dinesh@uniraj.ac.in

A. Kumar Alok · R. Sharma
Indian Institute of Technology Jodhpur, Jodhpur 342011, India
e-mail: akalok@iitj.ac.in

R. Sharma
e-mail: sharma.16@iitj.ac.in

J. Kumar
Department of High Energy Physics, Tata Institute of Fundamental Research, Mumbai 400005, India
e-mail: jka@tifr.res.in

A. Giri and R. Mohanta (eds.), *16th Conference on Flavor Physics and CP Violation*, Springer Proceedings in Physics 234,
https://doi.org/10.1007/978-3-030-29622-3_60

at the level of $\sim$4 σ. The measurement of $R_K \equiv \Gamma(B^+ \to K^+ \mu^+ \mu^-)/\Gamma(B^+ \to K^+ e^+ e^-) = 0.745^{+0.090}_{-0.074}$ (stat) $\pm$ 0.036 (syst) [2] in the low dilepton invariant mass-squared q^2 range ($1.0 \le q^2 \le 6.0\,\mathrm{GeV}^2$) deviates from the SM prediction $\simeq$1, by 2.6 σ which also supports this measurement. Very recently, the LHCb Collaboration has announced the measurement of $R_{K^*} \equiv \Gamma(B^0 \to K^{*0}\mu^+\mu^-)/\Gamma(B^0 \to K^{*0}e^+e^-)$ [3]:

$$R_{K^*}^{[0.045,1.1]} = 0.660^{+0.110}_{-0.070}\,(\text{stat}) \pm 0.024\,(\text{syst}), \tag{60.1}$$

$$R_{K^*}^{[1.1,6.0]} = 0.685^{+0.113}_{-0.069}\,(\text{stat}) \pm 0.047\,(\text{syst}). \tag{60.2}$$

which is measured in two different dilepton invariant mass-squared q^2 range. These measurements differ from the SM prediction, which is $\simeq$1, by 2.2$-$2.4σ in the low-q^2 region and by 2.4$-$2.5σ in the central-q^2 region. Apart from these, there are other measurements, all in the $b \to s\,\mu^+\,\mu^-$ sector, which show discrepancies with the SM.

We consider the VB model and three leptoquark models and perform different types of fits in order to constrain the NP couplings. The four fermion operator is not present in the three leptoquarks models at the tree level so these models donot contribute to $B_s{}^0 - \bar{B}_s^0$ mixing. We found from the global fit of these models that VB models is inconsistent with the present data and the three leptoquark models cannot accommodate the $R_{K^{(*)}}$ and $R_{D^{(*)}}$ simultaneously. We also study observables such as $B(B_c \to \tau\nu)$, $R_{J/\psi}$, τ polarization and D^* longitudinal polarization in $B \to D^*\tau\bar{\nu}$, branching ratios of the exclusive decays $B \to (K,\ K^*)\,\tau^+\tau^-$, $B_s \to \phi\,\tau^+\tau^-$ and $B_s \to \tau^+\tau^-$ induced by the quark level transition $b \to s\,\tau^+\,\tau^-$ and branching ratios of the lepton number violating decays $B \to (K,\ K^*)\mu\,\tau$ at the best fit points.

60.2 Methodology

In order to obtain constraints on NP couplings, we perform three kinds of fit,

1. Fit 1: Global fit
2. Fit 2: Fit with excluding $b \to c\tau\bar{\nu}$ data
3. Fit 3: Fit with only clean observables.

In fit 1, We perform the global fit by taking all the relevant data where these models give the NP contribution. We then remove the $R_{D^{(*)}}$ data from the fit 1 to perform second kind of fit. Fit 2 would enable us to know that how well these models can explain the anomalies in $b \to s$ sector. In the third kind of fit, we only consider clean observables such as $R_{K^{(*)}}$, $R_{D^{(*)}}$, $B_s \to \mu^+\mu^-$ and $B_s^0 - \bar{B}_s^0$ mixing. We perform this fit to see up to what extent our conclusions of the global fits are dependent on hadronic uncertainties.

We do a χ^2 fit using CERN minimization code `MINUIT`. The χ^2 function is defined as

$$\chi^2(C_i) = (\mathcal{O}_{\rm th}(C_i) - \mathcal{O}_{\rm exp})^T \mathcal{C}^{-1} (\mathcal{O}_{\rm th}(C_i) - \mathcal{O}_{\rm exp}) . \qquad (60.3)$$

$\mathcal{O}_{\rm th}(C_i)$ are the theoretical predictions of the observables which are calculated using `flavio`. $\mathcal{O}_{\rm exp}$ are the experimental measurements of the observables used in the fit. We obtain the total covariance matrix $\mathcal{C}$ adding the individual theoretical and experimental covariance matrices. We closely follow the methodology for global fits discussed in [4, 5]. We obtain the χ^2/dof for these four models in various fits which is helpful to know about the goodness of fit [6].

The list of observables, which is being used to constrain the new physics couplings, is following: (i) The measurement of $R_{D^{(*)}}$ and $R_{K^{(*)}}$, (ii) The branching ratio of $B_s^0 \to \mu^+\mu^-$, (iii) the differential branching ratio of $B^0 \to K^{*0}\mu^+\mu^-$ and $B^+ \to K^{*+}\mu^+\mu^-$ measured by LHCb, (iv) the CP-averaged differential angular distribution for $B^0 \to K^{*0}(\to K^+\pi^-)\mu^+\mu^-$, (v) the differential branching ratio of $B^0 \to K^0\mu^+\mu^-$ and $B^+ \to K^+\mu^+\mu^-$ measured by LHCb and CDF, (vi) the differential branching ratio of $B_s^0 \to \phi\mu^+\mu^-$ by LHCb and CDF and the angular observables measured by LHCb, (vii) the differential branching ratio of $B \to X_s\mu^+\mu^-$ measured by BaBar, (viii) the recent data by ATLAS and CMS for the angular observables in $B^0 \to K^{*0}\mu^+\mu^-$ decay, (ix) Mass difference ΔM_s in $B_s - \bar{B}_s^0$ mixing.

60.3 New Physics Models

The gauge invariant operators [7]

$$\mathcal{L}_{\rm eff} = \frac{G_1^{ijkl}}{\Lambda^2}(\bar{Q}_L^{\prime i}\gamma_\mu Q_L^{\prime j})(\bar{L}_L^{\prime k}\gamma^\mu L_L^{\prime l}) + \frac{G_2^{ijkl}}{\Lambda^2}(\bar{Q}_L^{\prime i}\gamma_\mu \sigma^I Q_L^{\prime j})(\bar{L}_L^{\prime k}\gamma^\mu \sigma^I L_L^{\prime l}). \qquad (60.4)$$

relates the $R_{K^{(*)}}$ and $R_{D^{(*)}}$. In the mass basis, the new physics couplings can be written as,

$$G_{(1,2)}^{ijkl} = g_{(1,2)} X^{ij} Y^{kl}, \qquad (60.5)$$

The matrices X and Y are the function of the rotation angles. For example, for $b \to s\ell\ell$ transitions, these are given by,

$$X = \begin{bmatrix} 0 & 0 & 0 \\ 0 & \sin^2\theta_{bs} & -\sin\theta_{bs}\cos\theta_{bs} \\ 0 & -\sin\theta_{bs}\cos\theta_{bs} & \cos^2\theta_{bs} \end{bmatrix}, \qquad (60.6)$$

$$Y = \begin{bmatrix} 0 & 0 & 0 \\ 0 & \sin^2\theta_{\mu\tau} & -\sin\theta_{\mu\tau}\cos\theta_{\mu\tau} \\ 0 & -\sin\theta_{\mu\tau}\cos\theta_{\mu\tau} & \cos^2\theta_{\mu\tau} \end{bmatrix}. \quad (60.7)$$

We consider one vector boson model and three leptoquark models which contribute both in $b \to s\mu^+\mu^-$ and $b \to c\tau\nu$. These models are [8]:

- SM-like vector boson model (**1**, **3**, **1**)
- a scalar $SU(2)_L$ singlet LQ S_1 (**3**, **1**, **−2/3**)
- a scalar triplet LQ S_3(**3**, **3**, **−2/3**) and
- a vector singlet LQ U_1(**3**, **1**, **4/3**).

In the gauge basis, The interaction of SM-like vector boson with the fermions is given by,

$$\Delta\mathcal{L}_V = g_{qV}^{33}\,(\overline{Q}'_{L3}\gamma^\mu\sigma^I Q'_{L3})V_\mu^I + g_{lV}^{33}\,(\overline{L}'_{L3}\gamma^\mu\sigma^I L'_{L3})V_\mu^I. \quad (60.8)$$

This generates only one operator when heavy vector boson is integrated out i.e $g_1 = 0$, $g_2 = -g_{qV}^{33}g_{\ell V}^{33}$. For simplicity, we set these couplings to a fixed value as $g_{qV}^{33} = g_{lV}^{33} = \sqrt{0.5}$.

The interaction Lagrangian in the gauge basis for three leptoquarks U_1, S_3 and U_3 can be written as, respectively

$$\Delta\mathcal{L}_{U_1} = g_{U_1}^{33}(\overline{Q}'_{L3}\gamma^\mu L'_{L3})U_{1\mu} + \text{h.c.}, \quad (60.9)$$

$$\Delta\mathcal{L}_{S_3} = g_{S_3}^{33}(\overline{Q}'_{L3}\sigma^I i\sigma^2 L'^{c}_{L3})S_3^I + \text{h.c.}, \quad (60.10)$$

$$\Delta\mathcal{L}_{U_3} = g_{U_3}^{33}(\overline{Q}'_{L3}\gamma^\mu\sigma^I L'_{L3})U_{3\mu} + \text{h.c.} \quad (60.11)$$

The couplings g_1 and g_2 in these leptoquark models can be identified as

$$U_1: \; g_1 = g_2 = -\frac{1}{2}|g_{U_1}^{33}|^2 < 0, \quad (60.12)$$

$$S_3: g_1 = 3g_2 = \frac{3}{4}|g_{S_3}^{33}|^2 > 0, \quad (60.13)$$

$$U_3: g_1 = -3g_2 = -\frac{3}{2}|g_{U_3}^{33}|^2 < 0, \quad (60.14)$$

We set the coupligs $g_{U_1}^{33}$, $g_{S_3}^{33}$ and $g_{U_3}^{33}$ to be one. The new physics contributions to the Wilson coefficients is given by,

$$C_9^{\mu\mu} = -C_{10}^{\mu\mu} = -\frac{\pi}{\sqrt{2}\alpha G_F V_{tb}V_{ts}^*}\frac{(g_1+g_2)}{\Lambda^2}\left(\sin\theta_{bs}\cos\theta_{bs}\sin^2\theta_{\mu\tau}\right), \quad (60.15)$$

$$C_V^{ij} = -\frac{1}{2\sqrt{2}G_F V_{cb}}\frac{2g_2}{\Lambda^2}\Big(-V_{cs}\sin\theta_{bs}\cos\theta_{bs} + V_{cb}\cos^2\theta_{bs}\Big)Y^{ij}, \quad (60.16)$$

$$C_L^{ij} = -\frac{\pi}{\sqrt{2}\alpha G_F V_{tb}V_{ts}^*}\frac{(g_1-g_2)}{\Lambda^2}\left(\sin\theta_{bs}\cos\theta_{bs}\right)Y^{ij}. \quad (60.17)$$

Table 60.1 Best fit values of the mixing angles and the pull obtained in vector boson model. For Fit 1, $\chi^2/dof = 120.77/114$

Observable		$R_{K[1.0-6.0]}$	$R_{K^*[0.045-1.1]}$	$R_{K^*[1.1-6.0]}$	$P'_{5[4.0-6.0]}$	R_D^{ratio}	$R_{D^*}^{\text{ratio}}$	$\frac{B_K^{SM+NP}}{B_K^{SM}}$	$\frac{B_{K^*}^{SM+NP}}{B_{K^*}^{SM}}$	$\tau \to 3\mu$
Measurement		0.75 ± 0.09	0.66 ± 0.09	0.69 ± 0.10	-0.30 ± 0.16	1.29 ± 0.17	1.21 ± 0.06	≤ 4.0	≤ 2.9	$< 2.1 \times 10^{-8}$
Standard model		1.0	0.93	0.99	−0.82	1.0	1.0			
$\theta_{\mu\tau}$ (rad.)	θ_{bs} (rad.)	Fit 1 (Global fit)								
0.655 ± 0.147	0.0046 ± 0.0016	0.75	0.88	0.75	−0.70	1.02	1.02	0.86	0.86	6.18×10^{-7}
χ^2		0.001	5.57	0.47	4.91	2.48	9.77	–	–	
$\theta_{\mu\tau}$ (rad.)	θ_{bs} (rad.)	Fit 2								
1.496 ± 1.502	0.002 ± 0.001	0.71	0.87	0.72	−0.68	0.94	0.94	0.94	0.94	1.06×10^{-7}
χ^2		0.16	5.10	0.11	4.59	4.29	20.59	–	–	
$\theta_{\mu\tau}$ (rad.)	θ_{bs} (rad.)	Fit 3								
0.594 ± 0.142	0.006 ± 0.002	0.72	0.87	0.73	−0.69	1.02	1.02	0.82	0.82	4.06×10^{-7}
χ^2		0.06	5.10	0.22	4.72	2.48	9.77	–	–	

Table 60.2 Best fit values of the mixing angles and the pull obtained in U_1 leptoquark model. For fit 1, $\chi^2/dof = 117.70/113$

Observable		$R_{K[1.0-6.0]}$	$R_{K^*[0.045-1.1]}$	$R_{K^*[1.1-6.0]}$	$P'_{5[4.0-6.0]}$	R_D^{ratio}	$R_{D^*}^{\text{ratio}}$
Measurement		0.75 ± 0.09	0.66 ± 0.09	0.69 ± 0.10	-0.30 ± 0.16	1.29 ± 0.17	1.21 ± 0.06
Standard model		1.0	0.93	0.99	−0.82	1.0	1.0
$\theta_{\mu\tau}$ (rad.)	θ_{bs} (rad.)	Fit 1 (Global fit)					
0.355 ± 0.215	0.008 ± 0.009	0.72	0.87	0.73	−0.69	1.04	1.04
χ^2		0.06	5.10	0.22	4.72	2.15	7.97
$\theta_{\mu\tau}$ (rad.)	θ_{bs} (rad.)	Fit 2					
0.099 ± 0.182	0.099 ± 0.365	0.72	0.87	0.73	−0.68	0.92	0.92
χ^2		0.06	5.10	0.22	4.59	4.64	22.73
$\theta_{\mu\tau}$ (rad.)	θ_{bs} (rad.)	Fit 3					
0.365 ± 0.206	0.008 ± 0.009	0.70	0.87	0.71	−0.68	1.04	1.04
χ^2		0.23	5.10	0.07	4.59	2.15	7.97

Table 60.3 Best fit values of the mixing angles and the pull obtained in S_3 leptoquark model. For fit 1, $\chi^2/dof = 120.75/113$

Observable		$R_{K[1.0-6.0]}$	$R_{K^*[0.045-1.1]}$	$R_{K^*[1.1-6.0]}$	$P'_{5[4.0-6.0]}$	R_D^{ratio}	$R_{D^*}^{\text{ratio}}$	$\frac{B_K^{SM+NP}}{B_K^{SM}}$	$\frac{B_{K^*}^{SM+NP}}{B_{K^*}^{SM}}$
Measurement		0.75 ± 0.09	0.66 ± 0.09	0.69 ± 0.10	-0.30 ± 0.16	1.29 ± 0.17	1.21 ± 0.06	≤ 4.0	≤ 2.9
Standard model		1.0	0.93	0.99	−0.82	1.0	1.0		
$\theta_{\mu\tau}$ (rad.)	θ_{bs} (rad.)	Fit 1 (Global fit)							
1.50 ± 2.03	-0.0010 ± 0.0002	0.71	0.87	0.72	−0.68	1.04	1.04	1.03	1.03
χ^2		0.16	5.10	0.11	4.59	2.15	7.97	–	–
$\theta_{\mu\tau}$ (rad.)	θ_{bs} (rad.)	Fit 2							
0.23 ± 0.42	-0.018 ± 0.067	0.72	0.87	0.73	−0.69	0.96	0.96	1.91	1.91
χ^2		0.06	5.10	0.22	4.72	3.75	17.26	–	–
$\theta_{\mu\tau}$ (rad.)	θ_{bs} (rad.)	Fit 3							
1.50 ± 2.02	-0.0011 ± 0.0003	0.69	0.86	0.70	−0.68	1.04	1.04	1.04	1.04
χ^2		0.41	4.84	0.02	4.59	2.15	7.97	–	–

Table 60.4 Best fit values of the mixing angles and the pull obtained in U_3 leptoquark model. For fit 1, $\chi^2/dof = 119.01/113$

Observable		$R_{K[1.0-6.0]}$	$R_{K^*[0.045-1.1]}$	$R_{K^*[1.1-6.0]}$	$P'_{5[4.0-6.0]}$	R_D^{ratio}	$R_{D^*}^{ratio}$	$\frac{B_K^{SM+NP}}{B_K^{SM}}$	$\frac{B_{K^*}^{SM+NP}}{B_{K^*}^{SM}}$
Measurement		0.75 ± 0.09	0.66 ± 0.09	0.69 ± 0.10	-0.30 ± 0.16	1.29 ± 0.17	1.21 ± 0.06	≤ 4.0	≤ 2.9
Standard model		1.0	0.93	0.99	−0.82	1.0	1.0		
$\theta_{\mu\tau}$ (rad.)	θ_{bs} (rad.)	Fit 1 (Global fit)							
1.50 ± 0.60	0.0010 ± 0.0002	0.71	0.87	0.72	−0.68	1.07	1.07	1.15	1.15
χ^2		0.16	5.10	0.11	4.59	1.70	5.56	–	–
$\theta_{\mu\tau}$ (rad.)	θ_{bs} (rad.)	Fit 2							
1.25 ± 2.15	0.001 ± 0.002	0.73	0.88	0.74	−0.69	1.04	1.04	1.15	1.15
χ^2		0.01	5.57	0.32	4.72	2.15	7.97	–	–
$\theta_{\mu\tau}$ (rad.)	θ_{bs} (rad.)	Fit 3							
1.50 ± 0.61	0.0011 ± 0.0003	0.71	0.87	0.72	−0.68	1.07	1.07	1.15	1.15
χ^2		0.16	5.10	0.11	4.59	1.70	5.56	–	–

60.4 Results and Discussion

The fit results for vector boson model are presented in Table 60.1. We can easily see that VB model evades the upper bound on $B(\tau \to 3\mu)$ and this results also holds on removing $b \to c\tau\bar{\nu}$ data from the fit. We conclude from this that VB model is not consistent with the present data in case of new physics coupling with third generation only in the weak basis.

Also the fit results for the U_1, S_3 and U_3 models are presented in Tables 60.2, 60.3 and 60.4. We note from the tables that the global fit of these three models can explain the $R_{K^{(*)}}$ measurement and these models can also reduce the tesnsion in $R_{D^{(*)}}$. In fit 2 and fit 3, we find that the tesnsion in R_{D^*} get worse than SM. We also find that these models cannot enhance the branching ratio of the $b \to s\tau\tau$ processes if the NP couplings are assumed to be only with third generation.

In summary, we look for the simultaneous explanations of the $R_{K^{(*)}}$ and $R_{D^{(*)}}$ in the VB, U_1, S_3 and U_3 models. We found that VB models evades the upper bound on the branching ratio of $\tau \to 3\mu$ whereas the three leptoquark models gives best fit to the data in the global fit but none of them can accomodate the $R_{D^{(*)}}$ data.

References

1. Y. Amhis et al. [HFLAV Collaboration], Eur. Phys. J. C **77**(12), 895 (2017) https://doi.org/10.1140/epjc/s10052-017-5058-4. arXiv:1612.07233 [hep-ex]
2. R. Aaij et al. [LHCb Collaboration], Phys. Rev. Lett. **113**, 151601 (2014). arXiv:1406.6482 [hep-ex]
3. R. Aaij et al. [LHCb Collaboration], JHEP **1708**, 055 (2017). arXiv:1705.05802 [hep-ex]
4. A.K. Alok, B. Bhattacharya, A. Datta, D. Kumar, J. Kumar, D. London, Phys. Rev. D **96**(9), 095009 (2017). arXiv:1704.07397 [hep-ph]
5. A.K. Alok, B. Bhattacharya, D. Kumar, J. Kumar, D. London, S.U. Sankar, Phys. Rev. D **96**(1), 015034 (2017). arXiv:1703.09247 [hep-ph]
6. A.K. Alok, D. Kumar, J. Kumar, R. Sharma, arXiv:1704.07347 [hep-ph]
7. B. Bhattacharya, A. Datta, D. London, S. Shivashankara, Phys. Lett. B **742**, 370 (2015). arXiv:1412.7164 [hep-ph]
8. B. Bhattacharya, A. Datta, J.P. Guvin, D. London, R. Watanabe, JHEP **1701**, 015 (2017). arXiv:1609.09078 [hep-ph]

Chapter 61
A Nonstandard Solution of the Fermionic Mass Hierarchy

Gauhar Abbas

Abstract We present a non-conventional solution of the fermionic mass hierarchy among and within the three fermionic families of the standard model and quark mixing using discrete abelian $\mathcal{Z}_2$ symmetries and gauge singlet scalar fields. An ultraviolet completion with vector-like fermions is also presented.

61.1 Introduction

One of the important questions which the standard model (SM) does not answer is the fermionic mass hierarchy among and within the three generations of the SM fermions. This problem is so compelling that even Setven Weinberg wants to see its solution in his lifetime [1, 2].

In addition to the mass hierarchy among and within the three fermionic generations, there is a hierarchy among quark-mixing angles, i.e. $\sin\theta_{12} \gg \sin\theta_{23} \gg \sin\theta_{13}$. It is expected that these three different hierarchies are intimately related.

There are several proposals to address this problem in literature [1–39], and one of the standard solutions is the celebrated Froggatt-Nielson mechanism [4]. In this paper, we present an incongruous solution which deviates from the standard Froggatt-Nielson approach by using only dimension-5 operators, which are next to Yukawa operator, for generation of masses of all fermions [1]. The main idea is that the Yukawa Lagrangian is forbidden by discrete $\mathcal{Z}_2$ type symmetries, and hence dimension-5 operators provide masses to fermions [40–42].

It turns out that one can create two models using the above discussed idea. The first model provides an origin of the fermionic mass hierarchy among and within the three fermionic families along with an explanation for quark-mixing pattern, and second model can explain the mass hierarchy among the three fermionic families

G. Abbas (✉)
Department of Physics, Indian Institute of Technology (BHU), Varanasi 221005, India
e-mail: gauhar.phy@iitbhu.ac.in
URL: https://www.iitbhu.ac.in/dept/app/people/gauharapp

A. Giri and R. Mohanta (eds.), *16th Conference on Flavor Physics and CP Violation*, Springer Proceedings in Physics 234,
https://doi.org/10.1007/978-3-030-29622-3_61

[1]. In this article, we only discuss the first model. The details about second model can be found in [1].

For achieving our goal, we need to employ three discrete symmetries and gauge singlet scalar fields. As we will show, an ultra-violet (UV) completion of these models may come from vector-like fermions.

The organization of this work is in the following pattern: In Sect. 61.2, we present the model providing an explanation for the fermionic mass hierarchy among and within the three fermionic families along with the quark-mixing. A discussion is provided in Sect. 61.3.

61.2 Model

We discuss now model which can provide an origin of the fermionic mass hierarchy among and within the three fermionic families along with the quark-mixing. This is achieved by introducing six real singlet scalar fields क_i [1] $(i = 1 - 6)$ to the SM. Their behaviour under the SM symmetry $SU(3)_c \otimes SU(2)_L \otimes U(1)_Y$ is given by,

$$\text{क}_i : (1, 1, 0). \tag{61.1}$$

Moreover, the symmetry of the SM is extended by imposing discrete symmetries $\mathcal{Z}_2$, $\mathcal{Z}_2'$ and $\mathcal{Z}_2''$ on the right handed fermions of each family and the scalar fields क_i as shown in Table 61.1.

It should be noted that the Yukawa Lagrangian is completely forbidden by the symmetries $\mathcal{Z}_2$, $\mathcal{Z}_2'$ and $\mathcal{Z}_2''$, and mass of fermions are now generated by dimension-5 operators. Thus mass Lagrangian can be written as,

$$\begin{aligned}
\mathcal{L}_{mass} &= \frac{1}{\Lambda}\Big[\Gamma_1 \bar{\psi}_L^1 \tilde{\varphi} u_R \text{क}_1 + \Gamma_2 \bar{\psi}_L^2 \tilde{\varphi} c_R \text{क}_3 + \Gamma_3 \bar{\psi}_L^3 \tilde{\varphi} t_R \text{क}_5 + \Gamma_1' \bar{\psi}_L^1 \varphi d_R \text{क}_2 \\
&+ \Gamma_2' \bar{\psi}_L^2 \varphi s_R \text{क}_4 + \Gamma_3' \bar{\psi}_L^3 \varphi b_R \text{क}_6\Big] + \frac{1}{\Lambda}\Big[\Gamma_4 \bar{\psi}_L^1 \varphi e_R \text{क}_2 + \Gamma_5 \bar{\psi}_L^2 \varphi \mu_R \text{क}_4 \\
&+ \Gamma_6 \bar{\psi}_L^3 \varphi \tau_R \text{क}_6\Big] + \frac{c}{\Lambda} \bar{l_L^c} \tilde{\varphi}^* \tilde{\varphi}^\dagger l_L + \text{H.c.}
\end{aligned} \tag{61.2}$$

We observe that the fermionic mass pattern of the three families of the SM, which is $m_d > m_u$, $m_c \gg m_s$, $m_t \gg m_b$, $m_\tau \gg m_\mu \gg m_e$, $m_b \gg m_s \gg m_d$, and $m_t \gg m_c \gg m_u$, can be explained by assuming the vacuum-expectation values(VEVs) of the singlet scalar fields in such a way that $\langle \text{क}_2 \rangle > \langle \text{क}_1 \rangle$, $\langle \text{क}_3 \rangle \gg \langle \text{क}_4 \rangle$, $\langle \text{क}_5 \rangle \gg \langle \text{क}_6 \rangle$, $\langle \text{क}_6 \rangle \gg \langle \text{क}_4 \rangle \gg \langle \text{क}_2 \rangle$, and $\langle \text{क}_5 \rangle \gg \langle \text{क}_3 \rangle \gg \langle \text{क}_1 \rangle$. Neutrino masses can be recovered from the Weinberg operator via celebrated seesaw mechanism.

---

[1] Consonant letter "क" (kə) is taken from the Devanagari script. It is pronounced as "Ka" in Kashmir [1].

Table 61.1 The charges of right-handed fermions of three families of the SM and singlet scalar fields under $\mathcal{Z}_2$, $\mathcal{Z}_2'$ and $\mathcal{Z}_2''$ symmetries. We show flavour of right-handed fermion by superscript

Fields	$\mathcal{Z}_2$	$\mathcal{Z}_2'$	$\mathcal{Z}_2''$
ψ_R^u	+	+	−
क$_1$	+	+	−
$\psi_R^{d,e}$	−	−	+
क$_2$	−	−	+
ψ_R^c	+	−	−
क$_3$	+	−	−
$\psi_R^{s,\mu}$	+	−	+
क$_4$	+	−	+
ψ_R^t	−	+	−
क$_5$	−	+	−
$\psi_R^{b,\tau}$	−	+	+
क$_6$	−	+	+

It is important to see if the model can provide an explanation to the observed pattern of quark-mixing. For this purpose, we write, for instance, down type mass matrix which approximately reads,

$$\mathcal{M}_\mathcal{D} = \begin{pmatrix} \frac{\Gamma_{11}^d v v_2}{2M} & \frac{\Gamma_{12}^d v v_4}{2M} & \frac{\Gamma_{13}^d v v_6}{2M} \\ \frac{\Gamma_{21}^d v v_2}{2M} & \frac{\Gamma_{22}^d v v_4}{2M} & \frac{\Gamma_{23}^d v v_6}{2M} \\ \frac{\Gamma_{31}^d v v_2}{2M} & \frac{\Gamma_{32}^d v v_4}{2M} & \frac{\Gamma_{33}^d v v_6}{2M} \end{pmatrix}. \tag{61.3}$$

This matrix can be diagonalized using bi-unitary transformation,

$$U_\mathcal{D}^\dagger \mathcal{M}_\mathcal{D} V_\mathcal{D} = diag(m_d, m_s, m_b). \tag{61.4}$$

In the limit $v, v_2, v_4 \ll v_6, M$, the following three mixing angles can be used to parametrize the matrix $V_\mathcal{D}$:

$$\begin{aligned} \tan\theta_{12}^d &\approx \frac{v_2 C_{12}^d}{v_4} + \mathcal{O}(\frac{v^3}{v_6^3}) + \mathcal{O}(\frac{v_2^2}{v_6^2}) + \mathcal{O}(\frac{v_4^2}{v_6^2}), \\ \tan\theta_{23}^d &\approx \frac{v_4 C_{23}^d}{v_6} + \mathcal{O}(\frac{v^3}{v_6^3}) + \mathcal{O}(\frac{v_2^2}{v_6^2}) + \mathcal{O}(\frac{v_4^2}{v_6^2}) \\ \tan\theta_{13}^d &= \frac{v_2^2 v_4 C_{13}^d}{v_6^3} + \mathcal{O}(\frac{v^3}{v_6^3}) + \mathcal{O}(\frac{v_2^2}{v_6^2}) + \mathcal{O}(\frac{v_4^2}{v_6^2}). \end{aligned} \tag{61.5}$$

Similarly the matrix $V_{\mathcal{U}}$ can be parameterized by the following three mixing angles:

$$\tan\theta_{12}^{u} \approx \frac{v_1 C_{12}^{u}}{v_3} + \mathcal{O}(\frac{v^3}{v_5^3}) + \mathcal{O}(\frac{v_1^2}{v_5^2}) + \mathcal{O}(\frac{v_3^2}{v_5^2}), \tag{61.6}$$
$$\tan\theta_{23}^{u} \approx \frac{v_3 C_{23}^{u}}{v_5} + \mathcal{O}(\frac{v^3}{v_5^3}) + \mathcal{O}(\frac{v_1^2}{v_5^2}) + \mathcal{O}(\frac{v_3^2}{v_5^2}),$$
$$\tan\theta_{13}^{u} = \frac{v_1^2 v_3 C_{13}^{u}}{v_5^3} + \mathcal{O}(\frac{v^3}{v_5^3}) + \mathcal{O}(\frac{v_1^2}{v_5^2}) + \mathcal{O}(\frac{v_3^2}{v_5^2}).$$

The Cabibbo-Kobayashi-Maskawa matrix can be obtained by $V_{CKM} = V_{\mathcal{U}}^{\dagger} V_{\mathcal{D}}$, and the three mixing angles in the standard parameterization reads,

$$\tan\theta_{12} \approx \frac{v_2 C_{12}}{v_4} + \mathcal{O}(\frac{v_1}{v_5}), \quad \tan\theta_{23} \approx \frac{v_4 C_{23}}{v_6} + \mathcal{O}(\frac{v_1}{v_5}), \tag{61.7}$$
$$\tan\theta_{13} \approx \frac{v_2^2 v_4 C_{13}}{v_6^3} + \mathcal{O}(\frac{v_1}{v_5}).$$

We observe that the quark-mixing angles turn out to be $\sin\theta_{12} \gg \sin\theta_{23} \gg \sin\theta_{13}$ in the limit $v_2 \ll v_4 \ll v_6$.

We now discuss a UV completion of the model discussed in this article. This can be achieved by introducing one vector-like isosinglet up type quark, one vector-like isosinglet down type quark, and one isosinglet vector-like charged lepton. Vector-like fermions transform in the following way under the SM symmetry $SU(3)_c \otimes SU(2)_L \otimes U(1)_Y$,

$$Q = U_{L,R} : (3, 1, \frac{4}{3})), D_{L,R} : (3, 1, -\frac{2}{3})), L = E_{L,R} : (1, 1, -2)); \tag{61.8}$$

and they behave trivially under the discrete $\mathcal{Z}_2$, $\mathcal{Z}_2'$ and $\mathcal{Z}_2''$ symmetries.

The masses of vector-like fermions are given by the following Lagrangian:

$$\mathcal{L}_V = M_U \bar{U}_L U_R + M_D \bar{D}_L D_R + M_E \bar{E}_L E_R + \text{H.c.} \tag{61.9}$$

Moreover, their interactions with the the SM fermions, for instance for quarks, can be written as,

$$\mathcal{L} = Y_1 \bar{q}_L^1 \varphi Q_R + Y_2 \bar{q}_L^2 \varphi Q_R + Y_3 \bar{q}_L^3 \varphi Q_R + \text{H.c}, \tag{61.10}$$

and

$$\mathcal{L} = C_1 \bar{Q}_L q_R^1 \,\text{क}_{1,2} + C_2 \bar{Q}_L q_R^2 \,\text{क}_{3,4} + C_3 \bar{Q}_L q_R^3 \,\text{क}_{5,6} + \text{H.c}, \tag{61.11}$$

where q_L is quark doublet of the SM. We can write a similar Lagrangian for leptons.

61.3 Discussion

We have discussed an alternative and a nonstandard solution of the fermionic mass hierarchy of the SM. This solution relies on the hierarchical VEVs of the singlet scalar fields. This will destabilize the scalar potential of the model. A cure of this is discussed in [1] which uses three discrete symmetries $\mathcal{Z}_3$ imposed on right-handed fermions and singlet scalar fields. This mechanism forbids dangerous quartic terms in scalar potential, thus providing a stability to the scalar sector. For more details, see [1].

We note that an explanation for hierarchical VEVs can come from a strongly interacting sector. For instance, six singlet scalar fields can be obtained by extending the minimal composite Higgs model $SO(5)/SO(4)$ [43–45] to $SO(5)/SO(4) \otimes SO(4)'$ where the SM fermions are singlet under the global $SO(4)'$, and isosinglet vector-like fermions behave non-trivially under the global $SO(4)'$. A qq condensate could be responsible for breaking of the global $SO(4)'$ such that six Nambu-Goldstone bosons become pseudo-Nambu-Goldstone bosons, providing six gauge singlet scalar fields having hierarchical VEVs. A phenomenological investigation of this model is under progress.

References

1. G. Abbas, Int. J. Mod. Phys. A **34**(20), 1950104 (2019). https://doi.org/10.1142/S0217751X19501045
2. G. Abbas, arXiv:1710.03582 [hep-ph]
3. G. Abbas, arXiv:1807.05683 [hep-ph]d
4. C.D. Froggatt, H.B. Nielsen, Nucl. Phys. B **147**, 277 (1979). https://doi.org/10.1016/0550-3213(79)90316-X
5. B.S. Balakrishna, Phys. Rev. Lett. **60**, 1602 (1988). https://doi.org/10.1103/PhysRevLett.60.1602
6. M. Cvetic, Phys. Rev. D **32**, 1214 (1985). https://doi.org/10.1103/PhysRevD.32.1214
7. A. Davidson, L. Michel, M.L. Sage, K.C. Wali, Phys. Rev. D **49**, 1378 (1994). https://doi.org/10.1103/PhysRevD.49.1378
8. M. Leurer, Y. Nir, N. Seiberg, Nucl. Phys. B **420**, 468 (1994). https://doi.org/10.1016/0550-3213(94),90074-4[hep-ph/9310320]
9. N. Haba, T. Kondo, Y. Shimizu, Phys. Lett. B **535**, 271 (2002). https://doi.org/10.1016/S0370-2693(02)01774-4 [hep-ph/0202191]
10. K.S. Babu, S. Nandi, Phys. Rev. D **62**, 033002 (2000). https://doi.org/10.1103/PhysRevD.62.033002 [hep-ph/9907213]
11. G.G. Ross, L. Velasco-Sevilla, Nucl. Phys. B **653**, 3 (2003). https://doi.org/10.1016/S0550-3213(03)00041-5 [hep-ph/0208218]
12. G.F. Giudice, O. Lebedev, Phys. Lett. B **665**, 79 (2008). https://doi.org/10.1016/j.physletb.2008.05.062 [arXiv:0804.1753 [hep-ph]]
13. F. Bazzocchi, S. Morisi, M. Picariello, E. Torrente-Lujan, J. Phys. G **36**, 015002 (2009). https://doi.org/10.1088/0954-3899/36/1/015002. arXiv:0802.1693 [hep-ph]
14. S.F. King, JHEP **1401**, 119 (2014). https://doi.org/10.1007/JHEP01(2014)119. arXiv:1311.3295 [hep-ph]

15. H. Ishimori, S.F. King, H. Okada, M. Tanimoto, Phys. Lett. B **743**, 172 (2015). https://doi.org/10.1016/j.physletb.2015.02.027. arXiv:1411.5845 [hep-ph]
16. F. Hartmann, W. Kilian, Eur. Phys. J. C **74**, 3055 (2014). https://doi.org/10.1140/epjc/s10052-014-3055-4. arXiv:1405.1901 [hep-ph]
17. C.E. Daz, S.F. Mantilla, R. Martinez, arXiv:1712.07263 [hep-ph]
18. H. Fritzsch, Phys. Lett. **73B**, 317 (1978). https://doi.org/10.1016/0370-2693(78)90524-5
19. S.M. Barr, Phys. Rev. D **21**, 1424 (1980). https://doi.org/10.1103/PhysRevD.21.1424
20. Y. Koide, Phys. Rev. D **28**, 252 (1983). https://doi.org/10.1103/PhysRevD.28.252
21. N. Arkani-Hamed, M. Schmaltz, Phys. Rev. D **61**, 033005 (2000). https://doi.org/10.1103/PhysRevD.61.033005 [hep-ph/9903417]
22. K. Yoshioka, Mod. Phys. Lett. A **15**, 29 (2000). https://doi.org/10.1142/S0217732300000062, 10.1016/S0217-7323(00)00006-2 [hep-ph/9904433]
23. E.A. Mirabelli, M. Schmaltz, Phys. Rev. D **61**, 113011 (2000). https://doi.org/10.1103/PhysRevD.61.113011 [hep-ph/9912265]
24. K.S. Babu, E. Ma, J.W.F. Valle, Phys. Lett. B **552**, 207 (2003). https://doi.org/10.1016/S0370-2693(02)03153-2 [hep-ph/0206292]
25. H. Georgi, A.E. Nelson, A. Manohar, Phys. Lett. **126B**, 169 (1983). https://doi.org/10.1016/0370-2693(83)90584-1
26. B.S. Balakrishna, A.L. Kagan, R.N. Mohapatra, Phys. Lett. B **205**, 345 (1988). https://doi.org/10.1016/0370-2693(88)91676-0
27. K.S. Babu, R.N. Mohapatra, Phys. Rev. Lett. **64**, 2747 (1990). https://doi.org/10.1103/PhysRevLett.64.2747
28. Z.G. Berezhiani, R. Rattazzi, Nucl. Phys. B **407**, 249 (1993). https://doi.org/10.1016/0550-3213(93)90057-V [hep-ph/9212245]
29. J.K. Elwood, N. Irges, P. Ramond, Phys. Lett. B **413**, 322 (1997). https://doi.org/10.1016/S0370-2693(97)01103-9 [hep-ph/9705270]
30. J.R. Ellis, S. Lola, G.G. Ross, Nucl. Phys. B **526**, 115 (1998). https://doi.org/10.1016/S0550-3213(98)00423-4 [hep-ph/9803308]
31. R. Barbieri, L.J. Hall, A. Romanino, Nucl. Phys. B **551**, 93 (1999) https://doi.org/10.1016/S0550-3213(99)00215-1 [hep-ph/9812384]
32. I. Dorsner, S.M. Barr, Phys. Rev. D **65**, 095004 (2002). https://doi.org/10.1103/PhysRevD.65.095004[hep-ph/0201207]
33. M. Bauer, M. Carena, K. Gemmler, JHEP **1511**, 016 (2015). 10.1007/JHEP11(2015)016. arXiv:1506.01719 [hep-ph]
34. R.S. Hundi, S. SenGupta, J. Phys. G **40**, 075002 (2013). https://doi.org/10.1088/0954-3899/40/7/075002. arXiv:1111.1106 [hep-th]
35. J.B. Dent, R. Feger, T.W. Kephart, S. Nandi, Phys. Lett. B **697**, 367 (2011). https://doi.org/10.1016/j.physletb.2011.02.028. arXiv:0908.3915 [hep-ph]
36. M. Bauer, M. Carena, K. Gemmler, Phys. Rev. D **94**(11), 115030 (2016). https://doi.org/10.1103/PhysRevD.94.115030. arXiv:1512.03458 [hep-ph]
37. K. Huitu, N. Koivunen, arXiv:1706.09463 [hep-ph]
38. F.J. Botella, G.C. Branco, M.N. Rebelo, J.I. Silva-Marcos, Phys. Rev. D **94**(11), 115031 (2016). https://doi.org/10.1103/PhysRevD.94.115031. arXiv:1602.08011 [hep-ph]
39. M. Bordone, C. Cornella, J. Fuentes-Martin, G. Isidori, Phys. Lett. B **779**, 317 (2018). https://doi.org/10.1016/j.physletb.2018.02.011. arXiv:1712.01368 [hep-ph]
40. G. Abbas, Phys. Rev. D **95**(1), 015029 (2017). https://doi.org/10.1103/PhysRevD.95.015029. arXiv:1609.02899 [hep-ph]
41. G. Abbas, arXiv:1706.01052 [hep-ph]
42. G. Abbas, Phys. Lett. B **773**, 252 (2017). https://doi.org/10.1016/j.physletb.2017.08.028. arXiv:1706.02564 [hep-ph]
43. D.B. Kaplan, H. Georgi, Phys. Lett. **136B**, 183 (1984). https://doi.org/10.1016/0370-2693(84)91177-8
44. D.B. Kaplan, H. Georgi, S. Dimopoulos, Phys. Lett. **136B**, 187 (1984). https://doi.org/10.1016/0370-2693(84)91178-X
45. K. Agashe, R. Contino, A. Pomarol, Nucl. Phys. B **719**, 165 (2005). https://doi.org/10.1016/j.nuclphysb.2005.04.035[hep-ph/0412089]

Chapter 62
$b \to cl\nu$ Anomalies in Light of Vector and Scalar Interactions

Aritra Biswas

Abstract We perform a model independent analysis of the charged current $b \to cl\nu$ anomalies under the presence of scalar and vector interactions. The analysis is carried out in two stages: (a) under the presence of both (left-handed) vector and scalar interactions and (b) under the presence of scalar interactions alone. We find that even after stringent bounds from similar quark-level processes such as $B_c \to \tau\nu$, such scenarios have the potential to explain the aforementioned anomalies. Contrary to the general notion, we show that even scalar interactions alone can explain such anomalies, provided they are complex. However, extended scalar sector models are unable to comply with these anomalies to $\sim 3\sigma$. We further illustrate our results with the help of three benchmark models corresponding to the presence of (i) both scalar and vector (ii) real scalar and (iii) complex scalar interactions.

62.1 Introduction

Over the past few years, there have been constant and consistent reports from experimental collaborations such as LHCb, Belle and BaBar about flavour observables with deviations of more than 3σ in exclusive $B \to D^*$ and $B \to J/\psi$ transitions. Both of these exclusive processes have the underlying sub-quark transition $b \to cl\nu$. These results are believed to be the hints of lepton-flavour universality violating (LFUV) type new physics (NP). We investigate the prospect of scalar and vector type NP's in explaining such deviations. We initially work from a model independent perspective and then illustrate our results further using the models: (i) Non-minimal universal extra dimensions (NMUED) for the case with one scalar and one vector NP operator, (ii) Goergi-Michacek (GM) model for the case of a single scalar NP operator preceded by a real Wilson coefficient (WC) and (iii) Leptoquark (LQ) model for the a single scalar NP operator preceded by a complex WC.

A. Biswas (✉)
Indian Association for the Cultivation of Science, Kolkata, India
e-mail: tpab2@iacs.res.in
URL: http://inspirehep.net/author/profile/Arita.Biswas.1

A. Giri and R. Mohanta (eds.), *16th Conference on Flavor Physics and CP Violation*, Springer Proceedings in Physics 234,
https://doi.org/10.1007/978-3-030-29622-3_62

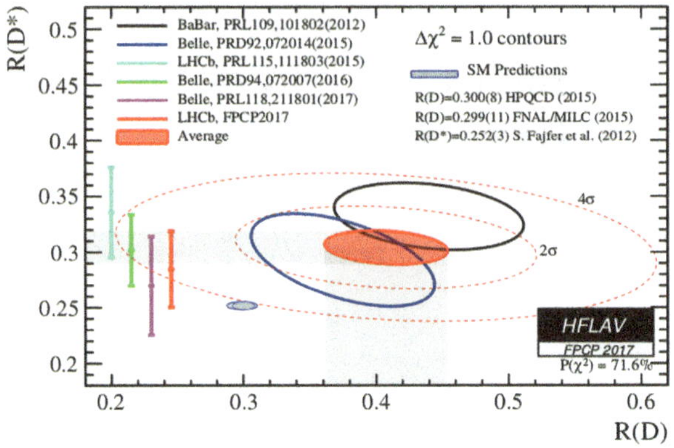

Fig. 62.1 Global average for $\mathcal{R}(D)$, $\mathcal{R}(D^*)$ and $P_\tau(D^*)$ and the deviation from the SM result

Table 62.1 Present status (both theoretical and experimental) of $\mathcal{R}(D)$, $\mathcal{R}(D^*)$ and $P_\tau(D^*)$. First uncertainty is statistical and the second one is systematic. The first row lists the arithmetic mean for the SM calculations reported in HFLAV

	$\mathcal{R}(D)$	$\mathcal{R}(D^*)$	Correlation	$P_\tau(D^*)$	$\mathcal{R}(J/\psi)$
SM	0.299(3)	0.258(6)		−0.491(25)	0.249(42)(LFCQ) 0.289(28)(PQCD)
Babar	$0.440(58)_{st.}(42)_{sy.}$	$0.332(24)_{st.}(18)_{sy.}$	−0.27		
Belle (2015)	$0.375(64)_{st.}(26)_{sy.}$	$0.293(38)_{st.}(15)_{sy.}$	−0.49		
Belle (2016)-I	–	$0.302(30)_{st.}(11)_{sy.}$			
Belle (2016)-II	–	$0.270(35)_{st.}\,{}^{+0.028}_{-0.025}$	0.33	$-0.38(51)_{st.}\,{}^{+0.21}_{-0.16}$	
LHCb (2015)	–	$0.336(27)_{st.}(30)_{sy.}$			
LHCb (2017)	–	$0.286(19)_{st.}(25)_{sy.}(21)$			
World avg.	$0.407(39)_{st.}(24)_{sy.}$	$0.304(13)_{st.}(7)_{sy.}$	0.20		$0.71(17)_{st.}(18)_{sy.}$

62.2 Current Status: Theory and Experiment

The present global average for the $\mathcal{R}(D^{(*)})$ anomalies are about 4σ away from the corresponding SM results. Figure 62.1 and Table 62.1 summarize the current theoretical and experimental status for these anomalies. The SM average is the arithmetic mean of the results from [1–4].

62.3 Formalism

The most general effective Hamiltonian describing the $b \to c\tau\nu$ transitions, with all possible four-fermi operators in the lowest dimension (with left-handed neutrinos) is given by:

$$\mathcal{H}_{\rm eff} = \frac{4G_F}{\sqrt{2}} V_{cb}\Big[(1 + C^{\ell}_{V_1})\mathcal{O}_{V_1} + C^{\ell}_{V_2}\mathcal{O}_{V_2} + C^{\ell}_{S_1}\mathcal{O}_{S_1} + C^{\ell}_{S_2}\mathcal{O}_{S_2} + C^{\ell}_{T}\mathcal{O}_{T}\Big], \quad (62.1)$$

where the operator basis is defined as

$$\begin{aligned}
\mathcal{O}_{V_1} &= (\bar{c}_L\gamma^{\mu}b_L)(\bar{\tau}_L\gamma_{\mu}\nu_{\tau L}),\\
\mathcal{O}_{V_2} &= (\bar{c}_R\gamma^{\mu}b_R)(\bar{\tau}_L\gamma_{\mu}\nu_{\tau L}),\\
\mathcal{O}_{S_1} &= (\bar{c}_L b_R)(\bar{\tau}_R\nu_{\tau L}),\\
\mathcal{O}_{S_2} &= (\bar{c}_R b_L)(\bar{\tau}_R\nu_{\tau L}),\\
\mathcal{O}_{T} &= (\bar{c}_R\sigma^{\mu\nu}b_L)(\bar{\tau}_R\sigma_{\mu\nu}\nu_{\tau L}),
\end{aligned} \quad (62.2)$$

and the corresponding Wilson coefficients are given by $C_X(X = V_1, V_2, S_1, S_2, T)$.

We are interested in the new scalar interaction $\mathcal{O}_{V_1}$ and $\mathcal{O}_{S_1}$, and thus we turn all other Wilson Coefficients to zero for this analysis.

Subject to the above hamiltonian, one can construct the differential decay rate for a particular exclusive decay, involving the NP WC's, the CKM elements and the corresponding hadronic form factors. The measurable observables are ratios fo these integrated decay rates with different leptons in the final states. The ratio cancels uncertainties due to the CKM elements completely, and also those due to the form factors to a large extent. For the theoretical details regarding the observables, the corresponding form factors and the constraints, the interested reader can look into [5, 6] and the references therein.

62.4 Analysis

The results for our fits with a single vector and scalar type NP are displayed in Fig. 62.2 and Table 62.2. In what follows $C_{S_1} = -C_H\, m_b\, m_\ell$. The WC's are considered to be real. It is clear that for all combinations of results shown in Fig. 62.2, there is a two-fold ambiguity in the best-fit results. One of these points is closer to SM than the other and this is the one that is important in constraining NMUED. We also note that while the results from Belle and LHCb are consistent with SM within 3σ, for any and all other combination of results, the SM is away from the best fit point by more than 3σ in the $C_W - C^{\tau}_{H}$ plane.

The results for our fits with a single vector and scalar type NP are displayed in Fig. 62.3 and Table 62.3, assuming c_H to be complex.

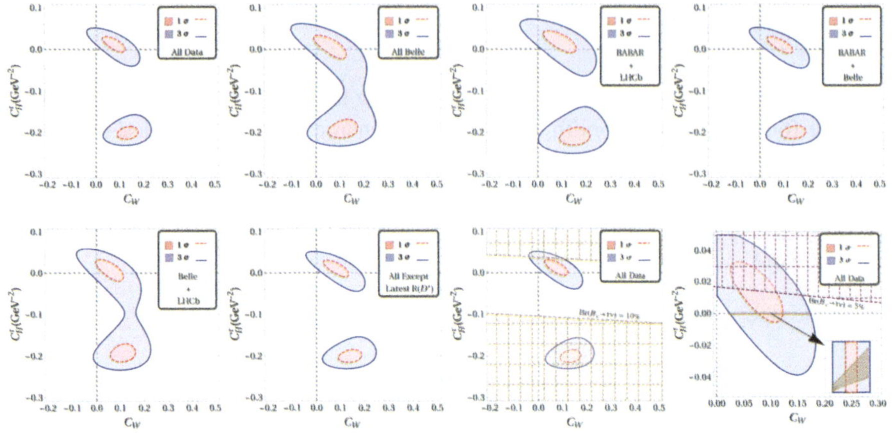

Fig. 62.2 $\mathcal{R}(D^{(*)})$ fit results corresponding to separate fits listed in Table 62.2 for the case with both O_{V_1} and O_{S_1}. Red (dotted) and blue (solid) lines enclose 1σ ($\Delta\chi^2 = 2.30$) and 3σ ($\Delta\chi^2 = 11.83$) regions respectively. Only the gridlines corresponding to $C_W = C_{V_1}$ and $C_H^{\tau} = 0$ are shown, such that there intersection point represents SM. The hatched regions in the last two figures show the constraints coming from $B_c \to \tau\nu$

62.5 Models

The model independent scenarios described in the previous sections are further illustrated with the help of benchmark models in this section. Details about the model parameters and their relations with C_{S_1} and C_{V_1} can be found in [5, 6].

62.5.1 *Real C_{V_1} and C_{S_1}: Non-minimal Universal Extra Dimension (NMUED)*

See Fig. 62.4.

62.5.2 *Real C_{S_1}: Georgi Michacek Model (GM)*

See Fig. 62.5.

62.5.3 *Complex C_{S_1}: Leptoquark Model (LQ)*

See Table 62.4.

Table 62.2 Results of fits for different combinations of experimental data-points along with our predictions for the charged current observables for the case with both O_{V_1} and O_{S_1}. The last two columns are obtained from fits without treating $\mathcal{R}_{J/\psi}$ as a data point

	Without $\mathcal{R}_{J/\psi}$		With $\mathcal{R}_{J/\psi}$				Fit results		Observable values	
			PQCD		LFCQ					
Datasets	$\chi^2_{\min}$/DoF	p-value (%)	$\chi^2_{\min}$/DoF	p-value (%)	$\chi^2_{\min}$/DoF	p-value (%)	Re(C_H) (GeV^{-2})	Im(C_H) (GeV^{-2})	$\mathcal{R}(D^*)$	$\mathcal{R}(D)$
All data	9.22/8	23.72	11.86/9	15.76	12.38/9	13.51	−0.031(8)	0.000(73)	0.2746(25)	0.448(42)
Belle	1.71/4	63.54	4.39/5	35.63	4.89/5	29.83	−0.023(11)	0.000(87)	0.2674(33)	0.406(60)
Babar+LHCb	6.42/3	4.03	9.00/4	2.92	9.54/4	2.29	−0.042(11)	0.000(84)	0.2764(34)	0.508(58)
Babar+ Belle	6.71/6	24.31	9.35/7	15.48	9.87/7	13.03	−0.030(8)	0.000(74)	0.2724(25)	0.445(43)
Belle+LHCb	4.70/6	45.41	7.37/7	28.82	7.88/7	24.72	−0.025(11)	0.000(78)	0.2700(34)	0.414(59)
All $\mathcal{R}_{D^*}$	2.37/5	66.78	4.31/6	50.53	4.99/6	41.67	–	–		
No $P_\tau(D^*)$	9.21/7	16.23	11.84/8	10.58	12.36/8	8.92	−0.031(8)	0.000(72)	0.2746(25)	0.448(42)

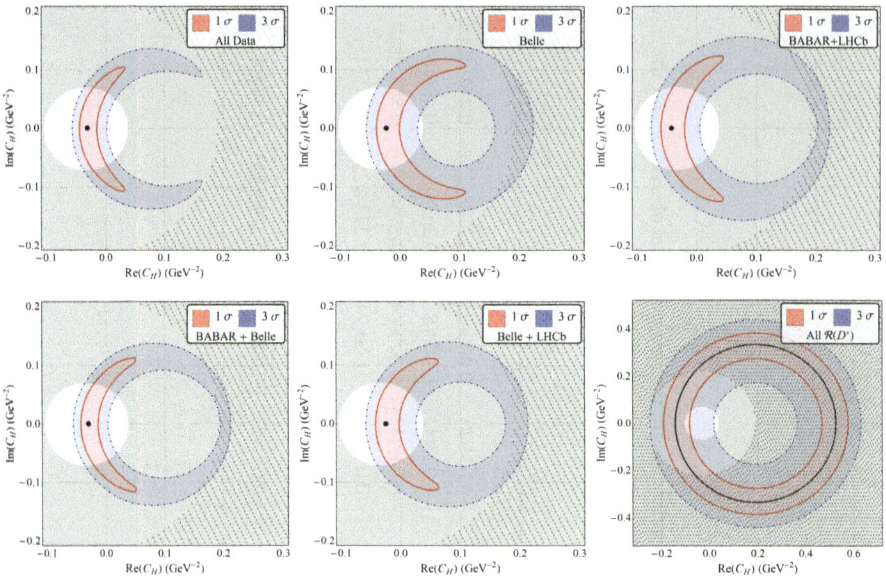

Fig. 62.3 Fit results in terms of the fixed $\Delta\chi^2$ contours representing 1σ (red, solid) and 3σ (blue, dot-dashed) confidence levels respectively, in the $\mathrm{Re}(C_H)$ and $\mathrm{Im}(C_H)$ parameter-space for the case with O_{S_1} only. The diagonally hatched region is ruled out from the B_c life-time constraint and the gray-shaded region is disallowed by the constraint $\mathcal{B}(B_c \to \tau\nu) < 10\%$

Table 62.3 Results of fits with different combinations of experimental data-points for the case with only O_{S_1}

Datasets	Without $\mathcal{R}_{J/\psi}$		With $\mathcal{R}_{J/\psi}$				Fit results	
			PQCD		LFCQ			
	χ^2_{min}/DoF	p-value (%)	χ^2_{min}/DoF	p-value (%)	χ^2_{min}/DoF	p-value (%)	$\mathrm{Re}(C_H)$ (GeV^{-2})	$\mathrm{Im}(C_H)$ (GeV^{-2})
All data	9.22/8	23.72	11.86/9	15.76	12.38/9	13.51	−0.031(8)	0.000(73)
Belle	1.71/4	63.54	4.39/5	35.63	4.89/5	29.83	−0.023(11)	0.000(87)
Babar+LHCb	6.42/3	4.03	9.00/4	2.92	9.54/4	2.29	−0.042(11)	0.000(84)
Babar+Belle	6.71/6	24.31	9.35/7	15.48	9.87/7	13.03	−0.030(8)	0.000(74)
Belle+LHCb	4.70/6	45.41	7.37/7	28.82	7.88/7	24.72	−0.025(11)	0.000(78)
All $\mathcal{R}_{D^*}$	2.37/5	66.78	4.31/6	50.53	4.99/6	41.67	–	–
No $P_\tau(D^*)$	9.21/7	16.23	11.84/8	10.58	12.36/8	8.92	−0.031(8)	0.000(72)

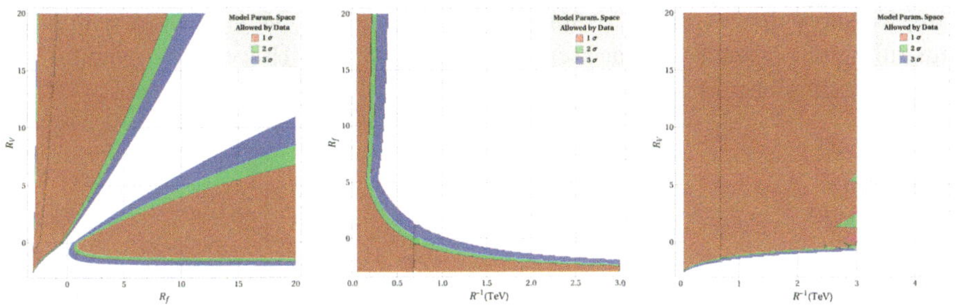

(a) R_V vs. R_f (b) R_f vs. R^{-1} (c) R_V vs. R^{-1}

Fig. 62.4 Regions in the NMUED model parameter space, allowed by $C_W - C_H^{\tau}$ fit of $\mathcal{R}(D^{(*)})$ data

Fig. 62.5 v_χ versus $m_{H_3^\pm}$ parameter space excluded by all Belle and LHCb data at 2σ (orange, solid) and 3σ (blue, dashed) confidence levels. Regions above the lines are excluded

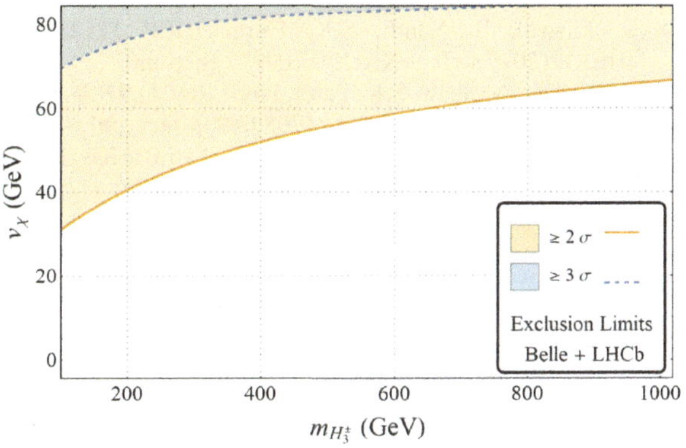

Table 62.4 Allowed values of the product of the couplings (both real and imaginary) of the chosen Leptoquark model involved with the Wilson coefficient $C_{S_1}^l$

Data	Re $(g_{2L}^{33} g_{2R}^{23*})$	Im $(g_{2L}^{33} g_{2R}^{33*})$
All data	−0.250(64)	0.0(6)
Belle	−0.186(90)	0.0(7)
Babar+LHCb	−0.338(89)	0.0(7)
Babar+Belle	−0.245(65)	0.0(6)
Belle+LHCb	−0.198(88)	0.0(6)
No P_τ (D^*)	−0.250(64)	0.0(6)

62.6 Conclusion

We have explored the scope of vector and scalar type NP in explaining the charged current anomalies, primarily from a model independent point of view, further illustrated by specific models. For the vector+scalar case, we show that the Non minimal universal extra dimensional model can explain such anomalies. For the single scalar case, we show that a real WC is unable to accommodate the anomalies within 3σ, thus ruling out extended scalar sector models in general. However, a complex WC can still do the job and is illustrated by a particular Leptoquark scenario.

References

1. D. Bigi, P. Gambino, Phys. Rev. D **94**(9), 094008 (2016). https://doi.org/10.1103/PhysRevD.94.094008. arXiv:1606.08030 [hep-ph]
2. F.U. Bernlochner, Z. Ligeti, M. Papucci, D.J. Robinson, Phys. Rev. D **95**(11), 115008 (2017). Erratum: [Phys. Rev. D **97**(5), 059902 (2018)]. https://doi.org/10.1103/PhysRevD.95.115008. https://doi.org/10.1103/PhysRevD.97.059902. arXiv:1703.05330 [hep-ph]
3. D. Bigi, P. Gambino, S. Schacht, JHEP **1711**, 061 (2017). https://doi.org/10.1007/JHEP11(2017)061. arXiv:1707.09509 [hep-ph]
4. S. Jaiswal, S. Nandi, S.K. Patra, JHEP **1712**, 060 (2017). https://doi.org/10.1007/JHEP12(2017)060. arXiv:1707.09977 [hep-ph]
5. A. Biswas, A. Shaw, S.K. Patra, Phys. Rev. D **97**(3), 035019 (2018). https://doi.org/10.1103/PhysRevD.97.035019. arXiv:1708.08938 [hep-ph]
6. A. Biswas, D.K. Ghosh, A. Shaw, S.K. Patra, arXiv:1801.03375 [hep-ph]

Chapter 63
Exploring Lepton Non-universality in $b \to c l \bar{\nu}_l$ Decay Modes in the Light of Recent Experimental Data

Suchismita Sahoo, Rukmani Mohanta and Anjan K. Giri

Abstract The observation of lepton non-universality in B sector has become a challenge for universality lovers. Recently, the LHCb Collaboration has added a new $R_{J/\psi}$ ratio into this challenging area, which has a 2σ deviation from its standard model prediction. In this regard, we present a model independent analysis of $b \to c l \bar{\nu}_l$ decays in order to scrutinize the presence of lepton universality violation in B sector. We first perform a global fit to $R_{D^{(*)}}$ data. We then estimate the lepton non-universality parameters of $B_c \to (\eta_c, J/\psi) l \nu_l$ and $B_s \to D_s^{(*)} l \bar{\nu}_l$ decay processes.

63.1 Introduction

Though the study of the semileptonic B decays with the tau-lepton in the final state is quite challenging experimentally, they are quite interesting from the theoretical point of view, due to the presence of several observables besides branching fractions, such as decay distributions and tau polarizations, which are quite sensitive to new physics (NP). In addition, the ratio of branching fractions of $\bar{B} \to D^{(*)} \tau \bar{\nu}_\tau$ decay to the semileptonic decay modes with light charged leptons (e, μ) in the final state, reduces the uncertainties significantly. The experimental measured values of $R_{D^{(*)}}$ [1]

$$R_D = \frac{\mathrm{Br}\left(\bar{B} \to D \tau \bar{\nu}_l\right)}{\mathrm{Br}\left(\bar{B} \to D l \bar{\nu}_l\right)} = 0.397 \pm 0.040 \pm 0.028, \tag{63.1}$$

S. Sahoo (✉)
Physical Research Laboratory, THEPH Division, Ahmedabad 380009, India
e-mail: suchismita8792@gmail.com

R. Mohanta
School of Physics, University of Hyderabad, Hyderabad 500046, India
e-mail: rukmani98@gmail.com

A. K. Giri
Department of Physics, IIT Hyderabad, Kandi 502285, India
e-mail: giria@iith.ac.in

A. Giri and R. Mohanta (eds.), *16th Conference on Flavor Physics and CP Violation*, Springer Proceedings in Physics 234,
https://doi.org/10.1007/978-3-030-29622-3_63

$$R_{D^*} = \frac{\text{Br}\left(\bar{B} \to D^* \tau \bar{\nu}_l\right)}{\text{Br}\left(\bar{B} \to D^* l \bar{\nu}_l\right)} = 0.316 \pm 0.016 \pm 0.010, \tag{63.2}$$

have about 1.9σ and 3.3σ deviations from their corresponding standard model (SM) predictions $R_D^{\text{SM}} = 0.300 \pm 0.008$ [2] and $R_{D^*}^{\text{SM}} = 0.252 \pm 0.003$ [3]. Apart from these anomalies, recently the LHCb Collaboration has observed discrepancy of 2σ in the $R_{J/\psi}$ parameter, $R_{J/\psi}^{\text{Expt}} = 0.71 \pm 0.17 \pm 0.18$ [4] from its SM result [5]. Keeping this in mind, in this work, we investigate the $B_c \to (\eta_c, J/\psi) l \bar{\nu}_l$ and $B_s \to D_s^{(*)} l \bar{\nu}_l$ processes in a model independent way.

The paper is organised as follows. We discuss the effective Lagrangian and the numerical fit to new parameters in Sect. 63.2. Section 63.3 contains the discussion on lepton non-universality (LNU) parameters of $b \to c l \bar{\nu}_l$ decay modes.

63.2 Effective Lagrangian and the Numerical Fit

Considering only the left handed neutrinos, the effective Lagrangian of $b \to c \tau \bar{\nu}_l$ process, is given as [6]

$$\mathcal{H}_{\text{eff}} = \frac{4G_F}{\sqrt{2}} V_{cb} \Big[(\delta_{l\tau} + V_L) \mathcal{O}_{V_1}^l + V_R \mathcal{O}_{V_2}^l + S_L \mathcal{O}_{S_1}^l + S_R \mathcal{O}_{S_2}^l + T_L \mathcal{O}_T^l \Big], \tag{63.3}$$

where G_F is the Fermi constant, V_{cb} is the CKM matrix element, $\mathcal{O}_x^l$ ($x = V_{1,2}$, $S_{1,2}$, T) are the effective operators and V_x are their corresponding Wilson coefficients. We neglect the tensor coefficient for simplicity.

After getting idea on the new parameters, we first perform a χ^2 test to measure the disagreement of SM with the data. We have assumed Gaussian errors for all the observables. The χ^2 is defined as

$$\chi^2 = \sum_i \frac{(\mathcal{O}_i^{\text{th}} - \mathcal{O}_i^{\text{exp}})^2}{(\Delta \mathcal{O}_i^{\text{exp}})^2}, \tag{63.4}$$

where $\mathcal{O}_i^{\text{exp}}$ represents the measured central value of the observables and $\Delta \mathcal{O}_i^{\text{exp}}$ represents the corresponding 1σ uncertainty. Here, $\mathcal{O}_i^{\text{th}}$ represents the theoretical prediction of the observables. We include $R_{D^{(*)}}$ measurements for the evaluation of χ^2. We assume that the NP contributions are coming from an operator characterized by a single NP Wilson coefficient only, which is considered to be imaginary in this analysis. Using (63.4) and the required input parameters from [7, 8] the contour plots for only V_L (top-left panel), V_R (top-right panel), S_L (bottom-left panel) and S_R (bottom-right panel) new coefficients are presented in Fig. 63.1. The best fit values of the real and imaginary parts of all the new parameters, their corresponding 1σ range and the minimum values of χ^2 are listed in Table 63.1.

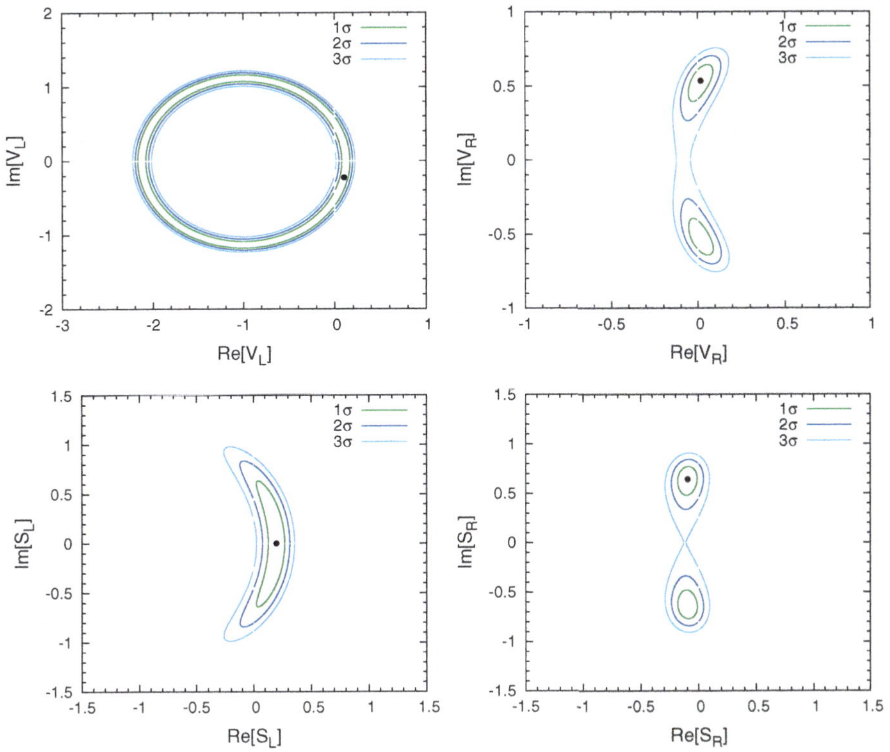

Fig. 63.1 Contour plots of Re[V_x] $\sim$ Im[V_x] coefficients obtained from $R_{D^{(*)}}$ parameters

Table 63.1 Bestfit values and 1σ range of new coefficients

NP WC	Bestfit values (Re, Im)	1σ range (Re, Im)	$\chi^2_{\rm min}$
V_L	(0.1026, −0.225)	([0.0026, 0.13], [−0.34, 0.34])	0.15
V_R	(0.0183, 0.535)	([−0.02, 0.06], [−0.59, −0.46])	9.92×10^{-30}
S_L	(0.2, -0.85×10^{-9})	([0.145, 0.245], [−0.32, 0.32])	0.125
S_R	(−0.095, 0.6414)	([−0.15, −0.06], [−0.72, −0.66])	8.63×10^{-31}

63.3 $B_{c,s} \to (P, V)l\bar{\nu}_l$ Processes

In this section, we study the impact of these new coefficients on the rare $B_{c,s} \to (P, V)l\bar{\nu}_l$ processes, where $P = \eta_c, D_s$ are the pseudoscalar mesons and $V = J/\psi, D_s^*$ are the vector mesons. We mainly focus on the $R_{P(V)}$ LNU parameters, which are defined as

$$R_{P(V)} = \frac{\mathrm{Br}(B_{c,s} \to P(V)\tau\bar{\nu}_\tau)}{\mathrm{Br}(B_{c,s} \to P(V)l\bar{\nu}_l)}, \quad l = e, \mu. \tag{63.5}$$

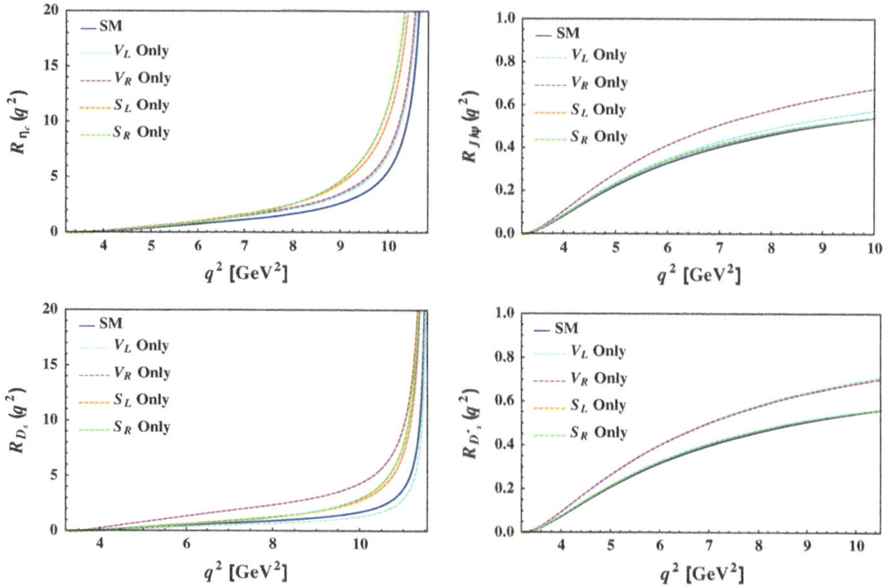

Fig. 63.2 The q^2 variation of R_{η_c} (top-left), $R_{J/\psi}$ (top-right), R_{D_s} (bottom-left) and $R_{D_s^*}$ (bottom-right) LNU parameters in the presence of one coefficient at a time

Table 63.2 Numerical values of the $R_{P(V)}$ parameters in the presence of new coefficients

Parameters	SM	V_L	V_R	S_L	S_R
R_{η_c}	0.417	0.528	0.552	0.624	0.62
$R_{J/\psi}$	0.283	0.3	0.354	0.289	0.29
R_{D_s}	0.34	0.257	0.805	0.45	0.437
$R_{D_s^*}$	0.259	0.328	0.325	0.264	0.26

For numerical analysis, the particle masses and the lifetime of $B_{c,s}$ mesons are taken from [7] and the form factors from [8]. Using the bestfit values of $V_{L,R}$ and $S_{L,R}$ coefficients from Table 63.1, we show the plots for the LNU parameters, R_{η_c} (top-left panel), $R_{J/\psi}$ (top-right panel), R_{D_s} (bottom-left panel) and $R_{D_s^*}$ (bottom-right panel) in Fig. 63.2. Here the blue solid lines represent the SM contribution, the cyan, magenta, orange and green color dotted lines stand for the contribution from the additional V_L, V_R, S_L and S_R coefficients, respectively. We found that these LNU parameters deviate significantly from their corresponding SM values in the presence of new coefficients. In Table 63.2, we report the numerical values of all these LNU parameters.

Acknowledgements SS and RM acknowledge the financial support from SERB, Govt. of India.

References

1. Y. Amhis et al., Heavy flavor averaging group. Eur. Phys. J. C **77**, 895 (2017). https://hflav.web.cern.ch
2. H. Na et al., Phys. Rev. D **92**, 054410 (2015). https://doi.org/10.1103/PhysRevD.92.054510
3. S. Fajfer, J.F. Kamenik, I. Nisandzic, J. Zupan, Phys. Rev. Lett. **109**, 161801 (2012). https://doi.org/10.1103/PhysRevLett.109.161801
4. R. Aaij et al., LHCb. Phys. Rev. Lett. **120**, 121801 (2018). https://doi.org/10.1103/PhysRevLett.120.121801
5. W.-F. Wang, Y.-Y. Fan, Z.-J. Xiao, Chin. Phys. C **37**, 093102 (2013). https://doi.org/10.1088/1674-1137/37/9/093102
6. V. Cirigliano, J. Jenkins, M. Gonzalez-Alonso, Nucl. Phys. B **830**, 95 (2010). https://doi.org/10.1016/j.nuclphysb.2009.12.020
7. C. Patrignani et al., Particle data group. Chin. Phys. C **40**, 100001 (2016). https://doi.org/10.1088/1674-1137/40/10/100001
8. M. Tanaka, R. Watanabe, Phys. Rev. D **87**, 034028 (2013). https://doi.org/10.1103/PhysRevD.87.034028; F. Wang, Y.Y. Fan, Z.J. Xiao, Chin. Phys. C **37**, 093102 (2013). https://doi.org/10.1088/1674-1137/37/9/093102; C.J. Monahan, H. Na, C.M. Bouchard, G.P. Lepage, J. Shigemitsu, Phys. Rev. D **95**(11), 114506 (2017). https://doi.org/10.1103/PhysRevD.95.114506

Chapter 64
Rare Semileptonic $b \to d\mu^+\mu^-$ Transitions and Effects of Z' in the 331 Model

Soram Robertson Singh and Barilang Mawlong

Abstract We study the rare semileptonic $B \to \rho\mu^+\mu^-$ decay mediated by $b \to d\mu^+\mu^-$ transition. Differential branching ratio, longitudinal lepton polarization and forward-backward asymmetry of $B \to \rho\mu^+\mu^-$ are analyzed within the Standard Model in the low q^2 region. We study the effects of Z' in the 331 or $SU_C(3) \times SU_L(3) \times U_X(1)$ model on this decay mode by varying the model parameter β at $M_{Z'} = 1,\ 3$ TeV. The couplings of Z' are constrained by flavor observables of $B_d - \bar{B}_d$ mixing. We find that the observables of interest are sensitive to Z' effects and we observe deviations from the SM for some values of β.

64.1 Introduction

Standard Model has been successful in explaining many experimental observations. But there are some observed discrepancies that SM cannot explain. Semileptonic B decays are good testing grounds for SM and models beyond it as these decays are highly suppressed in the SM. The $b \to sll$ modes have been studied extensively [1, 2] and it is important that we discuss the complementary $b \to dll$ modes as well as they have similar properties. The observables of $b \to dl^+l^-$ transitions are more CKM-suppressed than those of $b \to sl^+l^-$ transitions because the CKM factor $|\lambda_i^s|$ is greater than $|\lambda_j^d|$ where $\lambda_i^s = V_{\text{ib}}^* V_{\text{is}}$, $\lambda_j^d = V_{\text{jb}}^* V_{\text{jd}}$ with $i = c,\ t$ and $j = u,\ c,\ t$ and deviation of the observables from their SM values could hint the possibility of New Physics. In this paper, we thus study the $B \to \rho ll$ mode mediated by the $b \to dll$ transition in the SM and also in the 331 Z' model that allows $b \to dll$ FCNC transitions at the tree level. The effects of such FCNC Z' mediated transitions on these modes are analyzed here.

S. Robertson Singh (✉) · B. Mawlong
School of Physics, University of Hyderabad, Hyderabad 500046, India
e-mail: robsoram@gmail.com

B. Mawlong
e-mail: barilang05@gmail.com

A. Giri and R. Mohanta (eds.), *16th Conference on Flavor Physics and CP Violation*, Springer Proceedings in Physics 234,
https://doi.org/10.1007/978-3-030-29622-3_64

64.2 Standard Model Framework

In the SM framework, the effective Hamiltonian describing the quark level transition $b \to dl^+l^-$ is given in [3, 4] as

$$\mathscr{H}_{\text{eff}} = -\frac{G_F}{\sqrt{2}}\left[V_{\text{tb}}V_{\text{td}}^* \sum_{i=1}^{10} C_i(\mu)O_i(\mu) + V_{\text{ub}}V_{\text{ud}}^* \sum_{i=1}^{2} C_i(\mu)\left(O_i(\mu) - O_i^{(u)}(\mu)\right)\right], \tag{64.1}$$

where G_{F} is the Fermi coupling constant, $C_i(\mu)$ are the Wilson coefficients and $O_i(\mu)$ are local operators as defined in [2].

64.3 331 Model Framework

It is a model based on $SU(3)_C \otimes SU(3)_L \otimes U(1)_X$ guage group wherein the electric charge Q is defined as a linear combination of the diagonal generators of the group

$$Q = T^3 + \beta T^8 + XI, \tag{64.2}$$

where β is a key parameter that gives rise to different variants of the 331 model. The generators $T^3 = 1/2\ \text{diag}(1, -1, 0)$ and $T^8 = (1/2\sqrt{3})\ \text{diag}(1, 1, -2)$ are two of the SU(3) generators and X is the generator of $\text{U}(1)_{\text{X}}$. Details of the model can be found in [5, 6].

The NP contributions C_9^{NP} and C_{10}^{NP} are given in [5] as

$$C_9^{\text{NP}} = -\frac{1}{g_{\text{SM}}^2 M_{Z'}^2 \sin^2\theta_W}\frac{\Delta_L^{db}(Z')\Delta_V^{\mu\bar{\mu}}(Z')}{V_{td}^* V_{tb}}, \tag{64.3}$$

$$C_{10}^{\text{NP}} = -\frac{1}{g_{\text{SM}}^2 M_{Z'}^2 \sin^2\theta_W}\frac{\Delta_L^{db}(Z')\Delta_A^{\mu\bar{\mu}}(Z')}{V_{td}^* V_{tb}}, \tag{64.4}$$

where $\Delta_V^{\mu\bar{\mu}}(Z')$, $\Delta_A^{\mu\bar{\mu}}(Z')$ are lepton couplings of Z' and their values are taken from [7]. The quark couplings $\Delta^{bd}(Z')$ are given by $\Delta_L^{db}(Z') = -0.489 s_{13} e^{i\delta_1}$ for $\beta = \pm 2/\sqrt{3}$. The parameters (s_{13}, δ_1) are constrained by the flavor observables of $B_{\text{d}} - \bar{B}_{\text{d}}$ mixing and the obtained constraint regions are shown in Fig. 64.1. In this paper, we have presented the analysis of Z' effects for the value of $\beta = -2/\sqrt{3}$. The constraint regions are divided into 4 scenarios for $M_{Z'} = 1, 3$ TeV and they are given in Table 64.1.

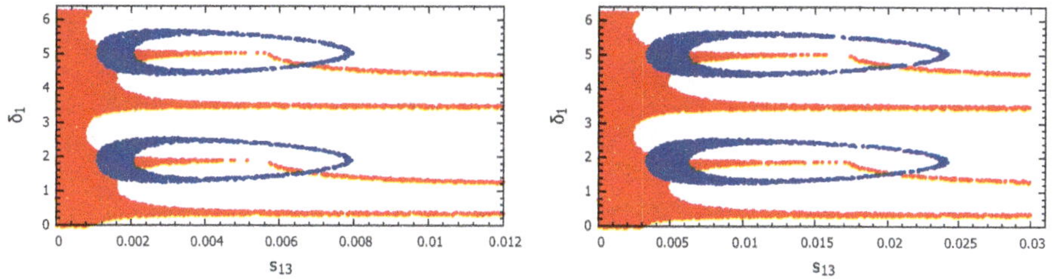

(a) $M_{Z'} = 1$ TeV (b) $M_{Z'} = 3$ TeV

Fig. 64.1 The constraint regions for s_{13} and δ_1 obtained from the observables of $B_d - \bar{B}_d$ mixing

Table 64.1 Constraint scenarios (s_{13} in units of 10^{-3} and δ_1 in radians)

	Scenario 1		Scenario 2		Scenario 3		Scenario 4	
$M_{Z'}$	s_{13}	δ_1	s_{13}	δ_1	s_{13}	δ_1	s_{13}	δ_1
1 TeV	1.08–2.06	1.41–2.10	1.08–2.08	4.54–5.28	6.61–6.82	1.58–1.57	6.60–6.80	4.70–4.73
3 TeV	3.30–6.56	1.41–2.10	3.32–6.39	4.55–5.27	20.13–20.72	1.58–1.60	20.32–20.66	4.72–4.78

64.4 Results

The differential branching ratio, longitudinal lepton polarization asymmetry and forward-backward charge asymmetry (A_{FB}) of $B \to \rho\mu^+\mu^-$ are computed in the SM and the effects of Z' boson arising from 331 model on these observables are analyzed. For scenarios 1 and 2, we have plotted bands for the observables. Since scenarios 3 and 4 are small, we have plotted only the lower curves for these two scenarios. The plots for the differential branching ratio, longitudinal lepton polarization asymmetry and forward-backward charge asymmetry are shown in Figs. 64.2, 64.3 and 64.4,

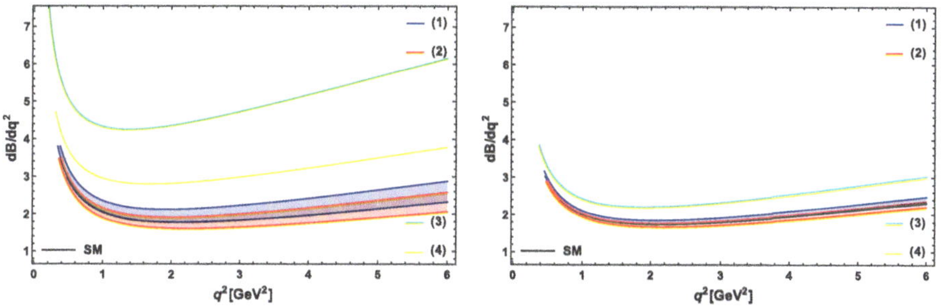

(a) $M_{Z'} = 1$ TeV (b) $M_{Z'} = 3$ TeV

Fig. 64.2 Variation of the differential branching ratio (in units of 10^{-9}) with q^2 both in the SM and in the 331-Z' model

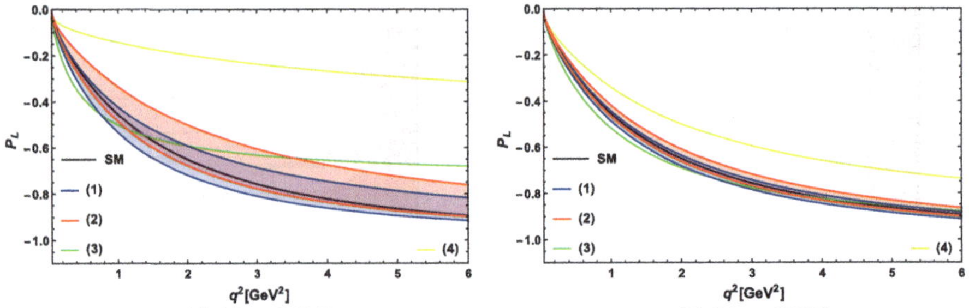

(a) $M_{Z'}$ = 1 TeV (b) $M_{Z'}$ = 3 TeV

Fig. 64.3 Variation of the longitudinal lepton polarization asymmetry with q^2 both in the SM and in the 331-Z' model

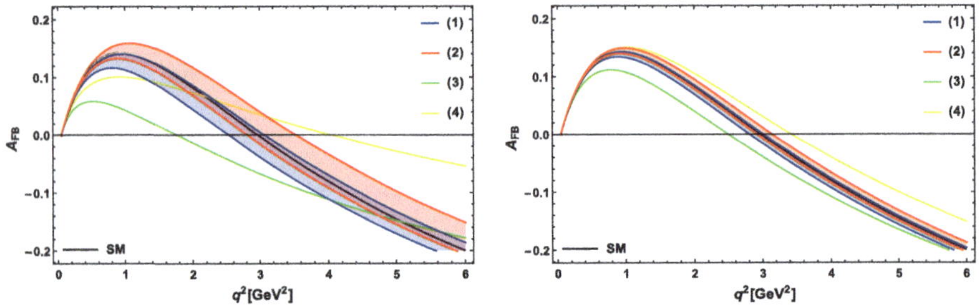

(a) $M_{Z'}$ =1 TeV (b) $M_{Z'}$ =3 TeV

Fig. 64.4 Variation of the forward-backward asymmetry with q^2 both in the SM and in the 331-Z' model

respectively. It can be seen from these figures that observables are sensitive to Z' effects and they deviate from SM prediction. The deviations are found to be larger for $M_{Z'} = 1$ TeV than $M_{Z'} = 3$ TeV and this is due to the dividing factor of $M_{Z'}^2$ in the $C_9^{\rm NP}$, $C_{10}^{\rm NP}$ as expressed in (64.3) and (64.4).

Acknowledgements S. Robertson Singh would like to acknowledge the financial support from DST(Inspire), Government of India.

References

1. A. Ali, P. Ball, L.T. Handoko, G. Hiller, Phys. Rev. D **61**, 074024 (2000)
2. D. Melikhov, N. Nikitin, S. Simula, Phys. Rev. D **57**, 6814 (1998)
3. R.N. Faustov, V.O. Galkin, Eur. Phys. J. C **74**, 2911 (2014)
4. G. Buchalla, A.J. Buras, M.E. Lautenbacher, Rev. Mod. Phys. **68**, 1125 (1996)
5. A.J. Buras, F. De Fazio, J. Girrbach, M.V. Carlucci, JHEP **02**(2013), 023 (2013)
6. R.A. Diaz, R. Martinez, F. Ochoa, Phys. Rev. D **72**, 035018 (2005)
7. A.J. Buras, F. De Fazio, J. Girrbach, JHEP **02**(2014), 112 (2014)

Chapter 65
Resolving R_D and R_{D^*} Anomalies

Suman Kumbhakar, Ashutosh Kumar Alok, Dinesh Kumar and S. Uma Sankar

Abstract The current world averages of the ratios $R_{D^{(*)}}$ are about 4σ away from their Standard Model prediction. These measurements indicate towards the violation of lepton flavor universality in $b \to c\, l\, \bar{\nu}$ decay. The different new physics operators, which can explain the $R_{D^{(*)}}$ measurements, have been identified previously. We show that a simultaneous measurement of the polarization fractions of τ and D^* and the angular asymmetries $A_{\rm FB}$ and $A_{\rm LT}$ in $B \to D^*\tau\bar{\nu}$ decay can distinguish all the new physics amplitudes and hence uniquely identify the Lorentz structure of new physics.

65.1 Introduction

In recent years, the evidence for charged lepton universality violation is observed in the charge current process $b \to c\tau\bar{\nu}$. The experiments, BaBar, Belle and LHCb, made several measurements of the ratios

$$R_D = \frac{\Gamma(B \to D\,\tau\,\bar{\nu})}{\Gamma(B \to D\,\{e/\mu\}\,\bar{\nu})}, \quad R_{D^*} = \frac{\Gamma(B \to D^*\,\tau\,\bar{\nu})}{\Gamma(B \to D^*\,\{e/\mu\}\,\bar{\nu})}. \tag{65.1}$$

The current world average of these measurements are about 4σ away from the Standard Model (SM) predictions [1].

S. Kumbhakar (✉) · S. U. Sankar
Indian Institute of Technology Bombay, Mumbai 400076, India
e-mail: suman@phy.iitb.ac.in

S. U. Sankar
e-mail: uma@phy.iitb.ac.in

A. K. Alok
Indian Institute of Technology Jodhpur, Jodhpur 342011, India
e-mail: akalok@iitj.ac.in

D. Kumar
University of Rajasthan, Jaipur 302004, India
e-mail: dinesh@uniraj.ac.in

A. Giri and R. Mohanta (eds.), *16th Conference on Flavor Physics and CP Violation*, Springer Proceedings in Physics 234,
https://doi.org/10.1007/978-3-030-29622-3_65

All the meson decays in (65.1) are driven by quark level transitions $b \to cl\bar{\nu}$. These transitions occur at tree level in the SM. The discrepancy between the measured values of R_D and R_{D^*} and their respective SM predictions is an indication of presence of new physics (NP) in the $b \to c\tau\bar{\nu}$ transition. The possibility of NP in $b \to c\mu\bar{\nu}$ is excluded by other data [2]. All possible NP four-Fermi operators for $b \to c\tau\bar{\nu}$ transition are listed in [3]. In [2], a fit was performed between all the $b \to c\tau\bar{\nu}$ data and each of the NP interaction term. The NP terms, which can account for the all $b \to c\tau\bar{\nu}$ data, are identified and their Wilson coefficients (WCs) are calculated. It was found that there are six allowed NP solutions. Among those six solutions, four solutions are distinct with a different Lorentz structure. In [4] it was found that the tensor NP solution could be distinguished from other possibilities provided $\langle f_L \rangle$, the D^* polarization fraction can be measured with an absolute uncertainty of 0.1.

Here, we consider four angular observables, $P_\tau(D^*)$ (τ polarization fraction), f_L (D^* polarization fraction), $A_{\rm FB}$ (the forward-backward asymmetry), $A_{\rm LT}$ (longitudinal-transverse asymmetry) in the decay $B \to D^*\tau\bar{\nu}$. Note that these asymmetries can only be measured if the momentum of the τ lepton is reconstructed. We show that a measurement of these four quantities can uniquely identify the Lorentz structure of the NP operator responsible for the present discrepancy in R_D and R_{D^*} [5].

65.2 Distinguishing Different New Physics Solutions

The most general effective Hamiltonian for $b \to c\tau\bar{\nu}$ transition can be written as

$$H_{\rm eff} = \frac{4G_F}{\sqrt{2}} V_{cb} \left[O_{V_L} + \frac{\sqrt{2}}{4G_F V_{\rm cb}} \frac{1}{\Lambda^2} \left\{ \sum_i \left(C_i O_i + C_i^{'} O_i^{'} + C_i^{''} O_i^{''} \right) \right\} \right], \quad (65.2)$$

where G_F is the Fermi coupling constant, $V_{\rm cb}$ is the Cabibbo-Kobayashi-Maskawa (CKM) matrix element and the NP scale Λ is assumed to be 1 TeV. We also assume that neutrino is always left chiral. The effective Hamiltonian for the SM contains only the O_{V_L} operator. The explicit forms of the four-fermion operators O_i, $O_i^{'}$ and $O_i^{''}$ are given in [3]. The NP effects are encoded in the NP WCs C_i, $C_i^{'}$ and $C_i^{''}$. Each primed and double primed operator can be expressed as a linear combination of unprimed operators through Feirz transformation.

The values of NP WCs which fit the data on the observables R_D, R_{D^*}, $R_{J/\psi}$, $P_\tau(D^*)$ and $\mathscr{B}(B_c \to \tau\bar{\nu})$, have been calculated previously [2]. Here $R_{J/\psi}$ is the ratio of $\mathscr{B}(B_c \to J/\psi\tau\bar{\nu})$ to $\mathscr{B}(B_c \to J/\psi\mu\bar{\nu})$ [6]. The results of these fits are listed in Table 65.1. This table also lists, for each of the NP solutions, the predicted values of the polarization fractions and the angular asymmetries in $B \to D^*\tau\bar{\nu}$ decay.

Here we compute $A_{\rm FB}(q^2)$ and $A_{\rm LT}(q^2)$ in $B \to D^*\tau\bar{\nu}$ decay, as functions of $q^2 = (p_B - p_{D^*})^2$, where p_B and p_{D^*} are the four momenta of B and D^* respectively.

Table 65.1 Best fit values of NP WCs at $\Lambda = 1$ TeV, taken from table IV of [2]. We provide the predictions of $\langle P_\tau(D^*)\rangle$, $\langle f_L\rangle$, $\langle A_{\rm FB}\rangle$ and $\langle A_{\rm LT}\rangle$ in decay $B \to D^*\tau\bar{\nu}$ with their uncertainties for each of the allowed solutions

NP WCs	Fit values	$\langle P_\tau(D^*)\rangle$	$\langle f_L\rangle$	$\langle A_{\rm FB}\rangle$	$\langle A_{\rm LT}\rangle$
SM	$C_i = 0$	-0.499 ± 0.004	0.45 ± 0.04	-0.011 ± 0.007	-0.245 ± 0.003
C_{V_L}	0.149 ± 0.032	-0.499 ± 0.004	0.45 ± 0.04	-0.011 ± 0.007	-0.245 ± 0.003
C_T	0.516 ± 0.015	$+0.115 \pm 0.013$	0.14 ± 0.03	-0.114 ± 0.009	$+0.110 \pm 0.009$
C''_{S_L}	-0.526 ± 0.102	-0.485 ± 0.003	0.46 ± 0.04	-0.087 ± 0.011	-0.211 ± 0.008
(C_{V_L}, C_{V_R})	$(-1.286, 1.512)$	-0.499 ± 0.004	0.45 ± 0.04	-0.371 ± 0.004	$+0.007 \pm 0.004$
(C'_{V_L}, C'_{V_R})	$(0.124, -0.058)$	-0.484 ± 0.005	0.45 ± 0.04	-0.003 ± 0.007	-0.243 ± 0.003
(C''_{S_L}, C''_{S_R})	$(-0.643, -0.076)$	-0.477 ± 0.003	0.46 ± 0.04	-0.104 ± 0.005	-0.202 ± 0.002

The predictions for $P_\tau(D^*)$, f_L and $A_{\rm FB}$ are calculated using the framework provided in [7] and for $A_{\rm LT}(q^2)$ we follow [8, 9].

The $B \to D^{(*)}\, l\, \bar{\nu}$ decay distributions depend upon hadronic form-factors. The form factors for $B \to D$ decay are well known in lattice QCD [10] and we use them in our analyses. For $B \to D^*$ decay, the HQET parameters are extracted using data from Belle and BaBar experiments along with lattice inputs. In this work, the numerical values of these parameters are taken from [1, 11].

This table lists six different NP solutions but only the first four solutions are distinct [2]. Thus we have four different NP solutions with different Lorentz structures. We explore methods to distinguish between them.

65.3 Results and Discussions

The average values of $P_\tau(D^*)$ and f_L for all six NP solutions are given in Table 65.1. Not surprisingly, there is a large difference between the predicted values for O_T solution and those for other NP solutions. If either of these observables is measured with an absolute uncertainty of 0.1, then the O_T solution is either confirmed or ruled out at 3σ level.

We now show that the angular asymmetries $A_{\rm FB}$ and $A_{\rm LT}$ have a good discrimination capability between the three remaining NP WCs. The plots for $A_{\rm FB}$ and $A_{\rm LT}$ as a function of q^2 are shown in the bottom row of Fig. 65.1 and their average values are listed in Table 65.1. We see that the plots of both $A_{\rm FB}(q^2)$ and $A_{\rm LT}(q^2)$, for (O_{V_L}, O_{V_R}) solution, differ significantly from the plots of all other NP solutions as

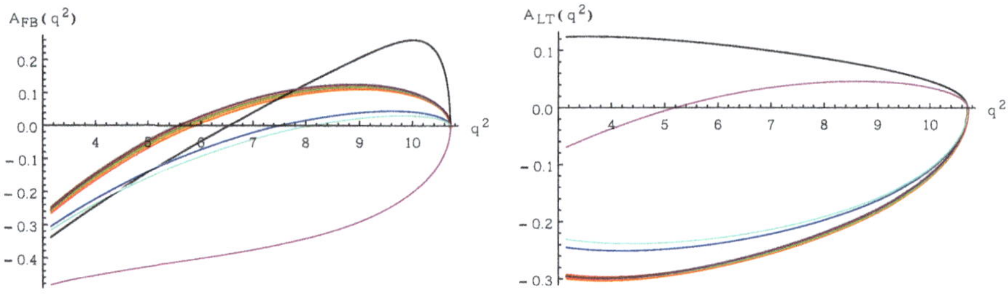

Fig. 65.1 Left and right panels correspond to $A_{\rm FB}(q^2)$ and $A_{\rm LT}(q^2)$, respectively for the $B \to D^* \tau \bar{\nu}$ decay. Red curves with yellow band corresponds to SM predictions. The band, representing 1σ range, is mainly due to the uncertainties in various hadronic form factors and is obtained by adding these errors in quadrature. In each panel, the color code for the NP solutions is: $C_{V_L} = 0.149$ (green curve), $C_T = 0.516$ (black curve), $C''_{S_L} = -0.526$ (blue curve), $(C_{V_L}, C_{V_R}) = (-1.286, 1.512)$ (magenta curve), $(C'_{V_L}, C'_{V_R}) = (0.124, -0.058)$ (purple curve), $(C''_{S_L}, C''_{S_R}) = (-0.643, -0.076)$ (cyan curve)

do the average values. If either of these asymmetries is measured with an absolute uncertainty of 0.07, then the (O_{V_L}, O_{V_R}) solution is either confirmed or ruled out at 3σ level.

So far we have identified observables which can clearly identify the O_T and the (O_{V_L}, O_{V_R}) solutions. As we can see from Table 65.1, one needs to measure $\langle A_{FB} \rangle$ with an absolute uncertainty of 0.03 or better to obtain a 3σ distinction between O_{V_L} and O''_{S_L} solutions. However, this ability to make the distinction can be improved by observing q^2 dependence of $A_{\rm FB}$ for these solutions. We note that $A_{\rm FB}(q^2)$ for O_{V_L} solution has a zero crossing at $q^2 = 5.6$ GeV^2 whereas this crossing point occurs at $q^2 = 7.5$ GeV^2 for O''_{S_L} solution. A calculation of $\langle A_{FB} \rangle$ in the limited range 6 GeV$^2 < q^2 < q^2_{max}$ gives the result +0.1 for O_{V_L} and +0.01 for O''_{S_L}. Hence, determining the sign of $\langle A_{FB} \rangle$, for the full q^2 range and for the limited higher q^2 range, provides a very useful tool for discrimination between these two solutions.

Hence, we find that a clear distinction can be made between the four different NP solutions to the R_D/R_{D^*} puzzle by means of polarization fractions and angular asymmetries. Note that only the observables ($P_\tau(D^*)$ and f_L) isolating O_T do not require the reconstruction of τ momentum. The reconstruction of τ momentum is crucial to measure the asymmetries which can distinguish between the other three NP solutions.

References

1. Y. Amhis et al. HFLAV collaboration. Eur. Phys. J. C **77**(12), 895 (2017)
2. A.K. Alok, D. Kumar, J. Kumar, S. Kumbhakar, S.U. Sankar, JHEP **09**, 152 (2018)
3. M. Freytsis, Z. Ligeti, J.T. Ruderman, Phys. Rev. D **92**(5), 054018 (2015)
4. A.K. Alok, D. Kumar, S. Kumbhakar, S.U. Sankar, Phys. Rev. D **95**(11), 115038 (2017)
5. A.K. Alok, D. Kumar, S. Kumbhakar, S.U. Sankar, Phys. Lett. B **784**(16) (2018)

6. R. Aaij et al., LHCb collaboration. Phys. Rev. Lett. **120**(12), 121801 (2018)
7. Y. Sakaki, M. Tanaka, A. Tayduganov, R. Watanabe, Phys. Rev. D **88**(9), 094012 (2013)
8. A.K. Alok, A. Datta, A. Dighe, M. Duraisamy, D. Ghosh, D. London, JHEP **1111**, 121 (2011)
9. M. Duraisamy, P. Sharma, A. Datta, Phys. Rev. D **90**(7), 074013 (2014)
10. S. Aoki et al., Eur. Phys. J. C **77**(2), 112 (2017)
11. J. A. Bailey et al., Fermilab lattice and MILC collaborations. Phys. Rev. D **89**(11), 114504 (2014)

Chapter 66
Probing Signatures of Beyond Standard Model Physics Through $B_s^* \to \mu^+\mu^-$ Decay

Jyoti Saini, Dinesh Kumar, Shireen Gangal and Sanjeeda Bharti Das

Abstract We perform a model independent analysis to identify new physics operators which can enhance the branching ratio of $B_s^* \to \mu^+\mu^-$ above its Standard Model (SM) prediction. We find that none of the new physics operators which provide a good fit to $b \to s\mu^+\mu^-$ data can enhance the Br($B_s^* \to \mu^+\mu^-$) above its SM value.

66.1 Introduction

Several observables related to the decays of B meson do not agree with their Standard Model (SM) predictions. For e.g., the measurement of the ratios $R_{K^{(*)}}$, angular observable $P_5^{'}$ in $B \to K^*\mu^+\mu^-$ decay in the 4.3–8.68 q^2-bin, branching ratio of $B_s \to \phi\mu^+\mu^-$ do not agree with their SM value. All of these discrepancies are related to the $b \to s\mu^+\mu^-$ sector. This can be attributed to the presence of new physics (NP) in $b \to s\mu^+\mu^-$ transition.

In [1, 2], new physics in $b \to s\mu^+\mu^-$ decays were analysed by making use of an effective Hamiltonian with all possible Lorentz structures. It was shown that any large effects in $b \to s\mu^+\mu^-$ sector, in particular decays like $B \to K^*\mu^+\mu^-$ and $B_s \to \phi\mu^+\mu^-$, can only be due to new physics vector (V) and axial-vector operators (A). After the advent of R_{K^*} data, several groups performed global fits to identify

J. Saini (✉)
Indian Institute of Technology Jodhpur, Jodhpur 342037, India
e-mail: saini.1@iitj.ac.in

D. Kumar
Department of Physics, University of Rajasthan, Jaipur 302004, India
e-mail: dinesh@uniraj.ac.in

S. Gangal
Center for Neutrino Physics, Department of Physics, Virginia Tech, Virginia, Blacksburg 24061, USA
e-mail: shireen.gangal@gmail.com

S. Bharti Das
Department of Physics, Ramanujan Junior College, Nagaon 782001, India
e-mail: sanjeedabd0194@gmail.com

A. Giri and R. Mohanta (eds.), *16th Conference on Flavor Physics and CP Violation*, Springer Proceedings in Physics 234,
https://doi.org/10.1007/978-3-030-29622-3_66

the Lorentz structure of NP [3–11]. Many NP solutions, all in the form of vector V and A operators, were obtained. In order to discriminate between these NP solutions, one needs new observables. It would be desirable to have an access to observables which are theoretically clean.

The purely leptonic decay of B_s^* meson is such a decay channel [12]. In this work we perform a model independent analysis of $B_s^* \to \mu^+\mu^-$ decay to see whether $\text{Br}(B_s^* \to \mu^+\mu^-)$ can discriminate between various NP solutions which provide a good fit to the $b \to s\mu^+\mu^-$ data [13].

66.2 $B_s^* \to \mu^+\mu^-$ Decay

The effective Hamiltonian for the quark level transition $b \to s\mu^+\mu^-$ within the SM is given by

$$\mathcal{H}_{\text{SM}} = \frac{4G_F}{\sqrt{2}\pi} V_{\text{ts}}^* V_{\text{tb}} \Bigg[\sum_{i=1}^{6} C_i(\mu) O_i(\mu) + C_7 \frac{e}{16\pi^2} [\bar{s}\sigma_{\mu\nu}(m_s P_L + m_b P_R) b] F^{\mu\nu} + C_9 \frac{\alpha_{em}}{4\pi} (\bar{s}\gamma^\mu P_L b)(\bar{\mu}\gamma_\mu \mu) + C_{10} \frac{\alpha_{em}}{4\pi} (\bar{s}\gamma^\mu P_L b)(\bar{\mu}\gamma_\mu \gamma_5 \mu) \Bigg]. \tag{66.1}$$

Here G_F is the Fermi constant, V_{ij} are elements of the Cabibbo–Kobayashi–Maskawa (CKM) matrix and $P_{L,R} = (1 \mp \gamma^5)/2$. The effect of the operators O_i, i = 1–6, 8 can be included in the effective Wilson Coefficients by redefining $C_7(\mu) \to C_{\text{eff}}^7(\mu, q^2)$ and $C_9(\mu) \to C_{\text{eff}}^9(\mu, q^2)$.

To study NP effects in $B_s^* \to \mu^+\mu^-$ decay, we consider the addition of V, A, S and P operators to the SM effective Hamiltonian of $b \to s\mu^+\mu^-$

$$\mathcal{H}_{\text{eff}}(b \to s\mu^+\mu^-) = \mathcal{H}^{\text{SM}} + \mathcal{H}^{\text{VA}} + \mathcal{H}^{\text{SP}}, \tag{66.2}$$

where $\mathcal{H}^{\text{VA}}$ and $\mathcal{H}^{\text{SP}}$ are as

$$\mathcal{H}^{\text{VA}} = \frac{\alpha G_F}{\sqrt{2}\pi} V_{\text{ts}}^* V_{\text{tb}} \Big[C_9^{\text{NP}} (\bar{s}\gamma^\mu P_L b)(\bar{\mu}\gamma_\mu \mu) + C_{10}^{\text{NP}} (\bar{s}\gamma^\mu P_L b)(\bar{\mu}\gamma_\mu \gamma_5 \mu) + C_9^{'\text{NP}} (\bar{s}\gamma^\mu P_R b)(\bar{\mu}\gamma_\mu \mu) + C_{10}^{'\text{NP}} (\bar{s}\gamma^\mu P_R b)(\bar{\mu}\gamma_\mu \gamma_5 \mu) \Big] \tag{66.3}$$

$$\mathcal{H}^{\text{SP}} = \frac{\alpha G_F}{\sqrt{2}\pi} V_{\text{ts}}^* V_{\text{tb}} \Big[R_S (\bar{s} P_R b)(\bar{\mu}\mu) + R_P (\bar{s} P_R b)(\bar{\mu}\gamma_5 \mu) + R_S^{'} (\bar{s} P_L b)(\bar{\mu}\mu) + R_P^{'} (\bar{s} P_L b)(\bar{\mu}\gamma_5 \mu) \Big] \tag{66.4}$$

where $C_9^{\text{NP}}, C_{10}^{\text{NP}}, C_9^{'\text{NP}}, C_{10}^{'\text{NP}}, R_S, R_P, R_S^{'}, R_P^{'}$ are NP couplings.

We find that,

$$\langle 0|\bar{s}b|B_s^*\rangle = 0, \ \langle 0|\bar{s}\gamma^5 b|B_s^*\rangle = 0.$$

Hence the $\mathrm{Br}(B_s^* \to \mu^+\mu^-)$ is not affected by NP in the form of S and P operators. The decay rate including NP V and A contribution is obtained to be,

$$\Gamma(B_s^* \to \mu^+\mu^-) = \frac{G_F^2\alpha^2}{96\pi^3}|V_{\mathrm{tb}}V_{\mathrm{ts}}^*|^2 f_{B_s^*}^2 m_{B_s^*}^2 \sqrt{m_{B_s^*}^2 - 4m_\mu^2}\Bigg[\ \Bigg|C_9^{\mathrm{eff}}(m_{B_s^*}^2) + 2\frac{m_b f_{B_s^*}^T}{m_{B_s^*} f_{B_s^8}} C_7^{\mathrm{eff}}(m_{B_s^*}^2) + C_9^{\mathrm{NP}} + C_9^{'\mathrm{NP}}\Bigg|^2 + \Big|C_{10} + C_{10}^{\mathrm{NP}} + C_{10}^{'\mathrm{NP}}\Big|^2\Bigg]. \qquad (66.5)$$

As the total decay width of B_s^* meson, $\Gamma(B_s^{*\mathrm{tot}})$ is not yet known precisely, it is assumed that $\Gamma(B_s^{*\mathrm{tot}})$ is comparable to the dominant decay process $B_s^* \to B_s\gamma$ which is found to be $\Gamma(B_s^* \to B_s\gamma) = 0.10 \pm 0.05$ KeV [12].

66.3 Methodology

We perform a χ^2 fit to all CP conserving data $b \to s\mu^+\mu^-$ sector for all possible combinations of NP couplings. The observables used in the fit are listed in [13].

The χ^2 function is constructed as

$$\chi^2(C_i) = (O_{\mathrm{th}}(C_i) - O_{\mathrm{exp}})^T \mathscr{C}^{-1}(O_{\mathrm{th}}(C_i) - O_{\mathrm{exp}}). \qquad (66.6)$$

The total covariance matrix $\mathscr{C}$ is obtained by adding the individual theoretical and experimental covariance matrices.

66.4 Results and Discussions

The fit results for various new physics scenarios, along with the corresponding predictions for the branching ratio of $B_s^* \to \mu^+\mu^-$, are presented in Table 66.1. It is obvious from Table 66.1 that none of the new physics scenarios can provide large enhancement in the branching ratio of $B_s^* \to \mu^+\mu^-$ above its SM value. In scenarios where a good fit to the data is obtained, $Br(B_s^* \to \mu^+\mu^-)$ is seen to be suppressed as compared to the SM value. Hence, most likely, the future measurements are expected to observe $B_s^* \to \mu^+\mu^-$ decay with a branching ratio less than its SM prediction.

Table 66.1 Calculation of the branching ratios of $B_s^* \to \mu^+\mu^-$ for various new physics scenarios. Here $\Delta\chi^2 = \chi^2_{\rm SM} - \chi^2_{\rm bf}$ and $\chi^2_{\rm bf}$ is the χ^2 at the best fit points

Scenario	New physics couplings	$\Delta\chi^2$	Branching ratio
$C_i = 0(SM)$	–	0	$(1.23 \pm 0.48) \times 10^{-11}$
$C_9^{\rm NP}$	-1.24 ± 0.18	43.27	$(0.95 \pm 0.48) \times 10^{-11}$
$C_{10}^{\rm NP}$	0.91 ± 0.19	29.47	$(1.01 \pm 0.51) \times 10^{-11}$
C_9'	0.13 ± 0.16	0.66	$(1.30 \pm 0.65) \times 10^{-11}$
C_{10}'	-0.11 ± 0.13	0.68	$(1.29 \pm 0.65) \times 10^{-11}$
$C_9^{\rm NP} = C_{10}^{\rm NP}$	0.01 ± 0.18	0.001	$(1.26 \pm 0.64) \times 10^{-11}$
$C_9^{\rm NP} = -C_{10}^{\rm NP}$	-0.65 ± 0.11	43.04	$(0.89 \pm 0.45) \times 10^{-11}$
$C_9' = C_{10}'$	-0.04 ± 0.17	0.06	$(1.26 \pm 0.64) \times 10^{-11}$
$C_9' = -C_{10}'$	0.07 ± 0.08	0.81	$(1.30 \pm 0.65) \times 10^{-11}$
$[C_9^{\rm NP}, C_{10}^{\rm NP}]$	[–1.10, 0.33]	47.33	$(0.88 \pm 0.44) \times 10^{-11}$
$[C_9', C_{10}']$	[0.08, –0.07]	0.81	$(1.31 \pm 0.66) \times 10^{-11}$
$[C_9^{\rm NP} = C_{10}^{\rm NP}, C_9' = C_{10}']$	[–0.02, –0.02]	0.07	$(0.97 \pm 0.49) \times 10^{-11}$
$[C_9^{\rm NP} = -C_{10}^{\rm NP}, C_9' = -C_{10}']$	[–0.67, 0.16]	46.27	$(1.00 \pm 0.52) \times 10^{-11}$
$[C_9^{\rm NP}, C_{10}^{\rm NP}, C_9', C_{10}']$	[–1.31, 0.26, 0.34, –0.25]	56.04	$(1.00 \pm 0.52) \times 10^{-11}$

References

1. A.K. Alok, A. Datta, A. Dighe, M. Duraisamy, D. Ghosh, D. London, JHEP **1111**, 121 (2011), arXiv:1008.2367 [hep-ph]
2. A.K. Alok, A. Datta, A. Dighe, M. Duraisamy, D. Ghosh, D. London, JHEP **1111**, 122 (2011), arXiv:1103.5344 [hep-ph]
3. B. Capdevila, A. Crivellin, S. Descotes-Genon, J. Matias, J. Virto, arXiv:1704.05340 [hep-ph]
4. S. Descotes-Genon, J. Matias, J. Virto, Phys. Rev. D **88**, 074002 (2013), arXiv:1307.5683 [hep-ph]
5. W. Altmannshofer, D.M. Straub, Eur. Phys. J. C **75**(8), 382 (2015), arXiv:1411.3161 [hep-ph]
6. W. Altmannshofer, D.M. Straub, arXiv:1503.06199 [hep-ph]
7. T. Hurth, F. Mahmoudi, S. Neshatpour, Nucl. Phys. B **909**, 737 (2016), arXiv:1603.00865 [hep-ph]
8. B. Capdevila, S. Descotes-Genon, J. Matias, J. Virto, JHEP **1610**, 075 (2016), arXiv:1605.03156 [hep-ph]
9. W. Altmannshofer, C. Niehoff, P. Stangl, D.M. Straub, Eur. Phys. J. C **77**(6) 377 (2017), arXiv:1703.09189 [hep-ph]
10. A.K. Alok, B. Bhattacharya, D. Kumar, J. Kumar, D. London, S.U. Sankar, Phys. Rev. D **96**(1), 015034 (2017), arXiv:1703.09247 [hep-ph]
11. A.K. Alok, B. Bhattacharya, A. Datta, D. Kumar, J. Kumar, D. London, arXiv:1704.07397 [hep- ph]
12. B. Grinstein, J. Martin Camalich, Phys. Rev. Lett. **116**(14), 141801 (2016), arXiv:1509.05049 [hep-ph]
13. D. Kumar, J. Saini, S. Gangal, S.B. Das, Phys. Rev. D **97**(3), 035007 (2018), arXiv:1711.01989 [hep-ph]

Chapter 67
Probing New Physics in Semileptonic Λ_b Decays

Atasi Ray, Suchismita Sahoo and Rukmani Mohanta

Abstract In recent times, several hints of lepton non-universality have been observed in $b \to c l \bar{\nu}_l$ and $b \to sll$ decay processes. In this context, we perform a model independent analysis of the semileptonic baryonic $\Lambda_b \to (\Lambda_c, p) l \nu_l$ decay processes in order to probe the structure of new physics. We constrain the new parameter space by using the measured branching ratios of $B^+_{c,u} \to \tau^+ \nu_\tau$, $B \to \pi \tau \nu_\tau$ processes and the existing experimental results on $R_{D^{(*)}}$ and R^l_π parameters. Using the constrained new couplings, we estimate the branching fractions and forward-backward asymmetries of the $\Lambda_b \to (\Lambda_c, p) l \nu_l$ processes. As like B meson decays, we also examine the possible presence of lepton universality violation in these decay modes.

67.1 Introduction

Recently, the LHCb Collaboration has reported a spectacular discrepancy of 1.9σ (3.3σ) and 2σ on the lepton non-universality (LNU) parameters $R_{D^{(*)}} = \mathrm{Br}(\bar{B} \to \bar{D}^{(*)} \tau \bar{\nu}_\tau)/\mathrm{Br}(\bar{B} \to \bar{D}^{(*)} l \bar{\nu}_l)$ [1] and $R_{J/\psi} = \mathrm{Br}(B_c \to J/\psi \tau \bar{\nu}_\tau)/\mathrm{Br}(B_c \to J/\psi l \bar{\nu}_l)$ [2] respectively from their corresponding SM values. Analogous LNU parameters are also observed in $b \to sll$ processes i.e., $R_{K^{(*)}} = \frac{\mathrm{Br}(\bar{B} \to \bar{K}^{(*)} \mu^+ \mu^-)}{\mathrm{Br}(\bar{B} \to \bar{K}^{(*)} e^+ e^-)}$ with discrepancies of 2.6σ $(2.2 - 2.4)\sigma$. So the natural question arises is, whether such LNU parameters would also be present in baryonic decays. Around 20% of the total number of hadrons produced at LHCb are Λ_b baryons, so it is interesting to look for such effects in semileptonic Λ_b decay modes. Thus, as in B decays one

A. Ray (✉) · R. Mohanta
School of Physics, University of Hyderabad, Hyderabad 500 046, India
e-mail: atasiray92@gmail.com

R. Mohanta
e-mail: rukmani98@gmail.com

S. Sahoo
Theoretical Physics Division, Physical Research Laboratory,
Ahmedabad 380009, India
e-mail: suchismita8792@gmail.com

A. Giri and R. Mohanta (eds.), *16th Conference on Flavor Physics and CP Violation*, Springer Proceedings in Physics 234,
https://doi.org/10.1007/978-3-030-29622-3_67

can also scrutinize the presence of lepton universality violation in the corresponding semileptonic baryon decays $\Lambda_b \to (\Lambda_c, p)l\bar{\nu}_l$ to corroborate the results from B sector and hence, to probe the structure of New Physics (NP).

The outline of the paper is follows. In Sect. 67.2, we present the general effective Lagrangian of $b \to (u, c)l\nu_l$ processes in presence of NP, and the necessary theoretical framework for analysing these processes. The new coefficients are constrained in Sect. 67.3. In Sect. 67.4, we discuss the q^2 variation of branching fractions and lepton non-universality parameters of $\Lambda_b \to (\Lambda_c, p)l\bar{\nu}_l$ processes in presence of NP. Our findings are summarized in Sect. 67.5.

67.2 Theoretical Framework

The most general effective Lagrangian associated with $B_1 \to B_2 l\bar{\nu}_l$ decay processes, where $B_1 = \Lambda_b,\ B_2 = \Lambda_c, p$ mediated by the quark level transition $b \to ql\bar{\nu}_l,\ (q = u, c)$ is given by [3]

$$\mathcal{L}_{\rm eff} = -\frac{4\,G_F}{\sqrt{2}}\,V_{qb}\left\{(1+V_L)\,\bar{l}_L\,\gamma_\mu\,\nu_L\,\bar{q}_L\,\gamma^\mu\,b_L + V_R\,\bar{l}_L\,\gamma_\mu\,\nu_L\,\bar{q}_R\,\gamma^\mu\,b_R \right.$$
$$\left. +S_L\,\bar{l}_R\,\nu_L\,\bar{q}_R\,b_L + S_R\,\bar{l}_R\,\nu_L\,\bar{q}_L\,b_R + T_L\,\bar{l}_R\,\sigma_{\mu\nu}\,\nu_L\,\bar{q}_R\,\sigma^{\mu\nu}\,b_L\right\} + \text{h.c.}\,, \tag{67.1}$$

where G_F denotes the Fermi constant, V_{qb} are the CKM matrix elements and $q(l)_{L,R} = P_{L,R}\,q(l)$ are the chiral quark(lepton) fields with $P_{L,R} = (1 \mp \gamma_5)/2$ as the projection operators. Here $V_{L,R}, S_{L,R}, T_L$ represent the vector, scalar and tensor type NP couplings, which are zero in the SM. For simplicity, we switch off the tensor coupling contribution in this analysis. In the presence of NP, the double differential decay distribution for $B_1 \to B_2 l\bar{\nu}_l$ processes with respect to q^2 and $\cos\theta_l$ (θ_l is the angle between the directions of parent B_1 baryon and the l^- in the dilepton rest frame) is given as [4]

$$\frac{d^2\Gamma}{dq^2\,d\cos\theta_l} = N\left(1-\frac{m_l^2}{q^2}\right)^2\left[A_1 + \frac{m_l^2}{q^2}\,A_2 + 2\,A_3 + \frac{4\,m_l}{\sqrt{q^2}}\,A_4\right], \tag{67.2}$$

where A_1, A_2, A_3 are functions of helicity amplitudes, which are expressed in terms of the form factors and the new couplings. $N = \dfrac{G_F^2|V_{qb}|^2 q^2\sqrt{\lambda(M_{B_1}^2, M_{B_2}^2, q^2)}}{2^{10}\pi^3 M_{B_1}^3}$, with $\lambda(a, b, c) = a^2 + b^2 + c^2 - 2(ab + bc + ca)$. The lepton non-universality parameter is defined as

$$R_{B_2} = \frac{\text{Br}(B_1 \to B_2\tau^-\bar{\nu}_\tau)}{\text{Br}(B_1 \to B_2\,l^-\bar{\nu}_l)}, \quad l = e, \mu. \tag{67.3}$$

Table 67.1 Predicted allowed ranges of the new coefficients

Decay processes	New coefficients	Minimum value	Maximum value
$b \to u\tau\bar{\nu}_\tau$	(Re[V_L], Im[V_L])	$(-2.489, -1.5)$	(0.504, 1.48)
	(Re[V_R], Im[V_R])	$(-0.478, -1.185)$	(0.645, 1.198)
	(Re[S_L], Im[S_L])	$(-0.136, -0.396)$	(0.672, 0.398)
	(Re[S_R], Im[S_R])	$(-0.6743, -0.398)$	(0.1265, 0.398)
$b \to c\tau\bar{\nu}_\tau$	(Re[V_L], Im[V_L])	$(-2.224, -1.228)$	(0.225, 1.225)
	(Re[V_R], Im[V_R])	$(-0.129, -0.906)$	(0.173, 0.89)
	(Re[S_L], Im[S_L])	$(-0.116, -0.788)$	(0.474, 0.8)
	(Re[S_R], Im[S_R])	$(-1.076, -0.809)$	(0.06, 0.807)

67.3 Constraints on New Couplings

In this analysis, the new Wilson coefficients are considered as complex. We further assume that only one new coefficient present at a time and accordingly compute the allowed parameter space of these couplings. The constraints on new parameter space associated with $b \to (u, c)l\bar{\nu}_l$ transitions are computed from the experimental data on $R_{D^{(*)}}$, $R_{J/\psi}$, R_π^l, Br$(B_{c,u} \to \tau\bar{\nu}_\tau)$ and Br$(B \to \pi\tau\bar{\nu}_\tau)$ observables. The allowed range of new couplings are presented in Table 67.1.

67.4 Effect of New Physics

First we considered the effect of new physics contribution to the SM result is coming only from the coupling associated with the left-handed vector-like quark currents i.e., $V_L \neq 0$ and $V_R, S_{L,R} = 0$. In this scenario we considered the constrained parameter space of V_L for $b \to ul\bar{\nu}_l$ transition obtained from Table 67.1, and obtained the q^2 variation of the branching ratio and LNU parameter of $\Lambda_b \to pl\bar{\nu}_l$ in presence of the new coefficient. We repeat the same procedure and performed the analysis in presence of V_R, S_L and S_R coefficients considering one at a time. Similarly using the real and imaginary part of new coefficients for $b \to cl\bar{\nu}_l$ obtained from Table 67.1, we obtained the variation of differential branching ratio and LNU parameters in presence of individual new coefficients. the q^2 variation of LNU parameters are presented in Fig. 67.1.

67.5 Conclusion

We performed a model independent analysis of the rare baryonic $\Lambda_b \to (\Lambda_c/p)l\bar{\nu}_l$ processes. We considered the most general effective Lagrangian in the presence of new Physics, which provides additional new couplings to the standard model

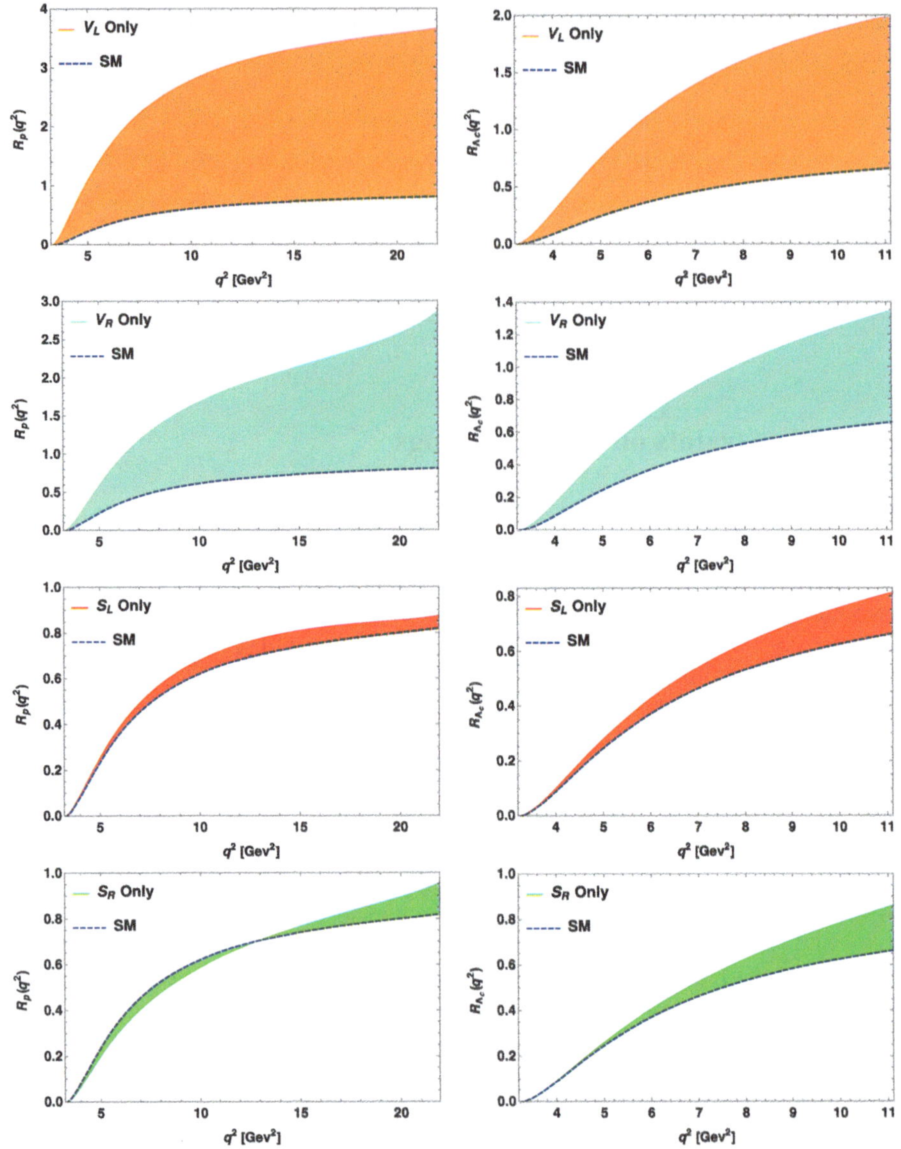

Fig. 67.1 The variation of R_p (left panel) and R_{Λ_c} (right panel) LNU parameters with respect to q^2 in the presence of the new coefficients

contributions. Using the experimental value of several branching ratios and LNU parameters, we have constrained the new couplings. We then computed the branching ratios and LNU parameters, in order to test the Lepton non-universality in Λ_b decay processes. The study of these modes are found to be very crucial in order to shed light on the nature of new physics.

Acknowledgements We acknowledge SERB for financial support.

References

1. R. Aaji et al., Measurement of the Ratio of Branching Fractions $\mathcal{B}(\bar{B}^0 \to D^{*+}\tau^-\bar{\nu_\tau})/\mathcal{B}(\bar{B}^0 \to D^{*+}\mu^-\bar{\nu_\mu})$. Phys. Rev. Lett. **115**, 111803 (2015). https://doi.org/10.1103/PhysRevLett.115.159901
2. R. Aaij et al., Measurement of the ratio of branching fractions $\mathcal{B}(B_c{}^+ \to J/\psi\tau^+\nu_\tau)/\mathcal{B}(B_c{}^+ \to J/\psi\mu^+\nu_\mu)$. Phys. Rev. Lett. **120**, 121801 (2018). https://doi.org/10.1103/PhysRevLett.120.121801
3. T. Bhattacharya, V. Cirigliano, S.D. Cohen, A. Filipuzzi, M. Gonzalez-Alonso, R. Graesser, Gupta, H.W. Lin, Probing novel scalar and tensor interactions from (Ultra) cold neutrons to the LHC . Phys. Rev. **D85**, 054512 (2012). https://doi.org/10.1103/PhysRevD.85.054512
4. S. Shivashankara, W. Wu, Datta, A.:$\Lambda_b \to \Lambda_c\tau\bar{\nu_\tau}$ Decay in the Standard Model and with New Phys. Rev. **D91**, 115003 (2015). https://doi.org/10.1103/PhysRevD.91.115003

Chapter 68
Signature of Lepton Flavor Universality Violation in $B_s \to D_s \tau \nu$ Semileptonic Decays

Rupak Dutta and N. Rajeev

Abstract Deviation from the standard model prediction is observed in many semileptonic B decays mediated via $b \to c$ charged current interactions. In particular, current experimental measurements of the ratio of branching ratio R_D and R_{D^*} in $B \to D^{(*)} l \nu$ decays disagree with standard model expectations at the level of about 4.1σ. Moreover, recent measurement of the ratio of branching ratio $R_{J/\Psi}$ by LHCb, where $R_{J/\Psi} = \mathcal{B}(B_c \to J/\Psi\, \tau\nu)/\mathcal{B}(B_c \to J/\Psi\, \mu\nu)$, is more than 2σ away from the standard model prediction. In this context, we consider an effective Lagrangian in the presence of vector and scalar new physics couplings to study the implications of R_D and R_{D^*} anomalies in $B_s \to D_s\, \tau\nu$ decays. We give prediction of several observables such as branching ratio, ratio of branching ratio, forward backward asymmetry parameter, τ polarization fraction, and the convexity parameter for the $B_s \to D_s\, \tau\nu$ decays within the standard model and within various new physics scenarios.

68.1 Introduction

Anomalies present in R_D, R_{D^*} and $R_{J/\Psi}$ challenged the lepton flavor universality. At present the deviation in R_D and R_{D^*} from the standard model (SM) expectation [1–6] is at the level of 4.1σ [7–12]. A similar deviation of 1.3σ has been reported by LHCb in the ratio $R_{J/\Psi}$ [13] as well. Inspired by these anomalies we study the corresponding $B_s \to D_s \tau\nu$ decay mode within the SM and within various new physics (NP) scenarios by using the B_s to D_s transition form factors obtained in lattice QCD of [14]. The $B_s \to D_s \tau\nu$ decays serve as a complementary decay channel to similar B decays mediated via $b \to cl\nu$ quark level transition. Again, in the limit of $SU(3)$ flavor symmetry $B \to Dl\nu$ and $B_s \to D_s l\nu$ decay modes should exhibit the similar properties.

R. Dutta · N. Rajeev (✉)
National Institute of Technology Silchar, Silchar 788010, India
e-mail: rajeev@rs.phy.student.nits.ac.in

R. Dutta
e-mail: rupak@phy.nits.ac.in

A. Giri and R. Mohanta (eds.), *16th Conference on Flavor Physics and CP Violation*, Springer Proceedings in Physics 234,
https://doi.org/10.1007/978-3-030-29622-3_68

Our main motivation here is to study the implication of R_D and R_{D^*} anomalies on $B_s \to D_s \tau \nu$ decay mode in a model independent way. We use the effective Lagrangian in the presence of NP couplings and give the predictions of various physical observables.

68.2 Theory

The effective Lagrangian for $b \to c l \nu$ quark level transition decays consisting of the SM and beyond SM operators is given by [15, 16]

$$\begin{aligned}
\mathcal{L}_{\text{eff}} = & -\frac{4\,G_F}{\sqrt{2}}\,V_{cb}\Big\{(1+V_L)\,\bar{l}_L\,\gamma_\mu\,\nu_L\,\bar{c}_L\,\gamma^\mu\,b_L + V_R\,\bar{l}_L\,\gamma_\mu\,\nu_L\,\bar{c}_R\,\gamma^\mu\,b_R \\
& +\widetilde{V}_L\,\bar{l}_R\,\gamma_\mu\,\nu_R\,\bar{c}_L\,\gamma^\mu\,b_L + \widetilde{V}_R\,\bar{l}_R\,\gamma_\mu\,\nu_R\,\bar{c}_R\,\gamma^\mu\,b_R + S_L\,\bar{l}_R\,\nu_L\,\bar{c}_R\,b_L \\
& +S_R\,\bar{l}_R\,\nu_L\,\bar{c}_L\,b_R + \widetilde{S}_L\,\bar{l}_L\,\nu_R\,\bar{c}_R\,b_L + \widetilde{S}_R\,\bar{l}_L\,\nu_R\,\bar{c}_L\,b_R\Big\} + \text{h.c.}\,,
\end{aligned} \tag{68.1}$$

where, G_F is the Fermi coupling constant and $|V_{cb}|$ is the CKM matrix element and the couplings such as V_L, V_R, S_L, S_R and $\tilde{V}_L$, $\tilde{V}_R$, $\tilde{S}_L$, $\tilde{S}_R$ denote the NP Wilson coefficients involving left handed and right handed neutrinos respectively. We investigate several q^2 dependent observables such as differential branching ratio DBR (q^2), ratio of branching ratio $R(q^2)$, lepton side forward backward asymmetry $A^l_{FB}(q^2)$, polarization fraction of the charged lepton $P_l(q^2)$, and convexity parameter $C^l_F(q^2)$ defined as,

$$\begin{aligned}
DBR(q^2) &= \frac{d\Gamma/dq^2}{\Gamma_{\text{Tot}}}, \qquad A_{FB}(q^2) = \frac{\left(\int_{-1}^{0} - \int_{0}^{1}\right) d\cos\theta \frac{d\Gamma}{dq^2\,d\cos\theta}}{\frac{d\Gamma}{dq^2}}, \\
R(q^2) &= \frac{\mathcal{B}(B_s \to D_s \tau \nu)}{\mathcal{B}(B_s \to D_s\, l\, \nu)}, \qquad P_l(q^2) = \frac{d\Gamma(+)/dq^2 - d\Gamma(-)/dq^2}{d\Gamma(+)/dq^2 + d\Gamma(-)/dq^2}, \\
C^l_F(q^2) &= \frac{1}{(d\Gamma/dq^2)} \left(\frac{d}{d\cos\theta}\right)^2 \left[\frac{d\Gamma}{dq^2\, d\cos\theta}\right]
\end{aligned} \tag{68.2}$$

We also give predictions on the average values of these observable by separately integrating the numerator and the denominator over q^2. A detailed discussion is reported in the [17].

68.3 Results and Discussions

We report in Table 68.1 the SM central values and the corresponding ranges of each observable for $B_s \to D_s l\nu$ decay mode. The SM central values are obtained by considering the central values of each input parameters and the ranges are obtained by including the uncertainties associated with $|V_{cb}|$ and the form factor inputs. The details are presented in the [17]. We notice that the branching ratio of $B_s \to D_s l\nu$ is of the order of 10^{-2} for the e mode and the τ mode. In SM, the observables A^l_{FB}, P_l, C^l_F for the e mode are observed to be quite different from the corresponding τ mode.

We now proceed to discuss the various NP effects in $B_s \to D_s \tau\nu$ decay mode. We consider four different NP scenarios each containing two NP couplings at a time: (V_L, V_R), (S_L, S_R), $(\widetilde{V}_L, \widetilde{V}_R)$ and $(\widetilde{S}_L, \widetilde{S}_R)$. To get the allowed NP parameter space we impose the 1σ constraints coming from the measured ratio of branching ratio R_D and R_{D^*} as well as the requirement of $\mathcal{B}(B_c \to \tau\nu) \leq 30\%$ from the LEP data [18]. In Fig. 68.1 we show the allowed ranges of each NP couplings and the corresponding $\mathcal{B}(B_s \to D_s \tau\nu)$ and $\mathcal{B}(B_c \to \tau\nu)$ allowed regions. It is observed that the $\mathcal{B}(B_c \to \tau\nu)$ put a severe constraint on the scalar NP couplings and hence in our present analysis we omit the related discussions. Table 68.2 reports the allowed ranges of each observables for $B_s \to D_s \tau\nu$ decay mode when (V_L, V_R) and $(\widetilde{V}_L, \widetilde{V}_R)$

Table 68.1 SM prediction of various observables for the e and the τ modes

Observables	Central value	Range	Observables	Central value	Range
$\mathcal{B}_e\%$	2.238	[2.013, 2.468]	$\mathcal{B}_\tau\%$	0.670	[0.619, 0.724]
P^e	−1.00	−1.00	P^τ	0.320	[0.273, 0.365]
A^e_{FB}	0.00	0.00	A^τ_{FB}	0.360	[0.356, 0.363]
C^e_F	−1.5	−1.50	C^τ_F	−0.271	[−0.253, −0.289]

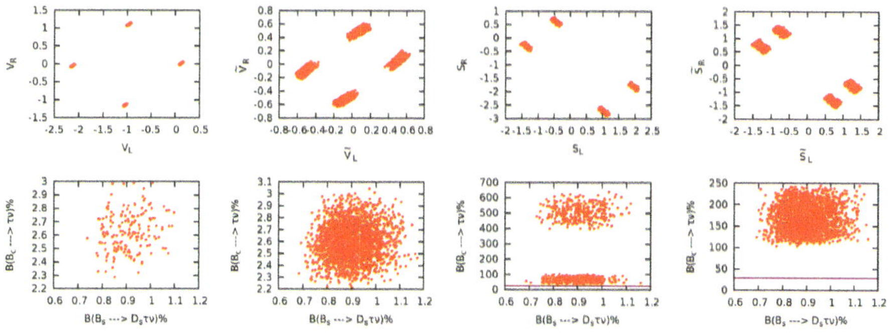

Fig. 68.1 Allowed ranges of (V_L, V_R), $(\widetilde{V}_L, \widetilde{V}_R)$, (S_L, S_R) and $(\widetilde{S}_L, \widetilde{S}_R)$ NP couplings are shown in the top panel once 1σ constraint coming from the measured values of R_D and R_{D^*} are imposed. We show in the bottom panel the allowed ranges in $\mathcal{B}(B_c \to \tau\nu)$ and $\mathcal{B}(B_s \to D_s \tau\nu)$ in the presence of respective NP couplings

Table 68.2 Allowed ranges of various observables with (V_L, V_R) and $(\widetilde{V}_L, \widetilde{V}_R)$ NP couplings

	$\mathcal{B}\%$	R_{D_s}	P^τ	A_{FB}^τ	C_F^τ
(V_L, V_R)	[0.733, 1.115]	[0.329, 0.496]	[0.234, 0.403]	[0.352, 0.364]	[−0.239, −0.305]
$(\widetilde{V}_L, \widetilde{V}_R)$	[0.684, 1.174]	[0.307, 0.519]	[0.064, 0.276]	[0.356, 0.363]	[−0.253, −0.289]

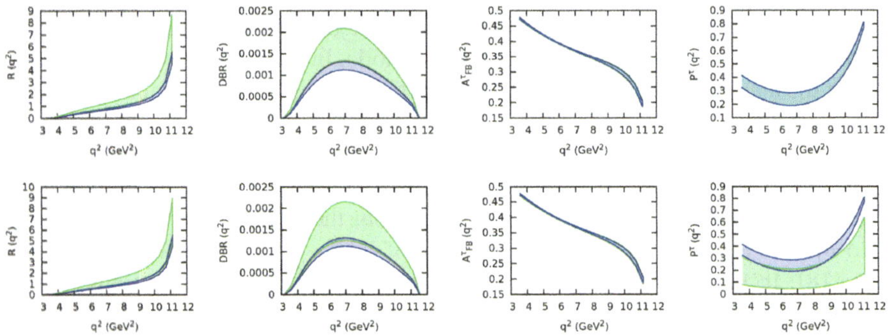

Fig. 68.2 $R(q^2)$, $DBR(q^2)$, $A_{FB}^\tau(q^2)$, $P^\tau(q^2)$ using (V_L, V_R) (top) and $(\widetilde{V}_L, \widetilde{V}_R)$ (bottom) NP couplings (green band). The corresponding 1σ SM range is shown with the blue band

NP couplings are present. We show the q^2 dependency of $R(q^2)$, $DBR(q^2)$, $A_{FB}^\tau(q^2)$, $P^\tau(q^2)$ using (V_L, V_R) (top) and $(\widetilde{V}_L, \widetilde{V}_R)$ (bottom) NP couplings in Fig. 68.2. In the presence of (V_L, V_R) NP couplings, we observe that $DBR(q^2)$ and $R(q^2)$ deviate considerably from the SM expectation whereas, no deviations are found in $A_{FB}^\tau(q^2)$, $P^\tau(q^2)$, $C_F^\tau(q^2)$. Similarly, in the presence of $(\widetilde{V}_L, \widetilde{V}_R)$ NP couplings the deviation is observed in $P^\tau(q^2)$ along with $DBR(q^2)$ and $R(q^2)$. Hence the polarization fraction of the charged lepton $P^\tau(q^2)$ can be used to distinguish between these two scenarios.

68.4 Conclusion

Based on the anomalies present in R_D and R_{D^*}, we study their implication on $B_s \to D_s\tau\nu$ decay mode within the SM and within various NP scenarios. We find that only vector type NP couplings satisfy the $\mathcal{B}(B_c \to \tau\nu)$ constraint whereas, the scalar type NP couplings are ruled out. Hence, studying the $B_s \to D_s\tau\nu$ decay mode will serve as an important stepping stone for R_D and R_{D^*} anomalies.

References

1. J.A. Bailey et al., [MILC Collaboration], Phys. Rev. D **92**(3), 034506 (2015)
2. H. Na et al., [HPQCD Collaboration], Phys. Rev. D **92**(5), 054510 (2015)
3. S. Aoki et al., Eur. Phys. J. C **77**(2), 112 (2017)
4. D. Bigi, P. Gambino, Phys. Rev. D **94**(9), 094008 (2016)
5. S. Fajfer, J.F. Kamenik, I. Nisandzic, Phys. Rev. D **85**, 094025 (2012)
6. D. Bigi, P. Gambino, S. Schacht, JHEP **1711**, 061 (2017)
7. J.P. Lees et al., [BaBar Collaboration], Phys. Rev. D **88**(7), 072012 (2013)
8. M. Huschle et al., [Belle Collaboration], Phys. Rev. D **92**(7), 072014 (2015)
9. Y. Sato et al., [Belle Collaboration], Phys. Rev. D **94**(7), 072007 (2016)
10. S. Hirose et al., [Belle Collaboration], Phys. Rev. Lett. **118**(21), 211801 (2017)
11. R. Aaij et al., [LHCb Collaboration], Phys. Rev. Lett. **115**(11), 111803 (2015)
12. Y. Amhis et al., [HFLAV Collaboration], Eur. Phys. J. C **77**(12), 895 (2017)
13. R. Aaij et al., [LHCb Collaboration], arXiv:1711.05623 [hep-ex]
14. C.J. Monahan, H. Na, C.M. Bouchard, G.P. Lepage, J. Shigemitsu, Phys. Rev. D **95**(11), 114506 (2017)
15. V. Cirigliano, J. Jenkins, M. Gonzalez-Alonso, Nucl. Phys. B **830**, 95 (2010)
16. T. Bhattacharya, V. Cirigliano, S.D. Cohen, A. Filipuzzi, M. Gonzalez-Alonso, M.L. Graesser, R. Gupta, H.W. Lin, Phys. Rev. D **85**, 054512 (2012)
17. R. Dutta, N. Rajeev, Phys. Rev. D **97**(9), 095045 (2018)
18. A.G. Akeroyd, C.H. Chen, Phys. Rev. D **96**(7), 075011 (2017)

The manufacturer's authorised representative in the EU is Springer Nature Customer Service Centre GmbH, Europaplatz 3, 69115 Heidelberg, Germany. If you have any concerns regarding our products, please contact ProductSafety@springernature.com

Printed and bound by CPI Group (UK) Ltd, Croydon, CR0 4YY

30/09/2026

02225228-0019